Uses of Immobilized Biological Compounds

NATO ASI Series

Advanced Science Institutes Series

A Series presenting the results of activities sponsored by the NATO Science Committee, which aims at the dissemination of advanced scientific and technological knowledge, with a view to strengthening links between scientific communities.

The Series is published by an international board of publishers in conjunction with the NATO Scientific Affairs Division

A Life Sciences **B Physics**	Plenum Publishing Corporation London and New York
C Mathematical and Physical Sciences **D Behavioural and Social Sciences** **E Applied Sciences**	Kluwer Academic Publishers Dordrecht, Boston and London
F Computer and Systems Sciences **G Ecological Sciences** **H Cell Biology** **I Global Environmental Change**	Springer-Verlag Berlin, Heidelberg, New York, London, Paris and Tokyo

NATO-PCO-DATA BASE

The electronic index to the NATO ASI Series provides full bibliographical references (with keywords and/or abstracts) to more than 30000 contributions from international scientists published in all sections of the NATO ASI Series.
Access to the NATO-PCO-DATA BASE is possible in two ways:

– via online FILE 128 (NATO-PCO-DATA BASE) hosted by ESRIN, Via Galileo Galilei, I-00044 Frascati, Italy.

– via CD-ROM "NATO-PCO-DATA BASE" with user-friendly retrieval software in English, French and German (© WTV GmbH and DATAWARE Technologies Inc. 1989).

The CD-ROM can be ordered through any member of the Board of Publishers or through NATO-PCO, Overijse, Belgium.

Series E: Applied Sciences - Vol. 252

Uses of Immobilized Biological Compounds

edited by

George G. Guilbault
Department of Chemistry,
University of New Orleans,
New Orleans, Los Angeles, U.S.A.

and

Marco Mascini
Institute of Analytical Chemistry,
University of Florence,
Florence, Italy

Springer-Science+Business Media, B.V.

Proceedings of the NATO Advanced Research Workshop on
Uses of Immobilized Biological Compounds
for Detection, Medical, Food and Environmental Analysis
Brixen, Italy
May 9–14, 1993

A C.I.P. Catalogue record for this book is available from the Library of Congress.

ISBN 978-94-010-4854-5 ISBN 978-94-011-1932-0 (eBook)
DOI 10.1007/978-94-011-1932-0

Originally published by Kluwer Academic Publishers in 1993
Softcover reprint of the hardcover 1st edition 1993

DEDICATION

To Susan and Maresa

Our Loves and Our Lives

TABLE OF CONTENTS

B. MEDICAL, CLINICAL AND PHARMACEUTICAL APPLICATIONS

C. ELECTROCHEMICAL BIOSENSORS

PREFACE

On May 9-14, 1993, a NATO Advanced Research Workshop on the Analytical Uses of Immobilized Biological Compounds was held in Bressanone, Italy. The Director of the Workshop was Professor George G. Guilbault of the University of New Orleans, and the Co-Director was Professor Marco Mascini of the University of Florence.

It was the purpose of this meeting to assemble scientists from all NATO Countries with an interest in immobilized biological compounds, to discuss:

- methods of immobilization of proteins and their properties;
- antibody-antigen reactions;
- enzyme electrodes and biosensors;
- optical devices utilizing immobilized enzymes;
- microsensors and microbial sensors and clinical uses of immobilized enzymes;
- flow injection analysis using enzymes;
- immobilized biological compounds in chemical defense detection;
- pharmaceutical analysis;
- uses in industrial analysis;
- enzyme reactors;
- air pollution detectors;
- disposable immunosensors;
- medical uses and applications;
- mass production of sensors.

Goals to be achieved by the conference were:

- to permit an exchange of views and experience in all these areas;
- to review and critically assess the state-of-the-art in these fields;
- to set guidelines for future research and establish collaborative projects between scientists in NATO laboratories in the above areas.

Fifty-four lectures were given by 53 speakers in all of the above areas. Sessions were devoted to (1) general aspects of immobilizing biological compounds; (2) medical, clinical and pharmaceutical applications; (3) electrochemical biosensors; (4) electrochemical and optical biosensors; (5) defense applications; (6) immunosensors and receptors; (7) food, environmental, clinical and analytical applications; (8) biotechnology and marketing.
Finally, two hours were devoted to an open discussion of future status, new directions and joint projects.
This book is a pubblication of most of the lectures given at this workshop.

We wish to thank for their financial support of this conference: Universita' di Padova, Universal Sensors (New Orleans) and especially NATO for the ARW grant that made the Conference possible. We also thank our organizing committee (Drs. Coulet, Weetall, Macca' and Palleschi) for the organization and local support.

CONCLUSIONS

In the last morning a round table discussion of selected speakers, the Directors and audience centered on the status of the field of Analytical Uses of Immobilized Biological Compounds, cooperation between labs and the future.

Several participants expressed appreciation to the Conference for the possibility to meet and develop contacts with scientists of other NATO Countries doing similar research. Collaborative projects were established between: (1) the Universities of Barcelona, Roma Tor Vergata, Coimbra, on Amperometric Based Biosensors Compatible with Mass Production Technologies for Pesticide Detection; (2) the Universities of Ioannina, Hacettepe, New Orleans and Coimbra on a CCMS Project on Management of Toxic Wastes. Their intention is to submit a project on Controlling Toxic Compounds with Biosensors; (3) the Universities of New Orleans and Quebec on New Biosensor Concepts; (4) the Universities of New Orleans and Bruxelles on Gas Phase Biosensors and PZ Antibodies; (5); the Universitites of New Mexico and Budapest on Stripping Analysis; (6) the Universitites of New Mexico and Milano on Food Biosensors; (7) the Universities of New Mexico and Bruxelles on Carbon Paste Biosensors; (8) the Universitites of New Orleans and Pavia on Toxin Biosensors; (9) the Universities of New Orleans, Florence and Roma Tor Vergata on PZ Immunosensors and Non-Invasive Sensors.

Discussion on new areas was devoted to (1) New Biological Materials, Receptors/Antibody Fragments, Toxin Probes; (2) Miniaturized Sensors, Electrochemical, Optical; (3) New Sampling Directions, Gas Phase, Non-Invasive; (4) Improved response/Market ability, Mediated Biosensors, New Directions, New Transducer, Optical, Arrays, Mass Production of Sensors, Screen Printing on Si, Al_2O_3 or Plastics; (5) Basic Studies, Importance of Position of Proteins and Antibodies, Studies with Microscopies like: AFM, TM, SEM, Thermodynamics Constants, Kinetics Equations, etc, CQN.

<u>**LIST OF CONTRIBUTORS**</u>

S. Alegret - Departamento de Quimica, Universitat Autonoma de Barcelona, E08193 Bellaterra, Catalonia, Spain

J. Alonso - Departamento de Quimica, Universitat Autonoma de Barcelona, E08193 Bellaterra, Catalonia, Spain

M. Alvarez-Icaza - Gesellschaft für Biotechnologische Forschung (GBF), Dept. of Enzyme Technology, 3300 Braunschweig, Germany

G. P. Anderson - Navy Research Labs, Washington, D. C. 20375

N. A. Anis - Biotechnology Division, U. S. Army Research and Development and Engineering Center, Edgewood, Maryland 21010

M. Y. Arica - Middle East Technical University, Department of Biological Sciences, 06531 Ankara, Turkey

S. Bacha - Laboratoire de Genie Chimique et Electrochimie, URA CNRS, Universite' Paul Sabatier, 31062 Toulouse, France

A. J. Bard - Department of Chemistry, University of Texas at Austin, Austin, Texas 78712

J. Bartroli - Departamento de Quimica, Universitat Autonoma de Barcelona, E08193 Bellaterra, Catalonia, Spain

T. Basinka - Center for Molecular and Macromolecular Studies, Polish Academy of Sciences, Sienkiewicza 112, 90-363, Lodz, Poland

A. Bergel - Laboratoire de Genie Chimique et Electrochimie, URA CNRS, Universite' Paul Sabatier, 31062 Toulouse, France

P. Bergveld - MESA Research Institute, University of Twente, 7500 AE Enschede, Netherlands

L. J. Blum - Laboratoire de Genie Enzymatique, E.P. CNRS, University Lyon 1, 69622 Villeurbanne, France

A. M. Oliveira Brett - Departamento de Quimica, University of Coimbra, 3049 Coimbra, Portugal

K. Cammann - Institut für Chemo- and Biosensorik und Lehrstuhl für Analytische Chemie, der Westfalischen Wilhelm Universitat, D-4400 Munster, Germany

L. Campanella - Dipartimento di Chimica, University of Rome I - La Sapienza, 00185 Roma, Italy

R. Carter - Universal Sensors, Inc.- 5258 Veterans Blvd., Suite D, Metairie, LA 70006

I. M. Christie - Department of Clinical Biochemistry, University of Manchester, Hope Hospital, Salford M6 8HD, United Kingdom

D. Compagnone - Distaam, Universita del Molise, Via Tiberio, 21 86100 Campobasso, Italy

M. Comtat - Laboratoire de Genie Chimique et Electrochimie, URA CNRS, Universite' Paul Sabatier, 31062 Toulouse, France

S. Cosnier - Laboratoire d'Electrochimie Organique et Photochimie Redox, URA CNRS, Universite' Joseph Fourier Grenoble, France

P. Coulet - Laboratoire de Genie Enzymatique, E.P. CNRS, Universite' Lyon 1, 69622 Villeurbanne, France

W. Cushley - Department of Biochemistry, University of Glasgow G12 8QQ, Scotland, United Kingdom

B. Danielsson - Technical University of Lund, Department of Pure and Applied Biochemistry, S-221 00, Lund, Sweden

D. Datta - Institute of Biotechnology, University of Cambridge, Cambridge CB2 1QT, United Kingdom

D. J. G. Davies, Centre for Drug Formulation Studies, School of Pharmacy and Pharmacology, University of Bath, BA2 7AY, United Kingdom

A. De Flora - Institute of Biochemistry, University of Genoa, 16132 Genoa, Italy

R. M. de La Rue - Department of Electronics and Electrical Engineering, University of Glascow, Glascow G12 8QQ, Scotland, United Kingdom

M. A. Desai - Department of Clinical Biochemistry, University of Manchester, Hope Hospital, Salford M6 8HD, United Kingdom

H. Durliat - Laboratoire de Genie Chimique et Electrochimie, URA CNRS, Universite' Paul Sabatier, 31062 Toulouse, France

R. A. Durst - Analytical Chemistry Laboratories, Department of Food Science and Technology, Cornell University, Geneva, New York 14456

A. T. Eldefrawi - Department of Phamacology and Experimental Therapeutics, School of Medicine, University of Maryland, Baltimore, MD 21201

M. Eldefrawi - Department of Phamacology and Experimental Therapeutics, School of Medicine, University of Maryland, Baltimore, MD 21201

M. Faridnia - Univesal Sensors, Inc.- 5258 Veterans Blvd., Suite D, Metairie, LA 70006

Z. Feher - Institut of General and Analytical Chemistry, Technical University of Budapest, Gellert ter, Budapest, Hungary

J. M. Fernandez-Romero - Department of Analytical Chemistry, Univesity of Cordoba, E-14004 Cordoba, Spain

M. Filipiak - Department of Biochemistry and Microbiology, Academy of Economics, Poznan, Poland

G. Frishman - Institute for Biological Research, Ness-Ziona, 70450, Israel

G. Gabor - Institute for Biological Research, Ness-Ziona, 70450, Israel

S. M. Gautier - Laboratoire de Genie Enzymatique, E.P. CNRS, Universite' Lyon 1, 69622 Villeurbanne, France

A. Gebbert - Gesellschaft für Biotechnologische Forschung (GBF), Department of Enzyme Technology, 3300 Braunschweig, Germany

S. Ghosh - Department of Clinical Biochemistry, University of Manchester, Hope Hospital, Salford M6 8HD, United Kingdom

J. P. Golden - Navy Research Labs, Center for Biomolecular Science and Engineering, Washington, D. C. 20375

C. Groom - Biotechnology Research Institute, National Research Council of Canada, Montreal, Quebec, Canada H4P 2R2

G. G. Guilbault, Department of Chemistry, University of New Orleans, and Universal Sensors, P. O. Box 736, New Orleans, LA 70148

C. E. Hall - Institute of Biotechnology, University of Cambridge, Cambridge CB2 1QT, United Kingdom

E. A. Hall - Institute of Biotechnology, University of Cambridge, Cambridge CB2 1QT, United Kingdom

E. Hansen - Chemistry Department A, Technical University of Denmark, DK-2800 Lyngby, Denmark

V. Hasirci - Middle East Technical University, Department of Biological Sciences, 06531 Ankara, Turkey

U. Hedberg - Technical University of Lund, Department of Pure and Applied Biochemistry, S-221 00, Lund, Sweden

S. P. J. Higson - Department of Clinical Biochemistry, University of Manchester, Hope Hospital, Salford, M68 HD, United Kingdom

C. Hödl - Zentrum fur Ultrastrukturforschung und Ludwig Boltzmann Institut für Molekulare Nanotechnology, A-1180 Wien, Austria

B. R. Horrocks - Department of Chemistry, University of Texas at Austin, Austin, Texas 78712

E. I. Iwuoha - School of Chemical Sciences, Dublin City University, Dublin 9, Ireland

S. Jackman - Sauflon Pharmaceutics Ltd., Ashford, Kent TN24 8EP, United Kingdom

M. N. Jones - Department of Clinical Biochemistry, University of Manchester, Hope Hospital, Salford, M68 HD, United Kingdom

M. I. Karayannis - Department of Chemistry, University of Ioannina, Greece

U. Karst - Institut für Chemo- and Biosensorik und Lehrstuhl fur Analytische Chemie, der Westfalischen Wilhelm Universitat, D-4400 Munster, Germany

I. Karube - Research Center for Advanced Science and Technology, University of Tokyo, 4-6-1 Komaba, Meguro-ku, Tokyo 153

J.-M. Kauffmann - University Libre de Bruxelles, Campus Plaine 205/6, 1050 Bruxelles, Belgium

J. B. Kay - Sauflon Pharmaceutics Ltd., Ashford, Kent TN24 8EP, United Kingdom

M.Kiremitci - Hacettepe University, Chemical Engineering Department, Beytepe, Ankara, Turkey

U. J. Krull - Department of Chemistry, University of Toronto, Erindale Campus, Mississanga, ON L5L 1C6, Canada

S. Küpcü - Zentrum fur Ultrastrukturforschung und Ludwig Boltzmann Institut fur Molekulare Nanotechnology, A-1180 Wien, Austria

P. Labbe - Laboratoire d'Electrochimie Organique et de Photochimie Redox, URA CNRS, Universite' Joseph Fourier Grenoble, France

P. J. Laybourn - Department of Electronics and Electrical Engineering, University of Glascow, Glascow G12 8QQ, Scotland, United Kingdom

A. Lewenstam - Department of Analytical Chemistry, Åbo Akademi University, Turku, Finland

F. S. Ligler - Center for Biomolecular Science and Engineering, Naval Research Laboratory, 4555 Overlook Ave., SW, Washington, D. C. 20375

G. J. Lubrano - Universal Sensors, Inc.- 5258 Veterans Blvd., Suite D, Metairie, LA 70006

J. H. Luong - Biotechnology Research Institute, National Research Council Canada, Montreal, Quebec, Canada H4P 2R2

M. D. Luque de Castro - Department of Analytial Chemistry, University of Cordoba, E-14004 Cordoba, Spain

A. Machado - Departamento de Quimica, University of Porto, P4000, Portugal

M. C. S. Magalhaes - Departamento de Quimica, University of Porto, P4000, Portugal

J. V. Magill - Department of Electronics and Electrical Engineering, University of Glascow, Glascow G12 8QQ, Scotland, United Kingdom

A. Makower - Max Delbuck Center for Molecular Medicine, 0-1115 Berlin, Germany

K. B. Male - Biotechnology Research Institute, National Research Council Canada, Montreal, Quebec, Canada H4P 2R2

S. Mannino - Department of Food Science and Technology, University of Milan, Via Celoria 2, 20133 Milano, Italy

N. Martens - Institute of Biotechnology, University of Cambridge, Cambridge CB2 1QT, United Kingdom

E. Martinez-Faregas - Departament de Quimica, Universitat Autonoma de Barcelona, E-08193 Bellaterra, Catalonia, Spain

M. Mascini - Dipartament de Sanita' Pubblica, Sezione di Chimica Analitica, University of Florence, 50121 Firenze, Italy

W. Matuszewske - Department of Chemistry, University of Warsaw, 02-093 Warsaw, Poland

J. Mauro - Geo-Centers, Inc., Fort Washington, MD 20744 and Naval Research Lab, 4555 Overlook SW, Washington, D. C. 20375

K. B. McDonnell - Chemical and Biological Defence Establishment, Porton Down, Salisbury, Wiltshire, SP4 OJQ, United Kingdom

C. McNeil - University of Newcastle, Newcastle upon Tyne, NE2 4HH, United Kingdom

C. McSharry - Department of Bacteriology and Immunology, Western Infirmary, Dumbarton Road, Glascow G11 6NT, Scotland, United Kingdom

B. J. Meakin - Centre for Drug Formulation Studies, School of Pharmacy and Pharmacology, University of Bath, BA2 7AY, United Kingdom

M. Montagne - Laboratoire de Genie Chimique et Electrochimie, URA CNRS, Universite' Paul Sabatier, 31062 Toulouse, France

D. Moscone - Dip. di Scienze e Tecnologie Chimiche, II University of Rome, 00173 Roma, Italy

M. N. Mustan - Institute of Biotechnology, University of Cambridge, Cambridge CB2 1QT, United Kingdom

A. Nagy - University Libre de Bruxelles, Campus Plaine 205/6, 1050 Bruxelles, Belgium

G. Nagy - Institute of General and Analytical Chemistry, Technical University of Budapest, Gellert ter, Budapest, Hungary

A. Neubauer - Zentrum fur Ultrastukturforschung und Ludwig Boltzmann Institut fur Molekulare Nanotechnologie, A-1180 Wien, Austria

D. Nikolelis - Department of Chemistry, University of Athens, 15771 - Athens, Greece

R. A. Ogert, Naval Research Labs, Washington, D. C. 20375

V. Owen - Sci Tec Management Consultants Ltd., Bletchley, Milton Keynes MK3 7QP, United Kingdom

G. Palleschi - Distaam, Universita del Molise, Via Tiberio, 21 86100 Campobasso, Italy

M. Pesavento - Dipartimento di Chimica Generale, University of Pavia, 27100 Pavia, Italy

Ch. Petit - University Libre de Bruxelles, Campus Plaine 205/6, 1050 Bruxelles, Belgium

D. Pfeiffer - Max Delbruck Center of Molecular Medicine, 0-1115 Berlin, Germany

A. K. Piskin - Hacettepe University, Department of Medical Biology, Sihhiye 06100 Ankara, Turkey

D. Pum - Zentrum fur Ultrastukturforschung und Ludwig Boltzmann Institut fur Molekulare Nanotechnologie, A-1180 Wien, Austria

E. Pungor -Institute of General and Analytical Chemistry, Technical University of Budapest, Gellert ter, Budapest, Hungary

M. Rank - Technical University of Lund, Department of Pure and Applied Biochemistry, S-221 00, Lund, Sweden

S. M. Reddy - Department of Clinical Biochemistry, University of Manchester, Hope Hospital, Salford M68 HD, United Kingdom

S. G. Reeves - Analytical Chemistry Laboratories, Department of Food Science and Technology, Cornell University, Geneva, New York 14456

R. Renneberg - Institut fur Chemo- and Biosensorik und Lehrstuhl fur Analytische Chemie, der Westfalischen Wilhelm Universitat, D-4400 Munster, Germany

M. A. Roberts - Analytical Chemistry Laboratories, Department of Food Science and Technology, Cornell University, Geneva, New York 14456

K. R. Rogers - U. S. Environmental Protection Agency, Environmental Monitoring Systems Laboratory, Las Vegas, Nevada 89193-3478

M. F. Rosenberg - Department of Clinical Biochemistry, University of Manchester, Hope Hospital, Salford M68 HD, United Kingdom

M. Sara - Zentrum fur Ultrastukturforschung und Ludwig Boltzmann Institut fur Molekulare Nanotechnologie, A-1180 Wien, Austria

F. Scheller - Max-Delbruck Center for Molecular Medicine, 0-1115 Berlin, Germany

R. D. Schmid - Gesellschaft fur Biotechnologische Forsschung (GBF), Department of Enzyme Technology, 3300 Braunschweig, Germany

H.-L. Schmidt - Technical Univesitat Munchen, Lehrstuhl fur Allgemeine Chemie und Biochemie, D-8050 Freising - Weihenstephan, Germany

F. Schubert - Physikolisch-Technische Bunde Sanstalt, W-1000 Berlin 10, Germany

L. C. Shriver-Lake - Naval Research Lab, 4555 Overlook SW, Washington, D. C. 20375

D. E. T. Sibley - Universal Sensors, Inc.- 5258 Veterans Blvd., Suite D, Metairie, LA 70006

S. A. Siebert - Analytical Chemistry Laboratories, Department of Food Science and Technology, Cornell University, Geneva, New York 14456

U. Sleytr - Zentrum fur Ultrastukturforschung und Ludwig Boltzmann Institut fur Molekulare Nanotechnologie, A-1180 Wien, Austria

S. Slomkowski - Center for Molecular and Macromolecular Studies, Polish Academy of Sciences, Sienkiewicza 112, 90-363, Lodz, Poland

M. Smyth - School of Chemical Sciences, Dublin City University, Dublin 9, Ireland

C. D. Stalikas - Department of Chemistry, University of Ioannina, Greece

B. Szczepanczyk - Department of Chemistry, University of Warsaw, Warsaw, Poland

M. Tomassetti - Dipartimento di Chimica, University of Rome I - La Sapienza, 00185 Roma, Italy

K. Toth - Institute of General and Analytical Chemistry, Technical University of Budapest, Gellert ter, Budapest, Hungary

P. H. Treloar - Department of Clinical Biochemistry, University of Manchester, Hope Hospital, Salford M68 HD, United Kingdom

M. Trojanowicz - Department of Chemistry, University of Warsaw, 02-093 Warsaw, Poland

A. P. F. Turner - Cranfield Biotechnology Center, Cranfield Institute of Technology, Cranfield, Bedford, MK43 OAL, England

S. M. Tzouwara-Karayanni - Department of Chemistry, University of Ioannina, Ioannina, Greece

P. Vadgama - Department of Clinical Biochemistry, University of Manchester, Hope Hospital, Salford M68 HD, United Kingdom

J. J. Valdes - Biotechnology Division, U. S. Army Research and Development and Engineering Center, Edgewood, Maryland 21010

F. Valdes-Perezgasga - Departament de Quimica, Universitat Autonoma de Barcelona, E-08193 Bellaterra, Catalonia, Spain

J. Wang - Department of Chemistry and Biochemistry, New Mexico State University, Las Cruces, NM 88003

A. Warsinke - Institut fur Chemo- and Biosensorik und Lehrstuhl fur Analytische Chemie, der Westfalischen Wilhelm Universitat, D-4400 Munster, Germany

H. Weetall - Biotechnology Division, National Institute of Standards and Technology, Gaithersburg, MD 20899

D. Wijesuriya - Geo Centers, Inc., 10903 Indian Head Highway, Ft. Washington, MD 20744

O. Wolfbeis - Karl Franzens University, Institute of Organic Chemistry, A-8010 Graz, Austria

U. Wollenberg - Fraunhofer Insstitut fur Siliziomtechnologie, W-1000 Berlin 33, Germany

S. Wood - Chemical and Biological Defense Agency, Edgewood Research Development and Engineering Center, Aberdeen Proving Ground, MD 21010

M. Wortberg - Institut fur Chemo- and Biosensorik und Lehrstuhl fur Analytische Chemie, der Westfalischen Wilhelm Universitat, D-4400 Munster, Germany

B. Xie - Technical University of Lund, Department of Pure and Applied Biochemistry, S-221 00, Lund, Sweden

K. Yokoyama - Research Center for Advanced Science and Technology, University of Tokyo, Japan

I. Zawicki - Institute of Biocybernetics and Biomedical Engineering, Polish Academy of Sciences, Warsaw, Poland

S. Zhao - Biotechnology Research Institute, National Research Council of Canada, Montreal, Quebec, Canada H4P 2R2

Y. Zhou - Department of Electronics and Electrical Engineering, University of Glascow, Glascow G12 8QQ, Scotland

IMMOBILIZATION OF BIOLOGICAL SPECIES ON BILAYER LIPID MEMBRANES

D. P. NIKOLELIS[1] AND U. J. KRULL[2]
[1] Department of Chemistry, University of Athens, Panepistimiopolis-Kouponia, 15771-Athens, GREECE
[2] Department of Chemistry, Erindale Campus, University of Toronto, 3359 Mississauga Road North, Mississauga, ON L5L 1C6, CANADA

Abstract. This work reports how bilayer lipid membranes (BLMs) can be used as the host matrices for the immobilization of hydrolytic enzymes and antibodies and how can be designed to provide maximized transduction of the activity of these biological species. A transient current signal is obtained as a result of the selective membrane incorporated protein/stimulant (analyte) interaction. The time appearance or magnitude of the transient charging signal is related to the concentration of substrate or antigen, respectively. The mechanism of signal generation is explored in the present paper by being related to rapid reorganization of the double layer and BLM structure.

I. INTRODUCTION

Natural chemoreception offers a glimpse of a practical system of biosensors. Chemoreceptive membranes consist of a bimolecular layer of lipids, which are physically associated with proteins that have molecular recognition properties, i.e. receptors. The ion channel system that is found in biological organisms is associated with rapid, reversible, sensitive and selective chemical signalling, with unique features such as the amplification and transduction of chemical information into an electrical pulse by switching of channel conductivity.

The implementation of chemoreceptive processes for biosensor development based on the use of artificial lipid membranes has received significant attention. The preparation of planar bilayer lipid membranes (BLMs) was first reported three decades ago by Mueller *et al.* [1]; Del Castillo and coworkers [2] were the first to point out that BLMs could be used as the basis for the development of electrochemical biosensors. Recent reports about electrochemical biosensors based on BLMs demonstrate the potential of these promising devices for applications which make use of selective proteins that are not molecular receptors[3-5]. The essential idea is that a protein which can selectively bind to a specific organic or biochemical species can be

G.G. Guilbault and M. Mascini (eds.), Uses of Immobilized Biological Compounds, 1–10.

incorporated into an ordered lipid membrane assembly so that selective binding events can lead to changes in the structure or electrostatic fields of the membrane (transduction). Since an artificial BLM has a thickness of a few nm, the response time for chemical interactions can be on the order of seconds. External energy can be stored across a BLM as electrochemical potential; a single selective binding event can generate a discharge of this energy. This is an intrinsic amplification step which provides high sensitivity. Finally, BLMs provide a host matrix which is conducive to the maintenance of activity of many biochemically selective proteins.

We are interested in developing an analogue of the response of an ion-channel, where the signal would be modulated by the concerted action of selective binding interactions, and the method of transduction would be suitable for the use of a wide variety of different proteins without implementation of ion-channel proteins from natural sources. Our approach begins with the immobilization of active proteins directly onto a carefully defined lipid mixture at an air/electrolyte interface, followed by a film casting technique for BLM formation. Subsequent experiments have shown that selective chemical reactions can be transduced as time-dependent singular transient ion current signals. The electrochemical event is due to the rapid reorganization of the double layer and structure of a BLM, and provides the basis for the construction of switchable biosensors with rapid response times and reversibility.

II. MATERIALS AND METHODS

Materials and Apparatus

Egg phosphatidylcholine (PC) was used as the structural lipid agent for BLMs, and was used in mixture with dipalmitoyl phosphatidic acid (DPPA). Protein studies included the use of the enzymes acetylcholinesterase (AChE), urease, and penicillinase supplied from Sigma Chemical Co. (St. Louis, MO.), and antibody from rabbit antiserum to thyroxine (T4), which was supplied from UCB Bioproducts S. A. (Braine-l'Alleud, Belgium).

A Saran-Wrap™ film (10 μm thickness) with an aperture of 0.32 mm diameter was used to separate two identical plexiglass solution chambers (each with a volume of ca. 10 ml and an air/water interface of 3 cm^2). The BLMs were located in the aperture and were supported in a 0.1 M KCl electrolyte solution. An external d.c. voltage of 25 mV was applied across the membrane between two Ag/AgCl electrodes. A digital electrometer (Model 614, Keithley Instruments, Cleveland, OH) was used as a current-to-voltage converter. The electrochemical cell and electronic equipment were isolated in a Faraday cage. All solutions were gently stirred and experiments were done at 25 ± 1^0 C.

BLM Formation
A revised monolayer folding method was used in our studies for the preparation of solventless BLMs [6]. A small amount (5-10 µl) of the lipid solution in *n*-hexane was placed (slowly, dropwise) onto the electrolyte surface in one of the cell compartments, followed by a delay to permit evaporation of the solvent. Over a period of a few seconds, the water level of the compartment was brought below the aperture and then raised again with a syringe. Over 95 % of the attempts to form BLMs were successful (assuming a freshly prepared dilute lipid solution) and these membranes were stable for periods of over 6 hours.

Protein immobilization
Membrane modification for the purpose of obtaining an analytical signal involves incorporation (immobilization) of the proteins into the lipid matrix as follows [5]:
1. Protein can be introduced into the lipid mixture before formation of BLMs or can be applied in a solvent to a lipid layer which is prepared at an air/electrolyte interface. This latter procedure was used in this work, and microliter volumes of enzyme or antibody solution were deposited onto the films of lipid at the air/water interface.
2. The protein can be delivered to a bilayer structure or a lipid film at an air/electrolyte interface by fusion of vesicles made with charged lipids (and containing the proteins) [4].
3. Membranes from natural sources which already contain the proteins can be directly isolated from tissues and patchclamped into an experimental configuration that is suitable for electrochemical study of BLMs.

III. RESULTS AND DISCUSSION

The planar BLMs used herein are unsuitable for practical biosensor implementation owing to the inherent mechanical fragility of these systems. Significant progress in the stabilization of lipid membranes by physical adsorption or chemical immobilization of lipids onto platinum electrodes has recently been reported [7, 8]. However, planar BLMs offer a degree of physical characterization and reproducibility that is difficult to achieve from immobilized membranes, and therefore the exploration of the basic physical chemistry and the utilization of sensing mechanisms of BLMs is essential for development of chemically selective configurations which may be practical.

BLMs prepared from mixtures of a neutral lipid such as PC and an ionizable lipid develop an electrical double layer at the membrane/electrolyte interface. DPPA has two potentially accessible ionizations with titration endpoints of about pH 4 and 7 [9], and mixtures of PC and DPPA are very sensitive to alterations of membrane surface charge [5].

Charged lipids such as DPPA can play an important structural role for the immobilization of biochemical species at BLMs by formation of hydrogen bonds with relevant amino acid side-chains of proteins. The formation of hydrogen bonding networks enhances the immobilization, and connects regions with differing proton or electron density to provide rapid conduction between such regions [9].

Transmembrane ion currents for membranes prepared from PC and DPPA measured at various values of pH clearly indicate the presence of different conductivity states which are determined directly by surface charge and also by phase domain structure [10, 11]. Recent work made use of rapid pH changes at only one side of a BLM, spanning the pH range required for the second ionization of DPPA [12]. This resulted in a rearrangement of membrane structure from one state of conductivity to another, which was accompanied by a transition in charging phenomena associated with restructuring of the double layer. The result of the relaxation of the surface charge was the evolution of a capacitive charging current which appeared as a fully reversible transient signal (seconds or less) once the electrostatic threshold responsible for phase redistribution was passed in either direction. These studies have therefore shown that it is possible to vary the surface electrostatic charge of one leaflet of an ionizable membrane without disturbing the structure of the opposite surface. Importantly, such dynamic modifications of surface charge have now been driven by appropriate hydrolytic enzyme reactions and by an immunochemical complexation at the surface of a BLM, to prepare a chemically-driven electrical switch. The electrochemical transient has an appearance similar to that observed during ion-channel gating [13] in that a single pulse of current of reproducible magnitude can be generated by selective chemical interactions.

Various orientations of proteins within BLMs have been proposed [9]. The partitioning of proteins into a lipid membrane is partially driven by electrostatic interactions and is therefore sensitive to the surface charge of a BLM. The surface charge is determined by the degree of ionization of the charged lipid, which can also can control the phase structure (largely dominated by hydrocarbon chain interactions) within a membrane. Proteins tend to partition into BLMs that are in a fluid state [9, 14]. Studies of probable orientations of AChE, urease and immunoglobulins have been presented [14-15]. The structure of these molecules is such that the hydrophobic portions are incorporated into BLMs, leaving the active site at the aqueous interface [15].

The proteins used in our studies were deposited directly onto the lipid mixtures at the air/water interface of one cell compartment to maximize the loading of these biological species in the BLMs. Experiments were done to estimate the maximum amount of the hydrolytic enzymes and anti-rabbit T4 that could be immobilized in the BLMs without inducing permeability changes. It was found that the minimum lipid-to-urease ratio could be about 140:1, and that smaller ratios resulted in the appearance of permanent conductivity increases and deviations from the expected residual (background) ion current. Similar experiments for penicillinase indicated that the minimum lipid-to-protein ratio which could be achieved without inducing permeability changes of BLMs was 210:1. A decrease of this ratio resulted in the appearance of multiple transient signals reminiscent of ion channel gating [13]. Table 1 summarizes the final lipid-to-protein ratio used for all enzyme-substrate experiments, the units of activity for enzymes immobilized in the BLM structure, and the related number of nmoles of acid or base per minute produced during the hydrolytic enzyme reaction at the BLM surface.

Table 1. Results of Optimized Amounts of Enzymes Immobilized in BLMs without Inducing Changes of Residual Ion Current

Enzyme reaction	Optimum Lipid-to-protein ratio	Enzyme activity (Units) at each surface of BLM	nmoles of acid or base produced/min
AChE/ACh	100:1	2×10^{-4}	0.2
Urea/urease	140:1	1.8×10^{-4}	0.32
Penicillin/penicillinase	210:1	2×10^{-4}	0.2

Transient or permanent ion current alterations (concentration dependent) were obtained when anti-rabbit T4 was incorporated into BLMs. Transient current signals observed for amounts of protein less than 5.2 µg decreased in magnitude and frequency over a period of not more than 20 minutes. This suggests that an aggregation of charged antibody molecules may occur, and that the interactions with the charged lipid component of the BLMs can induce electrostatic field gradients at a BLM surface. These processes terminated as the equilibrium concentration and distribution of protein was reached within 20 minutes. The results also indicate that it is unlikely that the protein spanned the membrane to form conductive pores since permanent alterations of ion current were not observed when using 5.2 µg or less of protein.

The difficulty of selecting an initial pH to optimize the sensitivity of the response to pH of negatively charged BLMs, and to concurrently optimize the activity of a hydrolytic enzyme, can be overcome by introducing Ca^{2+} to the bulk electrolyte solution [10, 11]. Adjustment of the intrinsic sensitivity of BLMs is achieved by selecting the lipid composition to contain concentrations of DPPA greater than 25 % w/w and by related adjustments of the pH and calcium ion activity [11]. The interaction of calcium ions with divalent negatively charged lipid (pH about 8) results in a phase separation of PC and DPPA and the formation of conductive zones in BLMs. These ion permeation zones can be switched on or off by selecting a pH value of the electrolyte solution below or above 7, respectively, in the presence of a limiting Ca^{2+} concentration which depends on the DPPA composition of BLMs.

The ACh/AChE, urea/urease and penicillin/ penicillinase hydrolytic enzyme reactions were examined at pH values of 8.0, 6.0 and 7.5, respectively, in the presence of calcium ions, as a compromise between optimum enzyme activity and enhancement of signal sensitivity. Figure 1 shows recordings of signals obtained at pH 8.0 for different concentrations of ACh. The transient responses appeared as singular events (no further transients were observed over periods of 10 minutes) which occurred in direct proportion to the reciprocal of time (Figure 2). The variability of response of the BLMs and the ability to reproducibly incorporate active protein is also indicated in Figure 2 (these results were obtained over a period of days, each result from a freshly prepared BLM). Similar transient current signals were obtained for the penicillinase catalysed hydrolysis of penicillin, whereas the signals for the urea/urease system were opposite in direction.

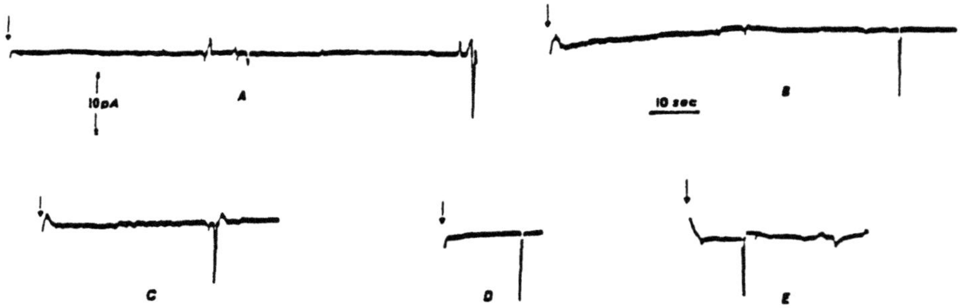

FIGURE 1. Experimental results obtained at pH 8.0 (0.1 M KCl, 10 mM HEPES and 1.0 mM Ca^{2+}) with BLMs consisting of 35 % w/w of DPPA when 3 μl of AChE (1.0 mg ml^{-1}) was co-deposited at the air/electrolyte interface. ACh concentrations are: (A): 2.20 μM; (B): 5.50 μM; (C): 11.0 μM; (D): 38.5 μM; and (E): 55.0 μM; the arrow indicates injection of substrate.

A number of control experiments using BLMs were completed to demonstrate that the transient signals were due to the enzyme reactions. These experiments included:
1. Addition of substrate to the electrolyte solution in the absence of enzyme, or in the presence of denaturated enzymes

(e.g. by heat) in the BLM structure. Transient current signals were not evident in such experiments.

2. Use of enzyme inhibitors; such studies involved e.g. the use of Neostigmine, a reversible inhibitor of the function of AChE, and Hg(II), a non-competitive urease inhibitor. Various amounts of inhibitors were added to one solution compartment after preparation of BLMs that contained active protein. Subsequent addition of substrate to the same solution compartment provided reduction or elimination of enzyme activity as controlled by the inhibitor concentration.

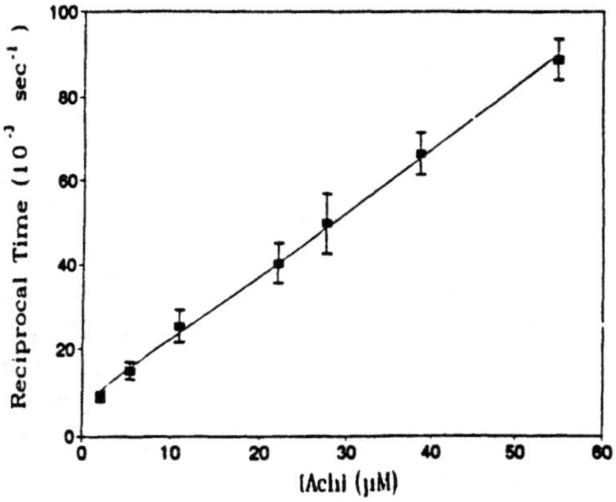

FIGURE 2. Calibration of the analytical signal from experiments as shown in Fig. 1. Each point represents the mean of 5 determinations.

The transient current signals are indicative of a rapid reorganization of surface charge of BLMs. Substantial pH alterations of the bulk solution were not observed or expected from the relatively small amount of enzyme in the cell compartment, however the local pH at the surface of a BLM should have varied dynamically as the enzyme converted substrate to product. The diffusion of substrate and product should control the ultimate pH value which would be expected to be different from that of the original buffered solution. As a result, substantial changes in the phase structure and/or electrostatic fields of BLMs must rapidly take place as the pH is altered. As the reactions proceed, the ionizable headgroups of the charged lipid in the membrane are dynamically titrated through an equivalence point with subsequent rapid reorganization of the double layer and BLM structure. Competition between the production or consumption of hydronium ions from the enzyme reaction with diffusion away from the membrane surface would set the detection limit for substrate determination, which can be e.g. 1 μM Ach (about 3 minutes were required before observation of the transient signal). The reliance on diffusion implies that strict attention must be given to stirring if calibration curves are to be reproducible.

The magnitude of the analytical signal was increased

when the amount or degree of ionization of DPPA was increased. An increase in the applied potential over a tested range of 25 to 100 mV resulted in a proportional increase of the magnitude of the charging signal. Alterations of the buffering strength of bulk electrolyte solution and of the amount of enzyme deposited in the BLM significantly affected the time required to produce a transient current signal and therefore the sensitivity of substrate determination. The experimental results offer exciting analytical sensing opportunities, where signal switching is not observed until a threshold substrate concentration is established and then a time-dependent signal burst of relatively constant magnitude appears.

Perhaps of greater significance is that similar signalling phenomena were observed for the antibody/antigen interactions involving modulation of surface charge by virtue of formation of the complex. The thyroxin/anti-rabbit T4 interaction was chosen as the immunological reaction owing to the clinical significance of T4. Figure 3 shows recordings of transient signals obtained at pH 6.0 (without Ca^{2+} in bulk solution) for different concentrations of T4 with BLMs composed of 15 % DPPA. The magnitude of these transient responses is in direct proportion to the logarithm of the antigen concentration in bulk solution. Control experiments demonstrated that these signals were due to selective interactions, and involved application of antigen to BLMs in the absence of antibody, and the use of non-selective protein in the presence of antibody.

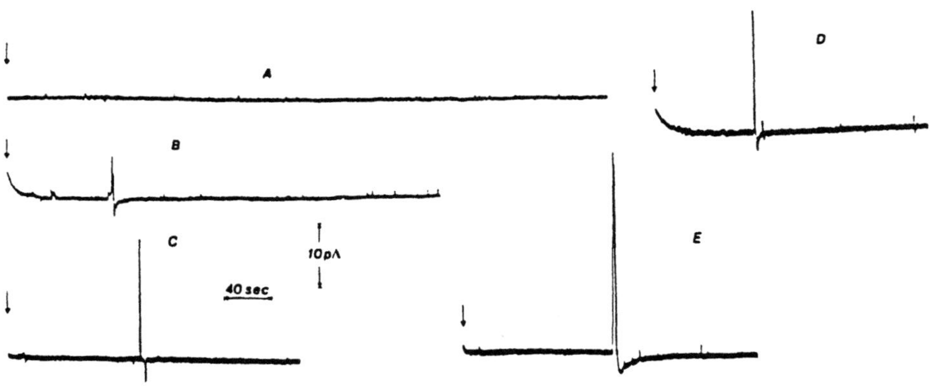

FIGURE 3. Experimental results obtained at pH 6.0 (0.1 M KCl, 10 mM HEPES and in the absence of Ca^{2+}) with BLMs consisting of 15 % w/w DPPA when 3 µl of anti-rabbit T4 solution (2.8 mg protein ml^{-1}) was co-deposited onto the air/water interface. Thyroxine concentrations: (A): 1.12 nM; (B): 11.2 nM; (C): 33.6 nM; (D): 112 nM; (E): 1.12 mM.

Rabbit anti-thyroxin antiserum contains antibodies with pI's between 6.3 to 9.0, whereas the pI of T4 is 5.0. Electrostatic forces are important in the mechanism of the formation of the antibody-antigen complex [16]. The kinetics of the immunological reaction and the electrostatics of this interaction on one side of a BLM would be expected to cause a slow non-linear double layer reorganization to occur. The appearance of a rapid transient current signal after substantial time delay (i.e. much longer than mixing or diffusion times) suggests that a capacitive charging current develops when some critical magnitude of surface charge is achieved, which results in a transition of membrane structure from one state to another (as in the case of hydrolytic enzymes). The speed of the signal suggests that alterations of antibody aggregations caused by the presence of antigen with concurrent increases of surface concentration of monovalent cations as given by Boltzman equation could be a possible mechanism of transient signal generation. Such a mechanism is reinforced by the semilogarithmic dependence of signal magnitude on T4 concentration.

The magnitude of the transient signals and therefore the sensitivity for antigen calibration is increased when the quantity of DPPA decreases to 15 %. A continuation of this trend is not evident for BLMs composed only of PC, which indicates that the phase structure of membranes is a predominant factor for the immobilization of immunoglobulins in BLMs [14] as these proteins prefer fluid areas enriched in DPPA. An increase of the pH of bulk solution (e.g. 8.0) which increases the degree of DPPA ionization (but decreases the number of hydrogen bonds between DPPA molecules) and reduces the positive charge of the antibody results in a decrease of the signal magnitude. The presence of calcium ions in bulk solution reduces signal magnitude and dramatically delays the time for the appearance of the transient signals. BLMs can therefore be designed to provide fast and sensitive transduction of immunological interactions; the previous example provided a detection limit of 1.0 nM of thyroxin (based on the experimental noise level of 0.7 pA) in a period of about 100 sec.

IV. SUMMARY/FUTURE PERSPECTIVES

Our results show that BLMs can be used as host matrices for the immobilization of biological species such as enzymes and antibodies without necessarily inducing permanent ion current alterations that would compromise the potential for transduction of selective reactions. The results further indicate that BLMs can be designed to provide fast, selective and sensitive generic transduction of hydrolytic enzyme reactions (fully reversible response) or interactions of charged antibodies with antigens. The possible uses of lipid membranes as transducers for biochemicals and species of environmental interest are now being explored. An improvement of the stability of the lipid membrane is a prerequisite for practical analytical use, but research is presently focused in this direction. This work suggests how the chemical composition of lipid membranes and their environment may be adjusted to optimize analytical response, and indicates some of the design criteria that will be required for development of stabilized biosensors based on the use of lipid membranes. Design criteria for achieving the goal of a biosensor based on lipid membranes are now becoming clear.

REFERENCES

1. P. Mueller, D. O. Rudin, H. Ti Tien and W. C. Wescott, Nature, 194, 979 (1962); Circulation, 26, 1167 (1962); J. Phys. Chem., 67, 534 (1963).
2. J. Del Castillo, A. Rodriguez, C. A. Romero and V. Sanchez, Science, 153, 185 (1966).
3. H. Minami, M. Sugawara, K. Odashima, Y. Umezawa, M. Uto, E.K. Michaelis, T. Kuwana, Anal. Chem., 63, 2787 (1991).
4. N. Sugao, M. Sugawara, H. Minami, M. Uto and Y. Umezawa, Anal. Chem., 65, 363 (1993).
5. D. P. Nikolelis and U. J. Krull, Electroanalysis, in press.
6. D. P. Nikolelis and U. J. Krull, Talanta, 39, 1045 (1992).
7. M. Otto, M. Snejdarkova and M. Rehak, Anal. Letters, 25 (4), 653 (1992).
8. J.D. Brennan, R.S. Brown, V. Ghaemagghami, K.M. Kallury, M. Thompson and U. J. Krull, Immobilization of Amphiphilic Membranes for Development of Optical and Electrochemical Biosensors in: Chemically-Modified Surfaces, H. Mottola and G. Steinmetz (Eds.), Elsevier, Amsterdam, 1992, p. 275.
9. P. Yeagle, The Structure of Biological Membranes, CRC Press Inc., Boca Raton, FL, 1992.
10. D. P. Nikolelis, J. D. Brennan, R. S. Brown and U. J. Krull, Anal. Chim. Acta, 257, 49 (1992).
11. D. P. Nikolelis and U. J.. Krull, Anal. Chim. Acta, 257, 239 (1992).
12. D. P. Nikolelis, M. G. Tzanelis and U. J. Krull, Anal. Chim. Acta, in press.
13. B. Hille, Ionic Channels of Excitable Membranes, Sinauer Associates Inc., Sunderland, MA, 2nd Ed., p. 4.
14. M. Vandenbranden, G. Kayser, S. Banerjee and J. M. Ruysschaert, Biochim. Biophys. Acta, 685, 177 (1982).
15. T. Weidmer, U. Brodbeck, P. Zahler and B. W. Fulpius, Biochim. Biophys. Acta, 506, 161 (1978).
16. Y. Endo, K. Miyai, N. Hata and Y. Iijima, Biotechn. Appl. Biochem., 9, 74 (1987).

POLYMERIC ENVIRONMENTS FOR PROTEINS AND WHOLE CELLS IN MOLECULAR SENSORS.

ELIZABETH A H HALL, CARL E HALL, NICOLAS MARTENS, M NABISAR MUSTAN AND DIPANKER DATTA.
Institute of Biotechnology, University of Cambridge, Tennis Court Road, Cambridge CB2 1QT, UK.

Abstract Methacrylate polymers are much used in microfabrication techniques. This paper examines their potential in the immobilisation of proteins and whole cells. A protein modified copolymer of glycidyl and methyl methacrylates is described and methacrylate-based emulsion polymers are discussed as a means of maintaining an isotonic environment for whole cells.

I INTRODUCTION

Biosensors have been investigated for a myriad of end uses but there have been few fundamentally new developments, and more alarmingly very few devices have made the transition from the research laboratory to the market place[1] The major reason for this is the basic incompatibility between the biological component of the sensor and bulk manufacturing techniques[2]. Within the laboratory environment the biological component can be carefully handled and immobilised in near ideal, aqueous conditions without undue loss of activity or specificity. However, bulk manufacturing techniques generally rely on the use of organic solvents and high processing temperatures to ensure rapid throughput of devices and therefore low manufacturing costs. Aqueous, low temperature immobilisation systems which are typically employed in biosensor fabrication techniques slow down the manufacturing process thereby making the devices uneconomical to produce or to market. Many immobilisation techniques have been employed however, which have included dialysis membranes, glutaraldehyde crosslinking, alginate entrapment, and entrapment in a wide variety of polymers [1,3-5].Some of these polymers such poly(vinyl alcohol) and poly(2-hydroxyethyl methacrylate) have been modified to make them patternable by photolithographic techniques[6,7]In general the polymers investigated have the advantage that they are cheap and readily available but they cannot be used within existing microcircuit fabrication techniques since they are water based and additionally in many cases, they require post-fabrication treatment such as crosslinking before use or are the designs which require direct human participation in the manufacturing process.

G.G. Guilbault and M. Mascini (eds.), Uses of Immobilized Biological Compounds, 11–21.

The aim of this paper is to explore the potential of polymeric materials which are routinely used in microstructure fabrication procedures, as materials for the immobilisation of biological recognition elements. The vast majority of polymers employed as photoresists are copolymers based on methylmethacrylate[8].The methacrylate monomer can be conveniently derivatised with functional groups, which will allow covalent attachment of other reagents (eg recognition molecules) and the copolymers can be cast from organic solvents. Ideally this polymer family should also include the capability to be fabricated under aqueous conditions, since in certain cases (eg the use of whole cells or cell components) maintainence of isotonic strength or just an aqueous environment is critical to the stability and function. Solution in both aqueous and non aqeous media for a single polymer is unlikely, without further derivatisation, however, a more satisfactory approach, and one which offers the opportunity for development in alternative printed fabrication techniques, is to exploit colloidal polymers, whose 'aqueous properties' can be tuned for both entrapment immobilisation (eg whole cells) and covalent immobilisation (eg enzymes, antibodies).

It is clear that this strategy offers a method of immobilisation appropriate to a considerable range of sensors. For example, the methacrylate polymer has dielectric properties suitable for use in contact with surfaces at which shear horizontal acoustic waves are launched [9] and behaves as a transparent material in the visible region of the spectrum (cf perspex). For the pupose of demonstration, this paper will consider some electrochemical models.

II ENZYME MODIFIED METHACRYLATE COPOLYMERS

Photoresist materials based on copolymers of methyl and glycidyl methacrylates[10-12] have been reported and significantly glycidyl groups have been employed for a number of years as a means of covalently binding enzymes[13] We have shown that copolymers of methyl and glycidyl methacrylate will bind enzyme (Scheme 1) in a heterogeneous aqueous reaction, and then these modified polymers can be cast from organic solvents in the same way as photoresist [14,15].

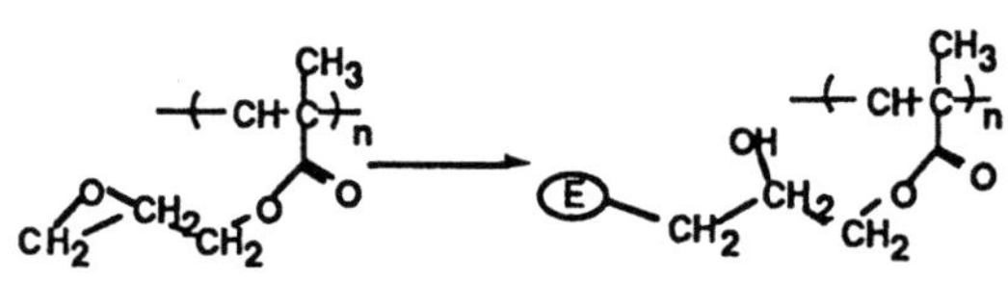

Scheme 1

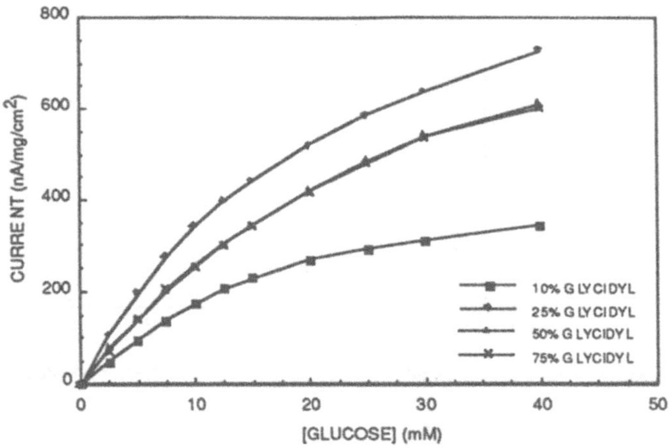

Figure 1: Influence of glycidyl ratio on current signal obtained from the enzyme modified copolymer.

Platinum electrodes coated with an oxidase modified polymer can be employed to create enzyme linked sensor devices. For example, figure 1 shows the electrochemical response to peroxide on a platinum electrode coated with a glucose oxidase modified polymer:

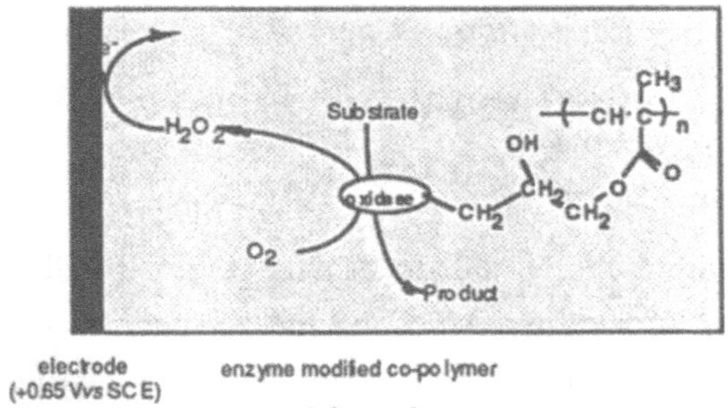

Scheme 2

As can be seen in the figure optimal response to the substrate, glucose is obtained with copolymer mixtures involving 25% glycidyl methacrylate; above this concentration binding becomes independent of epoxide concentration. Similar independence of functional epoxide group concentration has previously been observed by Nishikubo *et al*[10] Several other oxidases have also been employed; figure 2 shows the same polymer modified with lactate oxidase (LOX). This latter oxidase modified copolymer is a particularly useful result, since the enzyme is not especially stable, even in buffer solution and activity loss during other immobilisation processes or while in use, has been a point of concern[16].In this instance the modified polymer retained enzyme activity, even over a period of several months.

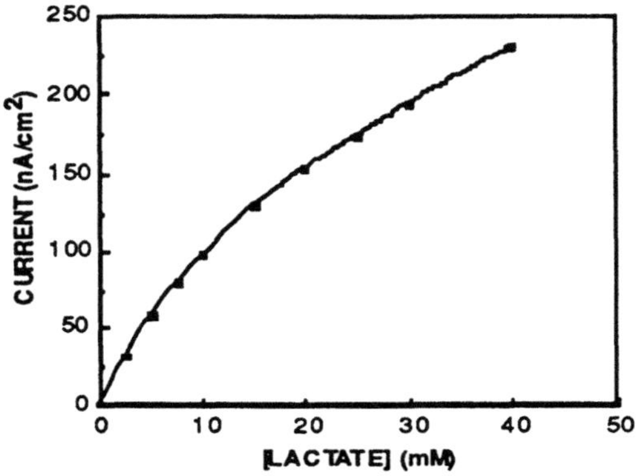

Figure 2 Response to lactate at a lactate oxidase modified methyl-glycidyl copolymer coated electrode.

Optimisation of the copolymer for enzyme immobilisation is not a simple matter of controlling the glycidyl ratio. The polymerisation involes a free radical initiation mechanism, where the relative importance of each pathway will depend upon the rates of each reaction and the concentration of the reacting species:

$$I_2 \rightarrow 2R\cdot \quad (1.)$$

$$R\cdot + M \rightarrow MR\cdot \quad (2.)$$

$$MR\cdot + M_{n-1} \rightarrow M_nR\cdot \quad (3.)$$

$$2R\cdot \rightarrow R_2 \quad (4.)$$

$$R\cdot + M_nR\cdot \rightarrow M_nR_2 \quad (5.)$$

$$2M_nR\cdot \rightarrow (M_nR)_2 \quad (6.)$$

Figure 3, for example shows the molecular weight distribution for copolymers initiated by benzoyl peroxide and AIBN and their respective current response, cast in an amperometric sensor, to glucose. The figure highlights the importance of molecular weight distribution. In general it has been found that narrow band molecular weight polymers can be better tuned to give optimal response.

Successful preparation of a narrow molecular weight copolymer must therefore consider the mechanism of polymerisation in order to direct the output. An initiator (I) with a long half-life and small k_d gives a slow generation of R·. The concentration of R· thus remains low and the reaction follows steps 1 - 3 of the reaction pathway without the termination reactions of steps 4 - 5, which would occur at high concentrations of radical species. With increase in n, the kinetics of step 3 decreases due to stabilisation of the radical $M_nR\cdot$; and step 6, terminating the polymerisation becomes significant as the concentration of $M_nR\cdot$ increases. As can be deduced from the indepenence of M_w on reaction time under these conditions, polymerisation

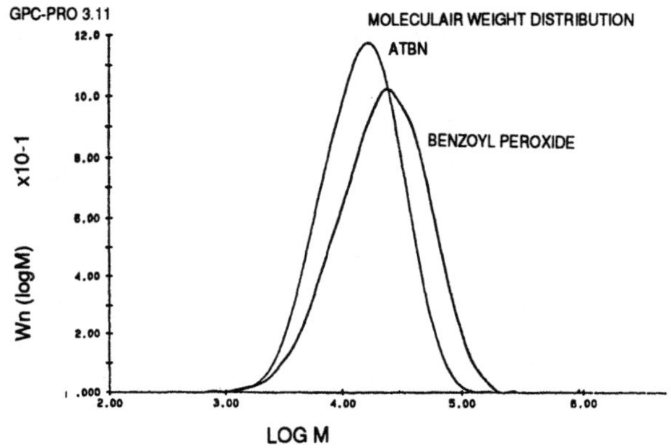

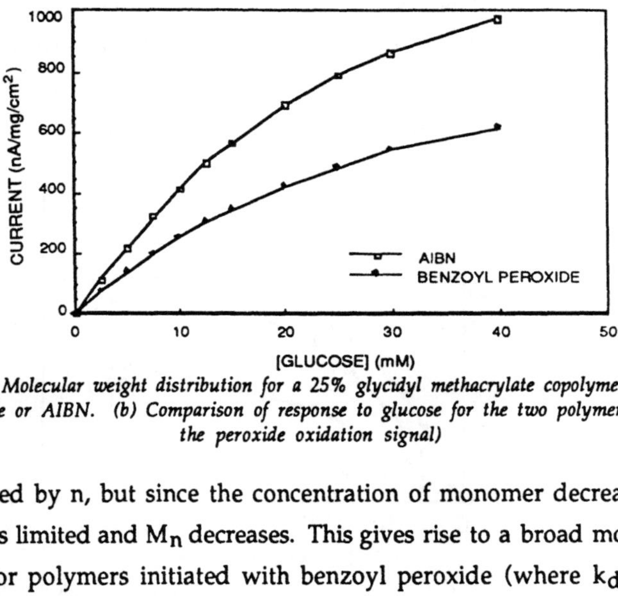

Figure 3: (a) Molecular weight distribution for a 25% glycidyl methacrylate copolymer, initiated by benzoyl peroxide or AIBN. (b) Comparison of response to glucose for the two polymers (measured via the peroxide oxidation signal)

length is limited by n, but since the concentration of monomer decreases with time, step 3 becomes limited and M_n decreases. This gives rise to a broad molecular weight distribution for polymers initiated with benzoyl peroxide (where k_d is small), but results in high yields.

If k_d is high (eg AIBN) the concentration of intiator radicals will be high and the termination steps 4-6 predominate over the propagation step 3. As a result of this the

Initiator Concn (Mol/l)	Polym. Time (hrs)	Polymer Yield %	Mw	Mn	Mw/Mn	Bound Enzyme* %(±0.07)
1.6×10^{-2}	7	52	9,450	6,100	1.55	0.51
8.0×10^{-3}	7	51	16,050	9,600	1.65	0.49
4.0×10^{-3}	7	49	23,000	12,800	1.8	0.43
4.0×10^{-3}	24	95	30,600	14,400	2.15	0.42

Table 1: Molecular weight and enzyme modification data for copolymers of 25%glycidyl/75% methyl methacrylate, prepared with AIBN.

(* calculated from the mass of enzyme consumed divided by mass of polymer used.)

molecular weight is lower and the distribution sharper. In this instance it is easier to control the molecular weight and distribution by control of the R· concentration through varying the initiator concentration. Table 1 shows the molecular weight data for some copolymers initiated with AIBN. For comparison a similar polymerisation carried out with benzoyl peroxide ($8x10^{-3}mol^{-1}$) gives a yield of 88% after 8 hours and M_w 87,350 with M_n 16,450. As would be expected from consideration of the mechanism, high concentrations of AIBN give the lowest molecular weight polymer (M_w 9,450) and as is observed from the results, this copolymer gave the highest signal, when modified with enzyme (figure 4). Indeed the differences in electrochemical response to substrate by an enzyme modified polymer, coated on an electrode has been unambiguously correlated with differences in molecular weight. Nevertheless, the ambiguity still remains as to the cause. The different particulate size in the copolymer must be important during the immobilisation step and incurs different properties on the cast film (particularly diffusion and water-uptake).

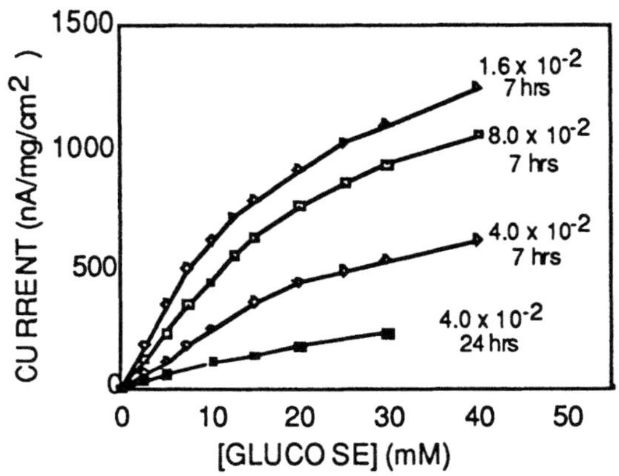

Figure 4: Plot of current due to H_2O_2 versus glucose concentration for a glucose oxidase modified copolymer initiated with different concentrations of AIBN. (time refers to polymerisation time)

In designing a modified polymer which would be suitable for fabrication purposes, the optimal signal conditions must sometimes be partially sacrificed for fabricatability. In the example cited here, extrapolation of the results would suggest that a molecular weight below 9000 might improve the signal further. However, this approaches the limits of usefulness, since the polymer must be precipitated.

III IMMUNOASSAY COMPATIBLE FORMATS

Electrochemically based immunoassay devices must be designed to include an electrochemical label or at least a label which gives rise to an electrochemical signal. Several labelling formats are possible; following on from the oxidase-linked system above, a glucose oxidase conjugate of the antibody can be employed which at constant glucose concentration can be assayed for total enzyme present, via the peroxide signal (Scheme 3).

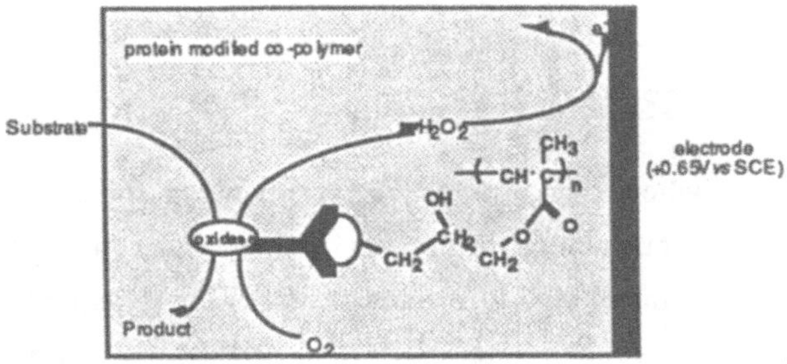

Scheme 3

In terms of the methacrylate copolymers, the glycidylmethacrylate polymer is now used to attach one element of the immunocomplex, the antibody say. In the example shown in Figure 5, the protein, anti-IgG has been bound to a glycidyl-methyl methacrylate copolymer and incubated with the IgG-GOD conjugate. It can be deduced from the data in this figure that the conjugate complexes mostly with the anti-IgG bound to the polymer near the sample interface, since the current response on addition of glucose decreases with polymer thickness. The enyme conjugate

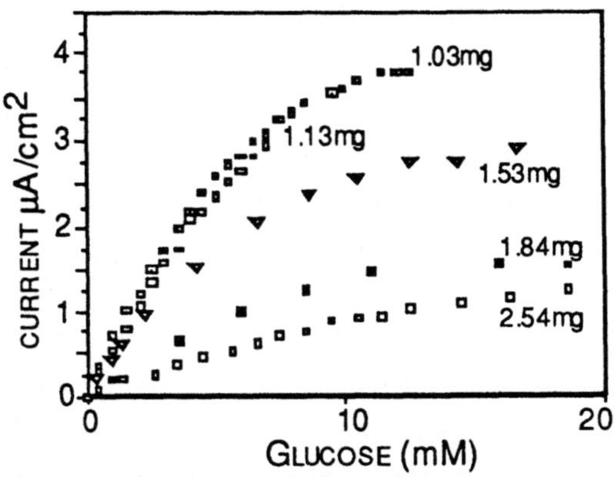

Figure 5: Current due to H_2O_2 versus glucose concentration from a GOD-IgG- conjugate, complexing with an anti-IgG modified polymer coated electrode, for different thicknesses of polymer coating.

behaves in an equivalent manner to the parent enzyme, but this copolymer-anti-IgG/IgG-GOD is ready for use as an immunosensor towards IgG, through a competitive displacement reaction. IgG added to the sample solution selectively displaces the enzyme conjugate .

This simple format can be manipulated for antibody or antigen assays and the enzyme label can be readily substituted for other enzymes if so wished, as long as an electrochemical functionality is retained. In all cases the assay will be related to the rate of displacement of conjugate into the bulk sample, rather than a steady state measurement as achieved in the oxidase-linked systems described in II.

IV WHOLE CELL ENVIRONMENTS

In the immobilisation of whole cells or intact cell components, the situation presents a need for even more rigorous control of the 'environment', since the phase must be isotonic with the cell if the membrane is to remain intact. These requirements appear incompatible with a non-aqueous immobilisation process, and indeed, existing methods have often been based on polysaccharides such as calcium alginate[17,18] and gelatine[19] but their commercial application is limited as they do not have the right hydrating or mechanical properties. They also lack the adhesion onto metal and many of the the surfaces required in sensor technologies. Synthetic aqueous support matrixes like acrylamide have also been explored, but they are liable to deactivate the cell due to the toxicity of the monomers[20] although this can be partially circumvented by using pre-polymerised linear functionalised polyacrylamides which can be cross-linked *in situ* , but this may still lead to some deactivation [21].

Members of the acrylate family described above have also found favour for the entrapment of cells and organelles in pre-polymerised photo crosslinkable resins based on hydroxyethylacrylate and poly(ethylene glycol) or poly (propylene glycol) [22]. The physico chemical nature of these matrixes can be altered via changes in the hydrophobic-hydrophilic balance and the ionic nature, but nevertheless, as the prepolymer is water soluble, cross linking of the matrix is necessary.

An alternative approach which utilises the acrylates is to use film forming emulsion polymers, as a single step immobilisation for whole cells. An emulsion is a stable suspension of particles, with diameters between 100nm and a few micron. In this instance the particles are a polymer, which incurs a film forming property to the emulsion. The polymerisation is activated by initiator and occurs in the monomer droplets or micelles which are formed on the addition of emulsifier (scheme 4), thus creating a complex microphase environment, where aqueous regions are trapped in micellular 'pockets'. Monomer, initiator and emulsifier will thus influence the nature of the resultant product and the fabrication protocol described can be tuned for ink-jet, screen print or other spray techniques, by the inclusion of 'bulking agents' if necessary, such as PEG or PVA or by varying the elements of the fundamental starting materials with different carbon chain lengths and side groups.

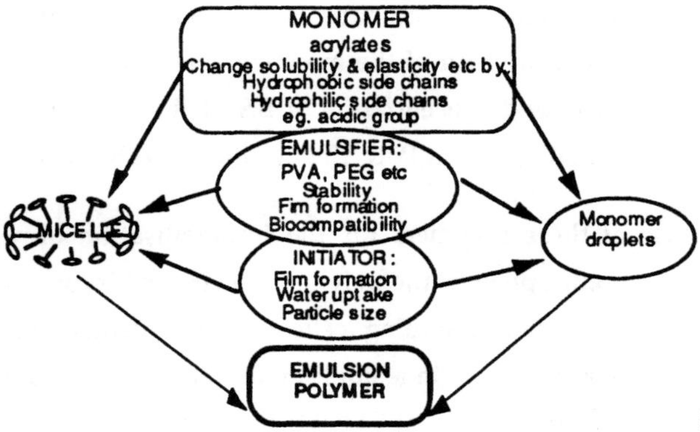

Scheme 4

Table 2 shows the composition of different emulsion polymers based on methylmethacrylate (MMA), methacrylic acid (MAA), sodium methacrylate (NaMAA) and butyl acrylate (BA). Polyvinyl alcohol is added as the emulsifier. The stability of the emulsion and the quality of the film formed after a planar glass surface is coated with the emulsion and subsequently dried are important criteria; the introduction of BA into the polymer for example, improves both the quality of the film and increases the adherence of the coatings to metal and glass surfaces. MAA and NaMAA are pH sensitive, such that at pHs where dissociation is complete the emulsion particle is strongly charged, leading to strong repulsion of the particles and inhibiting film formation. This prevents the use of these emulsions as matrixes for the immobilisation of whole cells. In the absence of pH sensitive monomers, the quality of the coatings improves and is not effected by the increase of the pH to physiological values.

	INITIAL CHARGE			MONOMER ADDITION				INIT-IATOR	STAB-ILITY	FILM FORM^N
	Solvent	Surfactant	Init·	MMA	MAA	NaMAA	BA			
Poly 4	100 ml	5% PVA	0.18 g	0.06 M (6.0 g)	0.02 M (1.72 g)	0.00 M (0.00 g)	0.00 M (0.00 g)	0.5 g	++	+ (pH dependent)
Poly 6	100 ml	2% PVA	0.18 g	0.06 M (6.0 g)	0.00 M (0.00 g)	0.02 M (1.72 g)	0.00 M (0.00 g)	0.5 g	+++	-
Poly 7	100 ml	2% PVA	0.18 g	0.055 M (5.5 g)	0.02 M (1.72 g)	0.00 M (0.00 g)	0.012 M (1.5 g)	0.5 g	+++	+ (pH dependent)
Poly 10	150 ml	2% PVA	0.18 g	0.055 M (5.5 g)	0.013 M (1.1 g)	0.00 M (0.00 g)	0.017 M (2.2 g)	0.5 g	+++	++ (pH dependent)
Poly 12	150 ml	2% PVA	0.18 g	0.055 M (5.5 g)	0.00 M (0.00 g)	0.00 M (0.00 g)	0.030 M (3.84 g)	0.5 g	+++	+++
Poly 14	150 ml	2% PVA	0.18g	0.030 M (3.00 g)	0.000 M (0.00 g)	0.000 M (0.00 g)	0.056 M (7.17 g)	0.5 g	+++	+++

Table 2: Emulsion polymers

The elasticity of the immobilisation matrix can also be modulated by altering the length of the hydrophobic carbon chain of the monomer. For example, Poly 12 and Poly 14 differ only in the increased concentration of BA of the later. Films cast from Poly14 appear soft whereas coatings from Poly 12 have more mechanical strength.

To demonstrate the use of these polymers for the immobilisation of whole cells, photosynthetic bacteria are entrapped in the polymer matrix. The incorporation of the bacterial cells was achieved by suspending cells in the aqueous emulsion and subsequent coating of a metal electrode in a single step procedure. The activity of the photosynthetic electron transport chain (PET) which requires an intact thylakoid membrane was monitored by measuring the photo current in an electrochemical cell. The charge transport between the electrode and the PET in the bacterial cell is realised via the redox mediator diaminodurene[23]. For example figure 6 shows the photo current measured with an electrode coated with films cast from two different emulsion polymers incorporating the cyanobacterium *Synechococcus sp.* and compares entrapment in an alginate matrix behind a dialysis membrane. It is reputed that alginate is an extremely gentle immobilisation procedure and in fact enhances the activity of the photosynthetic system in some cases[17] so this comparison is most encouraging.

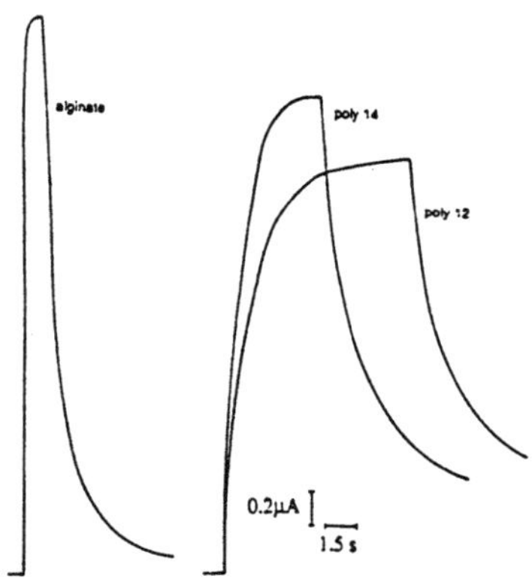

Figure 6: Comparison of the DAD mediated photocurrent obtained from cells of *Synechococcus sp.* immobilised in different matices.

REFERENCES

1. Griffiths D and Hall G. Biosensors-What real progress is being made. Trends in Biotechnology, 1993 *in press*.
2. Moore D F G. Molecular Sensor Technology Eurocourse, Cambridge Sept 1991.
3. Foulds N C and Lowe C R Enzyme entrapment in polypyrrole. Anal Chem 1988, 60, 2473.
4. Kennedy J F, White C A and Melo E H M. Enzyme Immobilisation. Chimicaoggi - Maggio 1988, 21.

5. Cooper J C and Hall E A H, Electrochemical response of an enzyme loaded polyaniline film. Biosensors and Bioelectronics, 1992, **7**, 473 - 485.
6. Urban G, Jobst G, Keplinger F, Ashauer E, Tilado O, Fasching R and Kohl F. Miniaturized multi-enzyme biosensors integrated with pH sensors on flexible polymer carriers for *in vivo* applications. Biosensors and Bioelectronics 1992, **7**, 733-739.
7. Vopel T, Ladde A and Müller. Amperometric glucose sensor with a photo-lithographically pattered enzyme membrane. Anal Chim Acta 1991, **251**, 117-120.
8. Reichmanis E and Thompson L F. Polymer materials for microlithography. Ann Rev Mater Sci 1987, **17**, 235.
9. Gizeli E, New Acoustic wave geometries. PhD Thesis University of Cambridge, 1993.
10. Nishikubo T, Iizawa T, Yamada M and Tsuchiya K Study of Photopolymer XVII. Synthesis of Novel Photosensitive Polymers with pendant photosensitive groups and photosensitizer groups. Polym. Sci. Polym. Chem. 1983, **21**, 1025.
11. Nishikubo T, Iizawa T, Takahashi E and Nono F. Novel Synthesis of self-sensitised photosensitive polymers by addition reaction of pendant epoxide groups in the polymer with some cinnamic esters. Polm Journal 1984, **16(4)**, 371.
12. Nishikubo T, Iizawa Novel synthesis and photochemical reaction of the polymers with pendant photosensitive and photosensitiser groups. Polym Prepr Am Chem Soc1984, **125**, 315.
13. Hannibal-Friedrich O, Chun M and Sernetz M. Immobilisation of β-galactoside, albumin and γ-globulin on epoxy activated acrylic beads. Biotechnol Bioeng 1980, **22**, 157.
14. Hall C E and Hall A A H Hall. The covalent immobilisation of glucose oxidase on methacrylate copolymers for use in an amperometric glucose sensor. Anal Chim Acta, *in press*.
15. Hall C E and Hall E A H. An evaluation of glucose oxidase modified methacrylate copolymers. Anal Chim Acta. *in press*.
16. Katakis I and Heller A. L-α-glycerophosphate and lL-lactate electrodes based on electrochemical "wiring" of oxidases. Anal Chem, 1992, **64**, 1008.
17. Hartmeier W. Co-Immobilization of Enzymes and Whole Cells. Food-Biotech. 1991 **4**, 1, 399-407
18. Kuwada Y. Ohta Y. Hydrogen production by an immobilised cyanobacterium, Lyngya sp. J. Ferment. Technol. **65**, 5, 597-602.
19. de-Alterilis E, Parascandola P, Pecorella M A, Scardi. V. Effect of gelatin-immobilisation on the catalytic activity of enzymes and microbial cells. Biotechnol. Tech. 1988, **2**, 3, 205-210.
20. Klein J. and Wagner F. Methods for the immobilisation of micorbial cells. *In* Immobilized Microbial Cells. (ed. Chibata I and Wingard L B) Academic Press, New York. 1983 pp 12-51.
21. Freeman A Y and Aharonowitz Y. Immobilisation of microbial cells in crosslinked, prepolymerized, liear polyacrylamide gels. Antibiotic production by immobilized Streptomyces clavuligerus. Biotechnol. Bioeng. 1981, **23** 2747-2759.
22. Fukui S. and Tanaka A, Application of Biocatalysts Immobilized by Prepolymer Methods. Advances in Biochemical Engineering?Biotechnology. Immobilized Biocatalysts Saccharomyces Yeasts Wastewater Treatment. (ed. Fiechter A) 1984 Springer Verlag Berlin.
23. Martens, N and Hall, E A H, Diaminodurene as a mediator of a photocurrent using intact cells of cyanobacteria. *Photochem. Photobiol.* 1993, *in press*.
24. Trevan M D and Mal A L, Immobilized algea and their potential for use as biocatalysts. Tibtech 1988, **6** 68-73.

USE OF RED BLOOD CELLS (RBC) AS CARRIERS OF BIOACTIVE COMPOUNDS

ANTONIO DE FLORA
Institute of Biochemistry, University of Genoa, Viale Benedetto XV, 1, 16132, Genoa, Italy.

Abstract Red blood cells (RBC) can be submitted to various procedures that induce the transient opening of pores across their membrane. These pores are large enough to allow internalization of a number of chemicals (up to proteins of 200 kDa) and can be closed again to yield loaded RBC having normal properties in terms of biochemistry, antigenicity, morphology and "in vivo" survival in circulation. Two main unusual functions can result from these treatments: 1) transport of molecules to be released either in the circulatory system or at targeted districts of the organism, 2) new performances as circulating bioreactors for several purposes including detoxification, intraerythrocytic production and release of drugs, vaccination. Several examples of exploitation of RBC as carriers of bioactive compounds are considered, including their use in human therapy.

I. INTRODUCTION

Red blood cells (RBC) from several mammalian species including human can be easily treated according to a variety of procedures in order to encapsulate a number of chemicals within them. Among the various procedures available, those based on a transient hypotonic shock in the presence of the compound to be entrapped, followed by isotonic resealing, prove to be the most versatile and conservative ones. Indeed, they yield normal RBC with respect to biochemical, morphological and antigenic properties and, importantly, the engineered RBC show a normal "in vivo" life-span when they are re-injected back intravenously to the same or to a compatible organism.

Due to these properties, RBC have been proposed as suitable systems for either delivery of chemicals (e.g. drugs) in circulation or display of new functional performances as bioreactors [1-5]. In both cases, the RBC loaded with a variety of compounds (ranging from low M_r molecules to macromolecules of up to 200 kDa proteins) can behave in a time- and site-programmed fashion. Modulation of chronology of release of chemicals depends on intrinsic properties such as the potential of encapsulation in RBC, the possible interaction of the internalized compound with the RBC metabolism, the mechanisms of release of the encapsulated compounds (e.g., diffusion versus mediated transport). Modulation of the site of action ("targeting") can be pursued through additional means of

G.G. Guilbault and M. Mascini (eds.), Uses of Immobilized Biological Compounds, 23–33.

perturbation of the RBC membrane that may be based on chemical treatments (e.g., with bifunctional reagents like glutaraldehyde [6]) or on more selective procedures exploiting mechanisms of biorecognition with target cells and tissues [7]. Targeting is easily affordable within tissues and organs having abundant reticulo-endothelial cells, like macrophages, in which erythrophagocytosis takes place, but it is still a challenging problem when other organs should be selectively reached by the engineered RBC.

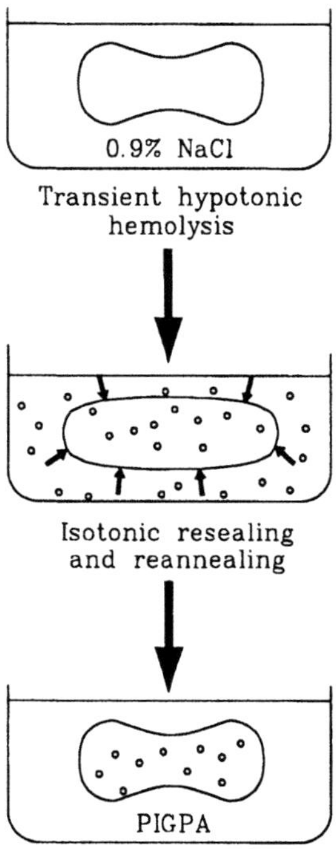

Figure 1. Scheme of the encapsulation procedures involving hypotonic shock-isotonic resealing of RBC.

II. THE RBC ENCAPSULATION TECHNOLOGY

Figure 1 represents a general scheme of the above mentioned procedures by which encapsulation is achieved in a two-step way, involving hypotonic shock-isotonic resealing. Osmotic stress of RBC leads to transient pores across the membrane allowing equilibration between internal and external compartments

8 and can be obtained by two general techniques, i.e. pre-swell and dialysis. Either technique can be used to yield engineered RBC, depending on the compound to be encapsulated and on subsequent utilization of the carriers. For instance, pre-swell seems to be preferable to dialysis if the antitumor drug methotrexate is to be internalized in RBC [9] Dialysis, however, proved to be in our laboratory more versatile and in addition it has been scaled up to meet the requirements (sterility, apyrogenicity, etc.) of transfusion technology in humans [10].

III. SELECTED EXAMPLES OF RBC AS CARRIERS AND BIOREACTORS

Although the difference between the two types of performances may be more semantic than real, "carriers" are meant as loaded RBC releasing their content either in circulation or at selected sites, while RBC can behave as "bioreactors" when encapsulation of a bioactive compounds confers unusual functions. These can be, for example, new metabolic properties following the entrapment of heterologous enzyme proteins, or the ability to release a bioactive compound originated from a RBC-entrapped precursor molecule, or optimization of delivery of Oxygen to peripheral tissues following encapsulation of inositol hexaphosphate [11-12]. The unifying methodology and strategy for RBC as carriers and bioreactors is therefore the internalization of a specific compound.

a) Encapsulation of enzyme proteins.

Sequestration of a protein in RBC provides a specific way of immobilization that can be exploited for a variety of purposes. Unless the intraerythrocytic functioning of an encapsulated heterologous enzyme is detrimental to the RBC survival (which may occur, e.g. following entrapment of the H_2O_2-producing enzyme Glucose oxidase [13]), the "bioreactor" will display unusual and potentially useful metabolic properties as compared to the unloaded RBC. Shielding of a heterologous enzyme protein from the many proteolytic enzymes present in plasma can preserve the protein itself since the triggering of resident proteases in the RBC requires abnormal events, like an enhanced Ca^{2+} influx through the membrane [14-15].The new features of the bioreactor so obtained depend on a number of parameters including permeation of both substrate(s) and product(s) across the RBC membrane and the kinetic properties of the encapsulated enzyme. Decrease of abnormally high blood metabolites or of abnormal molecules usually absent from plasma (e.g., xenobiotics) are typical performances of enzyme-loaded RBC, although not yet arrived at a clinical application. On the contrary, human RBC loaded with L-asparaginase (an enzyme protein purified from bacteria or fungi) have been successfully used in the therapy of hematopoietic diseases [16-17]. These "bioreactors" are efficient in decreasing plasma asparagine that is required for growth of blast cells and, in addition, they seem to prevent the anaphylactic reactions that are observed in some patients given the heterologous asparaginase intravenously.

An additional example of enzyme encapsulation in human RBC concerned the "in vitro" correction of genetic deficiency of Glucose 6-P dehydrogenase (G6PD) [18]. Deficiency is characteristically expressed, at various degrees of severity related to individual structural variants of G6PD, in the red cells that are unable to synthesize proteins. G6PD deficiency, in its Mediterranean variety, is particularly severe as the intrinsic levels of RBC G6PD activity do not exceed 0.15% of the activity measured in normal individuals having the wild-type G6PD (B type) [19]. Indeed, the Mediterranean variety of G6PD deficiency can cause acute hemolytic crises in the affected subjects upon exposure to triggering events such as ingestion of fava beans (favism), or administration of potentially oxidizing drugs [20].

G6PD B, purified to homogeneity from RBC of normal subjects was encapsulated, by the slow dialysis technique, in erythrocytes from G6PD-deficient individuals, up to restoring normal activity levels [18]. The G6PD-reconstructed RBC proved to acquire normal metabolic competence, quite comparable with that of normal RBC, even when they were exposed to oxidizing challenges. The results showed therefore that entrapment of a homologous enzyme protein in RBC involves normal intraerythrocytic functioning, this allowing to identify sites of metabolic control in RBC, i.e. bottlenecks in biochemical pathways. Thus, for instance, overloading normal RBC with hexokinase purified from human placenta showed an enhanced glucose consumption and a more efficient glycolytic pathway, as indicated by the rate of lactate output and by intraerythrocytic levels of intermediate metabolites [21]. Heterologous enzyme proteins, once encapsulated in RBC, may prove not to be regulated intracellularly like their homologous counterparts. This was found to be the case for yeast 6-phosphogluconate dehydrogenase, the enzyme involved with G6PD in the activity of the pentose phosphate pathway 18 Accordingly, the performances of RBC loaded with exogenous enzyme proteins should be accurately investigated under different conditions in order to validate the bioreactor so achieved.

b) Encapsulation of antigens

The natural termination of RBC in the spleen, with its antigen- processing macrophages, suggested to verify the potential of antigen-loaded RBC as transport systems for inducing an immune response. Indeed, RBC loaded with L-asparaginase had been found to elicit the production of antibodies against this heterologous enzyme protein, although no adverse reactions (e.g., anaphylactic shock) were observed because of unavailability of the circulating antibodies to the intraerythrocytic protein [17]. Use of mouse RBC loaded with three bacterial antigens of vaccinal interest (diphtheria, tetanus and pertussis toxoids) showed indeed that RBC are useful carriers for inducing an immune response [22]. This was investigated in terms of production of both specific and neutralizing antibodies (i.e., protecting from the toxic effects of the wild-type toxins). Titers of both antibodies were significantly higher upon administration of RBC-encapsulated diphtheria and tetanus toxoids than following an identical intravenous administration of the two free antigens [22]. The titers were conversely comparable for the RBC-encapsulated and the free pertussis toxoids. Thus, RBC seem to display adjuvant functions with some antigens only, while they ensure proper transport to the antigen-processing cells in all cases. Accordingly, use of antigen-loaded RBC may prove useful for vaccinations when fragile antigens (e.g., recombinant proteins), available in not too minute amounts, should be used.

c) Encapsulation of antitumor agents

These range from conventional cytotoxic drugs to the above mentioned case of L-asparaginase-based bioreactor. However, additional examples of RBC as useful bioreactors in oncology concern the bioconversion, catalyzed by endogenous erythrocyte enzymes, of encapsulated drug precursors to membrane-releasable active drugs. A brief account will be given here to cytotoxic drugs encapsulated in RBC as such and to those originated from RBC-encapsulated prodrugs.

d) Cytotoxic drugs

The case of doxorubicin is paradigmatic. This anthracycline molecule, widely used in the therapy of solid tumors, has a considerable limitation in the cumulative cardiotoxicity it produces in the treated patients. Use of carrier RBC, as well as of other transport systems like liposomes, was originally suggested by the requirement to develop a slow delivery system since sustained intravenous infusion of doxorubicin prevents or attenuates cardiotoxicity remarkably [23]. Doxorubicin features a very high encapsulation potential [24] and exposure of the loaded RBC to 0.1% - 0.3% glutaraldehyde resulted in two important effects: a) "in vitro", considerable slowing down of the drug, b) "in vivo", selective targeting of the carriers to liver, lungs and spleen, depending on the final concentration of the cross-linker. Both effects, observed with murine [7,25], canine [26-27], and human [28-29] RBC, were consistent with the feasibility of locoregional treatment with doxorubicin-loaded and glutaraldehyde-treated RBC. A murine model of metastatic tumor cells spreading to the liver was developed to this purpose and found to be reproducible and useful to assess the efficacy of the RBC-based therapeutic system [30]. Indeed, the effective dose of drug inhibiting at 50% the growth in liver of L1210 cells originally injected in the spleen was one order of magnitude lower in the RBC-encapsulated form than by intravenous injection of free doxorubicin [30]. Estimates of the therapeutic index of free versus RBC-encapsulated doxorubicin gave values of 1.8 and 4.2, respectively.

"In vivo" injection of doxorubicin-loaded and glutaraldehyde-treated RBC in healthy dogs revealed pharmacokinetic properties of considerable interest 27 notably a 5-fold higher systemic exposure than after intravenous administration of the free drug. Treatment of three dogs bearing spontaneous lymphomas with doxorubicin-loaded and glutaraldehyde-treated RBC determined tumor remission, with minimal acute toxicosis and no evidence of cardiotoxicity [31]. However, this treatment produced also remarkable, chronic myelosuppression, whose mechanisms are still completely undefined.

So far, only one human patient, bearing extensive hepatic metastases from colorectal cancer, was submitted to doxorubicin therapy with his glutaraldehyde-treated autologous RBC. Previous scintigraphic studies had confirmed prolonged persistence in the liver of the ^{99m}Tc-labeled RBC following exposure to glutaraldehyde. The autologous RBC could be infused locoregionally in the hepatic artery through an implanted catheter [28]. Distinctive cycles of chemotherapy were made, one with free doxorubicin and two with the RBC-encapsulated drug, at 0.15% and 0.3% glutaraldehyde, respectively, in order to compare the relevant pharmacokinetic parameters. Although these proved to be more favourable following administration of the RBC-encapsulated drug than after injection of free doxorubicin, and in spite of blockade of tumor progression [29] the patient developed grade IV pancytopenia and died after 4 months of therapy.

Development of myelosuppression is therefore a limit to this therapeutic strategy, although the pancytopenia leading the patient to death could have been the cumulative result of three previous and unsuccessful cycles of conventional chemotherapy [29]. Alternatively, myelotoxicity could be due to altered kinetics of release of blood cells from the bone marrow [31] or to some still unexplained reaction of glutaraldehyde with doxorubicin occurring in the loaded RBC. While treatment of additional patients awaits the unequivocal clarification of the underlying mechanisms of myelosuppression, a study is currently in progress for the scintigraphic screening of patients affected by various liver tumors (both primary and metastatic) and receiving their autologous ^{99m}Tc-labeled and glutaraldehyde-treated RBC through intravenous administration (rather than through the hepatic artery). Completion of such screening is expected to identify those patients in whom extensive hepatic targeting takes place and who might benefit from this chemotherapeutic approach.

e) Prodrug to drug conversion

Human RBC were shown to be efficient bioreactors catalyzing the conversion of encapsulated 5-F-2'-deoxyuridine monophosphate (FdUMP) to the membrane-diffusable drug 5-F-2'-deoxyuridine (FUdR) [32]. The latter is a highly toxic antitumor agent, particularly efficient against hepatic metastases from colorectal carcinoma. This type of bioconversion, whereby the prodrug-loaded RBC behave as miniaturized factories of specific drug molecules, is related to the intraerythrocytic activity of endogenous enzymes. Other examples of such unusual performances are available for the production and subsequent release of cytosine arabinoside [33] and of 2',3'-dideoxycytidine [34] from RBC-encapsulated prodrugs.

The potential of this approach requires careful evaluation of the properties of the bioreactor, in order to maximize its performances in terms of drug production and release. Specifically, attention should be paid to achieving time- and site-controlled patterns of delivery of the relevant drugs. Both issues are characterized by many problems. Targeting, as already mentioned, is not particularly difficult when the liver is the desired site of action. However, like in the case of intraerythrocytic FUdR production, liver involvement by primary or metastatic tumors could result in disordered anatomical structure (due to artero-venous shunts, or to portal hypertension, or to compressions on the vasculature, or to other reasons) and preclude accordingly access of the engineered RBC to the tumor itself. Indeed, a marked variability in the patterns and extent of targeting is emerging in the above mentioned scintigraphic screening of patients receiving their ^{99m}Tc-labeled and glutaraldehyde-treated autologous RBC. Targeting by means of alternative techniques (e.g., neuraminidase treatment of RBC [7]) should be considered in the future for human therapy.

The other hot topic concerning RBC as bioreactors is the time-dependent pattern of intraerythrocytic drug production and of subsequent release. Thus, for example, the hydrolase responsible for the FdUMP to FUdR conversion has a very high activity and the bioreactor would be too efficient to meet the pharmacokinetic requirements of sustained delivery of FUdR [32]. Although co-encapsulation of suitable enzyme inhibitors (e.g., ATP and GTP) is efficient "in vitro" in slowing down the process, it seems not to be adequate for "in vivo" applications [32]. Therefore, alternative strategies are being considered, and especially the design and synthesis of more metabolically remote precursors to be encapsulated in the RBC for a multi-step conversion to the desired drug. In the case of FUdR production, one of such compounds, i.e. a 5-F-2'-deoxyuridine dimer having a 3',5'-phosphodiester bridge between the two moieties, has been recently synthesized and found to be efficiently converted, when encapsulated in human RBC, to FUdR through FdUMP as an intermediate compound [35]. Therefore, in this case, the bioreactor is still an efficient producer and releaser of FUdR, although at a consistently lower rate than the single-step bioreactor already described. Other precursor molecules suitable to the purpose are under study.

IV. PROSPECTS

RBC seem to have inherent properties of natural, biocompatible and ubiquitous cells that make them ideal carriers and bioreactors. Their easy availability and simplicity of manipulation are additional advantages potentially exploitable for biomedical applications. Intensive studies are however required in order to elucidate those still unknown biochemical characteristics that represent bottlenecks in the use of engineered RBC. Specific subjects awaiting elucidation are a better characterization of RBC metabolism and its regulation, a deeper insight into structural and functional properties of the RBC membrane (and particularly analysis of receptorial molecules to be exploited for modulation of RBC metabolism and for triggering biorecognition mechanisms - e.g. by macrophages, lymphocytes, endothelial cells), studies on the process of opening and closure of pores across the RBC membrane during the various entrapment procedures. Without a careful consideration of these and of other basic properties of RBC and of their mechanisms of interaction with the organism, the RBC encapsulation technology would just stay as a versatile recipe, mostly to be used for model studies. Upgrading the thecnology through a better knowledge of mechanisms and processes and spreading information on its potential applications to clinicians, are compulsory steps toward practical exploitation in human and veterinary medicine.

ACKNOWLEDGMENTS

This study was supported by C.N.R. Target Projects " Biotechnology and Bioinstrumentation" and "A.C.R.O.", by the "Associazione Italiana per la Ricerca sul Cancro", by the "AIDS Project" of the Italian Ministry of Health, and by the "Biotechnology Action Programme" of EEC.

REFERENCES

1. J.R.. DeLoach and U. Sprandel (eds.), Red Blood Cells as Carriers for Drugs, Biblthca Haemat. vol.51, Karger, Basel, (1985).
2. C. Ropars, M. Chassaigne, C. Nicolau (Eds), Red blood cells as carriers for drugs. Potential therapeutic applications, Advances in Biosciences vol. 67, Pergamon Press, Oxford, (1987a).
3. R. Green and J.R. DeLoach, (Eds), Resealed erythrocytes as carriers and bioreactors. Advances in Biosciences, vol.81, Pergamon Press, Oxford, (1991).
4. M. Magnani, J.R. DeLoach, (Eds), The use of resealed erythrocytes as carriers and bioreactors. Advances in Experimental Medicine and Biology, vol. 326, Plenum Press, New York, (1992).
5. A. De Flora, The impact of biotechnology on diagnosis (Albertini et al., eds), Elsevier Science Publisher B.V., Amsterdam 223, (1985).
6. J.R. DeLoach, C. Barton, Am. J. Vet. Res., 43, 2210, (1982).
7. E. Zocchi, L. Guida, U. Benatti, M. Canepa, L. Borgiani, T. Zanini and A. De Flora, Biotechnol. Appl. Biochem., 9, 423, (1987).
8. J.F. Hoffman, J. Gen. Physiol., 42, 9, (1958).
9. C.A. Kruse, C.L. Freehauf, K.R. Patel, J.D. Baldeschwidler, Biotechnol. Appl. Biochem., 9, 123, (1987).
10. C. Ropars, M. Chassaigne, M.C. Villéreal, G. Avenard, C. Hurel and C.

Nicolau, in J.R. DeLoach and U. Sprandel (Eds), Biblthca Haemat. vol.51, 84, Karger, Basel, (1985).

11. M.C. Villéreal, C. Bailleul, M. Chassaigne and C. Ropars, in C. Ropars, M. Chassaigne and C. Nicolau (Eds), Advances in Biosciences, vol. 67, 55, Pergamon Press , Oxford, (1987).
12. B. Teisseire, C. Ropars, M.O. Vallez, R. Hérigault and C. Nicolau, J. Appl. Physiol., 58, 1810, (1985).
13 E. Zocchi, U. Benatti, L. Guida, M. Tonetti, G. Damonte and A. De Flora, in C.Ropars, M. Chassaigne and C.Nicolau, (Eds), Advances in Biosciences, Red Blood Cells as Carriers for Drugs, Potential Therapeutic Applications, Pergamon Press Oxford, 67, 95, (1987).
14. S. Pontremoli and E. Melloni, Annu. Rev. Biochem. 55, 455, (1986).
15. M. Grasso, A. Morelli and A. De Flora, Biochem. Biophys. Res. Commun., 138, 87 (1986).
16. R. Kravtzoff, I. Desbois, M. Chassaigne, J.P. Muh, J.P. Lamagnère, P. Colombat and C. Ropars, in R. Green and and J.R. DeLoach (Eds), Advances in Biosciences vol. 81, 127, Pergamon Press, Oxford, (1991).
17. R. Kravtzoff, I. Desbois, C. Dainel, P. Colombat, J.P Lamagnère , M. Chassaigne and C. Ropars, In M. Magnani and J.R. DeLoach (Eds), Advances in Experimental Medicine and Biology vol. 326, 175, Plenum Press, New York, (1992).
18. A. Morelli, U. Benatti, F. Salamino, B. Sparatore, M. Michetti, E. Melloni, S. Pontremoli and A. De Flora, Arch. Biochem. Biophys., 197, 543, (1979).
19. A. Morelli, U. Benatti, L. Lenzerini, B. Sparatore, F. Salamino, E. Melloni, M. Michetti, S. Pontremoli and A. De Flora, Blood, 54, 642, (1981).
20. A. Yoshida, E. Beutler, A. Motulsky, Bull. W.H.O., 45, 243, (1971).
21. M. Magnani, L. Rossi, M. Bianchi, G. Fornaini, U. Benatti, L. Guida, E. Zocchi and A. De Flora, Biochim. Biophys. Acta, 972, 1, (1988).
22. C. Polvani, A. Gasparini, U. Benatti, A. De Flora, S. Silvestri, G. Volpini and L. Nencioni, Biotechnol. Appl. Biochem., 14, 347, (1991).
23. S.S. Legha, R.S. Benjamin, B. Mackay, M. Ewer, S. Wallace, M. Valdivieson, S.L. Rasmussen, G.R. Blumenschein, E.J. Freireick, Ann. Intern. Med., 96, 133, (1982).
24. A. De Flora, U. Benatti, L. Guida, E. Zocchi, Proc. Natl. Acad. Sci.

U.S.A., 83, 7029, (1986)

25 E.Zocchi, M. Tonetti, C. Polvani, L. Guida, U. Benatti and A. De Flora, Biotechnol. Appl. Biochem., 10, 555, (1988).

26. M. Tonetti, B. Astroff, W. Satterfield, A. De Flora, U. Benatti and J.R. DeLoach, Biotechnol. Appl. Biochem., 12, 621, (1990).

27. M. Tonetti, B. Astroff, W. Satterfield, A. De Flora, U. Benatti and J.R. DeLoach, Am. J. Vet. Res., 52, 1630, (1991).

28. M. Tonetti, C. Polvani, E. Zocchi, L. Guida, U. Benatti, P. Biassoni, F. Romei, A. Guglielmi, C. Aschele, A. Sobrero and A. De Flora, Eur. J. Cancer, 27, 947, (1991).

29. M. Tonetti, E. Zocchi, L. Guida, C. Polvani, U. Benatti, P. Biassoni, F. Romei, A. Guglielmi, C. Aschele, A. Sobrero and A. De Flora, in M. Magnani and J.R. DeLoach (Eds), The Use of Resealed Erythrocytes as Carriers and Bioreactors, Plenum Publishing Corporation, vol. 326, 307, (1992).

30. E. Zocchi, M. Tonetti, C. Polvani, L. Guida, U. Benatti and A. De Flora, Proc. Natl. Acad. Sci. U.S.A., 86, 2040, (1989).

31 C.M. Matherne, W.C. Satterfield, A. Gasparini, A.B. Astroff, R.D. Schimdt,L.D. Rowe and J.R. DeLoach, Submitted for pubblication.

32 A. De Flora, E. Zocchi, L. Guida, C. Polvani, U. Benatti, Proc. Natl. Acad. Sci. U.S.A., 85, 3145, (1988).

33. J.R. DeLoach, J. Appl. Biochem., 4, 533, (1982).

34. M. Magnani, M. Bianchi, L. Rossi, V. Stocchi, Biochem. Biophys. Res. Commun., 164, 446, (1989).

35. M. Mazzei, T. Grandi, A. Balbi, A. De Flora, U. Benatti, A. Gasparini and G. Damonte, ECB6, 6th European Congress on Biotechnology, Florence, June 13-17 (1993) Abstract in press.

THEORETICAL AND EXPERIMENTAL ASPECTS FOR IMPROVEMENT OF ELECTROCHEMICAL BIOSENSORS BY VARIOUS KINDS OF IMMOBILIZATION

MAURICE COMTAT, HELENE DURLIAT, ALAIN BERGEL, SALIHA BACHA AND MARIELLE MONTAGNE
Laboratoire de Génie Chimique et Electrochimie, URA CNRS 192, Université Paul Sabatier, 118 Route de Narbonne, 31062 Toulouse Cedex France

Abstract : The performances of amperometric detection biosensors - response time, sensitivity, linearity range - are predicted by calculation for different kinds of immobilisation or confinement of the catalyst(s). The cases of monoenzyme electrodes with the enzyme in solution confined by a semi-permeable membrane near the electrode or grafted covalently to a membrane are considered. The calculations were extended to bienzyme sensors with the enzymes grafted to one side, or to both sides of a membrane. The experimental results obtained with L-lactate, D-lactate and L-glutamate specific electrodes are presented and used to qualify the model.

I INTRODUCTION

Amperometric detection biosensors are known for the diversity of substrates they are able to assay with good sensitivity and precision [1-2]. For a given substrate, numerous propositions are possible depending on whether the enzyme is confined in contact with the transducer by a semi-permeable membrane, immobilized in a gel or covalently grafted on to a membrane or on to the electrode itself.

Usually, optimization of these sensors is accomplished empirically depending on the goal of the application. In general, the performances sought are a short response time, high precision and sensitivity, associated with a wide response linearity range. Some theoretical studies are available concerning either potentiometric or amperometric detection sensors. However, it should be noted that very often the systems are treated in the steady state [3-9] and only rarely take into account transient phenomena [10-16]. It would nevertheless be useful to have a theoretical tool making it possible to select beforehand all the biochemical, geometrical and electrochemical parameters to

G.G. Guilbault and M. Mascini (eds.), Uses of Immobilized Biological Compounds, 35–45.

obtain the required performances. The purpose of this paper is to provide some elements for proposing such a strategy. The models considered concern mono- and bienzyme devices with catalysts positioned in different ways close to the transducer. The finite volume technique (FV) is used to solve the differential equations. The model is qualified by comparing the theoretical and the experimental results obtained in the case of L-lactate. Experimental results are also obtained for D-lactate and L-glutamate bienzyme biosensors for various kinds of immobilization.

II GENERAL PRINCIPLES

Figure 1 presents the various types of sensor studied. There are two kinds of sensors: those with the enzymes in solution in a reaction chamber whose volume is defined by the flat electrode surface and a semi-permeable membrane and those with the enzymes immobilized on a membrane in contact with the electrode. In all cases, the membrane is held in place with an O-ring. Figure 1 also reviews the parameters that condition sensor performance levels.

Electrode	Reaction chamber	Membrane	Analysed solution
Potential Nature of the material	Thickness Enzyme concentration	Thickness Permeability	Concentration of the different species Stirring

a

Electrode	Electrolyte film	Membrane	Analysed solution
Potential Nature of the material	Thickness	Position and Concentration of the enzymes	Concentration of the different species Stirring

b

FIGURE 1 : Scheme of the two types of considered sensors. The more important parameters are indicated. (a) the enzyme(s) is or are in solution; (b) the enzyme(s) is (or are) immobilized on a preactivated membrane.

Various types of reaction are involved. With a monoenzyme electrode, the substrate S reacts with the hexacyanoferrate III:

$$S + Fe(CN)_6^{3-} \rightarrow P + Fe(CN)_6^{4-}$$

Such is the case for L-lactate with baker's yeast L-lactate dehydrogenase.
With bienzyme sensors, the dehydrogenase catalyzes the following reaction:

$$S + NAD^+ \rightarrow P + NADH + X$$

NADH is electroactive on a platinum electrode at around 0.75 V versus saturared calomel electrode (sce). In order to reduce the important overvoltage corresponding to the direct detection of NADH through anodic oxidation, a second reaction is introduced, the oxidation of NADH catalyzed by a diaphorase:

$$NADH + Fe(CN)_6^{3-} \rightarrow NAD^+ + Fe(CN)_6^{4-} + H^+$$

These successive reactions are the same for the L- and D-lactate and L-glutamate ion assays.
In each case electrochemical detection is based on the $Fe(CN)_6^{4-}$ ion oxidation on a platinum electrode that is used as a transducer:

$$Fe(CN)_6^{4-} \rightarrow Fe(CN)_6^{3-} + e$$

On the second electrode occurs the reaction $Fe(CN)_6^{3-} + e \rightarrow Fe(CN)_6^{4-}$ from hexacyanoferrate III put in excess in the solution to be assayed.

The sensor response curve shows the electrolysis current variation as a function of time, with 80 mV imposed between the two electrodes.

The general principle is used for the three sensors presented :
- the specific L-lactate monoenzyme electrode implemented the following reaction:

$$CH_3CHOHCOO^- + 2\,Fe(CN)_6^{3-} \rightarrow CH_3COCOO^- + 2\,Fe(CN)_6^{4-} + 2H^+$$

- the oxidation reactions of D-lactate and L-glutamate by NAD^+ were as follows:

$$CH_3CHOHCOO^- + NAD^+ \rightarrow CH_3COCOO^- + NADH + H^+$$

$$^-OOCCH(NH_2)CH_2CH_2COO^- + NAD^+ + H_2O \rightarrow {}^-OOCCOCH_2CH_2COO^- + NADH + NH_4^+$$

III THE MODEL

The enzymatic reaction is : $S + Ox \xrightarrow{Enz} P + Red$

where S is the substrate to be assayed, Ox and Red are the oxidized and reduced forms of the cosubstrate. The kinetics of the reaction is assumed to be of "ping-pong" type, i.e. with a rate of the form:

$$\frac{k\,[Enz]}{1 + \frac{K_S}{[S]} + \frac{K_{Ox}}{[Ox]}}$$

Amperometric detection is based on the rapid oxidation of Red on the electrode Red $\rightarrow$ Ox + n e .

The mass balance equations written for the substrate S and the two forms of the Ox and Red cosubstrate fully describe the system.

Dimensionless parameters

All the concentrations and the kinetics parameters are divided by $[Ox]^b$ rather than $[S]^b$ (see list of symbols) because $[Ox]^b$ is an operating parameter which is maintained constant during the measurements, while $[S]^b$ varies from one assay to another.

- Concentrations: $S = [S]/[Ox]^b$; $Ox = [Ox]/[Ox]^b$; $Red = [Red]/[Ox]^b$
- Michaelis constants: $Km_S = K_S/[Ox]^b$; $Km_{Ox} = K_{Ox}/[Ox]^b$
- Diffusion coefficients: $Di_{S2} = D_{S2}/D_{R1}$; $Di_{S1} = D_{S1}/D_{R1}$; $Di_{R2} = D_{R2}/D_{R1}$
- Time and distance: $T = \dfrac{t\, D_{R1}}{\delta_m{}^2}$; $X = x/\delta_m$
- Thickness of the diffusion layer: $Lo = \delta/\delta_m$
- Current density: $\psi = \dfrac{i\, \delta_m}{n\, F\, D_{R1}\, [Ox]^b}$

- Catalytic activity: $Rm_{vol} = \dfrac{k\, [Enz]\, \delta_m{}^2}{D_{R1}\, [Ox]^b}$; $Rm_{sur} = \dfrac{k\, [Enz]\, \delta_m}{D_{R1}\, [Ox]^b}$

The relative catalytic activity Rm_{vol} is quite similar to the parameter σ^2 used in other works (7) $\sigma^2 = Rm_{vol}/Km_{Ox}$. It is directly related to the square of the Thiele modulus:

$$\phi^2 = \frac{Rm_{vol}\, [S]^b}{[S]^b(1+Km_{Ox}) + Km_S}$$

which gives the ratio of the potential rate of enzymatic reaction to the potential rate of diffusion. This notation is useful for physical interpretation of the results, but we keep here the parameter Rm_{vol} which allows an easier writting of the equations. Rm_{vol} is obviously equal to zero when the enzyme is only linked to the surface of the membrane.

The mass balance equations

These are:

$$\frac{\partial S}{\partial T} = Di_{S1} \frac{\partial^2 S}{\partial X^2} - \frac{Rm_{vol}}{1 + \dfrac{Km_S}{S} + \dfrac{Km_{Ox}}{Ox}}$$

$$\frac{\partial Ox}{\partial T} = \frac{\partial^2 Ox}{\partial X^2} - \frac{Rm_{vol}}{1 + \frac{Km_S}{S} + \frac{Km_{Ox}}{Ox}} \qquad [1]$$

$$\frac{\partial Red}{\partial T} = \frac{\partial^2 Red}{\partial X^2} - \frac{Rm_{vol}}{1 + \frac{Km_S}{S} + \frac{Km_{Ox}}{Ox}}$$

within the membrane if enzymatic reaction occurs, with the assumption of the equality of the diffusion coefficients of the two forms of the cosubstrate.

One have:

$$\frac{\partial S}{\partial T} = Di_{S2}\frac{\partial^2 S}{\partial X^2} \quad ; \quad \frac{\partial Ox}{\partial T} = Di_{R2}\frac{\partial^2 Ox}{\partial X^2} \quad ; \quad \frac{\partial Red}{\partial T} = Di_{R2}\frac{\partial^2 Red}{\partial X^2} \qquad [2]$$

within the diffusion layer. Balance of the mass flux densities at the membrane-diffusion layer interface should involve the enzymatic surface kinetics. No partition effect is taken into account:

$$Di_{S2}\frac{\partial S}{\partial X}\bigg|_{x=1+} = Di_{S1}\frac{\partial S}{\partial X}\bigg|_{x=1-} + \frac{Rm_{sur}}{1 + \frac{Km_S}{S} + \frac{Km_{Ox}}{Ox}}$$

$$Di_{R2}\frac{\partial Ox}{\partial X}\bigg|_{x=1+} = \frac{\partial Ox}{\partial X}\bigg|_{x=1-} + \frac{Rm_{sur}}{1 + \frac{Km_S}{S} + \frac{Km_{Ox}}{Ox}} \qquad [3]$$

$$Di_{R2}\frac{\partial Red}{\partial X}\bigg|_{x=1+} = \frac{\partial Red}{\partial X}\bigg|_{x=1-} - \frac{Rm_{sur}}{1 + \frac{Km_S}{S} + \frac{Km_{Ox}}{Ox}}$$

The boundary conditions

- at the electrode:

$$\frac{\partial S}{\partial X}\bigg|_{x=0} = 0 \quad ; \quad \frac{\partial Ox}{\partial X}\bigg|_{x=0} = -\frac{\partial Red}{\partial X}\bigg|_{x=0} \quad ; \quad Red\,|x=0 = 0 \qquad [4]$$

- current density is provided by:

$$\psi = \frac{\partial Red}{\partial X}\bigg|_{x=0} \qquad [5]$$

- at the diffusion layer-solution interface: Red = 0; $Ox^b = 1$; $S = S^b$ [6]

IV SOLUTION

The FV method is based on an integral form of the equations. The computational domain is divided into elementary volumes within which the integration is carried out (Fig.2a). Using the grid-point cluster shown in figure 2.b, we focused attention on the point located at distance X_i from the electrode. The dotted lines show the west (w) and east (e) frontiers of the cell centered in X_i. Let ΔX_i be the length of the cell and δX_i the distance between the points X_{i+1} and X_i. Integration of equation [1] over the cell provides:

$$\int_w^e \frac{\partial S}{\partial T} dX = Di_{S1} \int_w^e \frac{\partial^2 S}{\partial X^2} dX - \int_w^e \frac{Rm_{vol}}{1 + \frac{Km_S}{S} + \frac{Km_{Ox}}{Ox}} dX \qquad [7]$$

Every partial derivative equation is transformed into a set of algebric equations with a tridiagonal form.

The algebric systems obtained cannot be directly solved. The presence of a non-linear kinetic term introduces the concentration values at the n+1 iteration in the expression of the coefficients of the matrix. Iterations are carried out for every time steps.

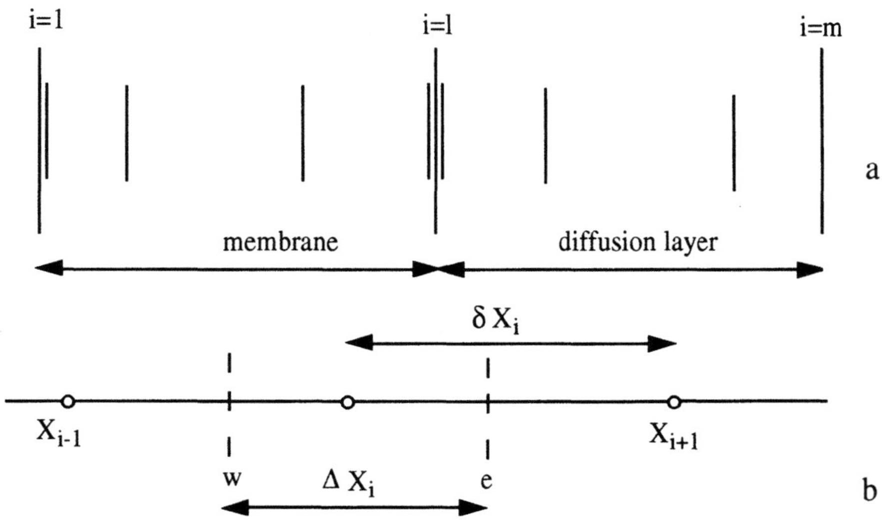

FIGURE 2 : Space discretization of the computational domain when m=10. a: For Finite Volume method, the physical domain: membrane and diffusion layer is subdivided into five subdomains. The space steps can be independently chosen within each. b: Example of grid-point cluster.

V EXPERIMENTAL METHOD

Apparatus

Spectrophotometric measurements were made on a Hewlett Packard 8451 spectrophotometer with a thermoregulated cell. A Tacussel PRT 20*2X potentiostat was used as a power supply to impose 80 mV between the enzyme electrode and the auxiliary electrode. A Sefram Luxytrace recorder was used to record the response curves.

Chemicals

Merck provided the pure salts used to prepare the solutions.

The following solutions were used:

- pH 7.0: mixture of monopotassium and disodium phosphate, 0.1 M phosphate
- pH 8.3: 0.1 M 3-[1,1-Dimethyl-2-hydroxyethyl)amino]-2-hydroxypropanesulfonic acid (AMPSO) acidified with HCl
- pH 9.1: 0.1 M AMPSO acidified with HCl or a mixture of mono- and disodium carbonate with a carbonate concentration of 0.1 M.

Sigma provided the biochemical reagents and the enzymes with the exception of the L-lactate dehydrogenase extracted from baker yeast, of the *Hansenula anomala* strain, that was supplied by Sétric Génie Industriel (Toulouse, France). The diaphorase E.C.1.8.1.4 was extracted from *Clostridium kluyveri*, the L-lactate dehydrogenase (L.LDH) E.C.1.1.1.27 was extracted from rabbit muscle, the D-lactate dehydrogenase (D.LDH) E.C.1.1.1.28 was extracted from *Leuconostoc mesenteroides*, the L-glutamate dehydrogenase (L.GDH) E.C.1.4.1.3 was extracted from beef liver.

Immobilization of the enzymes on the membrane

The preactivated membrane was of the Immunodyne TM type (Pall Industries, ref: BIA065HC5). Immobilization procedure was as follows: 10 μl of enzymatic solution were placed on the membrane during 10 minutes. Two rinse procedures were performed with the carbonate solution, the first rinse being performed in the presence of 1 M KCl.

With the membrane dipped into a solution with the required compounds, each of the enzymes present on the membrane catalyzed the specific reaction described above and it was possible to follow reaction progress using spectrophotometry. Regular stirring made it possible to assume that there was no concentration gradient in the solution. The quantity of enzyme fixed was correlated with the initial rate of the substrate disappearance reaction and this initial rate was measured each three hours. Between each measurement, the membrane was stored in the phosphate solution. A decrease in the initial rate of the reaction with time together with an increase in enzyme activity in the storage solution indicated the enzyme had dropped off the membrane.

VI RESULTS AND DISCUSSION

Previous work has already made it possible to determine several parameters [17]. The thickness of the reaction chamber for enzymes in solution should be to the order of 30 μm. Similarly, the semi-permeable membrane, that serves as a diffusion barrier for the substrates, is made from cellophane with a thickness of approximately 40 μm. Measurement temperature is regulated to 25 °C.

Various parameters have been studied.

- Stirring: figure 3 provides an example of response curves obtained with a D-lactate

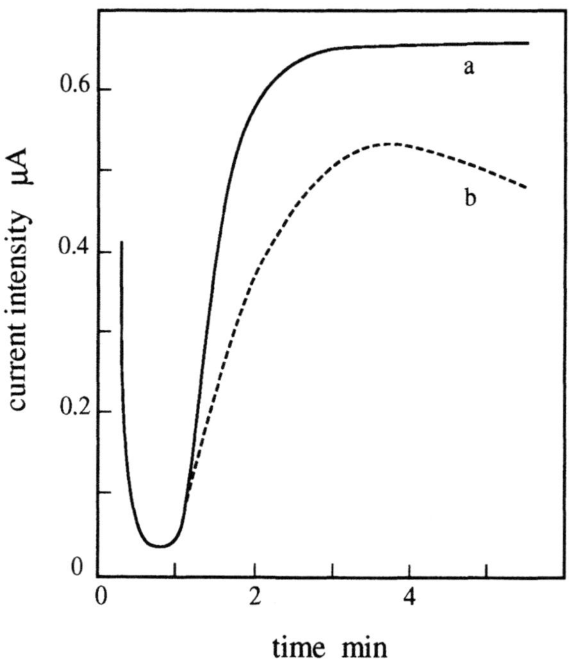

FIGURE 3 : Influence of stirring on the response curve. D-lactate dehydrogenase and diaphorase are in solution. D-lactate 0.5 mM. (a) stirred solution; (b) unstirred solution.

specific bienzymatic sensor, with the solution containing the substrate stirred (3a) or not (3b). In the first case, intensity becomes constant, whereas in the second case, intensity continously decreases when a diffusion layer forms between the membrane external surface and the solution to be assayed. The intensity considered as sensor response corresponds to the value where there is a plateau or a maximum. Stirring the solution also leads to an increase in the intensity measured and a slight decrease in the response time. All the experiments presented below involved stirring the solution, this was ensured using a magnetic stirrer with sufficient bar rotation rate to avoid concentration gradients near the membrane.

Disposition of the immobilized enzymes: figure 4 shows an example of response curves for a L-lactate specific bienzymatic sensor. The two enzymes are either fixed on the same side of the membrane (4a) or on each of the two sides (4b). For the two sensors proposed in this paper, optimal layout is co-immobilization on the same side of the membrane which should be opposite to the platinum disk. The layer of enzymes is protected since it is no longer in contact with the solution. Electrolyte thickness between the electrode and the membrane is the lowest possible in order to minimize the cell electric resistance.

This figure also shows the good agreement between the experimental and the calculated

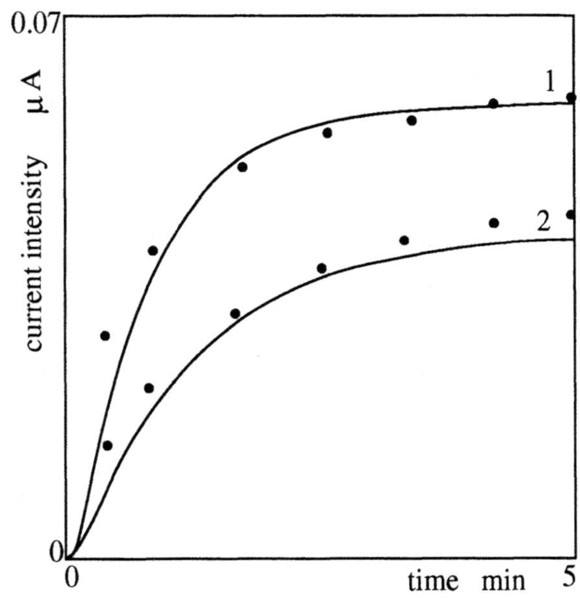

FIGURE 4 : Experimental (——) and calculated (...) response curves for the D-lactate bienzyme sensor.

1 : the enzymes are grafted on the same side of the membrane

2 : the enzymes are grafted on both sides of the membrane

results. Table 1 summarizes essential results obtained with these sensors, some of these results have already been published [17-19].

The reproducibility of the results depends of the number of enzymes used either in solution or immobilized on membrane. It is about 2% on the basis of 50 measurements in the case of the use of one enzyme and becomes 4% when two enzymes are used. In a general manner, response times vary between 0.5 and 5 minutes with the shortest times corresponding to highest enzyme activity levels. However, even the longest times are compatible with the applications planned for the sensors described here. Comparison of results obtained through spectrophotometric enzyme assay and through using the sensors shows a difference of less than 4 % on the basis of 50 measurements. The concentration ranges in which there is a linear response are sufficiently extensive (2 to 3 decades) for these sensors routine applications.

substrate	enzymes	mode of enzy-me confining	linearity range (mM)	stability (days)	initial slope (μA/mM)
L-lactate	yeast L-LDH	immobilized	2.10^{-3} - 5	60	1.1
	"	in solution	0.1 - 10	30	0.8
	muscle L-LDH				
	+ diaphorase	immobilized	5.10^{-3} - 0.15	2	0.2_5
	"	in solution	0.2 - 8	30	0.5
D-lactate	D-LDH				
	+ diaphorase	immobilized	5.10^{-3} - 0.15	2	0.2
	"	in solution	2.10^{-2} - 6	40	0.9
L-glutamate	L-GDH				
	+ diaphorase	immobilized	2.10^{-2} - 2	28	0.3

TABLE I : Specifications of the biosensors for L-lactate, D-lactate and L-glutamate.

NOMENCLATURE

Di, dimensionless diffusion coefficient; F, Faraday's constant; i, current intensity; k, rate constant; K, Michaelis constant; Km, dimensionless Michaelis constant; Li, dimensionless thickness of the electrolyte layer; Lo, dimensionless thickness of the diffusion layer; n, number of electron exchanged; Rm, dimensionless catalytic activity; t, time; T, dimensionless time; x, distance from the electrode surface; X, dimensionless distance from the electrode surface; b, value in the bulk solution
δ, diffusion layer thickness ; δi, electrolyte layer thickness; δm, membrane thickness; ψ, dimensionless current intensity
Subscripts : Ox, oxidized form of the cosubstrate; S, substrate; sur, relative to a surface; vol, relative to a volume; 1, in the membrane; 2, in the diffusion or the electrolyte layer

REFERENCES

1. A.P.F. Turner, I. Karube and G.S. Wilson, Biosensors: Fundamentals and Applications, Oxford University Press, Oxford, (1987).
2. F. Scheller and F. Schubert, Biosensors, Elsevier, New-York, (1990).
3. D.A. Gough and J.K. Leypoldt, Anal.Chem., 51, 439, (1979).
4. J.K. Leypoldt and D.A. Gough, Biotechnol.Bioeng., 24 2705, (1982).
5. J.K. Leypoldt and D.A. Gough, Anal.Chem., 56, 2896, (1984).
6. D.A. Gough, J.Y. Lucisano and P.H.S. Tse, Anal.Chem., 57, 2351, (1985).
7. P.N. Bartlett and R.G. Whitaker, J.Electroanal.Chem., 224, 27, (1987).
8. W.J. Albery , A.E.G. Cass and Z.X. Shu, Biosens.Bioelectron., 5, 367, (1990).
9. T.Tatsuma and T.Watanabe, Anal.Chem., 64, 625, (1992).
10. L.D. Mell and J.T. Maloy, Anal.Chem., 47, 299, (1975).
11. A. Bergel and M. Comtat, Anal.Chem., 56, 2904, (1984).
12. J.J. Kulys, V.V. Sorochinski and R.A. Vidziunaite, Biosensors, 2, 135, (1986).
13. P.H.S.Tse and D.A.Gough, Anal.Chem., 59, 2339, (1987).
14. J.Y.Lucisano and D.A.Gough, Anal.Chem., 60, 1272, (1988).
15. T.Tatsuma, T.Watanabe and Y.Okawa, Anal.Chem., 64, 630, (1992).
16. F.Battaglini and E.J.Calvo, Anal.Chim.Acta, 258, 151, (1992).
17. H. Durliat, M. Comtat and A. Baudras, Clin. Chem., 22, 1802, (1976).
18. H. Durliat, C. Causserand and M. Comtat, Anal. Chim. Acta, 231, 309, (1990).
19. M.Montagné, H.Durliat and M.Comtat, Anal.Chim.Acta, 278, 25, (1993).

DEVELOPMENT AND CHARACTERIZATION OF POLYMERIC BIOMATERIALS FOR BIOACTIVE AGENT IMMOBILIZATION

Menemşe KİREMİTÇİ
Hacettepe University, Chemical Engineering
Department, Beytepe, Ankara, TURKEY

Abstract. Polymeric biomaterials based on the poly(HEMA), polyurethane, and polystyrene structures have found a wide range of applications in biotechnology and biomedical engineering. Among the applications, immobilization of bioactive agents, i.e. enzymes, pharmaceutical agents, mammalian cells and microorganisms has particular interest in industrial and clinical fields. This paper describes the development of novel polymeric carriers by the incorporation of ionic comonomers and their immobilization characteristics for mammalian cells and plasma proteins.

1. INTRODUCTION

Many advanced biotechnological processes require that the immobilized bioactive agents (e.g. microorganisms, drugs, enzymes or proteins, plant or animal cells) because of their significant advantages, i.e. reusability, stability, continuous processing, reduced cost, long biological life, etc. [1-3]. There are also important in therapeutic and analytical applications including bioaffinity chromatography [4], biosensors [5], controlled release technology [6], and artificial organ systems [7]. There are a great number of alternative immobilization procedures, e.g. covalent coupling, adsorption, encapsulation, crosslinking, etc., to choose from when immobilization of bioactive agents is desired [8]. Important advances continue to be made in these techniques, the development and characterization of the carrier systems and potential applications. The carrier system may be a variety of solid supports, e.g. natural or synthetic polymers, glass, ceramics, etc. [9-11]. Among them synthetic polymers are widely used since they are available in a wide variety of compositions, properties and forms. They may be fabricated readily into complex shapes, e.g. hollow fibers, tubes, films, membranes, discs, micro-to-macro beads, fibers, etc.

G.G. Guilbault and M. Mascini (eds.), Uses of Immobilized Biological Compounds, 47–56.

Recently, we have developed a number of polymeric carriers both in membrane and bead forms with different surface properties such as chemical and physical structures, electrical charge, wettability, etc. which have been implicated as important parameters effective on immobilization characteristics of bioagents. These surfaces are poly(HEMA) and its copolymers with ionic comonomers, such as acrylic acid (AA) and dimethylaminoethyl-methacrylate (DMAEMA), polyetherurethanes (PU), crosslinked polystyrenes (PS) and their modified forms by glow-discharge treatment. Anchorage-dependent mammalian cells in different morphologies (i.e. fibroblastic BHK, epithelial MDBK cell lines and primary cells), microbial cells (i.e. C.albicans and E.coli), and plasma proteins (i.e. BSA and fibrinogen) are the bioactive agents used in our studies. Immobilization of these agents to the polymeric surfaces is achieved by adsorption process. This is physical adsorption occurred by the van der Waals forces, ionic and dispersion forces or other hydrophilic and hydrophobic attractions of agents to the solid support. In the case of cells this process is called "attachment" [12]. Attachment of cells to various inert supports is governed by both electrostatic and hydrophobic interactions [13]. A few solid support is available for cell attachment. The main reason is that solid materials carry generally a negative surface charge and the surfaces of cells is also negative at neutral pH, thus an electrostatic repulsion occurs. That is why, the surface properties must be modified in order to prevent cell-support electrostatic repulsion. These include coating the material with a positively charged layer, grafting amine groups or adsorbing polycations [14]. In addition, surface energy and microroughness may help attachment of cells. It is possible to modify surfaces by altering its energy [15].

This paper reviews preparation methods and characterization of polymeric carriers developed by us and their adsorption properties for mammalian cells and plasma proteins.

2. EXPERIMENTAL

2.1. Preparation and Characterization of Polymeric Supports

Functional poly(HEMA) beads

Acrylic polymer beads comprising combination of 2-hydroxyethylmethacrylate (HEMA) with acrylic acid (AA), dimethylaminoethylmethacrylate (DMAEMA) and methylmethacrylate (MMA) were prepared by suspension copolymerization technique of Mueller et al.[16] modified by Kiremitçi [17]. In this modified technique, polymerization was carried out in an aqueous phase containing MgO (BDH,UK) which was used as the stabilizing and salting-out agent. A three necked flask with a blade-type stirrer was used as the polymerization reactor. The mixture containing the appropriate amount of monomers (HEMA, Aldrich Chemical Co., or 25%(v/v) AA, BDH, UK or 25%(v/v)

DMAEMA, Aldrich Chem. Co, or 25%(v/v) MMA, BDH,UK), the crosslinker, ethyleneglycol-dimethacrylate (EGDMA, Merck, Germany), and the initiator, 2,2-azobis(2-methyl propionitrile) (AIBN, Merck, Germany) was added into the suspension medium which contained 0.015g MgO/ml aqueous phase. AIBN and EGDMA concentrations were kept constant at 0.0015 mol/mol HEMA, and 0.3 mol/mol HEMA, respectively. The monomer/aqueous phase ratio was 0.1(v/v) and stirring speed was 380 rpm. The mixture was maintained at 70°C for 3 h and then at 90°C for 1 h. After cooling, the polymer beads were washed with HCl solution and then water and dried in a vacuum desiccator. The resulting beads, poly(HEMA), poly(HEMA-AA), poly(HEMA-DMEAMA) or poly(HEMA-MMA), have a very narrow size distribution and excellent spherical shape.

The surface and bulk sturucture of the beads were characterized by using a multiprong technique including solid-state ^{13}C CP/MAS NMR, FTIR, ESCA, elemental analysis and DSC. Their molecular analysis was performed by buoyancy method and water transport properties were observed microscopically. The details of these methods and their discussion were reported in our previous publications [18,19].

Ionogenic poly(HEMA) membranes

Hydrophilic poly(HEMA) membranes were prepared by bulk polymerization of HEMA monomer in the presence of the initiator, AIBN and the crosslinker, EGDMA. In order to produce ionogenic poly(HEMA) membranes, acrylic comonomers, i.e. carboxyl group containing AA, and amino group containing DMAEMA were also added to the polymerization mixture. The casting solutions that contain proper amounts of the comonomers (5; 10; 20; 30% v/v), the crosslinker (0.04 ml/ml HEMA) and the initiator (0.002 g/ml HEMA) were casted between flat polypropylene plates and were heated at 60°C about 24 h for bulk copolymerization and crosslinking. The resulting membranes were approximately 300±25μ thick.

Polymer hydrophilicity was assessed through contact angle and equilibrium water content (EWC) measurements. Surface structure was investigated by SEM and ATR-FTIR. In addition, atomic composition of gels was determined by CNH elemental analysis. The details were given in elsewhere [20].

Polyurethane membranes

The PU membranes with different bulk and surface properties were prepared by using different solvent systems, i.e. dioxane (BDH, England), tetrahydrofurane (THF, Merck, Germany) and their mixture. Solutions (8%w/v) of polyurethane, Pellethane® 2363-80 A (Up John, The Netherlands) were cast on a glass surface, then the membranes were precipitated in an aqueous medium containing 20% glycerol (v/v) and they were extensively washed with distilled water and ethanol, then dried at 40°C under vacuum. The structural characterization methods described for poly(HEMA) membranes were applied to the PU membranes [21].

Polystyrene beads

Spherical crosslinked polystyrene beads (Biobeads SX2, styrene-2% divinyl benzene) having a diameter of 37-74 μ were purchased from BioRad Laboratories Inc. The surfaces of these beads were treated with alkylamines in a RF glow discharge apparatus. Two primary amines, i.e. allylamine (ALAM) and ethylenediamine (EDA) and a tertiary amine, triethylamine (TEA) were used as the modification agents. These amines were purchased from Aldrich Chem. Co. Treatment time and power of the plasma were varied while holding the initial monomer pressure and flow rates constant. All modified PS beads were characterized by ESCA [22].

2.2 Immobilization Procedure

Cell ımmobilization

The Baby Hamster Kidney cells (BHK21) which have fibroblast-like morphology and the Madine Darby Kidney cells (MDBK) which have epithelial morphology were used as model cell lines and BK cells isolated from bovine kidney were used as primary cells in order to investigate the effects of different cell morphologies on the immobilization procedure. The culture medium was Dulbecco's Modified Eagle's Medium (DME, Gibco, USA) for BHK cells and Eagle MEM (Gibco, USA) for MDBK and BK cells. These media were supplemented with pyruvate plus 10% fetal calf serum (FCS, DIFCO, USA) and 100μg/ml gentamycin (Gibco,USA).

Polymeric substrates sterilized by UV radiation and swollen in PBS were placed into six-well TCPS plates (Falcon 3046 35-mm, 6-well plates, Becton Dickinson, NJ) and cell culture studies were performed in static culture conditions at 37°C in a humidified 5% CO_2 atmosphere. The inoculation density was $2x10^5$ cells/ml for cell lines and $1x10^5$ cells/ml for BK cells. The number of attached cells to the polymer surfaces were determined by hemocytometric countings. That is why cells were harvested from surfaces by trypsinization procedure and then trypan blue exclusion method was used. The cells adhering to and spreading on the membranes were photographed with a light microscope (Olympus, CK).

Immobilization of plasma proteins

Immobilization of plasma proteins were performed with freshly prepared solutions of bovine serum albumin (BSA, Fraction V, M.W. 60000) and fibrinogen (M.W. 360000) in phosphate buffer solution (PBS, 0.019 M NaH_2PO_4, 0.0046 M $NaH_2PO_4.2H_2O$, pH 7.35). Proteins were supplied from Sigma (St. Louise, MO.). During these studies, buffer conditioned polymeric supports were immersed into the protein solution (concentration range: 10-200 μg/ml) at 37°C with gentle shaking and at specific time intervals samples were withdrawn from the reservoir and protein concentration was determined florimetrically at 342 nm emission wavelength for both of proteins. Fluorimetric measurements were taken in Jasco Model FP-550 Spectrofluorometer using 1 cm^2 quartz cells. The incubation times for immobilization were 24h. The adsorption studies were carried out comparatively with control protein solution which does not contain polymeric surface. The decrease in fluorescence intensity, ΔF, was followed as a function of time. It was observed that at the same environmental conditions, ΔF increases with the presence of membranes as compared with control solution. The difference between these two ΔF values refers the quantity of adsorbed protein to the surface. Here, the loss in protein from solution could be ascribed to surface adsorption.

3. RESULTS AND DISCUSSION

3.1. Surface and Bulk Properties of The Polymeric Carriers

Functional poly(HEMA) beads

Functional poly(HEMA) beads with very narrow size distribution (from 98μ to 2mm in different diameters) and with excellent spherical geometry were obtained by the suspension copolymerization tecnique as described before. Solid-state ^{13}C CP/MAS NMR results showed that, copolymerization and crosslinking reactions occur simultaneously and a highly crosslinked structure is obtained. Density and water content of the poly(HEMA) beads were determined as 1.29 g/ml and 44.7 % (on a dry basis), respectively. The presence of ionic comonomers caused lower values than that of poly(HEMA) beads (such as water content of MMA- containing beads was 25.8 % and their density was 1.19 g/ml). Further more, the beads were sterilizable as due to high degradation temperature (about 160° C).

Ionogenic poly(HEMA) membranes

Structural analysis has shown that, the incorporation of AA and DMAEMA in the bulk structure causes important variation in the EWC values of membranes. These values are greater than 40 % and they depend upon the pH of swelling medium. While AA-containing membranes swell significantly at pH values above 5, DMAEMA-containing ones swell at pH values below 7. Moreover, EWC values increase with increasing comonomer concentration up to 600 % in the case of 30 % AA-containing membranes.Phase transition behaviour of the membranes were discussed in elsewhere [23]. The surface free energy was found approximately same (~60-63 ergs/cm^2) for all surfaces as due to the high bulk water fraction of membranes. SEM study revealed uniform, smooth surfaces and ATR-FTIR studies showed that the presence of AA and DMAEMA in the structure leads to the different structures [20].

Polyurethane membranes

In our previous studies it was shown that, different microporous polymeric membranes can be obtained by changing type of solvent in the casting solution [21]. Here, PU-1, PU-2 and PU-3 surfaces prepared with pure THF, 50 % THF-50 % dioxane, and pure dioxane solutions, respectively, were used. According to the SEM study, in the case of dioxane, a symmetric, highly porous structure was obtained, while the presence of THF in the structure giving lowest porosity. The EWC values of PU membranes were changed between 2.6 and 23.7 % by using different solvent systems. The highest EWC was obtained for PU-3 membrane prepared with the mixture of THF-dioxane. The air contact angle (θ_{air}) of membranes roughly decreases from 76.1 to 51.0 degrees with increasing water content from 2.6 to 23.7 %.

Polystyrene beads

Exposure of the beads to the amine plasmas have resulted in the introduction of nitrogen on the surface of the PS beads. An increase in the treatment time (from 1 to 10 min.) results in an increase in surface nitrogen for all three monomers. The beads exposed to the EDA plasma for 10 min. have the largest amount of nitrogen (12.7 %) on their surfaces while the beads exposed to the ALAM and TEA plasmas contain 6.7 and 5.1 % nitrogen, respectively [22].

3.2. Cell Attachment Results

Attachment characteristics of fibroblasts and epithelial cells to the poly(HEMA) and PU based membranes were studied and the results were listed in Table 1 together with their average doubling time values. The values for attachment of cells were reported as a percentage of the total cells available for attachment during 2 h inoculation period. As shown in this table fibroblastic cell attachment is higher than the epithelial cells for both of the PU and poly(HEMA) membranes. BK cells show similar adhesion characteristics to BHK cells. The comparison of PU surfaces indicates that cell attachment is better on the PU-3 membrane than the others as due to the its porous and moderately wettable structure. In the series of poly(HEMA) based membranes the most pronounced cell attachment was seen on the positively-charged copolymer, and increasing DMAEMA concentrations within the structure caused increasing percent adhesion. However, as due to the toxic effects of higher DMAEMA concentrations than 10%, cell growth was not supported by 20% and 30% DMAEMA-containing surfaces. Similar results were obtained for functional poly(HEMA) beads [24].

Table 1. Attachment and growth characteristics of BHK21 and MDBK Cells to the PU and poly(HEMA) Based Membranes.

Membrane	BHK Cells		MDBK Cells	
	Attached cells (%)	Doubling time (h)	Attached cells (%)	Doubling time (h)
P(HEMA)	11.5±7.8	–	–	–
P(HEMA-10%DMAE)	78.4±5.1	14.50	52.5±6.0	28.29
P(HEMA-20%DMAE)	83.7±2.2	24.84	68.4±4.1	36.19
P(HEMA-30%DMAE)	89.3±4.1	–	80.3±2.8	–
P(HEMA-10%AA)	62.6±4.5	–	46.2±8.6	–
P(HEMA-20%AA)	66.3±6.9	–	48.3±8.6	–
P(HEMA-30%AA)	72.3±3.1	–	57.4±4.7	–
PU-1	69.9±3.7	nd	48.6±5.3	nd
PU-2	78.6±2.3	14.00	53.5±5.3	28.00
PU-3	85.6±3.2	nd	65.6±4.1	nd
PS(control)	100.0±3.8	12.00	100.0±3.5	13.20

(Inoculation period was 2h for cell attachment)
nd: not determined.

Adhesion experiments on alkylamine plasma modified PS beads were performed to determine the variability in attachment behaviour of BHK cells to these different surfaces. Table 2 shows the relation between surface nitrogen content and attached cell percent. In the case of unmodified PS beads no adhesion was seen. However, an increase in surface nitrogen content significantly increases the number of attached cells. All modified beads tested, showed a rapid attachment rate within 30 min after seeding and about 80% of the total cell attachment to the surfaces were achieved in this period.

Table 2. Relationship between surface nitrogen content and adhesion of BHK21 cells on the modified PS beads.

PS Beads	Surface Nitrogen (%)	Attached BHK21 Cells (%)
Unmodified PS	–	–
modified by		
ALAM;1min*	4.2	38.0±4.0
ALAM;10min	6.7	59.0±3.0
EDA;1min	2.5	70.0±2.5
EDA;10min	13.3	74.0±2.0
TEA;1min	4.5	46.0±1.5
TEA;10min	5.1	64.0±2.0

*treatment time

3.3. Adsorption Results of Plasma Proteins

The proteins used in this study are BSA and fibrinojen that have tryptophan residues which detectable by fluroscence, therefore adsorption to polymeric surfaces was followed by fluorescence spectroscopy. Table 3 indicates the adsorbed amounts of BSA and fibrinogen on ionogenic poly(HEMA) membranes at 30 min. and 1 h adsorption time. Although there is no important difference between the surface free energies of AA and DMAEMA containing membranes, they show extremely different adsorption kinetics. This result can be explained by means of different surface charges, charge densities and water sorption abilities of membranes. In the presence of DMAEMA the adsorption process is driven by the electrostatic interaction of the amino groups and the net negative charge of the BSA molecules (izoelectric point:4.8) at the adsorption pH (i.e. pH 7.0). Therefore larger amounts of BSA adsorbed on the DMAEMA-containing surfaces than the AA-copolymers and increasing DMAEMA concentrations increases both of the adsorption rate and adsorbed quantity of BSA. Because of high hydrophilicity of AA-copolymers, reversible protein adsorption occurs on these surfaces. Moreover, compared with BSA, the amount of fibrinogen adsorbed on DMAEMA-containing surfaces was high. This behaviour may be releated with molecular properties of BSA and fibrinogen possibly influencing their surface activity. In addition, it was shown that protein adsorption was greater at pH 7.0 compared to pH 3.0 and 5.0 in two different buffer solutions (Tris HCl and PBS) [25].

Table 3. Adsorbed quantities of BSA and Fibrinogen on the ionogenic poly(HEMA) membranes.

Membrane	Adsorbed BSA ($\mu g/cm^2$)		Adsorbed Fibrinogen ($\mu g/cm^2$)	
	1h	24h	1h	24h
P(HEMA)	–	–	0.07±0.05	0.08±0.03
P(HEMA-10%DMAE)	0.37±0.07	0.51±0.07	1.04±0.06	0.96±0.06
P(HEMA-20%DMAE)	0.69±0.06	1.16±0.06	1.45±0.04	1.07±0.04
P(HEMA-30%DMAE)	1.18±0.07	1.67±0.07	1.59±0.03	1.22±0.05
P(HEMA-10%AA)	0.02±0.07	0.25±0.08	0.32±0.03	0.30±0.04
P(HEMA-20%AA)	0.04±0.06	0.60±0.07	0.49±0.05	0.62±0.05
P(HEMA-30%AA)	0.06±0.06	0.85±0.07	0.69±0.02	0.82±0.06

Protein concentration:10μg/ml

Similar adsorption characteristics were observed for the functional poly(HEMA) beads. Figure 1 shows the adsorption isotherms of BSA onto beads. Measurements were made at seven albumin concentrations and the adsorbed quantities at the end of 1h were determined. This figure indicates a variation in surface concentration from about 0.5 to 4.5 mg/g beads for the poly(HEMA-DMAEMA) beads. Therefore this isotherm represents a multimolecular layer of BSA on the surface.

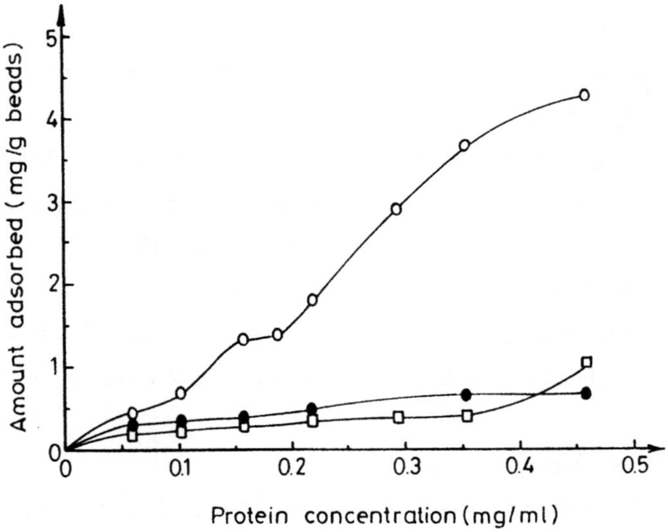

Figure1. Adsorption isotherms of BSA onto the functional poly(HEMA) beads.
(o: HEMA-DMAEMA; •: HEMA-MMA; □: HEMA-AA beads)

4. CONCLUSION

In literature, a number of modification methods for polymeric substrata have been described to increase the immobilization ability of bioactive agents. This study reviews modification of poly(HEMA), PU and PS surfaces by the incorporation of ionic comonomers in different routes. In addition, a correlation between bioagent (i.e. mammalian cells and plasma proteins) adsorption and the structural properties of substrata have been demonstrated. Thus, it was found that the presence of amino groups and pores in the structure significiantly increases adsorption abilities of anchorage- dependent mammalian cells and plasma proteins.

5. REFERENCES

1. O.R. Zaborsky, Immobilized Enzymes, CRC, Press, Cleveland, Ohio, U.S.A. ,1973.
2. B. Mattiasson, Immobilized Cells and Organelles, CRC, Press, Boca Raton, Florida, 1983.

3. N. Mozes, F. Marchal, M.P. Hermesse, J.L. Van Haecht, L. Reuliaux, A.J. Leonard and P.G. Rouxhet, Biotechnol. Bioeng., 30, 439 (1987).
4. P.D.G. Dean, W.S. Johnson, F.A. Middle, Affinity Chromatography, IRL Press, Oxford, 1985.
5. J. Schultz, Scientific American,August, 48 (1991).
6. R. Baker, Controlled Release of Bioactive Materials, Academic Press, New York, 1980.
7. C.D. Scott, Enzyme Microb. Technol., 9, 66 (1987).
8. G.G. Guilbault, in Enzyme Engineering, L.B. Wingard, Ed., Wiley, New York, 1972.
9. S. Reuveny, A. Mizrahi, M. Kotler, A. Freeman, Biotechnol. Bioeng., 25, 469 (1983).
10. O.H. Friedrich, M. Chun, M. Sernetz, Biotechnol. Bioeng., 22,157 (1980)
11. H.J. Lydon, T.W. Minett, B.C. Tighe, Biomaterials, 6, 396 (1985).
12. N. Mozes, P.G. Rouxhet, in Biofilms: Science and Technology, L.F. Melo, T.R. Bott, M. Fletcher, B. Capdeville (Eds.), Kluwer Academic Publishers, The Netherlands, p.125, 1992.
13. P.G. Rouxhet, N. Mozes, Wat. Sci. Tech., 22, 1 (1990).
14. S. Hattori, J.D. Andrade, J.B. Hibbs, D.E. Gregonis, R.N. King, J.Colloid Interface Sci., 104, 72 (1985).
15. J.L. Dewez, A. Doren, Y.L. Schneider,R. Legras, P.G. Rouxhet, in Interfaces in new materials , P. Grange and B. Delman (eds.), Elsevier Applied Science, London, p. 84 1991.
16. K.F. Mueller, S.J. Heiber, W.L. Plankl, US Pat. 4 224 427 (1980).
17. M. Kiremitçi, H. Çukurova, Polymer, 33, 15, 3257(1992).
18. M. Kiremitçi, H. Çukurova, Polymer International, 30, 357 (1993).
19. H. Çukurova, M. Kiremitçi, Doğa-Turkish Chemistry Journal, (1993) (in press).
20. M. Kiremitçi, M. Pulat, A. Peşmen, İ. Gürhan J.of Biomaterials Applications, 7, 250 (1993).
21. M. Kiremitçi, A.İ. Şerbetçi, R. Çolak, E. Pişkin, Clinical Materials, 8, 9 (1991).
22. E. Pişkin, M. Kiremitçi, E.B. Denkbaş, İ. Gürhan, W.R. Gombotz, A.S. Hoffman, Clinical Materials, 11, 171 (1992).
23. A. Peşmen, A. Gönen, M. Kiremitçi, Polymer Gels, 1993 (submitted).
24. M. Kiremitçi, İ. Gürhan, E. Pişkin, Biotechnol. and Appl. Biochem., 14, 170 (1991).
25. E. Gök, M. Kiremitçi, S. Ateş, Analytical Letters, 25 (1), 49 (1992).

TWO-DIMENSIONAL (GLYCO)PROTEIN CRYSTALS AS IMMOBILIZATION MATRIX FOR FUNCTIONAL MOLECULES

SETA KÜPCÜ, ANGELA NEUBAUER, CLAUDIA HÖDL, DIETMAR PUM, MARGIT SÁRA, UWE B. SLEYTR

Zentrum für Ultrastrukturforschung und Ludwig Boltzmann Institut
für Molekulare Nanotechnologie
Gregor-Mendel Straße 33, A-1180 Wien, Austria

Abstract: Crystalline cell surface layers (S-layers) represent the outermost cell envelope component in many bacteria. S-layers as isoporous two-dimensional protein or glycoprotein crystals are composed of identical subunits exhibiting free carboxyl, amino and hydroxyl groups in well defined positions and orientations. Due to these characteristics of the S-layer protein, the hexagonally ordered S-layer lattices from *Bacillus stearothermophilus* PV72 and *Clostridium thermohydrosulfuricum* L111-69 were used as immobilization matrix for a broad spectrum of enzymes and small nucleophilic compounds. For this purpose, the carboxyl groups of the S-layer protein were activated with carbodiimide. In the case of S-layer glycoproteins, the hydroxyl groups from the carbohydrate chains were either activated with cyanogen bromide or were converted into carboxyl groups by succinylation.
To demonstrate the application potential of S-layers as immobilization matrix, they were used for the development of amperometric and fiber optic biosensors. For this purpose enzymes such as glucose oxidase, ß-fructosidase and mutarotase were covalently attached to S-layer matrices after carbodiimide activation of the carboxyl groups from the S-layer protein. Sensing layers obtained that way were used in combination with an electrode as electrochemical sensor or an O_2-sensitive fluorescent dye as optical transducer in biosensor prototypes.

I INTRODUCTION

The immobilization of enzymes by fixation to an insoluble matrix provides a means of using enzymes repeatedly in a dynamic way as heterogeneous specific catalysts, thereby alleviating the problems of separation and cost. Furthermore, this method enables the use of enzymes in automated processes. Enzymes are termed "immobilized" if their mobility has been limited by widely differing methods of immobilization which are entrapment inside a molecular lattice, physical adsorption, extensive cross-linking or covalent attachment to an insoluble matrix. The last approach is certainly the most desirable. In order to characterize an enzyme reaction process by using immobilized enzymes, the basic chemical data on the matrix must be known. The density of functional groups and the structure, such as pore size and distribution, are essential characteristics for enzyme immobilization. The reproducible production of chemical polymeric matrices is a complicated task. Most chemical polymers are created by polymerizing a mixture of the monomers, thereby producing a distribution of chain lenghts and an approximately random sequence if more than one type of monomer is present.

In this paper we report about a novel type of immobilization matrix: crystalline surface layer

G.G. Guilbault and M. Mascini (eds.), Uses of Immobilized Biological Compounds, 57–66.

proteins from bacterial cells (S-layers). S-layers as isoporous two-dimensional monomolecular protein crystals are composed of identical subunits and their functional groups have a well defined position and orientation [1-4].

II CRYSTALLINE BACTERIAL CELL SURFACE-LAYERS (S-LAYERS)

Two-dimensional crystalline bacterial cell surface layers represent the outermost cell wall component in many prokaryotic cells [1] S-layer lattices can show oblique (p2), trigonal (p3), square (p4), or hexagonal (p6) symmetry. One morphological unit is composed of two (p2), three (p3), four (p4) or six (p6) identical protein or glycoprotein subunits. The center-to-center spacing of the morphological units is often a strain specific feature and lies between 3 nm and 35 nm. The molecular mass of the S-layer subunits have been shown to range from 40,000 to 200,000 on SDS-polyacrylamide gels. In S-layer proteins the content of acidic amino acids is high. Their pI is in the range of 3.5 to 5.5 and their lysine content is about 10%.

In this work the glycosilated S-layer lattice of *Clostridium thermohydrosulfuricum* L111-69 [5-7], the non glycosilated S-layer of *Bacillus stearothermophilus* PV72 [8] were used. Both are hexagonally ordered with centre-to-centre spacings of 14.2nm and 22.5nm and molecular weights of the constituent subunits of 120,000 and 130,00, respectively.

III THE S-LAYER LATTICE FROM *CLOSTRIDIUM THERMOHYDROSULFURICUM* L111-69 AND *BACILLUS STEAROTHERMOPHILUS* PV72 AS IMMOBILIZATION MATRIX

The outer face of the hexagonal S-layer lattice from *Clostridium thermohydrosulfuricum* L111-69 exhibits a net neutral surface charge, since carboxyl and amino groups are arranged in equimolar amounts. Labelling experiments with polycationized ferritin (PCF; a net positively charged topographical marker for electron microscopy) showed that the S-layer surface assumed a net negative charge upon cross-linking the subunits with glutaraldehyde [9] The S-layer protein of *C. thermohydrosulfuricum* L111-69 has also shown to be glycosilated [10] The carbohydrate chains, which have a molecular weight of approximately 20,000, are composed of dissacharide repeating [11] units of α-mannosyl and α-rhamnosyl residues and were found to be exposed on the S-layer surface [7] Therefore, the S-layer surface offers free and accessible functional groups such as carboxyl groups of aspartic and glutamic acid, amino groups of lysine and hydroxyl groups of the carbohydrate residue, which can be activated by different methods. The S-layer protein can be used as immobilization matrix in different forms. These are either whole bacterial cells completely covered with an S-layer lattice, cup-shaped cell wall fragments where both faces of the rigid cell wall layer are covered with an S-layer lattice, and S-layer ultrafiltration membranes where the S-layer material had been deposited on a microporous carrier (SUM) [12].

Activation of Functional Groups from the S-Layer Protein

For achieving chemical stability of the S-layer lattice towards acidic and alkaline pH conditions as well as against hydrogen bond breaking agents, the S-layer subunits are usually crosslinked with glutaraldehyde before the functional groups are activated. The free amino groups of the S-layer protein react under mild aqueous conditions with glutaraldehyde to form Schiff's bases as intermediate compounds, which are then reduced by sodium borohydride for giving stable alkylamine bond. Under standard conditions, at least 80% of the free amino groups of the non glycosilated S-layer protein from *Bacillus stearothermophilus* PV72 could react with glutaraldehyde. The degree of modification of the amino groups from the S-layer glycoprotein of *Clostridium thermohydrosulfuricum* L111-69 was 65-70% under the same conditions.

Activation of the Carboxyl Groups from the Protein Moiety of the S-Layer Protein

The low reactivity of carboxylic acids in water usually makes it difficult to use these groups to selectively modify proteins. The most important chemical modification reactions of carboxyl groups utilize a carbodiimide-mediated process.

In our experiments we have used the water-soluble 1-ethyl-3,3'(dimethylaminopropyl)-carbodiimide (EDC) [13] Quantitative modification of the carboxyl groups from the S-layer protein of *B.stearothermophilus* PV72 could be obtained if the activation with EDC was performed in the presence of highly concentrated hydrogen bond breaking agents, such as 5 M guanidine hydrochloride. In this case, glycine methyl ester was used as nucleophile. Amino acid analyses showed that under desintegration conditions 147 carboxyl groups per S-layer subunit could be modified. Only 99 carboxylic acid groups per S-layer subunit or 68% of the total number could be modified if the S-layer protein of the same organism was used in the form of cup shaped cell wall fragments where only the S-layer surface was accesible. Derived from the area covered by one morphological unit, nominally 1.35 carboxyl groups were arranged per nm^2 of the S-layer lattice.

Immobilization of macromolcules was either performed via spacer molecules, such as 4-amino butyric acid (4-ABA), 6-amino caproic acid (6-ACA) and poly-L-lysine, or directly to the S-layer lattice. The proteins used for immobilization were ferritin (M_r440,000), ß-glucuronidase (M_r280,000) and glucose oxidase (M_r160,000). Quantitative calculations showed that in the case of the hexagonally ordered S-layer lattice from *Clostridium thermohydrosulfuricum* L111-69, 700μg ferritin were bound per mg S-layer protein. Since the molecular size of ferritin (12 nm) corresponds approximately to the size of one morphological unit in the S-layer lattice of *C.thermohydrosulfuricum* L111-69 which is 14.2 nm, one ferritin molecule could bound per morphological unit. The immobilization density as calculated from quantitative measurements could be confirmed by freeze-etching of whole cells after immobilization of ferritin. Such preparations showed ferritin densely arranged on the outer face of the S-layer lattice.

Furthermore, thin-section preparations of cup-shaped cell wall fragments showed that the immobilized ferritin molecules were attached as a monolayer to the surface of the S-layer lattices. ß-glucuronidase was immobilized by the same procedure as applied to ferritin. In this case, 860μg ß-glucuronidase could be bound per mg S-layer protein. The enzyme retaind 80% of its activity. The immobilization of glucose oxidase was carried out either directly on the S-layer lattic or on the S-layer lattice after introducing spacer molecules. If glucose oxidase was immobilised directly on the S-layer lattice, 560μg enzyme protein were bound per mg S-layer protein. When using 4-ABA or 6-ACA as spacers, the amount of immobilized glucose oxidase was also 520-530μg enzyme protein per mg S-layer protein. If the immobilization was performed via poly-L-lysine, in which carboxyl groups were introduced by succinylation[14] of amino groups which were then activated with EDC, the values reached were in the range of 800μg enzyme protein per mg S-layer protein. There was a noticeble difference in the retention of enzyme activity. If the immobilization occured directly on the S-layer lattice, glucose oxidase retained 35% of its original activity. The highest enzyme activity of 60% was obtained if glucose oxidase was bound via the spacer molecules. Despite of the high amount of enzyme immobilized via poly-L-lysine, the retained activity of the enzyme proved to be only 50%.

Activation of the Hydroxyl Groups from the Carbohydrate Chains of the S-layer Glycoprotein

Cyanogen halide activation of hydroxyl groups is the most commonly used method for protein immobilization involving the use of polysaccharide carriers. Indeed, any polymeric matrices possesing cleavable vicinal hydroxyl groups can be utilized by this method. Cyanogen bromide reacts with vicinal hydroxyl groups to form a cyclic reactive imidocarbonate and carbamic acid ester. Imidocarbonates are susceptible to nucleophilic substitution by amino groups in proteins. As described before, the carbohydrate chains of the S-layer glycoprotein form *Clostridium thermohydrosulfuricum* L111-69 are composed of dissacharide repeatings of α-mannosyl and α-rhamnosyl residues. Due to the presence of vicinal hydroxyl groups, mannosyl residues are suitable for activation with cyanogen bromide [15].

The chemical basis of the activation process has been mainly studied in the case of Sephadex, but probably the same structures arise when the process is applied to other hydroxyl group containing matrices. Glutaraldehyde-treated cyanogen bromide activated cell wall fragments from *C.thermohydrosulfuricum* L111-69 could bind 150 μg ferritin per mg S-layer glycoprotein. Therefore, one ferritin molecule was bound per four morphological units each consisting of six identical subunits (Fig.1). Another possibility for immobilizing macromolecules to the carbohydrate chains was to convert the hydroxyl groups into carboxyl groups by succinylation with succinic anhydride. The introduced carboxyl groups were activated in a second stage with EDC, which due the lower pK of the introduced carboxyl groups was carried out at pH 3.75. In comparison to the immobilization via vicinal hydroxyl groups, the amount of covalently attached ferritin was significantly higher. The glutaraldehyde-treated, carboxy-amidated, succinylated and EDC activated S-layer glycoprotein could bind 1 mg ferritin per mg glycoprotein. Therefore, 1-2 ferritin molecules were bound per morphological unit. The different amounts of immobilized ferritin on the S-layer lattice must be considered when choosing among the different activation methods.

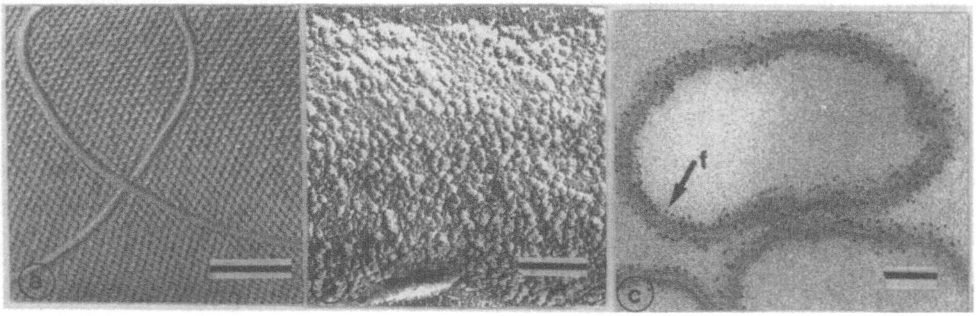

FIG.1 Freeze-etched preparation of glutaraldehyde-treated whole cells of *Clostridium thermohydrosulfuricum* L111-69 showing the hexagonal array of the S-layer a). After activation of the hydroxyl groups on the carbohydrate chains with cyanogen bromide, ferritin molecules were covalently bound to the S-layer surface b). Bars, 100 nm.

c) Thin-section preparation of glutaraldehyde-treated cell wall fragments of *C.thermohydrosulfuricum* L111-69 with covalently immobilized ferritin (f) after cyanogen bromide activation. Bars, 100 nm.

The greatest disadvantage of cyanogen bromide activation is the ensuing instability of imidoesters against hydrolysis. Imidoesters will convert to carbamates which do not react with nucleophiles. When both, succinylation of hydroxyl groups and activation of introduced carboxyl groups is performed, the immobilized ferritin quantity should be higher than the observed one. Derived from their molecular weight of about 20,000, it can be calculated that the carbohydrate chains are composed of approximately 30 repeating units of rhamnose and mannose. In the case of linear chain, this would correspond to a length of at least 30-40 nm. Therefore we expected to bind at least 3 ferritin molecules per morphological unit on the carbohydrate chain. As derived from the immobilized amount of ferritin, it seems that after binding of the first ferritin molecule at the outermost carboxyl groups of the carbohydrate chain, due to sterical hindrance the carboxyl groups in the inner regions are not accesible anymore.

IV CRYSTALLINE BACTERIAL SURFACE LAYERS AS IMMOBILIZATION MATRICES IN THE DEVELOPMENT OF BIOSENSORS

One of the features common to all biosensing systems is that the unique specifity of a biological element acting in close proximity with a signal transducer is utilized for the detection of a target species. For the immobilized biological recognition molecule (enzymes, antibodies, etc.) a high degree of specificity, a good stability under operating and storage conditions and high retention of activity in the immobilized state are required. Furthermore, the immobilized component should not be released from its support during use[16].Therefore, covalent binding of the biological component to the sensor surface is most frequently chosen among the various immobilization methods.[17].

Considering the qualities of bacterial surface layers (S-layers) as matrix for the covalent immobilization of functional molecules, it has become evident that 2-D crystalline protein arrays represent an interesting alternative to the immobilization matrices usually used in biosensors which all reveal a more or less amorphous structure and thus a random distribution and orientation of their functional groups.

In this section we will describe the construction principle and the main characteristics of single- and multienzyme biosensors as developed on the basis of S-layer technologies in our laboratory[18] (Table 1). Further, the application potential of S-layers as ultrathin immobilization matrices for the development of immunological sensors is shown.

Table 1: Amperometric Biosensors based on S-layer technology

Sensortype	Enzymes	Signal(μA*l/cm^2*mmol)	Linear Range	Application
glucose	glucose oxidase	15	0 - 20 mmol	blood analysis
ethanol	alcohol oxidase	2.5	0 - 7 mmol	blood analysis, microbiology, food technology
xanthine	xanthine oxidase	30	0 - 0.6 mmol	food technology
maltose	maltase + glucose oxidase	1.5	0 - 1.5 mmol	microbiology
sucrose	ß-fructosidase + mutarotase + glucose oxidase	1	0 - 12 mmol	microbiology
cholesterol	cholesterol esterase + cholesterol oxidase	6	0 - 0.5 mmol (cholesteryl palmitate)	blood analysis

For the fabrication of enzyme based S-layer-biosensors two different construction principles have been applied:

Single enzyme sensors, such as a glucose sensor, were produced by immobilizing densely packed monolayers of glucose oxidase (GOD) to the exposed S-layer surface of S-Layer-Ultrafiltration-membranes (SUMs) and employing them in close proximity to a transducer (a metal layer for amperometric or a layer containing an O_2-sensitive fluorescent dye for optical sensing methods). For amperometric measurements based on H_2O_2 oxidation, a 15 - 20 nm thick gold or platinum electrode was directly deposited on the enzyme layer by sputtering (Fig.2). In this way, a very close contact between the active center of the enzyme and the electrode was established with the effect of signals near maximum with respect to the amount of GOD immobilized at the sensor surface[19].

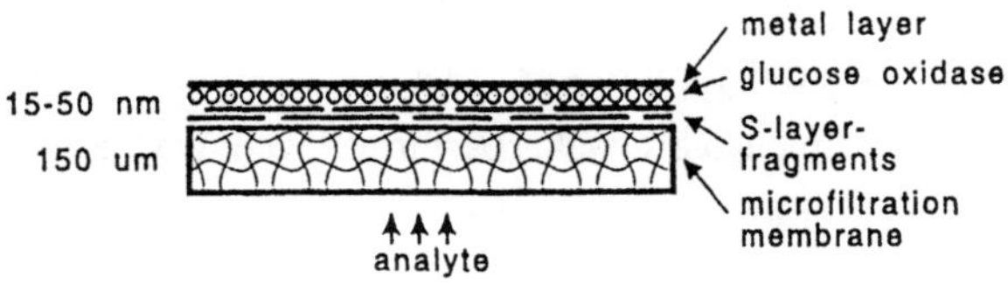

FIG.2 Schematic illustration of an amperometric single enzyme S-layer biosensor. The sensing layer consists of an S-layer ultrafiltration membrane with a monolayer of enzyme molecules immobilized on its surface. The electrical contact to the enzyme is established by depositing a metal electrode directly on the enzyme layer. The analyte reaches the sensing layer throuhg the open structure of the microfilter. (modified after [20]).

A rather different approach was developed for the construction of multienzyme sensors. In order to overcome the problem of uncontrollable competition for activated binding sites which usually occurs in simultaneous immobilization of different molecules, the enzymes were bound individually to the S-layer moiety of cup-shaped cell wall fragments in a first step and subsequently deposited and fixed on a porous support (Fig.3).

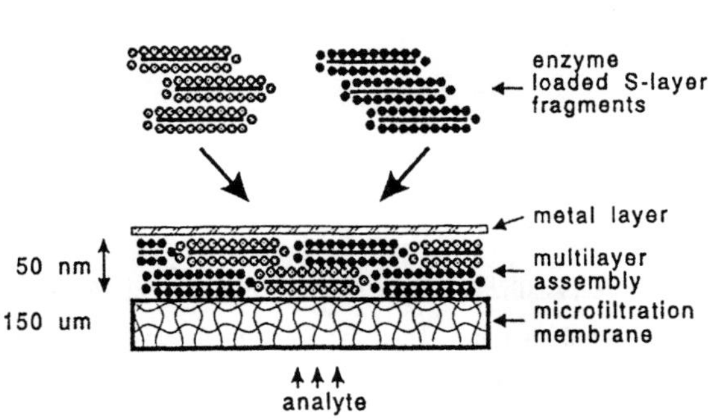

FIG.3 Schematic illustration of the fabrication process and construction principle of amperometric multienzyme S-layer biosensors: The multilayer assembly is produced by depositing the individual enzyme loaded S-layer fragments on the microporous support. The multilayer assembly is crosslinked with glutaraldehyde and subsequently coated with a metal layer. The analyte reaches the multienzyme assembly through the microfilter (modified after [20]).

In this way immobilization parameters could be optimized for each enzyme and the individual enzyme activities easily and accurately checked against one another. Sensing layers obtained by this method were well-structured sandwiches of monomolecular enzyme layers with a total thickness of about 50-100 nm.

On the basis of this technology several different multienzyme S-layer biosensors such as an amperometric sucrose sensor comprising a three enzyme cascade consisting of immobilized ß-fructosidase, mutarotase and glucose oxidase were developed (see Table 1).

Summarising the most remarkable features of S-layer biosensors, there are on the one hand the small thickness, isoporosity and high enzymatic activity of the biologically active layer which is reflected in high signal levels, high signal-to-noise ratios and short response times. On the other hand there is the possibility of creating a very close contact between a monomolecular layer of biologically active material and the transducer. Therefore, the transformation of the "biological message" into a physically exploitable signal is performed with maximum efficiency. Besides, in SUM based single enzyme sensors the molecular sieving properties of the active ultrafiltration layer of the SUM may be used as a selective barrier for the rejection of interfering macromolecules.

Recently, it has been demonstrated, that S-layers of numerous organisms are capable of recrystallizing on solid supports such as glass, carbon or synthetic polymers[21]or forming large scale crystals at interfaces[22] (Fig 4).

Transfer of S-layer protein monolayers formed either at the air/water interface or at phospholipidfilms onto solid surfaces can be carried out easily using standard LB-techniques.

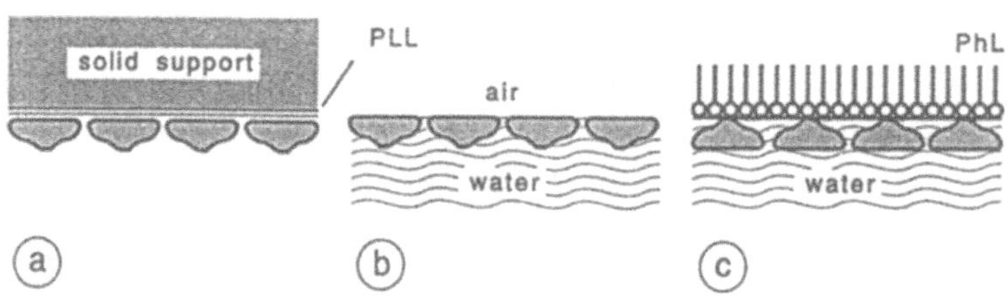

FIG.4 Schematic illustration of recrystallized S-layer sheets at different interfaces.
(a) on solid surfaces coated with poly-l-lysine, (b) at the air/water interface and (c) on (phospho)lipid films (DPPC; dipalmitoylphosphaditylcholine) (modified after[21]).

For the present, our efforts in biosensor development strongly focus on the utilization of that kind of recrystallized S-layer sheets for the fabrication of immunological sensors or miniaturized enzyme sensors. The idea is to cover an optical fiber, the surface of a semiconductor electrode, a piezo crystal or metal coated glass substrates as used for Surface Plasmon Resonance with a highly efficient but extremely thin immobilization matrix (a monolayer of S-layer protein is only a few nanometres thick with still plenty of precisely positioned functional groups available for chemical modification). As previously described S-layers on cup-shaped cell wall fragments are capable of binding IgG via Protein A in a high density [3].

Currently, we are investigating whether standard activation and immobilization techniques as developed and optimized for S-layer fragments in suspension or SUMs can be applied to 2-D-monolayers which are transferred to or recrystallized on solid surfaces.

V CONCLUSION

The aim of this work was to demonstrate that S-layers represent a completely new type of immobilization matrix in comparison to the amorphous polymers usually used as carriers for covalent attachment of macromolecules. With S-layer lattices isoporous immobilization matrices with a characteristic topography and a defined arrangement and orientation of functional groups became available for the first time. These characteristic features are seen as basic requirements for an immobilization matrix which allows a reproducible and geometrically defined binding of functional macromolecules.

Biosensors developed on the basis of S-layer technology yield high signal-to-noise-ratios, short response times and a good stability under working and storage conditions. In addition, the fabrication of multi-enzyme S-layer biosensors has shown that a controlled assembly of multifunctional devices can be realized by rather simple methods. Since S-layers are only several nanometres thick, the fabrication of extremely thin sensing layers is possible.

VI ACKNOWLEDGEMENTS

This work was supported by grants from the "Fonds zur Förderung der wissenschaftlichen Forschung in Österreich" Projekt S57/05 and S50/02, the Österreichisches Bundesministerium für Wissenschaft und Forschung and the Österreichische Bundeswirtschaftskammer.

V REFERENCES

1. U.B. Sleytr, P. Messner, Ann. Rev. Microbiol., 37, 311, (1983).
2. P. Messner, U.B. Sleytr, Adv. Microb. Physiol., 33, 213, (1992).
3. M. Sára, S. Küpcü, C. Weiner, S. Weigert, U.B. Sleytr, In: U.B. Sleytr, P. Messner, M. Sára, D. Pum, (eds)., Springer, New York, pp.71, (1993).
4. U.B. Sleytr, P. Messner, J. Bacteriol., 170, 2891, (1988).
5. U.B. Sleytr, K.J.I. Thorne, J. Bacteriol., 126, 377, (1976).
6. R.A. Growther, U.B. Sleytr, J. Ultrastruct. Res., 58, 41, (1977).
7. M. Sára, I. Kalsner, U.B. Sleytr, Arch. Microbiol., 149, 527. (1988).
8. U.B. Sleytr, M. Sára, Z. Küpcü, P. Messner, Arch. Microbiol., 146, 19, (1986).
9. M. Sára, S. Küpcü, U.B. Sleytr, Arch. Microbiol., 151, 416, (1989).
10. U.B. Sleytr, K.J.I. Thorne, J. Bacteriol., 126, 377, (1976).
11. R. Christian, P. Messner, C. Weiner, U.B. Sleytr, G. Schulz, Carbohydr. Res., 176, 160, (1988).
12. M. Sára, U.B. Sleytr, J. Membrane Sci., 33, 27, (1987).
13. M. Sára, U.B. Sleytr, Appl. Microbiol. Biotechnol., 30, 184, (1989).
14. A.D. Gounaris, G.E. Perlman, J. Biol. Chem., 292, 2739, (1967).
15. R. Axen, S. Ernback, Eur. J. Biochem., 18, 351, (1971).
16. E.A.H. Hall, Biosensors., Open University Press., (1990).
17. A.P.F. Turner, I.Karube, G.S. Wilson, Biosensors - Fundamentals and Application, Oxford University Press, (1987).
18. D. Pum, M. Sára, U.B. Sleytr, In: U.B. Sleytr, P.Messner, M. Sára, D. Pum, (eds), Springer, New York, pp.141, (1993).
19. A. Neubauer, D. Pum, U.B. Sleytr, Anal. Lett. (in press), (1993).
20. D. Pum, M. Sára, P. Messner, U.B. Sleytr, Nanotechnology, 2:196, (1991).
21. U.B. Sleytr, P. Messner, D. Pum, M. Sára, Journal of Applied Bacteriology (in press), (1993).
22. D. Pum, M. Weinhandl, C. Hödl, U.B. Sleytr, J. Bacteriol. (in press), (1993).

APPLICATION OF GRAPHITE-EPOXY COMPOSITES IN THE CONSTRUCTION OF ELECTROCHEMICAL SENSORS AND BIOSENSORS

SALVADOR ALEGRET, JULIAN ALONSO, JORDI BARTROLI,
ESTEVE MARTINEZ-FÀBREGAS AND FRANCISCO VALDÉS-PEREZGASGA[1]

Grup de Sensors i Biosensors, Departament de Química,
Universiat Autònoma de Barcelona, E-08193 Bellaterra, Catalonia, Spain

Abstract. O-3 graphite-epoxy composites have been developed and optimized as conducting plastic materials. These rigid and easily mechanized composites have been used in our laboratory to construct electrochemical sensors and biosensors in different sizes and shapes. All-solid-state potentiometric sensors and biosensors have been developed using grahite-epoxy conducting substrates. The procedure has been extended to CHEMFET and ENFET sensors, rendering hybrid solid-state sensing devices with long lifetimes. It have been found out that enzymes retain their bioactivity within the graphite-epoxy matrices. Different biocomposites have been prepared and applied as amperometric transducers. Biologically bulk-modified graphite-epoxy materials offer several advantages for amperometric sensing.

INTRODUCTION

Plastics are easily mouldable polymeric materials which have found many uses. New applications for plastics are being researched at present by modifying them with the addition of various fillers aiming at reducing their cost or, more interestingly, trying to alter or improve the bulk properties of the plastic, such as chemical, physical or mechanical stability, density, electrical or thermal conductivity, biocompatibility, etc.

A composite material can be simply defined as two or more dissimilar components combined so that each component keeps its

---

[1] Present address: Instituto Tecnológico de La Laguna, Cal. Cuauhtémoc y Boul. Revolución. Torreón. Coahuila. 2700 México

G.G. Guilbault and M. Mascini (eds.), Uses of Immobilized Biological Compounds, 67–79.

basic nature. The properties of the composite are greatly dependent on the arrangement of the individual phases. The arrangement of the individual materials in the composite is known as connectivity, defined as the dimensionality of the connections between individual phases [1,2]. A 1-3 connectivity, for exemple, would exist in an electrode formed by co-linear bundle of carbon fibers (dimensionality 1) bonded by a gluing polymeric matrix (dimensionality 3) [3].

The current work is about conducting 0-3 composites, formed by conducting particles dispersed in a polymeric matrix. This type of materials are often called conducting filled polymers [4]. When the volume fraction of the conductor (f) is small, the electric resistivity of the composite is high and very close to the value of the polymer bulk resistivity [ρ(polymer)](see Fig. 1). If f is raised further, a percolation threshold point (f_c) is reached where a first continuous conductive filament is formed by particle-particle contacts. When this point is reached, electrical conductivity raises abruptly once f is raised slightly. Eventually a limit is reached where resistivity is very close to that of the pure conductor [ρ(filler)].

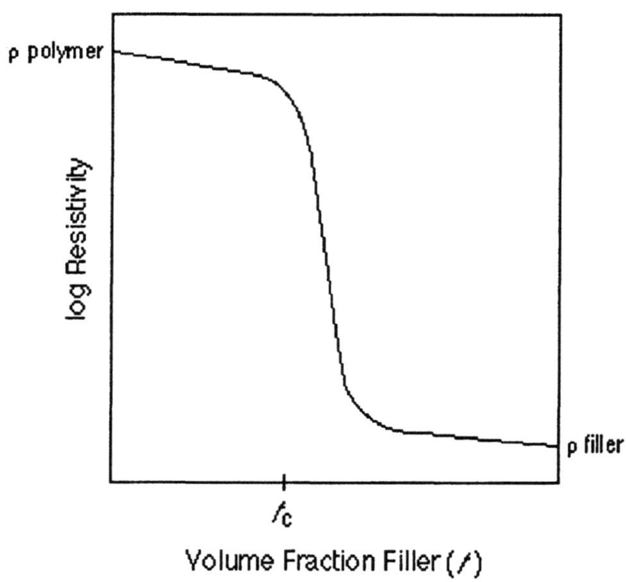

FIGURE 1. *Theoretical dependence of the resistivity (ρ) on the volum fraction of the conductive filler (f) in 0-3 composites based on conductive fillers and insulating polymer matrices*

0-3 Composites based on the dispersion of conductive loadings (V_2O_3, TiO, carbon black, etc.) in insulating polymeric matrices (polyethylene, polyurethane, polyvinyl alcohol, etc.) can be found in sensors [2,5]. When these materials are exposed to temperature, pressure or chemical environment they show a change in their electrical conductivity because the particle-particle contacts are disrupted or enhanced. Therefore, conducting 0-3 composites can be used to develope temperature, pressure or chemical sensors in a simple and inexpensive fashion.

One of the main drawbacks of most chemical sensors and biosensors is their high production costs because of the materials used or the technologies involved. Conducting composite materials can alleviate this problem since they can be prepared following very simple and economical procedures.

0-3 GRAPHITE-EPOXY COMPOSITES

For a long time, work in our laboratories has revolved around the use of graphite-epoxy composites for the development of electrochemical sensors and biosensors, based on potentiometric and amperometric transducers and ion-sensitive field-effect transistors (ISFETs).

The conductive composites are easily prepared by thoroughly mixing a non-conductive epoxy resin (formed by the resin itself and a hardener or cross-linking agent) and graphite powder (grain size below 50 μm). The graphite/ epoxy ratio depends on application (see Table 1). The resulting mixture is applied to the body of the sensing device (see Fig. 2). It is then hardened at a controlled temperature. The final product is rigid and hard and can be easily polished and mechanized.

An expensive, silver-loaded, epoxy resin (Epo-Tek 410, Epoxy Technology, Inc., Billerica, MA, USA) has been used as well to prepare conducting substrates for potentiometric transducers. The results obtained were identical to those yielded by the graphite-epoxy composite [9,10]. Wang *et al.* have used a commercially available graphite-epoxy resin (Dylon Co., Cleaveland, OH, USA) to build amperometric transducers [11].

TABLE 1. *Preparation of different conducting 0-3 graphite-epoxy composites used in the construction of electrochemical transducers*

Conductive substrate for	Epoxy resin	Graphite/ epoxy weight ratio	Curing Temp. ºC	Curing Time (h)	Ref.
Potentio-metric membrane	Araldit M/ Hardener HR (a)	1 : 2.5	40	24	[6]
id. ISFET	id.	1 : 1	60	16	[7]
Ampero-metric transducer	Epo-Tek H77/ Hardener HR (b)	1 : 4	50	48	[8]
id. bio-composite	id.	1 : 4	40	170	

a: Ciba-Geigy; b: Epoxy-Technology, Inc.,Billerica, MA,USA

One of the main advantages derived from the use of this type of composites in chemical sensors is that devices with a wide variety of shapes (planar, tubular, etc.) and size can be produced. This is because size and shape are determined by a highly mouldable substrate (see Fig. 2).

Potentiometric transducers. The use of conducting composites has permitted the production of potentiometric chemical sensors featuring a mobile carrier membrane in an all-solid-state (without internal reference solution) configuration [6, 10]. These sensors are produced by *in situ* casting and fixation of a PVC-based ion-selective membrane on the conducting substrate acting as an internal solid contact (see Fig. 2). The absence of an internal liquid contact enable these sensors to be used in any position (horizontally, upside-down, etc.) and situation (microgravidity, vibration, rotation, etc.). The building process is simple and economical and has been applied in the development of assorted ion-selective electrodes (IESs) featuring different types of mobile carriers (neutral and charged) and mediator solvents and designed in configurations suitable for flow-injection analysis (FIA) applications [6,10].

Generally speaking, potentiometric all-solid-state sensors based on 0-3 graphite-epoxy composites show a good response stability [12] and a longer durability when compared with sensors with a inner reference solution [13]. This can be attributed in part to the increased adhesion between the conducting substrate and the membrane. Additionally, the existence of a single solution-membrane interface reduces ionophore loss.

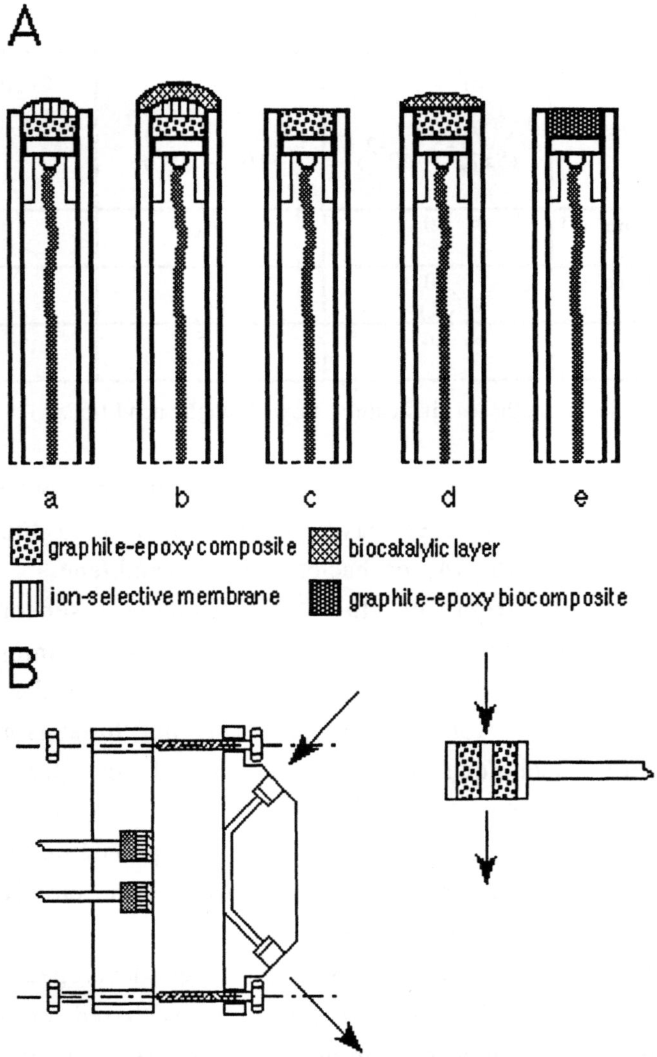

FIGURE 2. *Different configurations designed for electrochemical sensors based on 0-3 grahite-epoxy composites.*
A) Conventional configurations: potentiometric sensor (a) and biosensor (b); amperometric sensor (c) and biosensors (d,e). B) Flow-through configurations.

TABLE 2. *Characteristics of urea all-solid-state potentiometric biosensors based on an ammonium-selective membrane applied to a conducting 0-3 graphite-epoxy composite*

Urease immobilization	Nylon net	Glutaraldehyde spraying	Covalent bonding
Potential response (mV/decade)	58 (40-50)	55 (59)	50
Linear response range (M)	10^{-4} - 10^{-2} (5×10^{-5} - 10^{-2})	10^{-5} - 10^{-3} (10^{-5} - 10^{-2})	2×10^{-5} - 10^{-2}
Response time (s)	60	10	30
Lifetime (weeks)	50 (4)	5 (4)	3
pHbuffer solution	7.4	7.5	7.5

(The values in parentheses indicate flow-through conditions)

Ammonium potentiometric sensors, both in conventional or flow geometries, have been easily converted to urea biosensors by casting an enzyme [6, 14] or bacterial [15] membrane.

The combination of the potentiometric sensor building methodology described above and the enzyme immobilization technique reported by Tran-Minh *et al.* [16] results in a very powerful protocol for the fabrication of biosensors. This immobilization technique, originally conceived for glass electrodes, allows for the casting of very fine layers (1-2 μm) of cross-linked enzyme using an aerosol method. Using this procedure, we reported urea all-solid-state potentiometric biosensors that showed excellent response characteristics [17] (see Table 2).

A new type of urea biosensors has been developed following a similar technology. The building procedure is based on the *in situ* casting of an asymmetric ammonium-selective membrane on a conductive graphite-epoxy substrate. The asymmetric membrane is formed by a PVC membrane covered by layers of cellulose acetate which acts as a template for the *in situ* covalent bonding of the urease [18].

The immobilization methods described above comprise two innovations when applied to our potentiometric sensors: (a)

Potential mass production can be simplified by making the use of printing techniques feasible; and (b) the response of the resulting biosensor is made faster by speeding up the transfer of chemical species between the sample solution and the ion-selective membrane.

Solid state transducers. Ion-sensitive field-effect transistors (ISFETs) are recognized as very useful structures for the development of chemical sensors. ISFETs are small, robust and inexpensive devices that perform on-site impedance transformation.

The association of semiconductor-based chemical sensors with all-solid-state potentiometric sensors described above has yielded a new type of chemical sensor (CHEMFET or chemically sensitive field-effect transistor) based on an hybrid technology with several important features [19]. In this new design a graphite-epoxy composite layer is placed between the gate surface of the encapsulated ISFET and the ion-selective menbrane, as shown in Fig. 3. The resulting hybrid device shows response characteristics very similar to those of "classic" CHEMFETs as observed in repeated calibrations of ammonium- and potassium-selective CHEMFETs both with and without the graphite-epoxy composite layer present [7]. The drift observed in the hybrid CHEMFETs is lower than drift seen in sensors with "classic" CHEMFET construction.

The graphite-epoxy composite layer has a clear salutatory effect since it serves as a mask against light interference. CHEMFETs with optical sensitivity show an abrupt gate voltage change (in the order of $\pm$ 5 mV) when passing a hand between them and a light source for exemple. Once the conductive composite layer is in place, optical sensitivity disappears.

To work correctly, ISFET-based sensors have to be properly encapsulated. Should the aqueous solution come in contact with the electrical contacts or with the bulk of the chip, leakage currents will be present, resulting in sensor failure. The hybrid CHEMFETs reported here did not show leakage current after being immersed in an aqueous solution for long periods of time (> 5 months). The composite layer in conjunction with the epoxy encapsulant forms a continuous sealing surface, preventing the solution from acting on the encapsulant/chip interface (see Fig. 3). This eroding action is not easily preventable in conventional CHEMFETs. Leakage current prevention, together with an enhanced adhesion between the ion selective membrane and the surface of the composite substrate result in longer lifetimes for CHEMFETs reported here than those

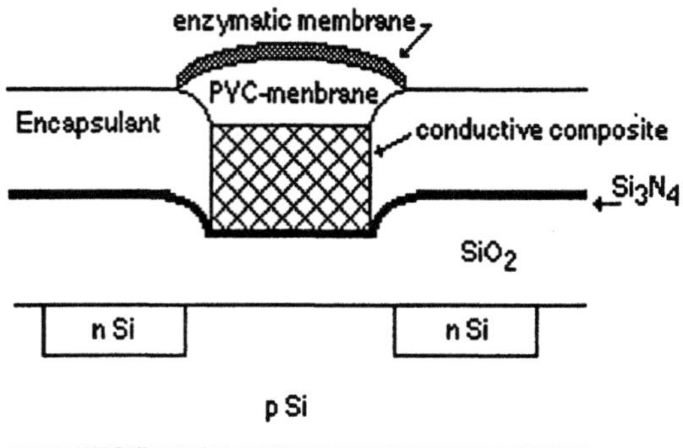

FIGURE 3. *Schematic representation of the gate modification performed on a chemically sensitive field-effect transistor (CHEMFET). A layer of conductive 0-3 graphite-epoxy composite is placed between the ISFET and the ion-selective membrane. The CHEMFET is converted in an ENFET by deposition of an additional enzyme layer*

observed for conventional CHEMFETs where the ion-selective membrane is cast directly on the gate surface.

Ammonium-selective CHEMFETs, built for conventional and flow applications using the described hybrid technology have been used to develope urea ENFETs by the deposition of very fine layers of reticulated enzyme as already mentioned, based on a spraying protocol [19] (see Fig. 3).

Amperometric transducers. Electrodes prepared from 0-3 graphite-epoxy composites (Fig. 2c) show voltammetric characteristics very similar to those shown by carbon paste electrodes. It has to be noted that carbon paste electrodes are mechanically soft and therefore not suitable for use in commercial sensors. Amperometric transducers based on a graphite-epoxy composite are mechanically robust, can be easily mechanized and their active surface is rejuvenated by a simple polishing procedure.

From an electrochemical point of view, these composite-based sensors show a higher signal to noise ratio and, as a consequence, lower detection limit compared to transducers featuring a single conducting phase (platinum, carbon, etc.). The surface of a composite-based electrode should be seen as formed by the ends of a bundle of microelectrodes, where each electrochemically active region measures a fews microns. Weber has discussed to some detail the origin of the signal and the noise in microelectrodes arrays and ensembles (*i.e.*, conducting composite surface) both in quiescent and in flowing solutions [20].

The feasibility of producing glucose amperometric biosensors based on a 0-3 graphite-epoxy composite has been demonstrated recently in our laboratories [8]. The resulting devices were of the first generation where a glucose oxidase membrane is applied to the surface of the composite transducer (see Fig. 2d). These sensors were previously characterized electrochemically for hydrogen peroxide detection.

BULK-MODIFIED 0-3-GRAPHITE-EPOXY COMPOSITES

The amperometric detection of various interesting chemical species is often restricted by the slow kinetics of the electron transfer on the surface of the electrode material. Therefore, surface modification techniques hold a great promise in the development of chemical sensors, allowing for the enhancement of the selectivity, sensitivity and application range of amperometric sensors.

The most common way of incorporating a modifier on the surface of an ampeometric transducer is by direct cast of a polymeric film or by covalent bonding fixation.

Another possibility is to blend the modifier into the bulk of the electrode matrix. It is well known that chemical and biological modifiers have been blended with carbon pastes in order to extend the scope of amperometric transducers. It is not surprising that some enzymes keep their bioactivity when confined to a non-polar environment, such as carbon paste, since there are plenty of references in literature on the performance of enzymes in hydrophobic organic phases.

In our laboratory soft carbon paste has been replaced by 0-3 graphite-epoxy composite as rigid support for the enzyme. The results are promising and the amperometric biosensors produced show the following characteristics:

- A rigid, machinable and polishable surface. The bulk of the electrode works as a reservoir of biocatalytic activity
- The homogeneous dispersion of the biological modifier in the graphite-epoxy matrix yields reproducible results after the surface has been polished
- The components of the active surface can be controlled. The preparation technique of the biocomposite permit a tight control of enzyme content in the bulk, and therefore on the surface, of the material. The same can be said of cofactors, redox mediators, catalysts, additional enzymes, etc.
- The resulting biosensor is inexpensive and can therefore be considered for single-use applications. The construction procedure is simple, compatible with thick film technology, and is carried out using dry chemistry methodologies

Table 3 shows the response characteristics of some glucose biosensors prepared in our laboratories based on graphite-epoxy biocomposites. These results are compared to those produced by conventional glucose composite biosensor (enzyme membrane cast on graphite-epoxy electrode).

Biocomposites-based glucose biosensors have been built where the composite matrix is modified in two different fashions.

In one case, the enzyme (glucose oxidase) and an hydrogen peroxide catalyst (Au-Pd) have been dispersed in the bulk of the graphite-epoxy matrix. The catalyst is reponsible for a lower working voltage and a shorter response time. Additionally, it helps produce a more stable analytical signal.

Yang *et al.* [21] proposed the surface modification of graphite rods using a gold-palladium sputtering technique. The mixing of these metals in the biocomposite is a very simple procedure in the technique reported here, with the additional advantage that a freshly bioactive surface can be produced with simple polishing manoeuvre.

Another modification procedure involves mixing a redox mediator (tetrathiafulvalene) and the enzyme (glucose oxidase) in the graphite-epoxy composite matrix. The resulting rigid material modulates the electron transfer between enzyme and the

TABLE 3. *Characteristics of different glucose biosensors based on bulk-modified and non- modified 0-3 graphite-epoxy composites*

Biosensor	Graphite-epoxy GOD	Graphite-epoxy Au-Pd-GOD	Graphite-epoxy TTF-GOD
Configuration	Membrane[a]	Biocomposite	Biocomposite
Detection	H_2O_2	H_2O_2	TTF
Working potential (mV)	1150	900	150
Linear response range (mM)	0.01 - 10	0.01 - 2	0.1 - 2
Sensitivity ($nA\ mM^{-1}$)	1400	1400	300
Response time (95%) (s)	32	6	11
K_m(app) (mM)	91	7.6	12

a) with an additional dialysis membrane

conductive graphite sites, producing a lower working potential than the one used with the biosensors mentioned above. The lowering of the working potential reduces interference problems that may arise from other electroactive species present.

CONCLUSIONS

Conducting 0-3 graphite-epoxy composites allow the easy construction of electrochemical sensors and biosensors with features similar to those obtained by different conventional techniques. The procedure is straightforward and inexpensive, making the use of thick film technology feasible, so that mass production of these type of sensors may well be considered. The procedure is particularly useful for preparation of sensors in uncoventional configurations (*e. g.*, flow-through geometries).

The construction procedure is suitable for the preparation of all-solid-state (without inner liquid contacts) potentiometric sensors and biosensors, rendering good response stability and long durability.

The procedure can be extended to CHEMFET and ENFET sensors. Covering the gate of an encapsulated ISFET with a layer of graphite-epoxy composite, the adhesivity of ion-selective PVC membranes improves, and therefore long lifetimes are achieved. Furthermore, the optical sensitivity is sharply reduced when compared to CHEMFETs accommodating transparent ion-selective membranes.

Biologically bulk-modified graphite-epoxy materials offer several advantages for amperometric sensing. Enzymes retain their bioactivity within the rigid graphite-epoxy matrices. Surface coverage can be controlled by varying the weight of the modifier (enzyme, electron transfer mediator, etc.) added to the matrix. Active surface can be easily renewed featuring good surface-to-surface reproducibility. And finally the resulting biocomposite-based biosensors are inexpensive, robust and easily mechanizable.

ACKNOWLEDGEMENTS

This work was supported in part by CICYT, Madrid, Spain

REFERENCES

1. R.E. Newnham, *Ferroelectrics*, **68** 1 (1986)

2. G.R. Ruschau, R.E. Newnham, J. Runt and B.E. Smith, *Sensors Actuat.* , **20** 269 (1989)

3. K.E. Creasy and B.R. Shaw, *Anal. Chem.*, **61** 1460 (1989)

4. W.J. Feast. In *Chemical Sensors*, T.E. Edmons, ed., Blackie & Son, Glasgow, 1988

5. P. Bergveld and B.H. van der Schoot, *Selective Electrode Rev.*, **10**, 5 (1988)

6. S. Alegret and E. Martínez-Fàbregas, *Biosensors,* **4** 287 (1989)

7. F. Valdés-Perezgasga, S.Alegret, J. Alonso and J. Bartrolí, *Sensors Actuat.*, 1993. *In press*

8. F. Céspedes, E. Martínez-Fàbregas, J. Bartrolí and S. Alegret, *Anal. Chim. Acta.*, **273** 409 (1993)

9. S. Alegret, J. Alonso, J. Bartrolí, J.M. Paulís, J.L.F.C. Lima and A.A.S.C. Machado, *Anal. Chim. Acta.*, **164** 147 (1984)

10. S. Alegret and A.A.S.C. Machado. In *Analytical Uses of Immobilized Biological Compounds,* G.G. Guilbault and M. Mascini, eds., D. Reidel, Dordrecht, 1988, pp. 309-317

11. J. Wang and K. Varughese, *Anal. Chem.*, **62** · 318 (1990)

12. S. Alegret and A. Florido, *Analyst*, **116** 473 (1991)

13. J.L.F.C. Lima and A.A.S.C. Machado, *Analyst*, **111** 799 (1986)

14. G. Palleschi, M. Mascini, E. Martínez-Fàbregas and S. Alegret, *Anal. Let.*, **21** 1115 (1988)

15. S. Alegret, E. Martínez-Fàbregas, A. Orellana, N. Gaju and R. Guerrero, *Quím. Anal.*, **10**, 93 (1991)

16. S. Kumaran, H. Meier, A.M. Danna and C. Tran-Minh, *Anal. Chem.*, **63** 1914 (1991)

17. D. Martorell, E. Martínez-Fàbregas, J. Bartrolí, S.Alegret and C. Tran-Minh, *Sensors Actuat.*, 1993. *In press*

18. M.H. Gil, A.P. Piedade, S.Alegret, J. Alonso, E. Martínez-Fàbregas and A. Orellana, *Biosensors Bioelectr.*, **7** 645 (1992)

19. S. Alegret, J. Bartrolí, C. Jiménez, D. Martorell and F. Valdés-Perezgasga, *Sensors Actuat.*, 1993. *In press*

20. S.G. Weber, *Anal. Chem.*, **61** 295 (1988)

21. X. Yang, G. Johansson and L. Gorton, *Mikrochim. Acta*, **1** 9 1989)

NOVEL PHEMA MEMBRANES FOR POTENTIAL BIOELECTRODE USE

M. YAKUP ARICA AND VASIF HASIRCI
Middle East Technical University, Department of Biological Sciences, Biotechnology Research Unit, 06531 Ankara, Turkey

Abstract Several types of PHEMA membranes were prepared via photopolymerization. In some applications, glucose oxidase (GOD) was entrapped in the bulk of the membrane or between layers. In others it was covalently bonded on the membranes.

The effect of chemical composition and various initiators on the entrapped GOD were determined. In the entrapment in the bulk it was found that as the monomer concentration in the polymerization medium increased, the GOD activity decreased. Among the initiators used, the highest enzyme activity was obtained with AIBN (71%).

Efficiency of immobilized enzyme activity was found to be 98.5 %, 93.7 % for entrapment, (in the bulk and between layers) and 17.4 % for covalent bonding.

A significant effect of the method used was observed on the stability of the enzyme upon use or storage. Although dependent on the operation mode (continuous or repeated use) and substrate concentration as well as on the entrapment method, activity retention to a level of 90% after 6 h continuous operation was possible.

The water contents of PHEMA membranes increased with ionic strength and decreased with the ionic radii of the ions in preparation medium. The highest permeability ($36.10 \times 10^{-8} cm^2.s^{-1}$) was obtained with the membrane containing GOD entrapped in the bulk.

I-INTRODUCTION

Biosensors and membrane bioreactors are advanced applications combining membrane technology and bioactive compounds. The resultant enzyme electrodes, that are capable of detecting biological reaction products[1,2], find application in biochemical and biomedical fields[3] and in the industry[4]

The preparation conditions of the membranes and the resultant properties such

G.G. Guilbault and M. Mascini (eds.), Uses of Immobilized Biological Compounds, 81–90.

as the water content and permeability are very important for the suitability of the membranes for the above uses. Poly(2-hydroxyethyl methacrylate) (PHEMA) is a biocompatible polymer whose most important application has been as a hydrogel in the biomedical field[5] and as a support material in enzyme immobilization[6]. It can be prepared in various forms, depending on the application, such as powder, bead, hydrogel, membrane as well as for coating on various materials.

In the present study, glucose oxidase (EC.1.1.3.4) has been immobilized on, in and between pHEMA layers to test their suitability for a wide range of applications. The effect of preparation conditions on the activity of the immobilized enzyme, and on the properties of resultant pHEMA membrane were investigated.

II-EXPERIMENTAL

Membrane preparation

HEMA polymerization was carried out in distilled water in a die at 25°C under nitrogen atmosphere by exposure to a long wave UV light for 30 min[7]. The following were added when and as required.

Effect of various initiator on GOD activity

In order to check the effect of initiator type on GOD activity various initiators such as α-α'-azo-isobutyronitrile (AIBN), ammonium persulfate (APS), benzoin and benzoyl peroxide were added to the solution containing the vacuum distilled monomer and glucose oxidase (5mg) (Type II Sigma Chemical Co., USA) in phosphate buffer (pH7, 0.1M) according to the compositions given in Table I.

Effect of HEMA and crosslinker concentration and type on entrapped glucose oxidase

The polymerization mixture contained HEMA monomer (30 %-50 % in aqueous medium), crosslinker N,N'-methylene bisacrylamide (bisAA) or ethylene glycol dimethacrylate (EGDMA),initiator AIBN, and GOD (5mg). The compositions are given in Table II and the polymerizations were carried out as above.

Entrapment of glucose oxidase between pHEMA layers

PHEMA membrane (N6) was prepared as described above, washed, and was sprayed with a solution of GOD (5mg/mL) to give a loading of 40 $\mu g/cm^2$. After a brief drying period at 4°C, the GOD/pHEMA membrane was sprayed with the membrane solution (to give a coat thickness of 65 μm).

Covalent immobilization of GOD on pHEMA membrane

The activation of alcoholic groups of pHEMA (N5) for covalent immobilization was achieved by reaction with epichlorohydrin by storing the membrane in its solution for 4h. The membrane was then washed with acetone and dried under reduced pressure at 50°C. Membrane was swollen in phosphate buffer (pH 7.5, 0.1 M) for 2h, immersed in enzyme solution, and then shaken for 24 h at 22°C.

Activity assay

Activity of soluble and immobilized glucose oxidase was determined according to the method reported by Malmstadt and Hicks[8] and was based on spectrophotometric determination of the amount of hydrogen peroxide formed.

Membrane permeability

Permeabilities of the membranes were determined at room temperature using a diffusion cell described elsewhere[9] which had two cylindrical plexiglass chambers (160 mL volume) and a connecting membrane holder (2cm i.d.). The amount of glucose in the effluent was determined an enzymatically using glucose oxidase and determining the peroxide.

Permeability was calculated from the following equation [10];

$$\ln M_{1,t} / M_{\infty} = P\, A_t / V_1\, l$$

where P is permeability($cm^2\, s^{-1}$), $M_{1,t}$ and M_∞ are solute concentrations in the upper chamber at time t and after equilibrium is reached respectively, A is membrane area(cm^2), V_1 is the volume of the upper chamber(mL), t is time(s) and l is membrane thickness (cm).

III-RESULTS AND DISCUSSION

Enzymes may be adversely affected by an increase of temperature or by chemical reaction during polymerization of HEMA. The effect of polymerization initiator on the activity of glucose oxidase was examined to explain their influence on the activity of the immobilized enzyme.

The highest enzyme activity was obtained with AIBN (6.1mM) and polymerization was completed within 2h (Table I). When AIBN concentration was doubled the polymerization time decreased from 2 to 1h and activity was decreased from 71.1 to 66.0 %. Shortest polymerization time was observed with benzoyl peroxide and benzoin (0.5h) but the activities decreased by about 10 % and 26.3 %, respectively. Under the same experimental conditions at 25 °C with APS as a initiator HEMA was not polymerized; only after the temperature was increased to 40 °C polymerization was completed (after 18h), with a modest entrapped enzyme activity of 39.6 %. At 25 °C AIBN (6.1 mM) and APS (1.8 mM) together required much less time (1h) to form a stable membrane with a slightly higher enzyme activity 43.2 %. Thus except APS all the initiators led to complete polymerization within a short time (0.5-2h) but this always was coupled with low enzyme activity. AIBN caused the least damage and decrease in its concentration led to higher activities.

Table I. Preparation conditions for pHEMA membranes with different initiators

Membrane No:	HEMA (M)	Initiator (mM)	Duration (h)	Temp. (°C)	Recovered* Activity(%)
M1	3.92	8.8(APS)	18.0	40	39.6
M2	3.11	12.2(AIBN)	1.0	25	66.0
M3	3.11	6.1(AIBN)	2.0	25	71.1
M4	3.11	8.3(BP)	0.5	25	10.4
M5	3.11	9.4(B)	0.5	25	26.3
M6	3.92	1.8(APS)+ 6.1(AIBN)	1.0	25	43.2

* With respect to equal quantity of free enzyme.

The dependence of glucose oxidase activity on the concentration of monomer and crosslinking agents (BisAA and EGDMA) was also studied (Table II). The activity was adversely affected by increases in monomer and crosslinker concentrations. The highest entrapped enzyme activity obtained with H_{30} membrane whereas, it did not posses good mechanical properties. The activity in the crosslinked membranes was always less than their uncrosslinked counterparts. This behaviour may be due to the increase in resistance to diffusion of the substrate into, and product out of the membrane pores upon crosslinkage. H_{40} membrane has good mechanical strength, and 71.1 % of enzyme activity when enzyme load is 1 mg GOD per gram of pHEMA.

Table II.Preparation conditions for pHEMA membranes with different crosslinker

Membrane No:	HEMA (M)	Crosslinker (mM)	Recovered GOD Activity (%)
H_{30}	2.34	-	78.3
H_{40}	3.11	-	71.1
H_{50}	3.92	-	69.3
H_{30X}	2.34	64(BisAA)	70.8
H_{40X1}	3.11	110(EGDMA)	67.3
H_{40X2}	3.11	220(EGDMA)	64.1
H_{40X3}	3.11	330(EGDMA)	57.6
H_{50X}	3.92	64(BisAA)	58.7

Each membrane preparation mixture contained initiator AIBN (6.1mM) and glucose oxidase (5mg) and it was carried out at 25°C for 2h.

Glucose oxidase was immobilized onto pHEMA membrane in two ways; by covalent bonding through epichlorohydrin, by entrapment within pHEMA matrix or between layers.

The optimization studies with GOD covalently bonded on the membrane surface led to an enzyme activity of 91 % although the immobilization efficiency was significantly lower (17.4 %) than the others. Similar studies with entrapment within matrix and between pHEMA layers yielded enzyme activities of 71 % and 86 % with almost 97 % and 90 % immobilization efficiency, respectively.

The effect of substrate concentration, pH and temperature of differently immobilized glucose oxidase was studied and compared with that of the free enzyme. The effect of substrate concentration was investigated in the range 5 to 100 mM D-glucose (Figure 1). With covalent bonded and free enzyme a linear increase in conversion rate was observed upto 20 mM glucose concentration. Beyond this glucose concentration, a steady decrease is obtained. The same behaviour is observed with the entrapped enzyme but the linear increase continued upto 50 mM glucose. The inactivation at high substrate concentration is due to poisoning of glucose oxidase by glucose.

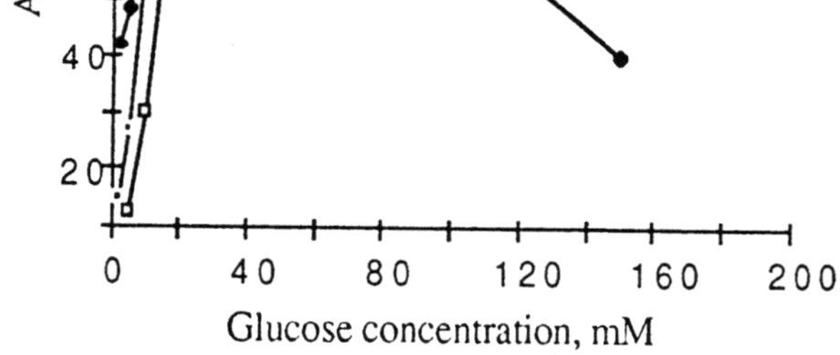

Figure 1. Dependence of substrate convertion rate on substrate concentration with free and immobilized glucose oxidase;

-♦-Free enzyme, -▪-Covalently bound enzyme, -□-Entrapped enzyme

In Figure 2, the relation between the conversion rates (at various glucose concentrations), and the duration of continuous operation for entrapped enzyme is presented. It is observed that although the enzymatic activity is not at its fullest at low concentrations (10 and 20 mM glucose), the enzyme maintains its activity (ca.90 %) for a duration of 6h.

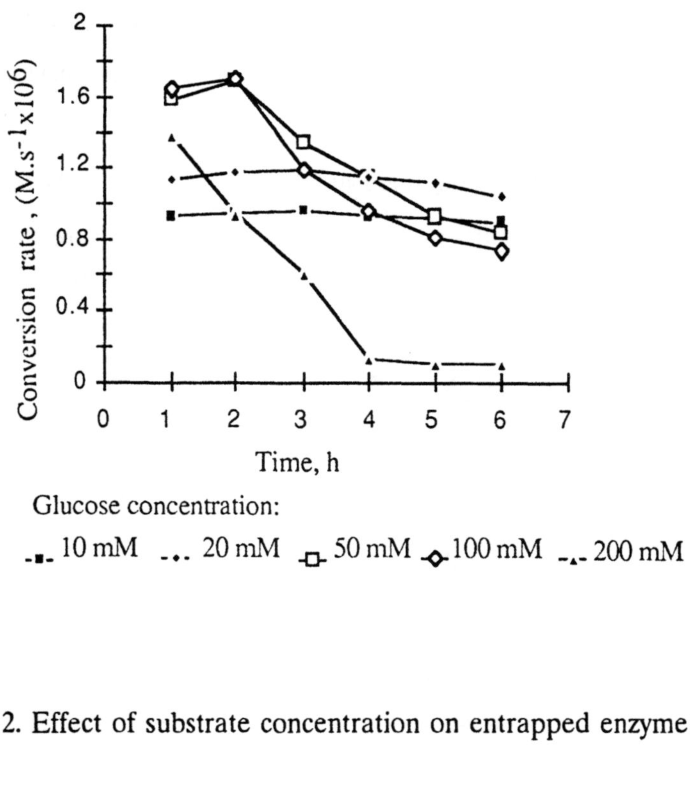

Figure 2. Effect of substrate concentration on entrapped enzyme activity

Figure 3 shows the effect of temperature (between 20-45 °C) on the activities of free and immobilized glucose oxidase. The data reveals a bell-shaped curve with optimum activity at 30 °C for both free and covalently bound, whereas it is 35 °C for entrapped enzyme.

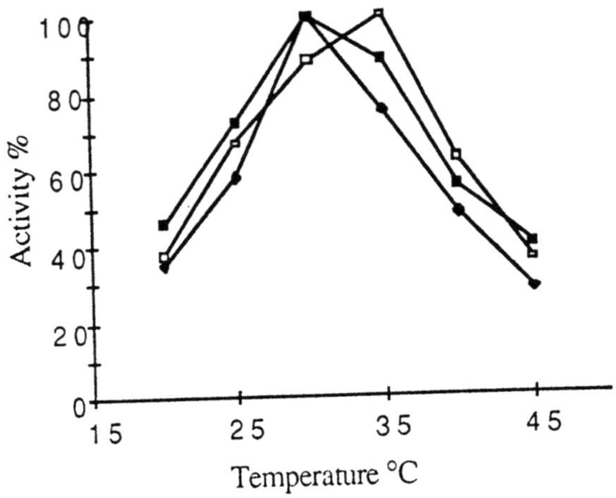

Figure 3. Effect of temperature on the activity of glucose oxidase
-♦-Free enzyme, -■-Covalently bound enzyme, -□-Entrapped enzyme

The optimum conversion was obtained at pH 5.0 for free, at pH 6.0 for covalently bound, and at pH 7.0 for both entrapped glucose oxidase (Figure 4).

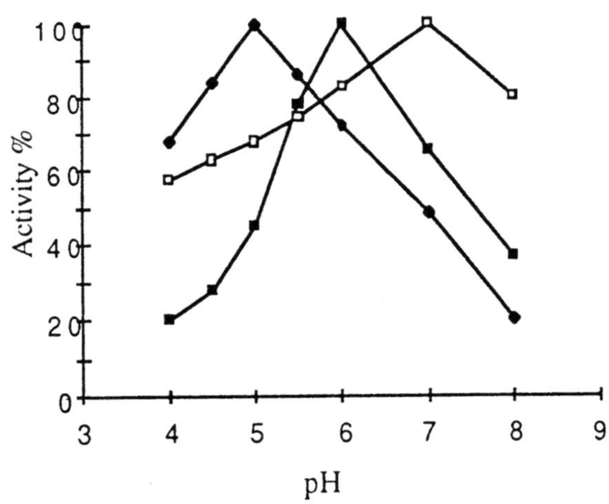

Figure 4. pH profiles of free and immobilized glucose oxidase
-♦-Free enzyme, -■-Covalently bound enzyme, -□-Entrapped enzyme

The similarities in the optimum pH and temperature between the free and covalently bound enzymes indicate that the microenvironment of the covalently bound enzyme is very similar to that of the free enzyme, whereas the entrapped seem more resistant to these environmental variations from ideal.

Physical properties of pHEMA membrane, such as water content, porosity, matrix reticulation and solute permeability was modified by varying ingredient composition and concentration (Table III). Increase of ion concentration (eg.from 0.112 M to 0.336 M for NaCl), led to an increase in the degree of porosity, water content and membrane permeability. When salts with cations of increasing atomic radii were used to create the same ionic strength the result was decreasing water contents (Sn^{+4}>Mg^{+2}>Na^{+}). When glucose oxidase, (a poly electrolyte which can not escape the matrix and thus lead to an osmotic gradient) was introduced in the bulk of the membrane, it was found that the water content (70.2%) and membrane permeability ($36.10x10^{-8}$ $cm^2.s^{-1}$) was higher. On the other hand, only a pHEMA coat (with no GOD) on the surface of the membrane had a detrimental effect on the water content.

Table III. Properties of the membrane prepared in media containing various ions

Membrane No:	Membrane Composition	Ionic Strenght	Permeability to Glucose ($cm^2s^{-1}x10^8$)	Water Content(%)
N1	NaCl (0.112 M)	0.112	5.85	56.5
N2	NaCl (0.336 M)	0.336	10.60	59.2
N3	$MgCl_2$ (0.075 M)	0.224	8.41	55.7
N4	$SnCl_4$ (0.022 M)	0.224	10.41	45.5
N5	K_2HPO_4+NaH_2PO_4(0.1 M)	0.224	11.30	59.5
N6	K_2HPO_4+NaH_2PO_4(0.1 M) +GOD	0.224	36.10	70.2
N7	K_2HPO_4+NaH_2PO_4(0.1 M) +HEMA coat	0.224	7.48	57.3

Thus, the membrane properties modified with medium ionic strength and ion type, monomer, crosslinker and GOD content enable one to design a membrane for electrode use with the right property. A choice of immobilization sites on the membrane provide a further advantage on tailor-making the enzyme membrane for biosensor uses.

REFERENCES

1. G.G. Guilbault and G.J. Lubrano,Analytica Chemica Acta, **64**, 439, (1973).
2. M. Rahni, G.G. Guilbault and G.N. Oliveira,Analytica Chemica Acta, 8, 219, (1986).
3. S. Dimitriu, M. Popa, and M. Dimitriu, J. Bioactive Biocomp. Polym., 3, 243, (1988).
4. K. Burg, O. Mauz, S. Noetzel, and K. Sauber, Angew Makromol. Chem., 157, 105, (1988).
5. C.C. Robert, P.A. Buri, and N.A. Peppas, J. Control Rel., 5, 151, (1987).
6. M. Canterella, L. Cantarella and F. Alfani. British Polym. J., 8, 477, (1988).
7. Y. Arıca and V.N. Hasırcı, Biomaterials, 8, 489, (1987).
8. H.V. Malmstadt and G.P. Hicks,Anal. Chem., 32, 394, (1960).
9. V.N. Hasırcı and I. Kamel, J. Appl. Sci., 33, 925, (1987).
10. R.Y.M. Huang, N.R. Jarvis, Polm. Sci., 41, 117, (1973).

IMMOBILIZATION OF CELL CYCLE SPECIFIC DRUG DFMO FOR CHEMOTHERAPY

A. Kevser Pişkin
Hacettepe University
Faculty of Medicine
Department of Medical Biology
Sıhhiye 06100 Ankara/TURKEY

ABSTRACT

Cell cycle specific drugs have advantages over conventional toxic cancer drugs. Immobilization of these drugs aims at increasing their bioavailability and efficiency. DFMO (Difluoro-methyl-ornithine) inhibits polyamine biosynthesis by binding to the rate limiting enzyme ornithine decarboxcylase (ODC). However, the very short halflife of the enzyme requires a continuous delivery system for DFMO. We have immobilized DFMO in alginate, a natural polymer, to obtain sustained release and have tested its efficiency on cultured myeloma cells.

INTRODUCTION

Growth of cancer cells can be inhibited by a number of chemicals. Among these, cell cycle specific drugs are quite favorable as they are less toxic and more effective. Polyamines increase during rapid cell growth and have important functions in preparing the cell to divide. DFMO is the irreversible inhibitor of ODC, the key enzyme in polyamine synthesis. The half life of ODC is only a few minutes [1,2]. Therefore, a sustained release system is necessary to deliver DFMO. In most reported animal studies DFMO was given by continuous ingestion to increase its bioavailability [2]. We have immobilized DFMO by intrapment into alginate gel spheres. Sodium alginate is a biocompatible polysaccharide which gels in the presence of calcium ions. The release profile of DFMO from the gel under changing ionic compositions was determined and the efficiency of the system was tested on cultured myeloma cells.

G.G. Guilbault and M. Mascini (eds.), Uses of Immobilized Biological Compounds, 91–99.

THE CELL CYCLE

Proliferation is a common feature of cells in growing or regenerating tissues. In human body the majority of cells stop dividing and enter a quiescent phase (G_0) with the exception of few cells i.e. skin, bone marrow cells. Theoretically, proliferating cells should proceed through the phases of the cell cycle [4].

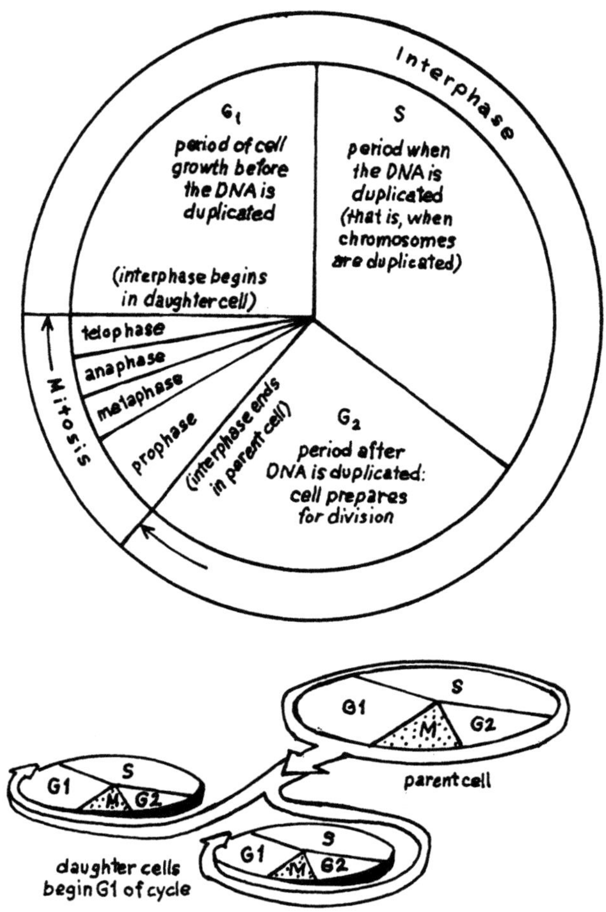

Figure 1. The Cell Cycle.

Cancer cells are transformed to show uncontrolled proliferation. Therefore, they go through the cycle most frequently. Molecules which have important functions in the cell cycle are targets for chemotherapy. Several inhibitors of these molecules are used and novel ones are in progress. These include mytomycin-C, 5-Fluorouracil and other base analogs.

POLYAMINES

Biological polyamines, are commonly used in general reference to putrescine, spermidine and spermine. The compounds are low molecular weight aliphatic amines which behave as organic cations under physiological conditions. They are capable of noncovalent interactions with biological macromolecules such as nucleic acids and regulate their functions. The need for polymines in order to maintain cell growth is firmly established [5,6]. Spermine and spermidine are synthesized from their precursor putrescine by the action of ODC.

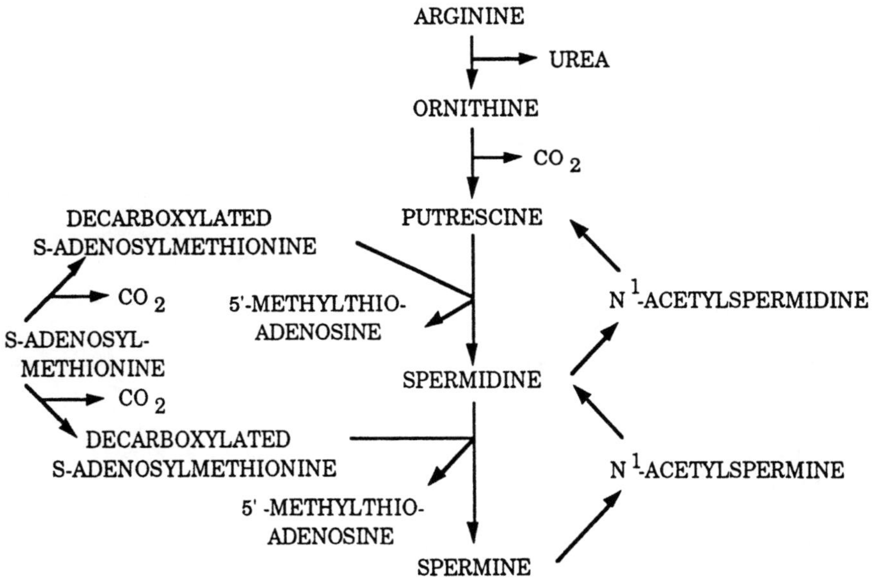

Figure 2. Polyamine Synthetic Pathway.

Ornithine decarboxylase is irreversibly inhibited by DFMO. The growth of hamster ductal adenocarcinoma and other mouse gastrointestinal carcinomas are inhibited by DFMO [7].

IMMOBILIZATION OF DFMO

Immobilization of DFMO was achieved by dropwise addition of alginate-DFMO mixture (30 mg DFMO in 500 µl of 1 % alginate) into 2 M $CaCl_2$ solution. Gel spheres were stabilized by treating with 0.1 M polylysine solution.

Release of DFMO: The amount of DFMO was measured by dansilation, thin layer chromatography and spectrofluorometry as reported elsewhere [8]. Table 1 and 2 show the effect of time and cations on DFMO release respectively.

Table 1. Release of DFMO,

Time	% Release
5 min	19
10 min	20
20 min	22
30 min	24
60 min	27
4 hr	38
24 hr	40
48 hr	43

Table 2. Effects of Cations on DFMO Release.

Incubation Medium	% Release
H_2O	38
$CaCl_2$	20
$MgCl_2$	6
$NiCl_2$	0
$BaCl_2$	2
$MnCl_2$	0
NaCl	38
KCl	38

Immobilization of DFMO in alginate provided a sustained release system, since 43 % of immobilized DFMO was released over 48 hours. Monovalent cations had no effect on the release, whereas most divalent cations decreased DFMO release. Effect of pH was analyzed in order to test the feasibility of using this system for gastrointestinal delivery. Table 3 shows the effect of incubating the spheres in buffers with low and high pH on DFMO release.

Table 3. The Effect of pH on DFMO Release.

Time	% Release	
	pH: 2	pH: 6.0
5 min	6	14
4 hr	16	35
24 hr	35	40
48 hr	38	43

Release of DFMO at pH:2 was lower especially during the first 4 hrs as compared to at pH:6. This may be due to the firming of the gel matrix under acidic conditions which limits the diffusion of DFMO. However, the analysis of sphere surfaces by scanning electrone microscopy before immobilization and after DFMO release and in the presence of DFMO indicates that DFMO does not simply diffuse out of the gel. As shown in Figure 3a and b the SEM photographs of the gel surfaces before DFMO immobilization and after its release were similar.

The SEM Photographs of DFMO-alginate gel sphere surface however was quite different (Figure 4). This indicates that DFMO is immobilized also by chemical means, effecting the gel structure.

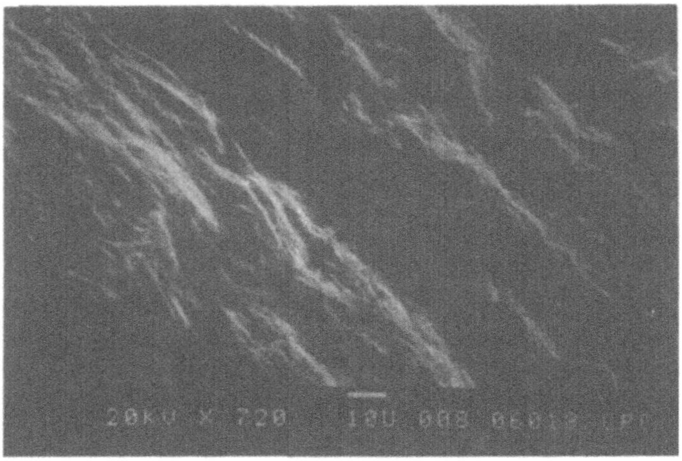

(a)

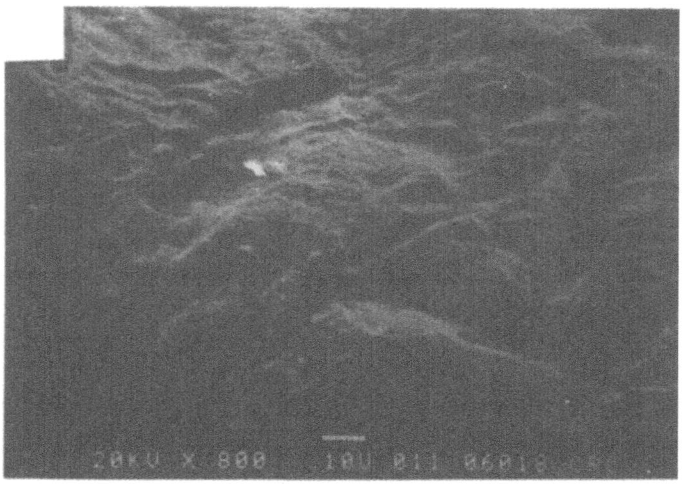

(b)

Figure 3. SEM Photographs of Gel Sphere Surfaces,
a. Before DFMO Immobilization
b. After DFMO Release.

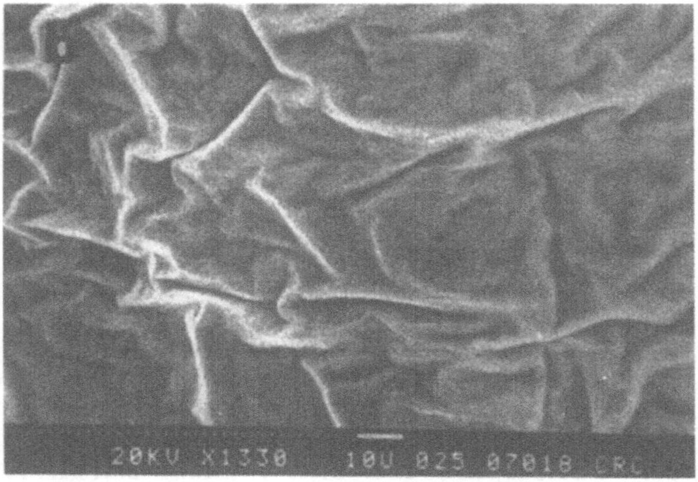

Figure 4. SEM Photograph of DFMO-Alginate Gel Surface.

Effect of DFMO on Myeloma Cells

Myeloma cells were grown in RPMI 1640 medium containing 10 % fetal calf serum. The number of cells was determined by taking into the account of living cells determined by trypan blue exclusion [9]. Alginate spheres containing 30 mg DFMO and the same amount of DFMO were added to 3 ml culture medium. Figure 5 shows the effect of free and immobilized DFMO on myeloma cell growth.

The exponential growth of control cells were inhibited by the action of DFMO, however, after 48 hours the number of cells started to increase. This indicates that single application of DFMO was not sufficient to inhibit polyamine synthesis so as to stop cell growth. As a result the uneffected cells were able to continue proliferating.

The continuous release of DFMO lead to a steady decrease in the number of cells. Within 5 days the number of living cells was reduced down to 15.5 %.

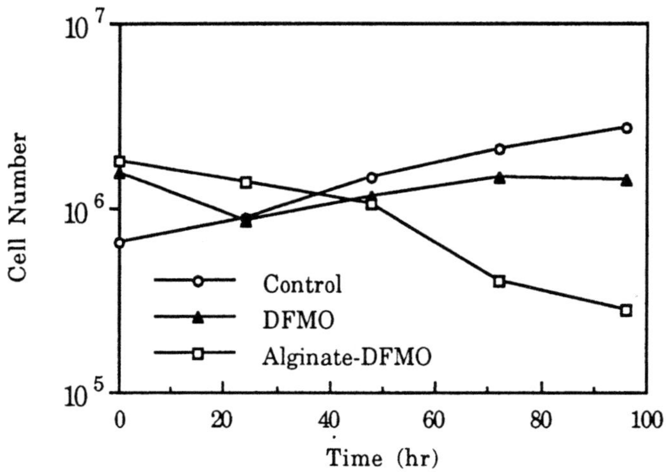

Figure 5. Effect of Free and Immobilized DFMO on myeloma Cell Growth.

In order to determine the polyamine levels in the cells same dansilation procedure used in DFMO analysis was applied. Figure 6 shows the final concentration of polyamines after three day application of DFMO or alginate-DFMO.

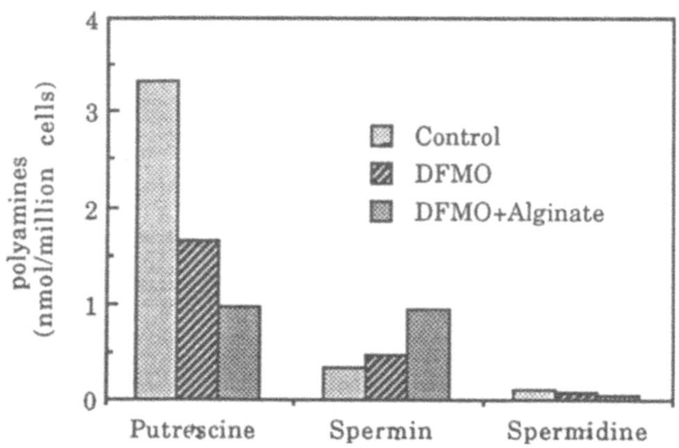

Figure 6. Polyamine Content of Myeloma Cells; Put: Putresine; Sper: Spermine; Sprd: Spermidine.

A decline in the levels of putrescine and spermidine was created by DFMO. However, spermin concentration was increased. This may be due to differentiation induced by DFMO. Continuous release of DFMO resulted in a greater decline in putrescine concentration and also increased spermin level.

CONCLUSION

Several different types of systems have been designed to improve the bioavailability of chemotherapeutic agents. Immobilization of DFMO into alginate spheres provides prolonged action of DFMO, thus, may permit the use of smaller doses than required for systemic administration. The system has the advantage of pH dependent drug release which makes it a potential orally administered drug conjugate. Further animal studies may better demonstrate and prove the feasibility of its chemotherapeutic application.

REFERENCES

1. Saydjari, R., Alexander, R.W., Upp, J.R., Barranco, S.C., Townsend, C.M., Thompson, J.C., Int. J. Cancer, 47: 44 (1991).
2. Heby, O., Adv. Enyme Regulation, 24: 103 (1986).
3. Nellikunja, J.P., Sunkara, P.S., Cancer Research, 43, 3192, (1983).
4. Alberts, B., Bray, D., Lewis, J., Raff, M., Roberts, K., Watson, J.D., The Molecular Biology of The Cell, p. 727, Gurland Publ. Co., N.Y., (1989).
5. Janne, J., Pösö, H., Raine, A., Biochimica et Biophysica Acta, 473: 241, (1978).
6. Pegg, A.E., Cancer Research, 48, 759, (1988).
7. Akar, C., Pişkin, A.K., Criss, W., Kanser, 20: 16 (1990).
8. Alper, Ö., M. Sc. Thesis, Hacettepe University, Ankara, (1993).

USE OF IMMOBILIZED BIOLOGICAL COMPOUNDS IN DRUG ANALYSIS

J-M Kauffmann, A. Nagy, Ch. Petit
Free University of Brussels (ULB)
Institute of Pharmacy Campus Plaine
205/6, Bd. du Triomphe, Brussels, BELGIUM

ABSTRACT. Improvement in sensor technology, better control and understanding of the immobilization process of biological molecules along with pressing demands from the market side have stimulated research towards the development of sensors allowing selective, fast, and simple assays of therapeutic drugs. After showing several application fields for drugs analysis, a review of recently developed devices comprising immobilized biological materials for drug analysis is presented.

1. INTRODUCTION.

With the commercial launching of drugs showing high activity and narrow ranges of therapeutic doses (Table 1) with the drastic purity requirements (e.g. optical purity [1]) along with the increasing demand for immediate results, drug analysis has reached a very active and high level of technology.
In addition to clinical laboratory instrumentation, there is a need for portable, automated and cost-effective detection devices for screening for drugs of abuse (cocaine...), for real time monitoring in hospital intense care units (anaesthetics, anticoagulants), for routine and emergency assays in physicians' office (patients wards) (e.g. salicylates, theophylline, digoxine, neuroleptics, opiates, antibiotics...), in non - clinical situation (in sports, employee testing...). With this respect, bedside testing of drugs has an obvious appeal to physicians. It allows great reduction for the turn around time for the assay and a better security with respect to activity versus toxicity.
Immobilized biological compounds integrated into sensing and reporting systems can be regarded as very attractive devices with regards to the above cited

G.G. Guilbault and M. Mascini (eds.), Uses of Immobilized Biological Compounds, 101–113.

2.DRY REAGENT CHEMISTRY [12-14]

The first clinical application of immobilized biological compounds dates back to the 1950s of Ames Clinistix urine reagent strips for testing urinary glucose. To date, more sophisticated dry chemistries for quantitative analysis of blood constituents and drugs are available or are under intense development. In all cases, a complete (bio)chemistry is miniaturized into a dispensable format. The sample is applied directly onto the strip and the analysis is usually monitored by instrumentation. The latter ranges from hand-held dedicated devices to large free standing instruments.
The chemical reactions on dry reagent elements are usually monitored by diffuse reflectance photometry [15] and, where appropriate, by front-face fluorescence [16]. The technology has been applied for the analysis of a variety of blood constituents including metabolites, enzymes, electrolytes, hormones (insulin, thyroxine) and therapeutic drugs [12-14]. In the latter, the reaction takes advantage of the high specificity of immunochemistry and in all cases the detection chemistry is based on an enzyme assay. The chemistry is generally a competitive protein-binding assay based on the substrate-labeled fluorescence immunoassay. Instrumentation and elements for fluorescence immunoassay and for apoenzyme reactivation immunoassay (ARIS) for therapeutic drugs have been described [12-14]. Several dry reagent chemistry strips have been described for the antiasthmatic drug theophylline in serum, for the determination of antibiotics (gentamicin, amikacin and tobramycin) and anticonvulsant drugs (carbamazepine, phenytoin, pirimidone, and phenobarbital) [12-14]. The performances of dry reagent chemistry analysis vis-a-vis established technologies have been reviewed [12-14].

3. ELECTROCHEMICAL BIOSENSORS

The biosensor technology has mainly focused on the development of sensors for the analysis of blood constituents. In the last 3 years new efforts have been made in the application of biosensor technology to the determination of drug substances (see also ref.17 for papers published before 1990).

3.1. Sensors based on enzymatic reactions

demands. Antibodies, antigens, enzymes and almost any biological component constituting cellular membranes, even whole cells have been combined with physical sensing systems. The latter may be electrical, optical, mass or heat sensitive [2-4].

The investigation of therapeutic drugs using immobilized biological compounds is of great pharmacological interest also from a mechanistic point of view. With this respect the development of biological receptors based transducers has already significantly contributed to the understanding of the processes governing drug-receptor phenomena [1,5,6]. It is well-known that drugs act on either receptors, enzymes or membranes. The interaction may be more or less biologically specific and is generally potent in very high dilution. High specificity of some drug receptors may be examplified with drugs existing in two enantiomeric forms, the two isomers exhibiting different activity [1]. By analogy with the enzyme active site, the drug receptor is thought to be part of a protein struture (e.g. cholinergic receptors). It is not, however, advisable to restrict the drug receptor to protein molecules but rather proteins and their surrounding lipidic matrix structure should be considered as a whole entity [7-9].

TABLE I. Examples of therapeutic doses (in serum) for drugs frequently analyzed in hospitals [10,11].

Drug	Concentration range
Carbamazepine	4 - 12 µg/ml
Digitoxine	10 - 30 ng/ml
Digoxine	0.8 - 2.4 ng/ml
Diphenylhydantoin	10 - 20 µg/ml
Valproic acid	50 - 100 µg/ml
Ethosuximide	40 - 100 µg/ml
Lidocaine	2 - 6 µg/ml
Procainamide	6 - 14 µg/ml
Lithium	0.6 - 0.8 mEq/l
Theophylline	7.5 - 15 µg/ml
Phenytoin	10 - 20 µg/ml
Phenobarbital	20 - 50 µg/ml
Salicylates	150 - 300 µg/ml
Quinidine	2 - 5 µg/ml
Gentamycin	1 - 10 µg/ml

In the following paper several examples of devices comprising immobilized biological compounds for therapeutic drug analysis will be reviewed.

The first type of biosensors operate by enzymatic action on the analyte to be detected. With electrochemical detectors, some product formation or reactant consumption is then monitored by either an amperometric or potentiometric electrode transducer. Within the last three years, several enzyme and cell (microbial cell, plant tissue) immobilized amperometric electrodes have been investigated as possible sensing devices for drug analysis as illustrated in Table 2. Depending on their configuration and shape (microsized) these sensors may be used for medical diagnosis as portable instruments or may see some useful application in the area of process control. The development of organic-phase enzyme biosensors [18-22] is particularly useful for challenging pharmaceutical products [23-26]. Many pharmaceutical formulations contain therapeutic drugs or preservatives which are not readily dissolved in aqueous media (phenol derivatives, peroxides) [26]. The development of biosensors for these substrates will be very attractive for quality-control and process monitoring in the pharmaceutical industry.

The development of disposable single-use electrochemical biosensors by printing technology is a recent very interesting and demanding area [27]. Since selectivity is the most important criteria when measurements of drugs have to be made in complex samples such as serum or whole blood, new strategies have been investigated by developing biosensors where the product of the enzyme reaction is detected at low potentials [28-33] or by using the enzyme (e.g. theophylline oxidase) in the presence of a suitable redox mediator [34,35]. Disposable sensor strips consisting of a carbon working electrode and an Ag-AgCl reference electrode have been fabricated for theophylline [30], salicylates [31], and acetaminophen [32,33].

TABLE II. Amperometric biosensors for drug analysis

A.Enzyme Immobilized Electrode

DRUG	SENSING ELEMENT	CHARACTERISTICS
Phenolic compounds	Carbon μelectrode HRP or Tyrosinase adsorbed	D.L.~20 μM [23]
Phenol,Peroxide Antiseptics	GCE/Tyrosinase in polymer (Eastman)	Linear: 0 -40μM F.I.A [24]
Theophylline	Carbon-strip/ Alkaline Phosphatase	Linear:0-300μM[30] Interferences: -Caffeine

		-Theobromine -Phenylalanine
Acetylsalicylic acid (in blood)	Carbon strip/NADH Salicylate hydroxylase/ benzoate	Linear:1-7mM [31] Stab.: 3 months
Acetaminophen (in blood)	Carbon-strip/aryl acylamidase	Linear:0-3 mM[32] Stab.:6months[33] Interfer.:thiols
Theophylline (in blood)	Pt/Theophylline Oxidase/Nafion Cyt.C (FeIII)	Linear:0-.3mM[34] D.L. = 2 μM Life Time:35 days
Theophylline (in serum)	NMP.TCNQ/adsorbed Theophyl.oxidase	Resp.Time: 60 s [35]
Phenol	TCNQ-graphite/ads. a) tyrosinase b) phenolase	[36] D.L. = 0.2 μM
Glutathion	GCE/Glutathion Peroxidase(ads.)	Linear:0.2-2.4mM [37]
Insulin (serum)	Pt/anti-Insulin IgG GA/Protein A/membrane	Linear:2-200μIU [38]
L-Aspartate	Pt/Immobilon-AV BSA-GA/Aspartate aminotransferase+ Glutamate oxidase	[39] Linear:1μM-0.2mM Life Time:2month
L-Amino Acids	Pt/L-AA oxidase on Polyamide	Linear: 1μM-1mM D.L.: 1 μM [40] Life Time:1month
Phenothiazines (body fluids)	Au/LB film,Tyrosine hydroxylase	Linear: 0-10 mM Ads.voltamm.[41]
B. Tissue and Microbial Cells		
Catechol	CPE/Eggplant Tissue (Polyphenoloxidase)	Linear:5μM-50μM D.L. :1μM [42] Life Time:23days
Catechol	Oxygen electrode Potato Slice (Polyphenoloxidase)	Linear:25μM-.2mM D.L.:10 μM [43] Interfer:DOPA,AA
L-Proline	Oxygen electrode PVA gel(*Pseudomonas*)	Linear: 0.01-1mM [44]

ACh+Choline	Pt/Pall Immuno Nylon ChOx + ACh esterase	Linear: 0-20 μM D.L.= 50 nM [49] Resp.Time= 2 min
ACh+Choline	μC/ Enzymes on activated PVA	Linear:.2-10μM[50] Resp.Time:0.8 min
ACh+Choline	GCE/ChOx+Ach.esterase +HRP/PVP-Os(bpy)2Cl	D.L. ~ 1μM [51] Resp.Time < 30 s

Proton sensitive electrochemical sensors such as the glass pH electrode or ISFETs have been preferentially used for penicillin determination by immobilizing penicillinase (eq.1) or penicillin amidase (eq.2). The biosensors were comprised in off-line flowing systems allowing penicillin monitoring and its determination in fermentation broths, (Table 3 and ref.52 for a thorough review).

$$\text{Penicillin} + H_2O \longrightarrow \text{Penicilloate}^- + H^+ \quad (1)$$

$$R'\text{-CO-NH-}R'' + H_2O \longrightarrow RCOOH + H_2N\text{-}R'' \quad (2)$$

TABLE III. Potentiometric enzyme immobilized biosensors

Biological component	Immobilization	Characteristics
Glass pH electrode		
Recombinant E.Coli	Adsorption on acetylcellulose	Linear:5-30 mM [54] Resp.Time: 8 min Life Time:13 days
Penicillinase	Cross-linking with spray GA	Linear:1.5-15mM[55] Res.Time: < 10 s Life Time:12 days
ISFET		
Penicillin-amidase	Cross-linking with BSA & GA	Linear:1.5-25mM[53] Resp.Time: 90 s
Penicillin-amidase	Cross-linking with GA	Linear:0.3-30mM[56] Res.Time: 99 s Life Time:100 days
A:Penicillinase B:Penicillin-amidase	Cross-linking with BSA + GA	Linear:0.5-10mM[57] Life Time: A: 4 months B: 3 months

Phenol Chlorophenol Dichlorophenol	Graphite/Mushroom or Banana Tissue adsorbed and pressed	Linear: 0 - 5mM D.L.: 30 μM [25] Resp.Time:8-20 s

D.L.= Detection Limit, PVA = polyvinyl alcohol, CPE = carbon paste electrode, GCE = glassy carbon electrode TCNQ = Tetracyanoquinodimethane, NMP = N-methyl phenazinum, GA = glutaraldehyde, BSA= bovine serum albumin, Linear=linear range,Interfer= interferences

Insulin was determined in serum using an antibody immobilized electrode with immunodetection in the presence of glucose oxidase-labelled insulin [38]. Immobilization of plant tissues has been successfully realized for the assay of some phenolic xenobiotics[25,42,43]. When intended for use in the treatment or prevention of diseases, L-amino acids may be defined as drugs (*Laws Enforced by the U.S. Food and Drug Administration* HHS FDA 80-1051. Washington, D.C.: U.S. Government Printing Office, 1981). Recent L-amino acid biosensors have been constructed by optimizing the immobilization of the enzyme [39,40] or by proper selection of microbial cells [44].
Acetylcholine is only rarely employed as therapeutic drug (cholinergic in surgery of eyes) but the determination of this neurotransmitter in physiological fluids is of great pharmacological significance. The enzymes acetylcholinesterase and choline oxidase have been successfully immobilized in micro-columns connected to flow-injection systems [45] or to HPLC columns [46], the liberated hydrogen peroxide being detected amperometrically at a platinum electrode (eqs.3,4). Several amperometric electrodes comprising both enzymes and detecting hydrogen peroxide or a suitable mediator (eq.5) have also been developed as reported in Table 4.

$$\text{Acetylcholine} \longrightarrow \text{Choline} + \text{Acetate} \quad (3)$$
$$\text{Choline} + 2O_2 + H_2O \longrightarrow \text{Betaine} + H_2O_2 \quad (4)$$
$$Fe(CN)_6^{3-} + e^- \longleftrightarrow Fe(CN)_6^{4-} \quad (5)$$

Table IV. Amperometric Acetylcholine and Choline Biosensors

Drug	Sensing element	Characteristics
Choline or Ach+Choline	ChOx and ACh esterase Adsorption/Nafion/GCE Hexacyanoferrate(III)	Linear: 0-5mM D.L. 10 μM [47]
ACh	Pt/PVA + ChOxidase + ACh esterase	Linear: 0-5 mM D.L.= 10 μM [48] Life Time 3 months

A unique and recently developed biosensor used a light addressable potentiometric sensor (silicon based device) to measure the rate at which immobilized cells (adsorbed or entrapped within a fibrin clot) acidify their environment [58]. Actually, this biosensor monitors extracellular acidification, thus allowing indirect detection of any metabolism change in cell physiology in the presence of xenobiotics. Cells, whether eukaryotic or prokaryotic, have been retained in a microflow chamber in aqueous diffusive contact with the pH-sensitive surface (Si-nitride). The light addressable nature of the sensor allowed the measurement of subtle changes in cell physiology. Authors showed that this "silicon microphysiometer" is particularly useful for screening for new receptor ligands, new therapeutic drugs and also for toxicology [59,60].

Several optical (fiber optics) transducers with penicillinase and a pH-sensitive fluorescent dye immobilized in a suitable polymer (polyacrylamide) have been developed for batch, continuous flow and flow injection analysis and the penicillin sensors have been reviewed [52].

A thermal biosensor (heat generated in the enzyme column) was constructed for on-line monitoring of penicillin V production in a bioreactor [61].

Sensors (oscillating crystal) sensitive to mass changes have also been developed for determination of thyroxine and cocaine based on interaction between immobilized antibodies with antigen [see 17,62 and refs. cited therein]. Other piezoelectric crystals have been covered by a membrane of natural or synthetic lipids for quantifying the adsorption of many different bioactive compounds such as surfactants, odorants, anaesthetics and antibiotics [63 and references cited therein]. Actually, the immobilization of mono or multilayers of lipids onto electrodes may also be advantageously applied for the study of drugs which modulate ion transport or interact strongly with the lipidic part of biological membranes. [64-68].

3.2. Bioaffinity sensors

This second type of sensors is based on binding interactions between the immobilized biomolecule and the analyte of interest. The binding event itself is detected and readout is proportional to analyte-biomolecule binding. Other bioaffinity sensors may utilize detection of other events which come about or near the bioactive surface as the result of binding event (enzyme immuno biosensors). Antibody and receptor

based bioaffinity sensors for drug analysis, and the corresponding immobilization modes have been recently thoroughly reviewed [6,17].
Biochips with antibodies and receptors immobilized onto interdigited electrodes on an epoxy board able to be pluged into a portable field electronics unit have been constructed [69]. The receptor chips operated in a differential mode (automatic compensation of non specific binding), the output signal may be either a change in impedance, voltage, current or capacitance. Acethylcholine (ACh) and Opiate (Opr) receptors have been immobilized in a one step procedure in the presence of proteins, stabilizers, lipid and detergent. The reported sensitivities for cholinergic agents was in the range of 1.5-4 μM (or 1-3ng ACh in the 10-25 μl sample) and 0.1-25 μg/ml for opiates [6]. ISFETs using immobilized ACh receptors with similar sensitivities were also reported [70]. Optical transducers comprising ACh receptors adsorbed onto quartz fibers have been developed as well [71]. Changes in fluorescence is monitored when a suitable fluorescein-labeled molecule competes with the cholinergic analyte to be measured. Intact neuronal receptors in contact with a conventional electrophysiological electrode have been investigated for the detection of pyrazinamide and selected local anaesthetics by monitoring the action potential spikes [72].
Finally, surface plasmon resonance (SPR) which is a label-free immunoassay that measure the changes that occur in the refractive index when the ligand analyte binds to its antibody or receptor immobilized on a thin metal film [73] may see some very useful applications in drug-receptor interaction studies. Actually due to the low molecular weigth of drugs, the detected signal is to low for confident direct measurements thus indirect binding techniques have been developed for example for theophylline measurements in serum samples [74].

4. ACKNOWLEDGEMENT

Thanks are expressed to the European Economic Community for financial assistance as part of a Community Mobility Action (Mrs A.Nagy) - PECO contract N° ERB-CIPA-CT 922015 686.

5. REFERENCES

1. Albert, A., *Selective Toxicity*, 1st ed. Methuen and Co, Ltd, London and John Wiley & Sons, N.Y. 1951; 7th ed., Chapman and Hall, London, 1985.

2. G.G.Guilbault, M.Mascini (Eds), *Analytical Uses of Immobilized Biological Compounds for Detection, Medical and Industrial Uses*, NATO ASI Series, Reidel Pub.Co Dordrecht Holland 1988.
3. C.H.Suelter, L.Kricka (Eds), *Bioanalytical Applications of Enzymes*, vol 36 John Wiley & Sons, Inc. N.Y. 1992.
4. R.F. Taylor (Ed.), *Protein Immobilization, Fundamentals and Applications*, Marcel Dekker, Inc. N.Y. 1991.
5. G.M.Eckert, I.R.Neering, 'Physiological Processes with an Electrochemical Basis:Mechanism of Action of Drugs and Electrical Alternatives which Modify these Processes', in Electropharmacology , (G.M. Eckert, F. Gutmann, H. Keyzer, Eds.) CRC Press, Boca Raton, Fl. 1990, p 93-106.
6. R.F. Taylor, 'Immobilized Antibody and Receptor Based Biosensors' in *Protein Immobilization, Fundamentals and Applications*, Marcel Dekker, Inc. N.Y. 1991, p.263-303.
7. E.Goormaghtigh, J-M Ruysschaert, Biochim. Biophys. Acta, 779, 271 (1984).
8. A.Sautereau, M.Betermier, A.Altibelli, J.F.Tocane Biochim.Biophys.Acta, 978, 276 (1989).
9. P.G.Green,J.Hadgraft,Int. J.Pharm., 37, 251 (1987)
10. *Répertoire Commenté des Médicaments*, Centre Belge d'Information Pharmacothérapeutique, Brussels,1992
11. G.F.Johnson, 'The Drug Assay Laboratory:Clinical Utility' in *The Scientific Basis of Clinical Pharmacology*, R. Spector Ed., Little Brown & Co Boston, 1986, p 131-152.
12 B.Walter, Anal.Chem. 55, 4894 (1983).
13. B.Walter, 'Fundamentals of Dry Reagent Chemistries The Role of Enzymes 'in *Bioanalytical Applications of Enzymes*, C.H.Suelter, L.Kricka (Eds), vol.36 John Wiley & Sons, Inc., 1992 p. 35-62.
14. K.Emancipator,D.G.Deutsch,'Dry Reagent Chemistries in *Analytical Aspects of Drug Testing*, D.G. Deutsch Ed., John Wiley & Sons, N.Y., vol 100, (1989) 203-228
15. G.Kortüm, *Reflectance Spectroscopy: Principles, Methods, Applications*. N.Y. Springer, 1969.
16. A.J.Pesce, C.G.Rosen, T.L.Pasby, *Fluorescence Spectroscopy*, New York: Dekker, 1971.
17. G.G.Guilbault, R.D.Schmid, Biotechnol.and Appl. Biochem. *14* , 133 (1991).
18. G.F.Hall, D.J.Best, A.P.F.Turner, Enzyme Microb. Technol. 10, 543 (1988)
19. G.F.Hall, D.J.Best, A.P.F.Turner, Anal.Chim.Acta 23, 113(1988).

20. S.Saini, G.F.Hall, M.E.Down, A.P.F.Turner, Anal. Chim.Acta, 249, 1 (1991).
21. G.F.Hall, A.P.F.Turner, Anal.Lett., 24,1375(1991).
22. J.Wang, L.H.Wu, L.Angnes, Anal.Chem., 63, 2993 (1991).
23. J.Wang, Y.Lin, Anal.Chim.Acta, 271,53 (1993).
24. J.Wang, Y.Lin, L.Chen, Analyst, 118, 277(1993).
25. J.Wang, N.Naser, H-S Kwon, M.Y.Cho, Anal.Chim.Acta 264,7(1992).
26. A. Nagy, Ch.Petit, A.Amine, J-M Kauffmann, Anal. Letters, submitted (1993).
27. P.I.Hilditch, M.J.Green, Analyst, 116, 1217(1991).
28. H.Bramwell, A.E.G.Cass, P.N.B Gibbs, M.Green, Analyst, 115, 185(1990).
29. J.E.Frew, N.C.Foulds, J.M.Wilshere, N.J.Forrow, M.Green, J.Electroanal.Chem., 266, 309 (1989).29.
30. N.C.Foulds, J.M.Wilshere, M.Green, Anal.Chim.Acta, 229, 57 (1990).
31. J.E.Frew, S.W.Bayliff, P.N.B.Gibbs, M.J.Green, Anal.Chim.Acta, 39, 224 (1989).
32. A.F.Jones, McAleer, R.A.Braithwaite, L.D.Scott, S.S. Brown, J.A.Vale, Lancet, 35 793(1990).
33. P.A.Vaughan, L.D.L.Scott, J.F.McAleer, Anal.Chim. Acta, 248, 361(1991).
34. J.Wang, E.Dempsey, M.Ozsoz, M.Smyth, Analyst, 116, 997(1991).
35. C.J.McNeil, J.M.Cooper, J.A.Spoors, Biosensor & Bioelectronics 7, 375(1992).
36. J.Kulys, R.Schmid, Anal.Letters, 23,589(1990)
37. C.Hua,M.Smyth, C.O'Fagain, Analyst, 116, 922(1991)
38. S.Kaku, S.Nakanishi, K.Horiguchi, Anal.Chim.Acta, 225, 283(1989).
39. R.L.Villarta, G.Palleschi, G.J.Lubrano, A.A. Suleiman, G.G.Guilbault, Anal.Chim.Acta, 245, 63 (1991).
40. G. Radu, P.Coulet, Analusis, 21, 101(1993)
41. J.Wang, Y.Lin, A.V.Eremenko, I.N.Kurochkin, M.F.Mineyeva, Anal.Chem;, 65, 513(1993).
42. A. Navaratne, M.S.Lin, G.A.Rechnitz, Anal. Chim. Acta, 237, 107(1990).
43. F.Botre, F.Mazzei, M.Lanzi, G.Lorenti, C. Botre, Anal.Chim.Acta, 255, 59 (1991).
44. A.L.Simonian, E.I.Rainina, V.I.Lozinsky, I.E. Badalian, G.E.Khachatrian, S.SH.Tatikian, T.A. Makhlis, S.D.Varfolomeyev, Applied Biochem & Biotechnol., 36, 199 (1992).
45 M.Masoom, Anal.Chim.Acta, 214, 173 (1988).
46. R.J.Storella, S.Begen, in *Current Separations*, Bioanalytical Systems Inc.USA Internal Review 12 N°1, 8 (1993).
47. B.L.Ruiz, E.Dempsey, C.Hua, M.Smyth, J.Wang, Anal. Chim.Acta, 273, 425(1993).

48. R.Rouillon, N.Mionetto, J-L Marty, Anal.Chim.Acta, 268, 347 (1992).
49. R.M.Morelis, P.R.Coulet, Anal.Chim.Acta, 231, 27 (1990).
50. E.N.Navera, K.Sode, E.Tamiiya, I.Karube, Biosensors & Bioelectronics, 6, 675(1991).
51. M.G.Garguilo, N.Huynh, A.Proctor, A.C.Michael, Anal.Chem., 65, 523(1993).
52. W.Höbel, A.Papperger, J.Polster, Biosensors 1 Bioelectronics, 7, 549(1992).
53. U. Brand, B.Reihnardt, F.Rüther, T.Scheper, K. Schügler 'Development of sensor system based on enzyme-modified field effect transistors for process control in biotechnology' in *Biosensors Applications in Medicine, Environmental Protection and Process Control* (GBF monographs) vol 13 (1989) p 127.
54. E.Galindo, D.Bautista, J.L.Garcia, R.Quintero, Enzyme Microbiol.Technol., 12, 642 (1990).
55. H.Meier, S.Kumaran, A.M.Danna, C.Tran-Minh, Anal. Chim.Acta, 249, 405(1991).
56. U.Brand, T.Scheper, K.Schügler, Anal.Chim.Acta, 226, 87 (1989).
57. U.Brand, B.Reinhardt, F.Rüther, T.Scheper, H-L Schmidt, Anal.Chim.Acta, 238, 201(1990).
58. H.M.McConnell, J.C.Owicki, J.W.Parce, D.L.Miller, G.T.Baxter, H.G.Wada, S.Pitchford, Science, 257, 1906 (1992).
59. H.M.McConnell, P.Rice, G.H.Wada, J.C.Owicki, J.W.Parce, Current Biology, 1, 647(1991).
60. H.M.McConnel, J.C.Owicki, J.W.Parce, D.L.Miller, G.T.Baxter, H.G.Wada, S.Pitchford, Science, 257, 1906(1992).
61. M.Rank, B.Danielsson, J.Gram, Biosensors and Bioelectronics 7, 631(1992).
62. J.Ngeh-Ngwainbi, A.A.Suleiman, G.G.Guilbault, Biosensors & Bioelectronics, 5, 13(1990).
63. Y.Okahata and H.Ebato, Anal.Chem., 63, 203(1991).
64. U.J.Krull, D.P.Nikolelis, J.D.Brennan, S.Brown, M.Thompson, V.Ghaemmaghami, K.M.Kallury, Anal. Proceedings, 28, 370(1991).
65. O.Chastel, J-M Kauffmann, G.J.Patriarche and G.D.Christian, Anal.Chem., 61,170(1989).
66. J.Wang, Z.Lu, Anal.Chem., 62, 826(1990).
67. J.Wang, M.Ozsoz, Analyst, 115, 831(1990).
68. J-M Kauffmann, O.Chastel, G.Quarin, G.J.Patriarche M.Khodari, Bioelectrochem.Bioenerg.,23, 167(1990).
69. R.F.Taylor, I.G.Marenchic, E.J.Cook, Anal. Chim. Acta, 213, 131(1988).
70. M.Gotoh, E.Tamiaya, I.Karube, J.Memb. Sci.,41 , 291(1989)

71. K.R.Rogers, J.J.Valdes, M.E.Eldefrawi, Anal. Biochem., 182, 353 (1989).
72. D.Leech, G.A.Rechnitz, Analytica Chim.Acta, 274, 25 (1993).
73. S.Löfas, M.Malmqvist, I.Rönnberg, E.Stenberg, B.Liedberg, I.Lundström, Sensors and Actuators B, 5, 79(1991).
74. Application Note 202,Pharmacia Biosensor AB.

IN VIVO MONITORING WITH MICRODIALYSIS PROBE

D. MOSCONE[1], M. MASCINI[2]

1 Dip. di Scienze e Tecnologie Chimiche, II Universita' degli Studi di Roma, Via Orazio Raimondo, 00173 Roma, Italy.
2 Dip. di Sanita' Pubblica, Epidemiologia e Chimica Analitica Ambientale, Sezione di Chimica Analitica, Via Gino Capponi 9, 50121 Firenze, Italy.

Abstract Using a sterilized hollow fiber as microdialysis probe, placed subcutaneously and a glucose biosensor, we were able to monitor the variation of glucose during a glucose load (in the bloostream with rabbit and oral with humans). The technique revealed good stability of the glucose sensitivity of the probe during the experiments (up to three hours) and a fair correlation of the glucose values in blood and subcutaneously. In few cases with humans (3 of 15) we found a delay of 30 min.

INTRODUCTION

One of the challenge of biosensors has been ever the in vivo measurement, especially for glucose, the key metabolite for diabetes control.
Different strategies have been tested from various groups and the keypoint of biocompatibility of the probe with the sampling point limited up to now reproducible behavior of such probes.
Microdialysis is an in vivo sampling technique for the continuous removing of small quantities of metabolites from the fluid compartments of the body, such as from brain tissue, the bloodstream, subcutaneous fluid, etc., without removing any liquid.
A thin dialysis tube in the range 100-200 μm of diameter is perfused with a physiological solution at a low flow rate in the range 2-30 μl/min.
Metabolites of small molecular weight diffuse in the direction of lowest concentration, the driving force being the concentration gradient.
This technique was firstly developed for measurement of neurotransmitters in the brain and has been coupled to liquid chromatography [1-2]; in our case the microdialysis probe is inserted subcutaneously and the probe is coupled to a microcell where a glucose electrochemical biosensor is inserted to obtain an apparatus for continuous in vivo glucose monitoring [3-4]. Therefore, it is possible in this particular case to think about the microdialysis as an in vivo biosensor technology with the only difference that substances are measured immediately outside the body instead of by a microsensor (like a needle biosensor) introduced into the body [5]. In this way several advantages are obtained: the sensitivity of the biosensor can be checked frequently; the hollow fiber used as microdialysis probe can

G.G. Guilbault and M. Mascini (eds.), Uses of Immobilized Biological Compounds, 115–122.

easily be sterilized and can be inserted in conscious animals and humans; the whole process of in vivo implantation can be much more closely controlled and reactions such as inflammation and clotting can be easily monitored and eliminated.
We report on this paper our results obtained in vivo on rabbits and human volunteers of continuous monitoring of glucose during a glucose load.

ASSEMBLING OF THE SENSOR

The wall-jet flow cell used in our work [6] includes three separate electrodes: the working electrode (platinum disk with diameter of 1.6 mm), the reference (Ag/AgCl) and the auxiliary electrode. A thin (20 μm) membrane of cellulose acetate was stretched over the platinum electrode surface: it removes the electrochemical interferences (uric acid, ascorbate etc.) with its nominal MWCO of 100 [7-9]; this has the effect that ascorbic and uric acid do not reach the electrode surface while hydrogen peroxide passes through easily. A nylon net (thickness 100 μm), with the immobilized glucose oxidase enzyme was placed over the electrode area.
Sleeve of suitable diameter was used to stretch both the membranes on the Pt electrode surface.

PROCEDURES

The flow system for in vivo experiments is shown in Figure 1. For in vitro experiments the microdialysis probe was immersed in a beaker with a physiological solution.
The peristaltic pump drives the carrier solution through the microdialysis probe immersed in glucose standard solutions at constant rate. A steady-state current is obtained. The standard solutions, where the microdialysis probe was immersed, were manually changed.
To place the microdialysis hollow fiber subcutaneously a sterilized needle was inserted transcutaneously through the skin for about 1 cm and the needle tip was pulled out. Then the sterilized fiber was inserted from the needle tip and the needle was taken out leaving the hollow fiber under the skin. The fiber was connected to nylon tubes and fixed with epoxy glue. For checking variations in sensitivity during the experiments, a flow injection system was used. An injection valve with a 20 μl sample loop was introduced in the flow system just after the microdialysis probe (Fig. 1). A glucose standard buffered solution filled the loop of the injection valve and flowed through the glucose biosensor. We obtained a current profile similar to a peak. Figure 2 shows an in vitro experiment. The microdialysis probe was immersed in different standard solutions of glucose giving a steady state signal and the sensitivity of the glucose biosensor was checked by the height of the peaks. The signal shows a very stable steady state and a high reproducibility of the peaks.

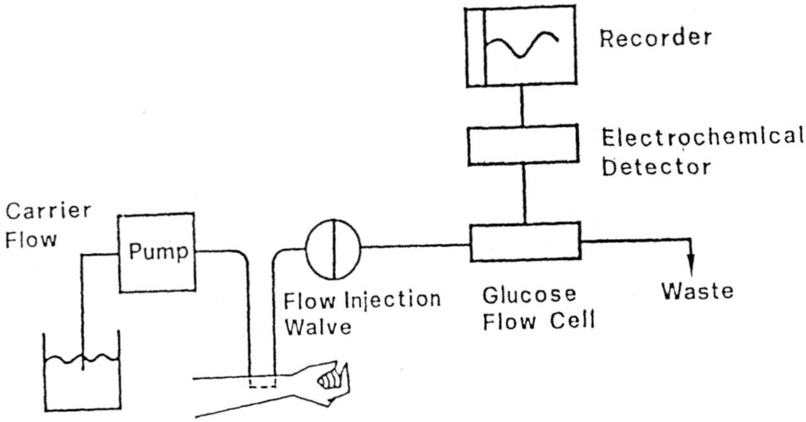

Fig. 1 Diagram of the flow system.

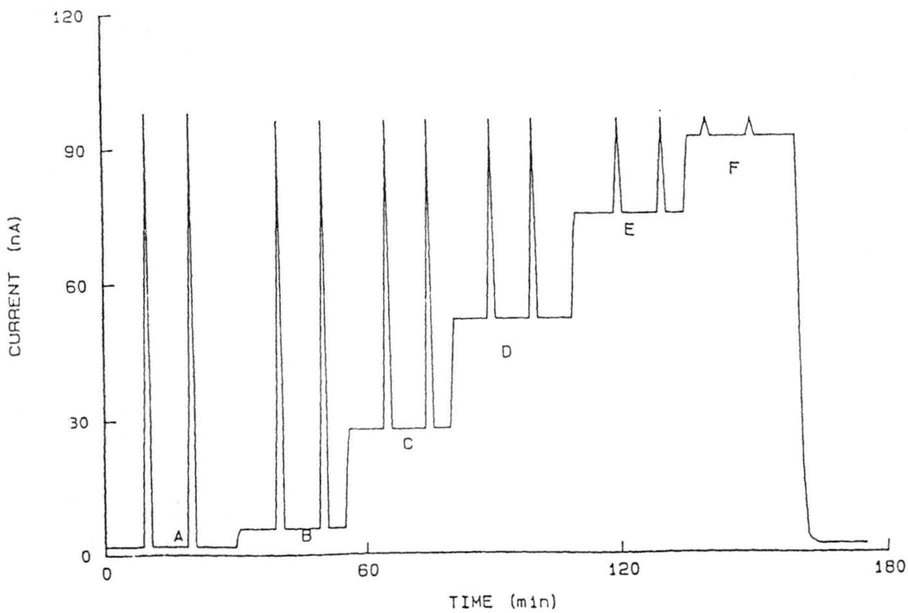

Fig. 2 In vitro experiment. The microdialysis probe was immersed in different standard glucose solutions: A, buffer; B, 1 mM; C, 5 mM; D, 10 mM; E, 15 mM; F, 20 mM. Peaks are obtained injecting through the sampling value a standard glucose solution of 2 mM.

RESULTS AND DISCUSSION

The geometry of the flow cell and the MWCO of the dialysis fiber is very important in in-vivo experiments. Recently we demonstrates how sensitivity could be stabilized with a suitable choice of such parameters [5].

Fig. 3 shows the experiment with the rabbit (3 Kg) where a glucose load of 2 g (30 ml of glucose solution 7%) was infused in the ear vein during about 8 minutes. After about 2-3 minutes from the start of the infusion procedure we could notice a variation of current measured by the cell. Such increase due to the subcutaneous glucose reached a maximum value after 15 minutes, then the current decreased reaching the basal value after about one hour.

During this time the sensitivity of the biosensor was checked and we could notice only a slight decrease less than 5% after the three hours of experiment.

During the experiment, blood aliquots were taken off by the ear of the rabbit and analyzed in the clinical laboratory for glucose.

Assuming that the constant value of current before the glucose load was proportional to the concentration measured in the blood, we could obtain the continuous monitoring of glucose concentration during the experiment (right side of the plot). Therefore, we assume a linear relation of biosensor output (current) and concentration in the range of interest (one point calibration). For comparison the plot of Fig. 3 also includes the concentrations of glucose in blood samples analyzed in laboratory as a control. It can be seen how this system follows with high accuracy the concentration value of blood in rabbits with a simple one point calibration. The response time of the overall system is of 2-3 minutes and the stability of the sensitivity of the biosensor perfectly acceptable for such glucose load experiments.

HUMAN VOLUNTEERS

In Fig. 4 we report 15 experiments on human volunteers submitted to glucose oral load; we follow them by continuous monitoring the subcutaneous glucose value and by normal clinical methods by taking blood samples every 30 minutes.

The glucose value of the blood was measured with clinical standard procedures. In only 3 cases (5,6,13) the microdialysis probe had to be replaced during the course of the experiment, due to accidental break of the probe.

To correlate the continuous monitoring of the current output with the glucose value we assume that the current value before the load corresponds to the glucose value of the first blood sample. (one point calibration). In few experiments (1,3,8,9) the glucose value for calibration has been chosen in the flat part of the curve.

The experiments reported as n° 2,14,15, did not perfectly correlate (Fig. 5) but the correlation increases (Fig. 5 at right) if we assume a delay between blood glucose value and

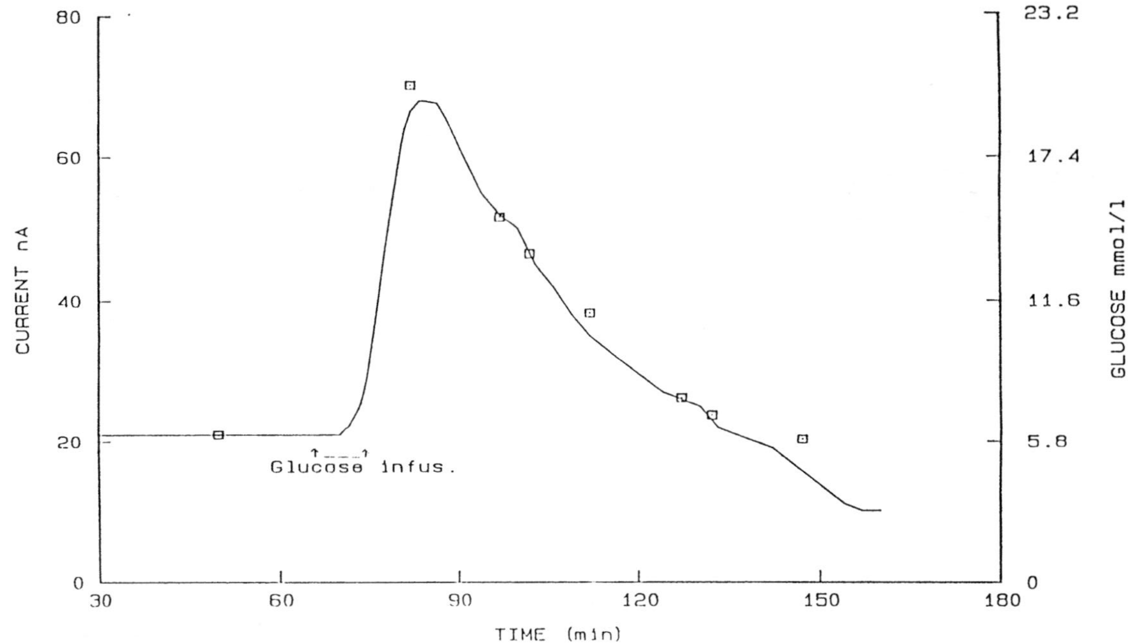

Fig. 3 In vivo glucose load with a 3 Kg rabbit. The glucose load was realized infusing a glucose concentrated solution in the ear vein during 8 minutes.
The hollow fiber has MWCO of 9.000 and was inserted subcutaneously.
By adjusting the stable value before the load with the clinical value we could plot the right ordinate (current proportional to concentration) according to the one point calibration.
The values plotted are the clinical values of blood taken out during the experiment (Correlation coefficient r= 0.993).

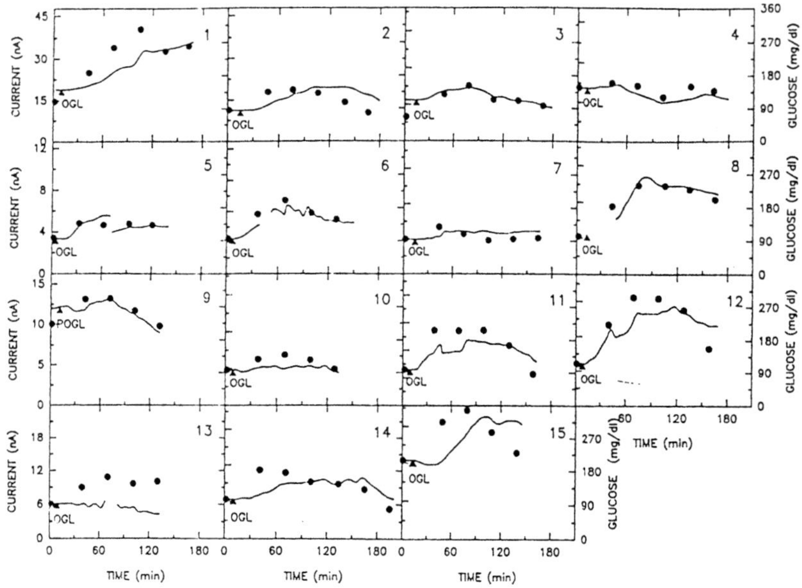

Fig. 4 Glucose oral load with 15 volunteers. The glucose load was realized by drinking 75 g of glucose at the arrow. The values obtained by clinical analyses are reported.

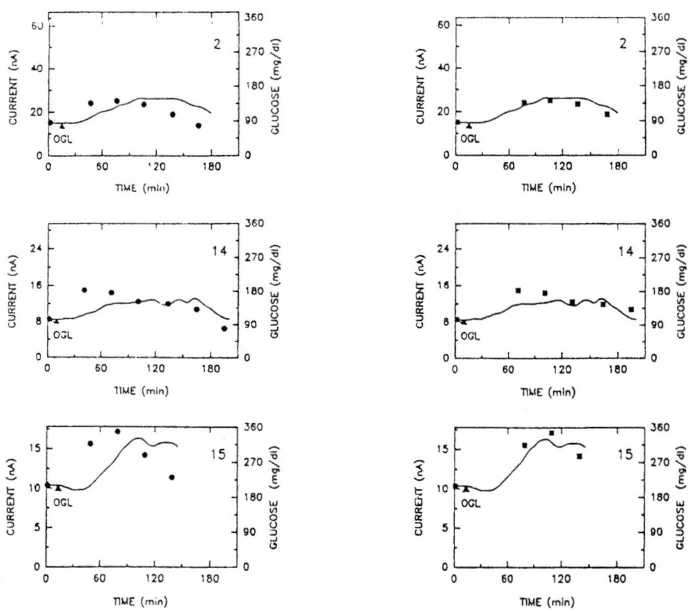

Fig. 5 The experiments n° 2,14,15 have been delayed in respect of their glucose value in blood of 30 min. Better agreement was then obtained.

subcutaneous values of 30 minutes.
In our experiments the microdialysis probe was located in the forearm or in the arm, but other sampling sites could be more suitable for a better correlation with the glucose values in blood.
This has been considered in a similar experiment [10] where up to 18 min. of delay was calculated between glucose blood and subcutaneous value.
Other three experiments (1,11,12) show a low correlation; the trend of glucose in blood seems at glance dissimilar or opposite to the subcutaneous values trend.
The experiments 3,4,7,8,9,10 seem correlate much better and give the hope for a real portable subcutaneous glucose monitoring.
The actual size of the apparatus does not permit to make experiments longer than 3-4 hours and a wearable system will be soon realized for extend the experiments.
During such experimental period we realize that the control of sensitivity was not necessary in the period of time up to 3-4 hours. Variation of the sensitivity of glucose probe was less than 10% which is fairly acceptable for a continuous monitoring.
One of the effect of the position of the microdialysis probe was the involuntary muscle contraction near the sampling point.
Often this contraction (generally when the blood sampling was taken for clinical analysis) lead to a sudden variation of the continuous monitoring due in our opinion to a collapse of the subcutaneous fiber.
This was really evident in Exp. n°6 where a series of small current peaks are evident on the continuous monitoring.
In other experiment (n°15) the patient was affected slightly by Parkinson disease and we obtained a series of unwanted peaks which have been smoothed out in the reported continuous monitoring.

CONCLUSIONS

The microdialysis coupled with a suitable flow cell equipped with a glucose biosensor gives a real in vivo sampling and continuous monitoring of glucose during glucose load experiment. One point calibration is perfectly acceptable with this approach.
Studies on sampling site reveal new problems, but a real continuous glucose monitoring can be performed on such principle.

ACKNOWLEDGMENT

This research was supported by the CNR Target Project on Biotechnology and Bioinstrumentation.

REFERENCES

1 Kissinger P.K.
Biomedical applications of liquid chromatography-electrochemistry
J. Chromatogr. 488(1989), 31

2 Hurd Y.L., Kehr J., Ungerstedt U.
In vivo microdialysis as a technique to monitor drug transport: correlation of extracellular cocaine levels and dopamine overflow in the rat brain
J. Neurochem. 51(1988), 1314

3 D. Moscone, M. Pasini and M. Mascini
Subcutaneous microdialysis probe coupled with glucose biosensor for in vivo continuous monitoring.
Talanta 39(1992) 8 1039-1044

4 Moscone D., Mascini M.,
Microdialysis and glucose biosensor for in vivo monitoring
Ann. Biol. Clin. 50(1992), 323

5 Vadgama P., Desai M.A.,
In vivo Biosensors from biosensors principles and applications (Blum L.J., Coulet P.R., eds) M. Dekker Inc, New York, 303-338 (1991)

6 Moscone D., Mascini M.,
Microdialysis coupled with glucose biosensor for subcutaneous monitoring
Analusis 21(1993), M40-M42

7 Mascini M., Mazzei F.
Amperometric sensor for pyruvate with immobilized pyruvate oxidase
Anal. Chim. Acta 192(1987), 9

8 Taylor P.J., Kmetec E., Johnson J.M.
Design, construction and applications of a galactose selective electrode
Anal. Chem. 49(1977), 789.

9 Palleschi G., Nabi Rahni M.A., Lubrano G.J., Ngwainbi J.N., Guilbault G.G.
A study of interferences in glucose measurements in blood by hydrogen peroxide based glucose probes
Anal. Biochem 159(1986), 114.

10 C. Meyerhoff, F. Bischof, F. Sternberg, H. Zier, E.F. Pfeiffer
On line continuous monitoring of subcutaneous tissue glucose in men by combining portable glucosensor with microdialysis
Diabet. 35(1992) 1087-1092.

IMMOBILISED CATALASE AND HYDROGEN PEROXIDE - A NOVEL CONTACT LENS DISINFECTION SYSTEM

SIMON A. JACKMAN, J. BRIAN KAY, BRIAN J. MEAKIN* AND DAVID J. G. DAVIES*
Sauflon Pharmaceuticals Ltd., 5, Mace Industrial Estate, Ashford, Kent TN24 8EP and *Centre for Drug Formulation Studies, School of Pharmacy & Pharmacology, University of Bath, Claverton Down, Bath BA2 7AY.

ABSTRACT Hydrogen peroxide has been used for several years as a disinfectant for contact lenses. This report details a novel neutralisation system for the peroxide, utilising catalase immobilised to the inner surface of a contact lens case by covalent attachment to a hydrolysed nylon surface.

I INTRODUCTION

Hydrogen Peroxide, at 3 % (w/v), was the first "cold" disinfection solution to be applied to soft contact lenses. It has broad spectrum activity against most pathogens and offers considerable safety providing that adequate neutralisation can be assured. Early aftercare regimes involved neutralisation of the peroxide with sodium thiosulphate, with later developments introducing both a platinum catalytic disc and soluble catalase to the range of possible neutralisers [1].

The naturally occurring enzyme catalase converts hydrogen peroxide to water and oxygen according to the equation :

$$2\ H_2O_2 \longrightarrow 2\ H_2O + O_2$$

This is a highly specific process that occurs at a very high catalytic rate under physiological conditions. When an excess of catalase, in the form of a tablet, is added to a 3 % hydrogen peroxide solution, neutralisation is effected within 10 minutes [2]. When this tablet is included as part of an aftercare regime, lenses must be rinsed, in order to remove the catalase, with sterile saline solution before applying them to the eyes.

If catalase can be immobilised onto a surface that is contacted with the peroxide solution, this may obviate the need for a final rinsing step, thereby producing a one-step system for both disinfection and subsequent neutralisation. Indeed, the resultant solution, once buffered to the correct pH and tonicity should contain an excess of oxygen, as produced from the catalase reaction, a situation that is advantageous for a contact lens that otherwise reduces the amount of oxygen available to the surface of the eye.

Catalase immobilisation has been studied for a number of decades in the field of removing hydrogen peroxide from treated food products [3] and as a component of glucose biosensor systems. A wide variety of surfaces and methods have been investigated, with detailed reports of the activity and stability of the immobilised enzyme having been published.

The requirements of a contact lens care system are such that the immobilisation matrix must be a polymer suitable for moulding into a contact lens case and the enzyme must be irreversibly fixed to this surface. In addition, the peroxide must be present for sufficient

G.G. Guilbault and M. Mascini (eds.), Uses of Immobilized Biological Compounds, 123–130.

length of time at sufficient concentration to enable complete disinfection of the lenses, and must then be neutralised within a convenient timescale. The upper threshold for tolerance of the eye to hydrogen peroxide is 60 ppm, therefore, in practice, neutralisation is considered achieved providing the concentration is reduced below this threshold.

It is worth mentioning, at this point, another facet of this system that has important bearing on the amount of catalase that is required to be immobilised. This is that catalase, alongside its decomposition of hydrogen peroxide, is itself inactivated by reaction with high concentrations of the substrate. This has led to problems in the food industry where deactivation must be overcome with sufficient quantities of stabilised enzyme [4]. The problem may be even greater when applied to the contact lens aftercare system because of the higher concentrations of peroxide used. As a result of this inactivation phenomenon, catalase does not behave according to standard Michaelis-Menton enzyme kinetics, but is, instead, consumed during the reaction. This is most clearly demonstrated in Fig. 1 where catalase, applied at different concentrations to irradiated polystyrene tubes, was immobilised, but its activity towards 3 % (w/v) hydrogen peroxide tailed off in each case. Overcoming such a "threshold" amount of catalase in order to completely neutralise the peroxide will require significantly greater amounts of catalase than would otherwise be needed for a straightforward enzyme reaction.

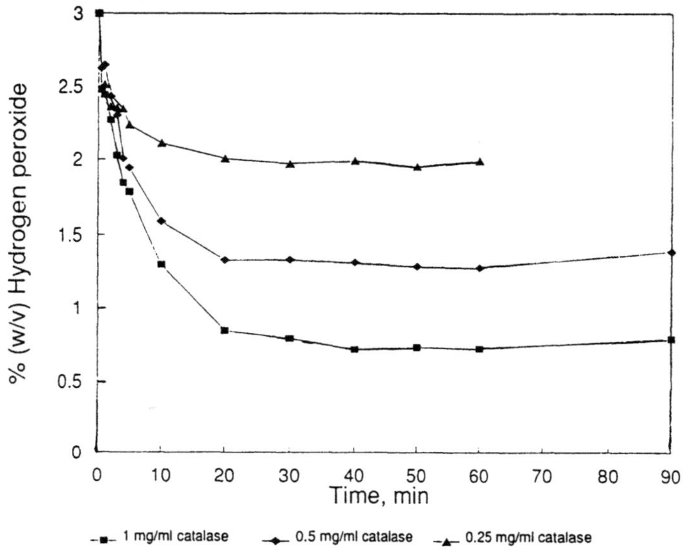

FIGURE 1 Activities of catalase immobilised onto irradiated polystyrene tubes, demonstrating the inactivation of the enzyme by peroxide.

We have approached all of the above goals and challenges with a multi-disciplinary approach of modifying published procedures and developing new techniques.

II INVESTIGATION OF IMMOBILISATION TECHNIQUES

Techniques available for immobilising enzymes include adsorption, entrapment in polymers, encapsulation and covalent cross-linking [5].

A. Adsorption.

The most straightforward technique available is adsorption, where a solution of the enzyme is simply applied to an activated form of the polymer. This technique has been used most successfully to produce antibody-based diagnostic kits in which the antibody is applied to an irradiated polystyrene vessel.

Aspergillus niger catalase was adsorbed onto the surface of polystyrene tubes that had been irradiated with a 2.5 Mrad dose of gamma-irradiation. This procedure was optimised by employing 10 mg/ml catalase in citrate buffer, pH 5. The adsorbed enzyme was capable of neutralising a 3 % solution of hydrogen peroxide. The most important analytical steps for this procedure were to determine whether the catalase could be desorbed from the surface and whether its activity could be retained during storage.

Due to the low levels of desorption expected from this system, the process was investigated using radiolabelled ^{125}I-catalase as opposed to attempting to measure the activity of the desorbed enzyme. Radioactive experiments demonstrated that, whilst the first five washes gave expected levels of soluble catalase from rinsing enzyme that was not firmly attached to the surface, a subsequent five washes also contained radioactivity and these at levels that were not decreasing (see Fig. 2). These experiments were corroborated by results from a 24-hour incubation of the immobilised catalase in a suitable buffer in which significant levels of both radioactivity and enzymic activity were measured.

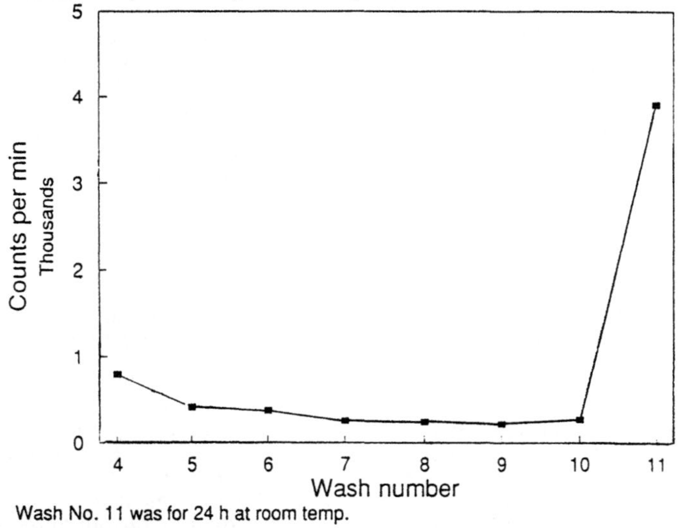

FIGURE 2 Radioactive content of wash solutions from irradiated polystyrene tubes to which ^{125}I-catalase has been bound.

Doubt concerning the suitability of this method has also been raised in the literature because it appears that enzymes adsorbed onto polystyrene are "inverted" and thereby inactivated due to a slow reaction of their hydrophobic cores with the styrene moeity of polystyrene[5].

B. Covalent binding.

Covalent binding of catalase to polymeric surfaces requires suitable reactive groups both on the surface of the enzyme and on the polymer surface. Polystyrene materials bearing amino groups have recently been developed by A/S Nunc [6] and those with amino or carboxyl groups by Costar Corporation. Nylon, being in nature a polyamide, is a good alternative polymer. It can be hydrolysed at its surface to yield exposed amino and carboxyl groups suitable for reaction with cross-linkers [7,8].

A number of amino acid residues on the enzyme surface are capable of being used for cross-linking. Catalase from *Aspergillus niger* is a glycoprotein. The attached polysaccharide residues confer increased stability upon the enzyme [9], but preliminary cross-linking experiments suggested that these may reduce the accessibility of reactive amino acid residues, and therefore the bovine liver enzyme, which is not a glycoprotein, was chosen for all subsequent covalent immobilisation work. The most commonly chosen reactive group on a protein is the sulphydryl, but analysis of bovine liver catalase demonstrated that no free groups were present. Therefore, surface amino groups were targetted, the majority of which are lysine Σ-amino groups.

The possible linkers available given the polymer and enzyme surfaces were homobifunctional amine cross-linkers and heterobifunctional amine-carboxyl linkers.

Results demonstrated that none of the polystyrene-based surfaces yielded significant enzyme binding, whereas binding to hydrolysed nylon was successful. Having selected the most suitable cross-linker, binding was characterised and optimised.

III CATALASE IMMOBILISATION EXPERIMENTS

Reagents

Catalase (E. C. 1.11.1.6) was from bovine liver, and was obtained from Biozyme Labs Ltd., Blaenavon, U. K. Hydrogen peroxide solution was 3 % (w/v) in sodium pyrophosphate buffer, pH 3.8. Potassium titanium (IV) oxalate was purchased from BDH Ltd., Poole, U. K. Stock titanium reagent was prepared by mixing 54.4 ml conc. sulphuric acid with approximately 60 ml ice-cold distilled water. 7.08 g Potassium titanium oxalate was dissolved in the acid solution and the volume was made up to 200 ml with distilled water. For measurement of hydrogen peroxide concentration this stock solution was diluted 5-fold in distilled water.

Enzyme immobilisation

Whilst it is not appropriate to disclose the details of the cross-linker and its conditions of use, the activation of the nylon surface and catalase immobilisation were performed under the following conditions :

Nylon contact lens cases were filled to a volume of 10 ml with 4.25 M hydrochloric acid and incubated at room temperature for 1 hour. The cases were then rinsed thoroughly with distilled water and the cross-linking reagent added. After the nylon surface was activated with cross-linker, a 1 mg / ml solution of catalase in phosphate-buffered saline (PBS), pH 7.4 (7.574 g Na_2HPO_4, 2.1 g $NaH_2PO_4.2H_2O$ & 4.4 g NaCl in 1 litre) was added to each case. The enzyme solution was applied to the activated surface at room temperature for a period of 1 hour. Following the reaction, the enzyme-coated surface was washed 5 times with 10 ml PBS containing 0.05 % (w/v) Tween-20. These washes were required to ensure that all non-covalently attached enzyme was removed from the surface of the case.

Measurement of catalase activity

The catalytic activity of immobilised catalase was determined by addition of 10 ml hydrogen peroxide solution to each case. Samples of 10 ul were taken at time intervals and added to 990 ul titanium reagent. The titanium reagent forms a complex with hydrogen peroxide which absorbs at 400 nm [10]. A standard curve for hydrogen peroxide (Fig. 3) demonstrates that the absorbance of the peroxide-titanium complex is proportional to the amount of peroxide up to 0.006 % (w/v) peroxide. This is equivalent to 0.6 % (w/v) peroxide in the lens case. Therefore, when peroxide concentrations of greater than 0.6 % were being measured, a 10-fold dilution of the initial peroxide-titanium solution was carried out.

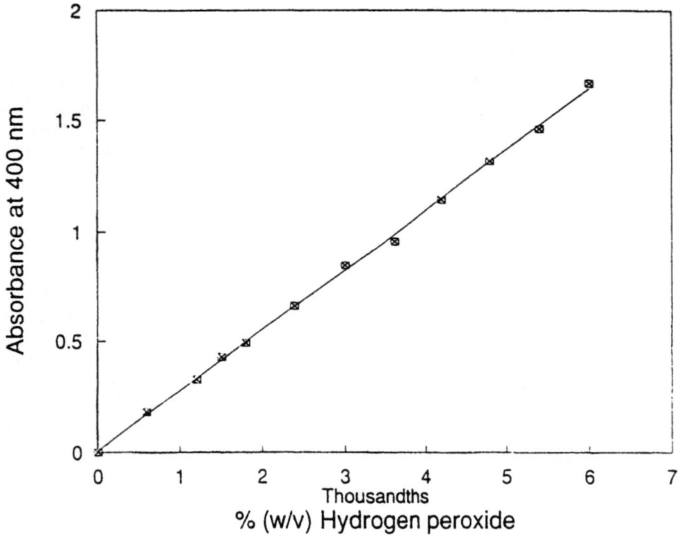

FIGURE 3 Standard curve for the absorbance of a hydrogen peroxide-titanium complex at 400 nm.

The results shown in Fig. 4 (steepest curve) portray the activity of catalase immobilised to nylon cases. Cases measured immediately after activation with catalase demonstrated very high activity, with the peroxide concentration being reduced to below 0.3 % (w/v) in under 10 minutes. After 2 hours, the concentration of hydrogen peroxide had been reduced below 25 ppm, as measured with Merckoquant peroxide test strips. This is sufficient for contact lenses to be placed directly into the eye after treatment.

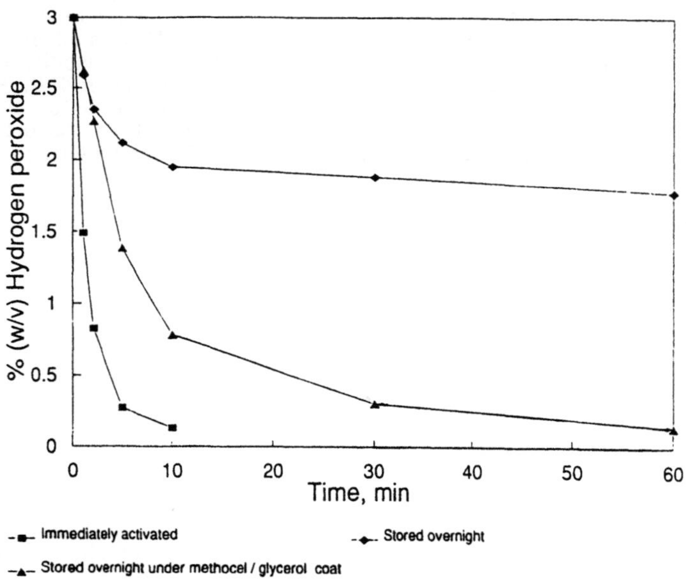

FIGURE 4 Activity of immobilised catalase stored with and without a methocel / glycerol coat.

IV STABILISATION OF IMMOBILISED CATALASE

Catalase that has been successfully linked to a nylon surface loses its catalytic activity rapidly upon drying, as demonstrated by the results from cases stored overnight under dessicated conditions (Fig. 4). In addition to this feature, immobilised catalase activity, as determined immediately after activation, is greater than required, with the result that the peroxide is removed before it has fully disinfected the system. It is therefore necessary to stabilise and coat the immobilised enzyme.

Coating

Stabilisation of immobilised enzymes through coating has been investigated using polyhydroxyls, polyelectrolytes and sugars [11]. No clear pattern emerges from the literature as to which combinations of reagents are most effective. Coating with 2 % methocel (methylcellulose) and 20 % glycerol in PBS led to considerable stabilisation of immobilised catalase through overnight storage (Fig. 4). This coating was carried out by simply pouring the coating solution into the lens case, and then removing it with a pasteur pipette. The residual coating solution was dried onto the lens case surface in dessicated conditions under a vacuum. Similar results were obtained when 20 % (w/v) lactose replaced glycerol.

Stabilisation through cross-linking

An alternative and possibly complementary method of enzyme stabilisation is through the cross-linking of individual sub-units to form a more stable oligomeric structure. The theory and methods involved in such a procedure have been reviewed recently by Wong & Wong [12]. It is currently thought that a substantial contribution to catalase instability comes from the dissociation of its four sub-units. Shaked & Wolfe [13] have cross-linked *Aspergillus niger* catalase sub-units with dimethyl adipimidate (DMA) and dimethyl suberimidate (DMS), with DMA cross-linking conferring considerable thermostability upon the enzyme.

For cross-linking of bovine liver catalase, we used DMA according to the procedure of Shaked & Wolfe [13], with the catalase being dissolved in 10 mM citrate buffer, pH 5 at 4°C at a concentration of 4 mg/ml. 16 mg DMA was dissolved in 2 ml methanol. The catalase solution was adjusted to pH 9.5-9.7 with NaOH, and DMA was added dropwise, maintaining the pH of the enzyme solution to between 9.5 and 9.7 by addition of NaOH. When all of the DMA had been added, the mixture was stirred for 1 hour at 4°C. Following this incubation, the catalase solution was passed down a Sepharose G-25 column pre-equilibrated with PBS, pH 7.4 in order to separate the enzyme from the cross-linker. Chromatographed enzyme was diluted to 1 mg/ml in PBS, pH 7.4 and added to activated nylon cases.

The results of this experiment are shown in Fig. 5.

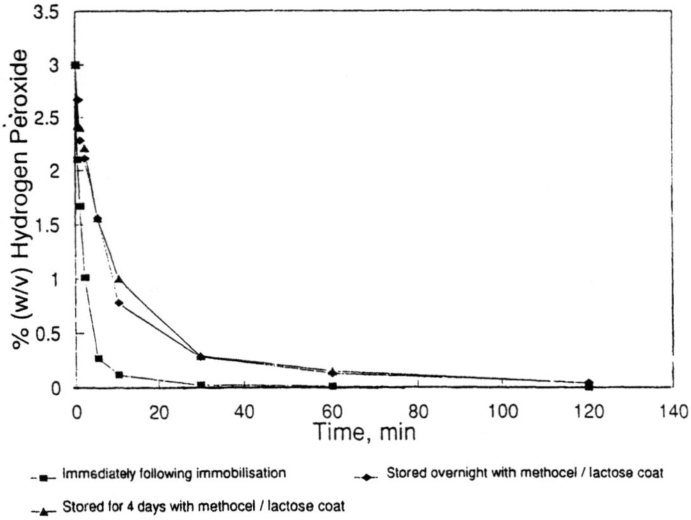

FIGURE 5 Activities of immobilised DMA-cross-linked catalase under different coating conditions.

Immobilised cross-linked enzyme stored overnight under methocel/lactose showed activity that was lower than that for enzyme assayed immediately after binding to the surface, but when the enzyme was assayed after 4 days of storage, the activity was at the same level, and this activity was significant enough to reduce the level of peroxide to below 60 ppm in 4 hours (see Fig. 6).

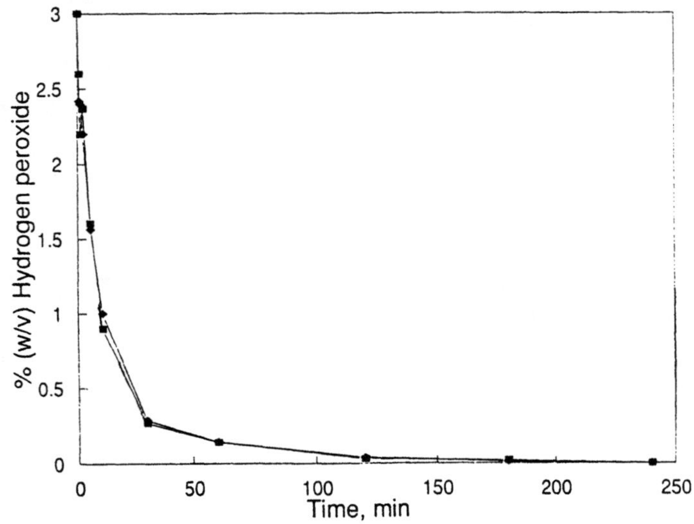

FIGURE 6 Activity of immobilised DMA-cross-linked catalase after storage under methocel / lactose coat for 4 days.

This is a significant improvement over results obtained with coating alone, where less activity was detected after overnight storage (see Fig. 4) and activity decreased dramatically for storage over 2 or more days.

V CONCLUSIONS

The results from immobilisation and coating of DMA cross-linked catalase are extremely promising as the enzyme appears to have sufficient activity for our purposes, and the stability data, although only preliminary, point to the possibility of reasonable storage stability over considerable timescales.

It now remains for us to extend the stability analysis of the enzyme and to investigate more precise coating techniques such that the enzyme can be stabilised further and its activity retarded more significantly. The extent and nature of the cross-linking by DMA must also be more carefully analysed through gel electrophoresis and other molecular studies.

In conclusion, a hydrogen peroxide-immobilised catalase contact lens aftercare system utilising the bovine liver enzyme covalently bound to nylon cases has been designed that will effect a one-step disinfection and neutralisation of contact lenses.

REFERENCES

1. M. J. Sibley Contact Lens Forum 57-63 (July 1987).

2. R. C. Courtney, N. Tarantino & P. Brown International Contact Lens Congress 17, 67-73 (1990).

3. B. P. Sharma & R. A. Messing in Immobilised enzymes for food processing (Pitcher W. H., ed.), 185-209, CRC Press (1980).

4. P. T. Vasudevan & R. H. Weiland Biotechnol. Bioeng. 36, 783-789 (1990).

5. W. H. Scouten Methods Enzymol. 137, 30-65 (1988).

6. S. E. Rasmussen Ann. Biol. Clin. 48, 647-650 (1990).

7. W. E. Hornby & L. Goldstein Methods Enzymol. 44, 118-134 (1976).

8. V. P. Sundaram in Biomedical applications of immobilized enzymes and proteins Volume 2 (T. M. S. Chang, ed.) pp. 317-326 (1977).

9. D. Scott & F. Hammer Enzymologia 22, 229-237 (1960).

10. R. M. Sellers Analyst 105, 950-954 (1980).

11. T. D. Gibson & J. R. Woodward in Biosensors and Chemical Sensors ACS Symposium Series 487 (P. J. Edelman & J. Wang, eds.), American Chemical Society, Washington DC, pp. 40-55 (1992).

12. S. S. Wong & L.-J. C. Wong Enzyme Microbiol. Technol. 14, 866-874 (1992).

13. Z. Shaked & S. Wolfe Methods Enzymol. 137, 599-615 (1988).

EX-VIVO SENSORS; MICROENCAPSULATION OF ENZYME LAYER IN BIOSENSORS

Paul H. Treloar, Séamus P.J. Higson, Mohamed A. Desai, Ian M. Christie, Sharmistha Ghosh, Mark F. Rosenberg, Subrayal M. Reddy, Malcolm N. Jones, Pankaj M. Vadgama.
Department of Medicine (Clinical Biochemistry), University of Manchester, Hope Hospital, Eccles Old Road, Salford, M6 8HD, United Kingdom.

ABSTRACT. Various solids, solid/liquid composites and quasi-liquid membranes have been developed to protect the enzyme layer and working electrode of enzyme electrodes. These include isopropyl myristate-containing porous membranes, liposomes and PVC with a high plasticiser content. The additional value of such membranes beyond their surface 'biocompatibility' has been their facility for controlling mass transport to the enzyme layer. In this way effective enzyme K_m values have been manipulated, the dependence on co-substrate has been reduced, and where appropriate the co-substrate has been retained with the enzyme without a specific need for a co-substrate chemical immobilisation so avoiding the inherent problems of co-substrate mobility. Specific examples of these new encapsulated systems are provided.

INTRODUCTION

Over the last three decades, biosensors have been the focus of much research activity in an effort to achieve simplified biomedical analyses [1,2]. A diversity of biomolecules have been utilised including antibodies[3], receptors[4] and polynucleotides[5]. However, enzymes have been the most intensively investigated, principally in electrochemically devices (enzyme electrodes)[6-9]. Glucose measurement has received particular attention due to a widespread demand for diabetic blood glucose monitoring together with some industrial demands, eg. fermentation monitoring[10]. Updike and Hicks[11] pioneering development of an enzyme electrode for glucose utilised a Clark oxygen electrode overlaid with immobilised glucose oxidase, though measurement of hydrogen peroxide product rather than oxygen consumption became popular subsequently[12]. Some form of immobilisation of an enzyme was recognised as being necessary for stabilisation, and for facilitating transduction. However, only more recently has it been appreciated how important is the precise mode of interfacing of enzyme, and indeed the external sample matrix. In particular, the additional use of covering (interfacing) membranes has proved vital as a protection from interferents and from surface active colloidal materials.

G.G. Guilbault and M. Mascini (eds.), *Uses of Immobilized Biological Compounds*, 131–140.

EXPLOITATION OF MEMBRANE TECHNOLOGY

One classical configuration has been enzyme immobilisation by covalent crosslinking of the enzyme, using a bifunctional coupling agent such as glutaraldehyde[13] typically to bovine serum albumin. The enzyme layer is entrapped between two supporting microporous membranes, eg. polycarbonate[9]. The enzyme laminate can then be readily handled and may be mounted over a permselective membrane (Fig. 1).

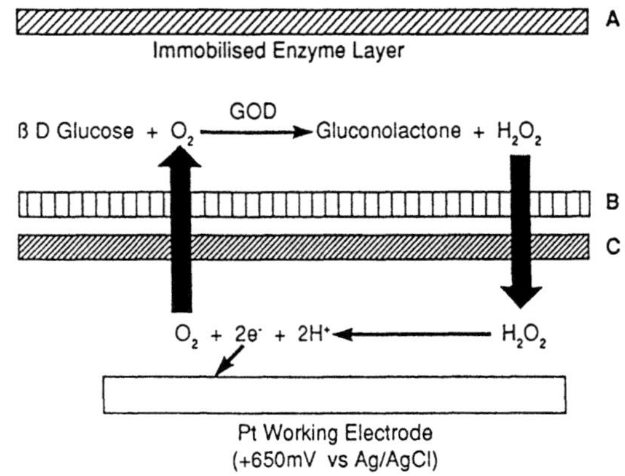

Fig. 1. Schematic diagram of a glucose enzyme electrode showing immobilised enzyme, sandwiched between outer (A) and inner (B) microporous membranes, mounted over a permselective membrane (C) protecting the polarised electrode.

Changes in the membrane laminate have allowed modification of behaviour. Thus, for example, as with other enzymic methods, the linear range is limited by the Michaelis constant. Therefore, to achieve a linear signal output and measurement one major objective has been to extend the working linear range. A key approach has been to introduce a diffusion limiting barrier between the sample and sensor enzyme layer[9-12,14]. Thus for an electrode incorporating glucose oxidase, the local, effective glucose concentration can be much reduced whilst maintaining supply of the co-substrate (oxygen). This overcomes co-substrate limitation at the higher bulk solution glucose concentrations giving much extended linearity.

Biofluid analysis is critically dependent upon an ability to control biofouling of the enzyme electrode surface. Fouling presents serious problems, especially during analysis of whole blood where the potentiated deposition of proteins, lipids as well as cells significantly hinders diffusion of substrate to the enzyme layer. In fabrication of enzyme electrodes, therefore, it is crucial that the outer membrane is sufficiently biocompatible and that permeability of substrate is not significantly marred.

Selectivity may pose further problems in electrochemical devices, which generally offer poorer selectivity than spectrophotometric techniques. With regard to hydrogen peroxide detection in oxidase electrodes, a range of opportunist species present in biological matrices can result in a high interferent signal. Design of a permselective membrane, between the working electrode surface and enzyme, selective on the basis of charge and molecular weight has been an important means of attenuating interferent effects at the polarised working electrode (Fig. 1).

The Outer Membrane

A variety of outer membranes have been employed in order to extend sensor response linearity. Microporous neutron track etched polycarbonate has been especially used[6,7,9]. This barrier operates by control of transmembrane flux by variation of membrane pore size[7,9,15]. Membrane permeability is essentially a function of available pore area. Model solutes with differing charge, polarity and functional groups display no specific transport behaviour indicative of selective partitioning or solute/membrane surface interaction[15]; however, permeability is inversely related to solute molecular weight. For polycarbonate, it is notable that oxygen permeability is approximately two fold greater than that for organic species. This indicates a trans-membrane + trans-pore pathway, clearly attractive for maintaining co-substrate supply to the enzyme, and is confirmed by data that shows that the oxygen permeability verses pore size relationship does not extent through the origin.

It has been demonstrated that decreasing the membrane pore size reduces the substrate/oxygen permeability ratio[9,15] eg. reduction of pore radius from 0.05 to 0.015 μm can extend the linear range for glucose analysis from 2 to 20mM. Also of importance to sensor performance is an ability to reduce absolute permeability such that electrode response is under diffusion control rather than following intrinsic enzyme kinetics. In this way, stir independence together with a lower dependence on temperature and pH variation has been possible[6,7].

In early studies, silanisation of the outer microporous membrane was explored. Organosilane polymerisation, eg. using di and trichloro- derivatives, was found to be effective in manipulating calibration linearity[6]. An optimised system was developed, involving treatment of polycarbonate with methyldichlorosilane, giving a 10 fold extension of linearity to 50mM for a glucose enzyme electrode while responses could still be obtained in 30-90s. This optimisation was possible as oxygen diffusion through the silane remained relatively unhindered due to its high oxygen permeability[16], while much reducing glucose transport. Silanisation also improved biocompatibility considerably, a property already used with silicon rubber, which is exploited as a biomaterial.

In a different approach, microporous membranes have been modified by coating with the lipid isopropyl myristate (IPM)[7]. A composite liquid membrane then results with IPM entrapped in the pores of the polycarbonate (Fig. 2). As with silane, the solubility of oxygen in lipid is exploited[17], while much restricting the diffusion of glucose, which of course does not partition into IPM. Increases in linearity for glucose have achieved up to 100mM[7]. An additional benefit of this composite was its rejection of electrochemically active interferents notably ascorbate and urate. The remaining interferent response was thereby $<$ 5% of the glucose signal in the clinical range, which obviated the need for an additional inner permselective membrane, enabling practical measurement and faster responses in whole blood.

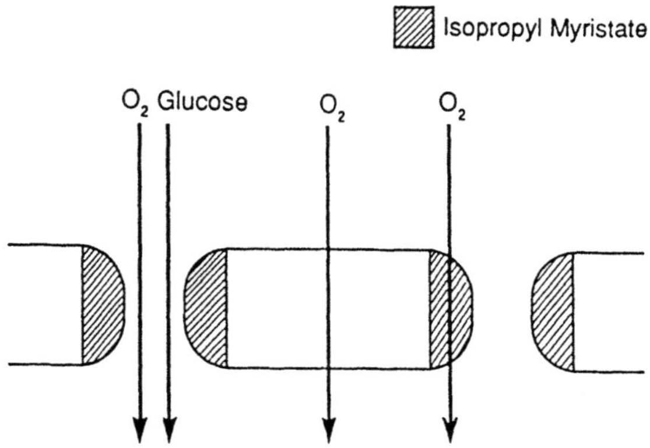

Fig. 2 Transport of oxygen and glucose across an IPM modified microporous polycarbonate membrane.

Recently, Diamond-Like Carbon (DLC) has been investigated as a novel material for coating polycarbonate membranes[9]. DLC is a dense, amorphous hydrocarbon derived polymer having properties which in many ways resemble those of diamond[18]. It has received attention recently as a coating for medical implants, since highly biocompatible surface films can be produced which minimise adverse cell and tissue effects[19]. Furthermore, DLC coatings are mechanically strong, flexible and chemically inert[19,20].

Methodology for DLC deposition involves first membrane cleaning by fast argon atom bombardment and then coating with DLC in a neutral saddle source[21]. Coating thickness can be controlled by simple variation of deposition time. For a given DLC layer, a decrease in permeability for both glucose and oxygen is found[9] and a progressively diminished glucose permeability achieved for a given membrane by DLC coating thickness. A key functional end result is that oxygen transport is differentially much less affected by DLC application than is the diffusion of glucose even at total pore occlusion. DLC coating is particularly attractive, moreover, since a hitherto unseen degree of control of membrane permeability is possible by fine tailoring of the deposition process. Though in principle, the greatest extended linear range for a glucose enzyme electrode is achieved by the highest ratio of glucose/oxygen permeability[6], membranes offering the greatest absolute resistance to the passage of glucose apparently imparted the greatest extensions in linearity. Thus, a linearity of > 80mM glucose was found using the highly impermeable 0.01μm pore size polycarbonate (Fig. 3), as compared with membranes of larger pore size with greater permeability ratios.

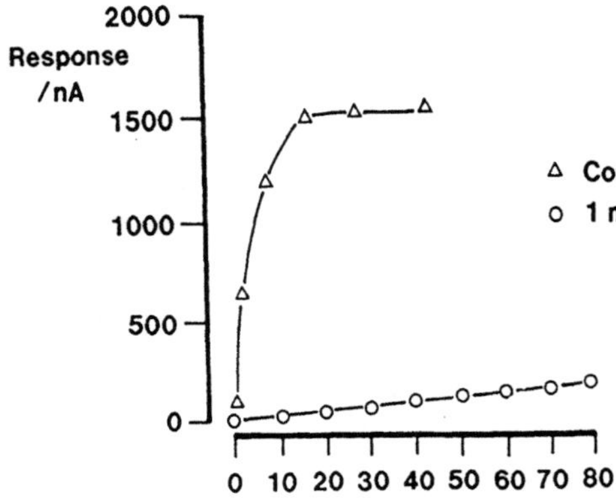

Fig. 3. Effect of DLC coating of the outer membrane on Glucose electrode calibration. (From Higson and Vadgama, 1993, with permission.)

The biocompatibility of DLC coated polycarbonate membrane has been evaluated following exposure to heparinised whole blood. Membrane pore size critically influences biofouling and response attenuation increases with an increase in pore size following blood exposure. DLC double sided coated membranes when used as upper covering membranes have enabled electrodes to withstand 30 min blood exposure with only $< 6\%$ loss of response. This is an improvement over most membranes, and is associated with membranes imparting greatest linear ranges. Two enzyme electrodes with outer DLC-coated polycarbonate membranes of either 0.01 or 0.1 μm pore size were calibrated following whole blood exposure and used to analyse blood samples of known glucose concentration. Whilst results with the 0.1μm pore size membrane gave a low bias, the results using the 0.01 μm pore size membrane showed very close agreement with the routine method ($r^2 = 0.99$). It was also of interest that electrochemically active interferents present in blood had little influence on responses with the low permeability DLC coated membranes. The membranes remain microporous, and permselectivity was not observed in diffusion chamber experiments. One possibility therefore was an intrinsically anionic environment offered by the enzyme matrix (A and B, Fig. 4) which possibly countered diffusion of the major anionic interferents. Exclusion of these species was not seen of course at the high fluxes and concentrations in the enzyme layer when less permeable microporous membranes were utilised. Supportive of this possibility is a loss of selectivity at high glucose levels. Under these conditions, a possible build up of gluconic acid in the enzyme layer following oxidation of glucose may increase the local enzyme protonation (C), eliminating the electrostatic repulsion of anionic interferents.

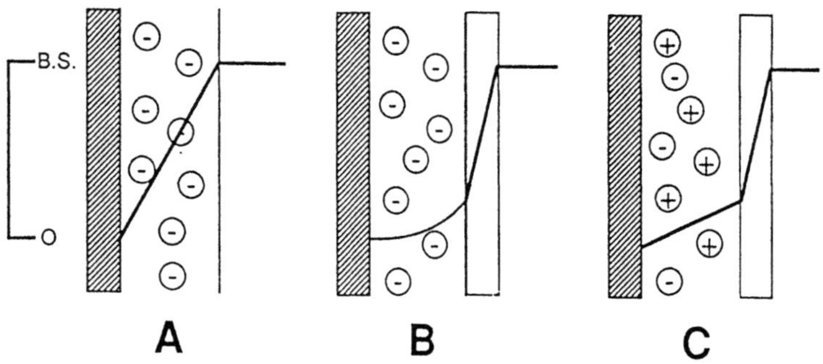

Fig. 4. Interferent flux profiles across a glucose enzyme electrode from bulk solution (B.S.). (A) high permeability outer membrane, (B) low glucose flux environment with a low permeability membrane; anionic charge exclusion, (C) high glucose flux environment with a low permeability membrane; diminished anionic charge exclusion.

Inner membranes

Ascorbate, urate and thiols are all potential interferents during electrochemical detection of hydrogen peroxide. These compounds may either cause false signals or a passivation of the electrode surface. An important strategy to reduce these problems has been to control transport of the species by protecting the electrode with a permselective membrane[22,23](Fig. 1). Unfortunately, this approach cannot easily be adopted when using mediator systems[24]due to the need for intimate contact of enzyme and working electrodes. Careful inner membrane design has also been highlighted for miniature polyurethane protected needle enzyme electrodes for invasive clinical monitoring[25].

Dense cellulose acetate membranes have been commonly used as inner membranes; a successful commercial glucose analyser utilises this permselective barrier[26]. Such membranes can be readily cast from acetone solution[27]. The result is a dense homogeneous membrane, therefore any permeation demands close molecular interaction between microsolute and polymer. Investigation of membrane permeability by mounting over an electrode has shown attenuation of responses to a variety of electrochemically active species as anticipated[27]. However, responses to hydrogen peroxide were found to be relatively maintained, while those to ascorbate were amongst the lowest. The very low permeability of ascorbate is probably due to a combination of charge and shape; phenols are able to traverse the membrane relatively freely[28].

By covering an electrode with cellulose acetate, biocompatibility is also improved. Exposure of glucose enzyme electrodes to whole blood for 120 min., when thus protected, produced relatively small losses of response[29]. Modification of cellulose acetate membranes by incorporation of detergent results in increased permeability of hydrogen peroxide. In this capacity, Tween 80 proved particularly effective in enhancing hydrogen peroxide signals (×3) without compromising ascorbate rejection.

Polyether sulphone (PES) has been employed as an internal membrane on needle electrodes to confer permselectivity[29]. PES can also markedly reduce loss of response due to biofouling. An explanation of this is likely to be reduced access to electrode surface passivating species of low molecular weight.

Plasticised PVC has been employed in the fabrication of enzyme electrodes as either an inner or outer membrane. For glucose enzyme electrodes, it has been employed as an inner membrane to impart high selectivity, eg. the ratio of hydrogen peroxide/ascorbate can be enhanced ~7 fold in comparison to with cellulose acetate, and once again there is much reduced biofouling[30].

A typical enzyme electrode with an inner 0.03μm polycarbonate membrane showing a linear range of ~3mM glucose can be manipulated with an inner PVC membrane such that the linear range is now 16mM. When glucose concentrations of whole blood samples are determined using this enzyme electrode the results show a close correlation (r = 0.978, Fig. 5) in comparison to those obtained by a routine clinical biochemistry laboratory method. It is possible that a high PVC permeability for oxygen is assisting maintenance of the pO_2 within the enzyme layer following the electrochemical conversion of hydrogen peroxide to oxygen (Fig. 1). Unplasticised PVC has also been employed as an outer membrane in the construction of an enzyme electrode for oxalate. In this case, selective transport of an ionic substrate was achieved over (more hydrophilic) interferents and acceptable calibration achieved.

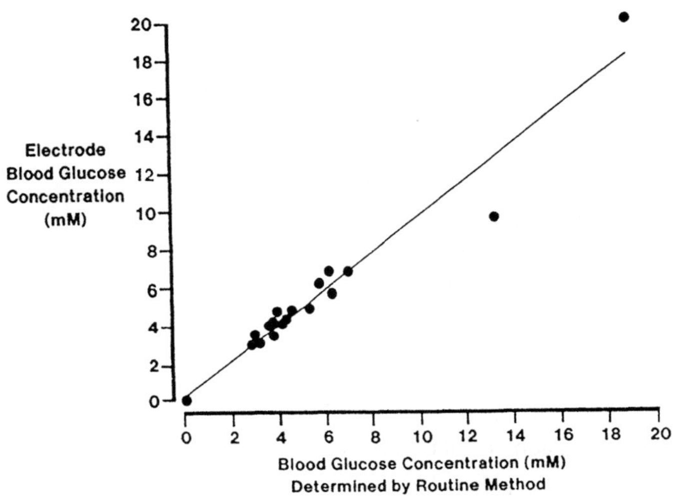

Fig. 5. Measurement of whole blood glucose using an enzyme electrode with outer 0.03μm polycarbonate and inner PVC membrane.

Liposome Encapsulation

A further method of increasing the ratio of co-substrate : substrate supply is to microencapsulate sensor enzyme with a liquid barrier membrane. This has been realised by the entrapment of glucose oxidase within liposomes[8]. The liposomes enclose an aqueous volume containing glucose oxidase with a bilayer lipid barrier (usually phospholipid). The latter membrane is particularly appropriate for the glucose enzyme electrode given that oxygen permeates lipid[17] while polar molecules such as glucose display very low permeability through liposomal membranes[31] (Fig. 6). By adsorbing liposome encapsulated enzyme onto a Millipore membrane, it has been possible to immobilise the liposomes adjacent to an amperometric electrode[8]. On exposure of the electrode to glucose, hydrogen peroxide generated by the enzyme then diffuses out of the liposome sufficiently to be determined at a polarised electrode. Liposome microencapsulation is particularly attractive as the composition of the bilayer membrane can be tailored to control solute permeation. In particular, the permeability of liposomes to glucose can be shown to be strongly dependent on the fluidity the bilayer membrane which can be modified on the basis of the lipid acyl chain length. Appropriate entrapment of the enzyme can increase the apparent K_m from 13mM for the free enzyme to 111mM[8].

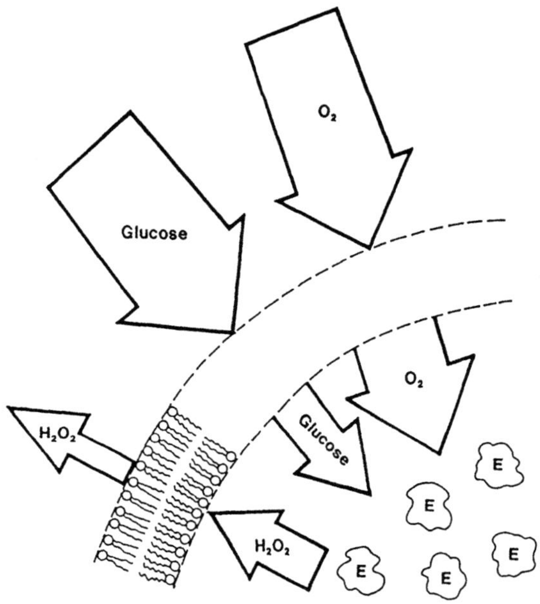

Fig. 6. Schematic cross section of liposome encapsulated glucose oxidase on exposure to glucose solution.

Membrane retention

Immobilisation of enzyme within a spherical vesicle is one obvious form of microencapsulation. However as an extension of this concept, we have developed a reagentless dehydrogenase enzyme electrode for ethanol with all the requisite components behind a plasticised PVC membrane (Fig. 7). Here, the enzyme substrate, ethanol (and product acetaldehyde), can readily permeate the lipophilic liquid membrane, but strong rejection of interferents in blood and urine is achieved[30]. Enzyme co-substrate NAD^+ retention is feasible because of its charged nature. This presents the possibility of wider exploitation of dehydrogenases in a viable, reagentless, biocompatible format. There is also the possibility of soluble electron transfer mediator retention behind plasticised PVC. Quinones are a particularly attractive choice in the case of NADH allowing the oxidation overpotential to be substantially lowered while avoiding problems of NADH electrochemistry [32]. Naphthoquinone sulphonic acid as mediator has thus allowed use of the electrochemical properties of a quinone, the highly charged sulphonic acid group preventing loss into bulk solution. Taken with the biocompatibility of plasticised PVC[30], PVC incorporated electrodes are an attractive development on existing membrane strategies.

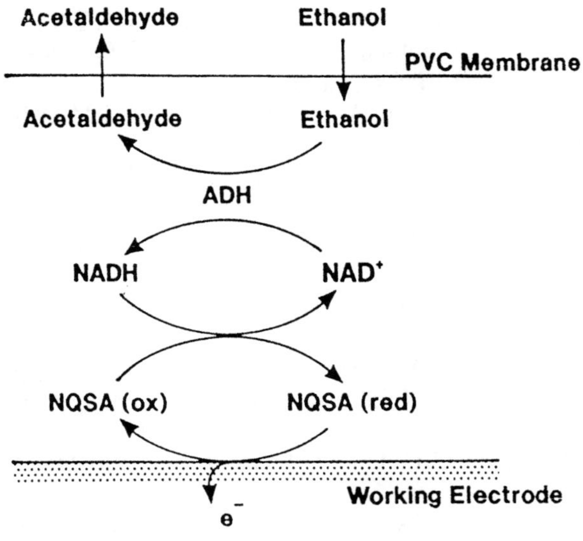

Fig. 7. Detection scheme for a reagentless enzyme electrode for ethanol showing membrane retention of alcohol dehydrogenase (ADH), NAD^+, and napthoquinone sulphonic acid (NQSA).

CONCLUSIONS

Enzyme electrodes present us with elegant analytical systems and a wide range of approaches are possible. Membrane entrapment or microencapsulation of the enzyme provide a key means of adapting these devices, with much reduced dependence on enzyme technology. Furthermore, more novel mediation and non-mediation electrode principles are now likely to be usable for practical analysis as more membrane materials become available.

REFERENCES

1. A.P.F. Turner, I. Karube, G.S. Wilson (eds.) Biosensors; fundamentals and applications, Oxford University Press, New York, 1987.
2. L.J. Blum, P.R. Coulet (eds.) Biosensor principles and applications, Marcel Dekker, New York, 1991.
3. M. Aizawa, S. Kato, S. Suzuki, Chem. Lett., 779, (1977).
4. M. Gotoh, E. Tamiya, M. Momoi, Y. Kagawa, I. Karube, Anal. Lett., 20, 857, (1987).
5. N.C. Fawcett, J.A. Evans, L.C. Chien, N. Flowers, Anal. Lett., 21, 1099, (1988).
6. W.H. Mullen, F.H. Keedy, S.J. Churchouse, Anal. Chim. Acta, 183, 59 (1986).
7. L.X. Tang, Z.B. Koochaki, P. Vadgama, Anal. Chim. Acta, 232, 357, (1990).
8. M.F. Rosenberg, M.N. Jones, P.M. Vadgama, Biochim. Biophys. Acta, 1115, 157, (1991).
9. S.P.J. Higson, P.M. Vadgama, Anal. Chim. Acta, 271, 125, (1993).
10. N. Cleland, S.-O. Enfors, Anal. Chem., 56, 1880, (1984).
11. S.J. Updike, G.P. Hicks, Nature, 214, 986, (1967).
12. I. Hanning, P. Vadgama, A.K. Covington, K.G.M.M. Alberti, Anal. Lett., 19, 461, (1986).
13. F.M. Richards, J.R. Knowles, J. Mol. Biol., 37, 231, (1968).
14. M.B. McDonnell, P.M. Vadgama, Select. Electrode. Rev., 11, 17, (1989).
15. Z. Koochaki, S.P.J. Higson, M. Mutlu, P.M. Vadgama J. Memb. Sci, 76, 261 (1993).
16. W.J. Roff, J.R. Scott, Fibres, Films, Plastic and Rubbers, Butterworths, London, 1971, p463.
17. G. Thews, Pflugers Arch. Gesamte Physiol. Menschen Tiere, 271, 227, (1960).
18. J.C. Angus, in P. Koidl and P. Oelhafen (Eds.), EMRS Symposia Proc., European Materials Research Society, Les Ulis, 17, 179, (1987).
19. L.A. Thompson, F.C. Law, J. Franks and N. Rushton, Biomaterials, 12, 37, (1991).
20. A.C. Evans, J. Franks, P.J. Revell, Med. Device Technol, May, 26, (1991).
21. J. Franks, J. Vac. Sci., A71, 2307, (1989).
22. T.Tuschuda, K. Yoda, Clin. Chem., 29, 837, (1983).
23. P. Vadgama, J. Spoors, L.X. Tang, C. Battersby, Biomed. Biochim. Acta, 48, 935, (1989).
24. A.E.G. Cass, G. Davis, G.C. Francis, H.A.O. Hill, W.J. Aston, I.J. Higgens, E.V. Plotkin, L.D.L. Scott, A.P.F. Turner, Anal. Chem., 56, 667, (1984).
25. M. Schichiri, R. Kawamori, N. Hakui, H. Abe, Lancet, ii, 1129, (1982).
26. K.S. Chua, I.K. Tan, Clin. Chem., 24, 150, (1978).
27. Z. Koochaki, I. Christie, P. Vadgama, J. Memb. Sci., 57, 83, (1991).
28. T. Matsuura, S. Sourirajan, J. Appl. Polym. Sci., 15, 2905, (1971).
29. M.A. Desai, S. Ghosh, P.W. Crump, Y. Benmakroha, P.M. Vadgama, Scand. J. Clin. Lab. Invest., 53 Suppl. 214, 53 (1993).
30. I.M. Christie, P.H. Treloar, P. Vadgama, Anal. Chim. Acta, 269, 65, (1992).
31. R.R.C. New, in Liposomes a practical approach, R.R.C. New (ed.), Oxford University Press, New York, 1990, p1.
32. L. Gorton, E. Csoregi, E. Dominguez, J. Emneus, G. Jonsson-Pettersson, G. Marko-Varga, B. Persson, Anal. Chim. Acta, 250, 203, (1991).

CLINICAL APPLICATION OF BIOSENSING WITH AMPEROMETRIC DETECTION OF AMMONIA-NITROGEN

MAREK TROJANOWICZ, WOJCIECH MATUSZEWSKI,
AND BEATA SZCZEPANCZYK
Department of Chemistry, University of Warsaw, Pasteura 1,
02-093 Warsaw, Poland

ANDRZEJ LEWENSTAM
Department of Analytical Chemistry, Abo Akademi University,
Turku, Finland

Abstract An interaction of polypyrrole with ammonia can be utilized for amperometric detection of ammonia using platinum working electrode, modified with electropolymerized pyrrole. This interaction leads however to permanent inactivation of detection at too high ammonia concentration and too large positive polarization of working electrode . In submillimolar concentration of ammonia interacting with polymer surface in short time interval in flow-injection measurements a satisfactory sensitivity and selectivity of amperometric response to ammonia is observed in the presence of a large excess of alkali and alkaline earth cations and numerous electroactive components of blood serum. A flow-injection system for amperometric detection of urea in blood serum using such a detection of ammonia has been developed.

INTRODUCTION

The most important analytical applications of polypyrrole are the use for the imobilization of enzymes in electrochemical biosensors design, as an electroactive material in ion-selective electrodes and as a gas-sensing material, which exhibits changes in its conductivity in different gas environments.

The first report about the possibility of applying of polypyrrole to ammonia detection has appeared in very early work of Kanazawa *et al.* [1], who reported that exposing polypyrrole to ammonia gas decreases the room temperature conductivity by factor of 10. The authors of the above mentioned work relate that ammonia interacts only weakly with polypyrrole and can be readily pumped off restoring the conductivity to the original value.

The sensitivity of the resistance of polypyrrole film to ammonia was demonstrated by Nylander *et al.* [2]. Gustafsson *et al.* [3] found that low concentra-

G.G. Guilbault and M. Mascini (eds.), Uses of Immobilized Biological Compounds, 141–150.

tions of ammonia produced a reversible increase in the resistance of polypyrrole, while concentrations of ammonia greater than 1 atm or the presence of water led to a permanent increase in resistance. That increase in resistance was attributed to nucleophilic attack by ammonia or hydroxide on the polymer leading to loss of conjugation and ring opening. It has been reported that electrochemically prepared polypyrrole film could detect 0.01% ammonia [4]. Also an ultrathin layer of polypyrrole coated on non-conducting substrates can be used as chemiresistor for sensitive detection of ammonia and hydrazine [5].

From the study involving measurements of polypyrrole resistance change and mass changes in the presence of selected gases and vapours, including ammonia, it was concluded that the response mechanism of polypyrrole sensitivity is due to a mixed response involving electronic effects and physical effects [6]. Mass changes of polypyrrole were utilized for the design of piezoelectric ammonia as sensors with polypyrrole deposited electrochemically on the surface of the gold electrode of a piezoelectric crystal [6,7].

The first application of polypyrrole modified electrode for amperometric detection of ammonia was described by Pandey and Mishra [8], who have used it in air-gap ammonia gas microelectrode for design of microsensor for urea. In this work neither an optimization of the conditions of amperometric ammonia detection nor the stability and selectivity of response were examined. The study of that was undertaken in this work in order to define a usefulness of such an amperometric detector of ammonia for application as an internal sensor in biosensors or flow-through systems for the determination of substrates converted enzymatically into ammonia.

PREPARATION OF POLYPYRROLE MODIFIED ELECTRODE

Platinum disk electrode before electropolymerization was polished with 1.0 and 0.3 μm alumina powder to obtain a smooth, shiny surface, then rinsed 2 min. in absolute methanol and dried. Electropolymerization was carried out 3 min. in three electrode system with auxiliary glassy carbon disk electrode and referen- ce Ag/AgCl electrode in 40 ml of non-stirrred 0.1 M pyrrole solution containing various background electrolytes at potential of working electrode +0.8 V vs. Ag/AgCl reference one. The electrolyte used was deaerated 20 min. with argon stream.

For prolonged use the modifed electrode was stored in 0.05 M borate buffer of pH 9.2. In such conditions *ca.* 1.5 μm layer of polypyrrole of good mechanical stability and shiny, black surface was formed.

INSTRUMENTATION

Basic electrochemical measurements for the study of modified electrode properties were made using voltammeter CV 37 from Bioanalytical Systems (West Lafayette, IN, USA) connected to x-y recorder 29000 from Bryans (Mildham, England) or to x-t recorder Laborgraph model E 586 from Metrohm (Herisau, Switzerland). In conventional, non-flow measurements electrodes and cell used were from Bioanalytical Systems.

Flow-injection measurements were performed in setup consisted of multichannel peristaltic pump MP13GJ4 from Ismatec (Zurich, Switzerland) or PLG 132100 from Desaga (Heidelberg, Germany), a rotary injection valve model 5020 from Rheodyne (Cotati, CA, USA) and a home-made large volume wall-jet cell. Flow part of the setup was assembled using 0.55 mm i.d. teflon tubing.

Configurations of the flow-injection systems used are shown in Fig.1.

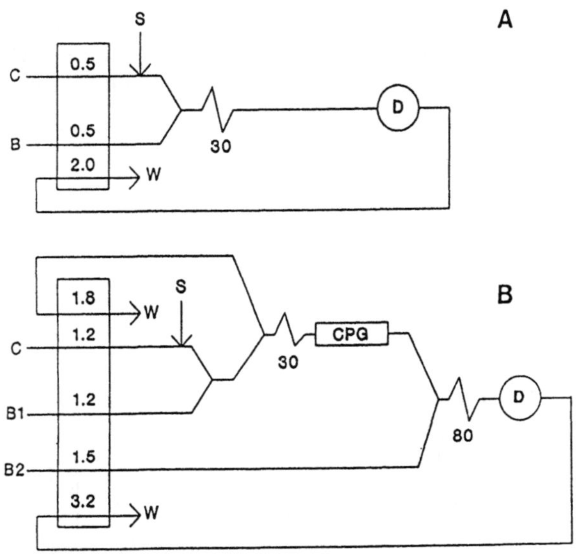

FIGURE 1 Schematic diagarms of flow-injection systems used for determination of ammonia (A) and urea (B). C - carrier stream of distilled water, B,B2 - 0.1 M borate buffer, pH 9.2, B1 - 0.01 M phosphate buffer , pH 7.1, containing 0.14 M NaCl, S - sample injection point, CPG - flow-through reactor with urease immobilized on controlled-pore glass, D - wall jet amperometric detector, W - waste.

Flow-injection measurements of ammonia were performed in the system A. System B with partial subtraction of sample zone before CPG enzyme column and detector was designed for measurements of urea using a significantly larger dispersion of sample segment in order to decrease the analyte concentration in the stream approaching the detector.

Enzyme reactor was prepared from urease of activity 88 units per mg from Serva (Heidelberg, Germany) by imobilization of 500 units on 500 mg of controlled porous glass of porosity 700 A from Cormay (Warsaw, Poland) according to procedure described earlier [9] and packing into a glass column of 4 mm i.d. and 80 mm long.

AMPEROMETRIC DETECTION OF AMMONIA

As it was reviewed above, the interaction of molecular, gaseous ammonia with polypyrrole, leading to the change of resistance of polymer and a change of its mass employed in piezoelectric detection is well documented. Those phenomena are associated also with a change of voltammetric behaviour of an inert electrode modified with a layer of electropolymerized polypyrrole. Addition of ammonia to the borate buffer solution, where approximately 50% of ammonia exists in nonprotonated form, distinctly changes cyclic voltammetric curves recorded in the anodic polarization range. Flow-injection response to ammonium-nitrogen shown in Fig.2 in two different buffers (taking into account

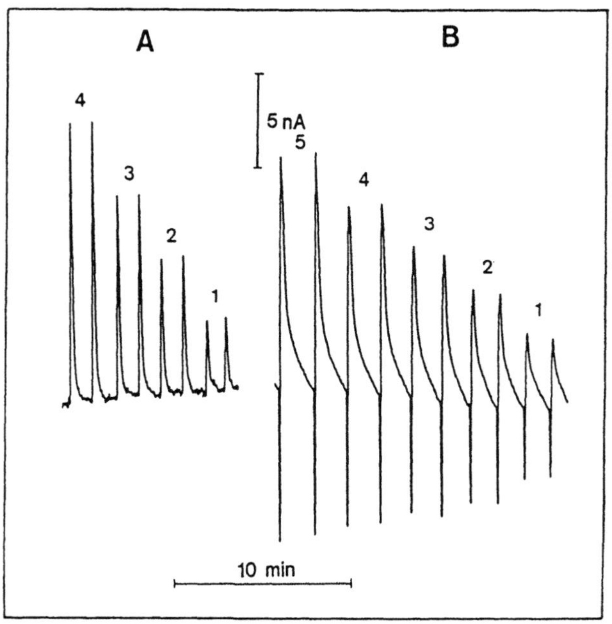

FIGURE 2 Flow-injection response of Pt disk electrode of Ø 1.5 mm covered with polypyrrole layer and polarized at +0.3 V obtained in the system shown in Fig.1A using 0.1M borate buffer, pH 9.2 (A) and 0.1 M Tris buffer, pH 7.8 (B) for 10 µl injections of (A) 0.025 (1), 0.05 (2), 0.075 (3), 0.1 (4) mM NH_4Cl solutions and (B) 2 (1), 4 (2), 6 (3) 8 (4) and 10 (5) mM NH_4Cl solutions.

different total concentrations of ammonia-nitrogen in recordings A and B) indicates amperometric sensitivity to molecular non-protonated ammonia and not to ammonium ions. The magnitude of response to ammonia does not depend significantly on the material of inner support , although among platinum, gold and glassy carbon the largest signals were obtained for platinum. Among several electrolytes examined for the electropolymerization of pyrrole the largest signal for ammonia was observed for the Pt electrode covered with polypyrrole layer from the sodium chloride solution containing 0.1 M pyrrole.

The most serious drawback of examined detection is a decrease of the signal magnitude in time as a result of an interaction of ammonia with polypyrrole. This can be observed either in consecutive cyclic voltammograms or in flow-injection response or in chronoamperometric measurements.

The change of current in time was observed for polypyrrole modified electrode, when at polarization at +0.6V vs. Ag/AgCl it was immersed in the borate buffer solution containing various concentration of ammonia in the milimolar range. Then the increase of ammonia concentration in solution, where the electrode has reached the steady current value, does not result in current increase, what indicates permanent deactivation of polypyrrole electrode after exposure to such a level of ammonia concentration. The efficiency of that deactivation process depends also on the polarizing voltage applied to the electrode For a smaller value of polarizing voltage applied, a smaller current response to ammonia is observed and a slower deactivation rate of electrode sensitivity was found.

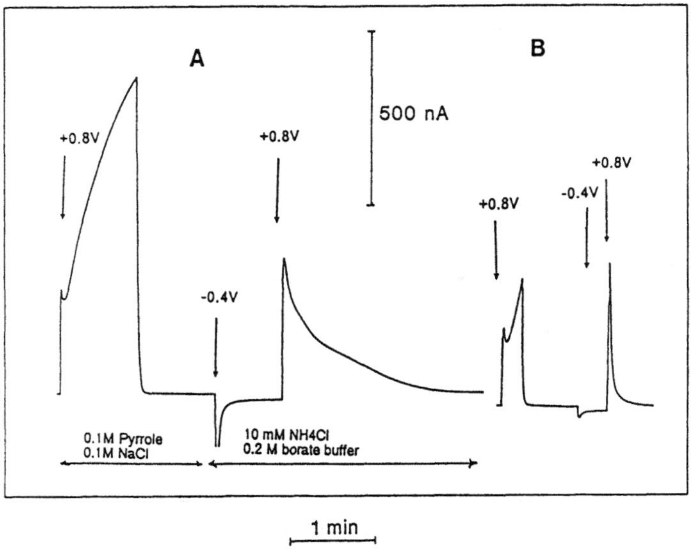

FIGURE 3 Chronoamperometric response of Pt disk electrode of Ø 0.125 mm polarized at +0.8 (points 1 and 3) and -0.4 (point 2) V vs. Ag/AgCl reference electrode obtained in non-stirred solutions of 0.1 M pyrrole containing 0.1 M NaCl (between points 1 and 2) and 10 mM ammonium chloride in 0.2 M borate buffer, pH 9.2 (between points 2 and 3) for 60 (A) and 15 (B) s electropolymerization time.

Several different chemical and electrochemical procedures were examined to adjust the electrode functioning to the initial state prior the exposure to ammonia. They included conditioning the deactivated electrode in borate buffer, in 0.5M sodium chloride without and with various negative polarization from -0.2 to -0.8 V, cyclic polarization in the range from +0.1 to +0.8 V in sodium chloride solution with and without pyrrole. None of them was effective. The polypyrrole layer exposed to a larger concentration of ammonia in aqueous environment can not be in those conditions regenerated to its initial sensitivity to ammonia of that of freshly formed polymer.

The same behaviour was observed in chronoamperometric measurements. In the first phase of experiment, the working Pt electrode was immersed in sodium chloride solution containing pyrrole and polarized 15 or 60 s at +0.8 V vs. Ag/AgCl to form a fresh polypyrrole layer. Then the electrode was transferred to the borate buffer solution and polarized at -0.4 V (point 2 in Fig.3). Finally ammonia solution was added to the borate buffer and polarization of the working electrode was changed again to +0.8 V (point 3). According to Bull *et al.* [10] the thickness of polypyrrole layer formed can be estimated. Those data allows also to estimate the charge transferred during electropolymerization and deactivation basing on the area of the surfaces under *i vs. t* curves recorded. Except the thickest polymer layer, for all other cases the ratio of charge passed during deactivation and electropolymerization was 0.5. It means that while the electropolymerization is two electron process, inactivation of the polymer is one electron process, which can result in addition of amine group to every radical-ion unit of polymer with or without pyrrole ring opening and the loss of condu-

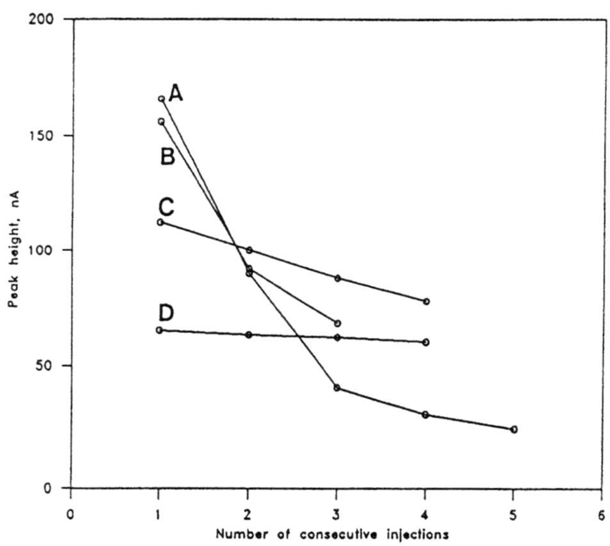

FIGURE 4 Effect of the polarizing voltage on the stability of flow-injection response obtained for injections of 10 mM ammonium chloride in the setup shown in Fig.1A using Pt disk electrode of Ø 1.5 mm for +0.8 (A), +0.6 (B), +0.4 (C) and +0.2 (D) V vs. Ag/AgCl. For each measurement freshly prepared polypyrrole electrode was used.

ctivity of the polymer. The electrode process taking place is irreversible. Further studies to confirm the mechanism of anodic deection of ammonia are in progress.

OPTIMIZATION OF FLOW-INJECTION AMMONIA DETECTION

Different rate of deactivation of the polypyrrole modified electrode at various polarizing voltage and ammonia concentration has to be taken into account in optimization of its use for analytical purposes. Effect of polarizing voltage and ammonia concentration on flow-injection response is shown in Figs.4 and 5. At +0.4 V for 10 μl of injected 25 μM ammonia solution a stable response of the polypyrrole electrode was observed (Fig.5).

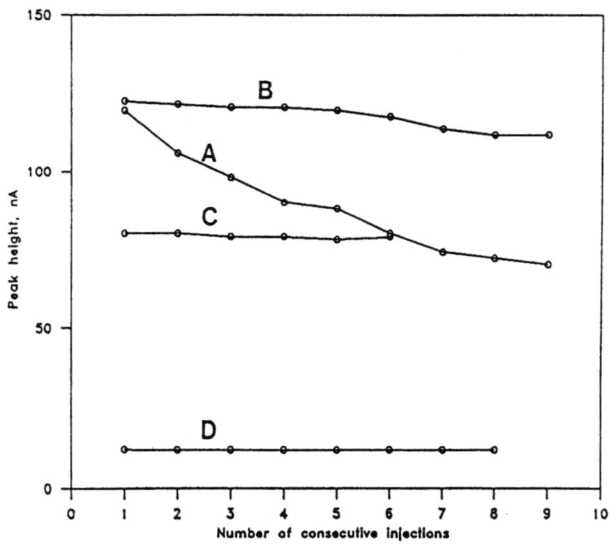

FIGURE 5 Effect of the concentration of ammonium chloride in injected solution on the stablility of flow-injection response obtained at +0.4 V polarizing voltage for 5.0 (A), 0.5 (B, 0.2 (C) and 0.025 (D) mM in flow-injection setup shown in Fig.1A.

As the extent of deactivation of polypyrrole depends on ammount of ammonia contacting the surface of polypyrrole layer, the test of continuous exposure of the electrode to the 10 and 100 mM ammonia solution was performed, showing that 50% decrease of detection sensitivity occurs after pumping 240 and 85 ml of above solutions, respectively. It corresponds to 240,000 or 8,500 injections of 10 μl ammonia solutions of the same concentration. It means that for series of 50 injections of 10 mM ammonia solution the observed loss of sensitivity should be practically neglegible. This was confirmed experimentally.

Between measurements the electrode was stored in the borate buffer, as for the electrode stored in air a worse stablility of base line and worse reproducibility of response was found.

The selectivity of the developed detection was measured in the flow-injection setup by injection of 5.0 mM ammonium chloride solution without and with addition of several possible interferents present in natural blood serum samples. As it is shown in Table 1, the most serious interference in the range of concentration encountered in natural serum is caused by the presence of 100 mg/l of ascorbic acid. No response to alkali metals and calcium ions was observed.

Table 1. Effect of interfering species on flow-injection determination of 5 mM ammonia.

Interferent	Concentration of interferent, mg/l	Error, %
Isoniazid	500	+43
	50	0
Haemoglobin	4000	-12
Salicylamide	500	0
	100	0
Creatinine	200	0
Ascorbic acid	400	+28
	200	+18
	100	+8.0
Paracetamol	500	+28
	50	0
Lactic Acid	1000	0
Glucose	2000	+2.0
Na-Salicylate	2000	0
K-Oxalate	2000	+2.0
0.2 M NaCl + 10 mM KCl + 10 mM $CaCl_2$ mixture		0

FLOW-INJECTION DETERMINATION OF UREA IN BLOOD SERUM

Flow-injection determination of urea with polypyrrole modified Pt electrode was performed in the system shown in Fig.1B with wall-jet detector and enzyme reactor with urease immobilized according to reported earlier procedure

[9]. The efficiency of enzymatic conversion of urea in CPG column was estimated as 98%. The working electrode was polarized at +0.3 V. The calibration plot obtained from data shown in Fig. 6A is linear up to 4 mM of urea. For 6 injections of the control serum sample the relative standard deviation was equal 1.8%. The correlation plot shown in Fig.6B demonstrates a satisfactory agreement of results obtained in comparison to determinations carried out using a commercial clinical analyzer PROGRESS from KONE (Finland) with correlation coefficient equal 0.9946 and equation describing the correlation plot as : $y = 0.8879x + 1.175$.

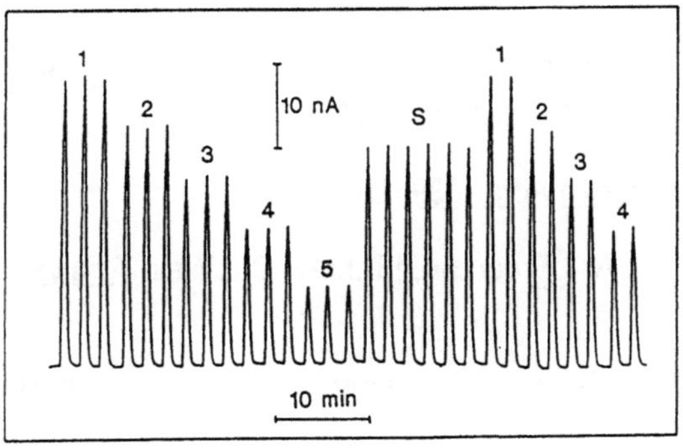

A

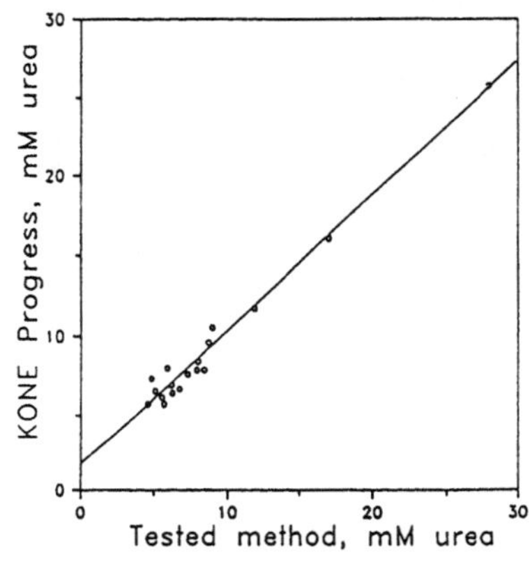

B

FIGURE 6 Flow-injection response to urea (A) and correlation plot for determination of urea in blood serum samples (B) obtained for Pt electrode of Ø 2.0 mm covered with polypyrrole layer and polarized at +0.4 V in the system shown in Fig.1B for injection volume 10 µl. Urea standard solutions 10 (1), 8 (2), 6 (2), 4 (4) and 2 (5) mM. S - control serum sample.

REFERENCES

1. K.K.Kanazawa, A.F.Diaz, R.H.Geiss, W.D.Gill, J.F.Kwak, J.A.Logan, J.F.Rabolt, G.B.Street, J.Chem.Soc., Chem.Commun., 854, (1979).

2. C.Nylander, M.Armargth, I.Lundstrom, in Proceedings of the 1st International Meeting on Chemical Sensors, Fukuoka, Japan, September 1983, Elsevier, Amsterdam 1986, p.203.

3. G.Gustafsson, I.Lundstrom, B.Liedberg, C.R.Wu, O.Inganas, O.Wennerstrom, Synth.Met., 31, 1989, (1989).

4. J.J.Miasik, A.Hooper, P.T.Moseley, B.C.Tofield, in Conducting Polymers, Ed.L.Alcacer, Reidel, Dordrecht 1987, p.189.

5. N.M.Ratcliffe, Anal.Chim.Acta, 239, 257, (1990).

6. J.M.Slater, E.J.Watt, N.J.Freeman, I.P.May, D.J.Weir, Analyst, 117, 1265, (1992).

7. S.J.Vigmond, K.M.R.Kallury, V.Ghaemmaghami, M.Thompson, Talanta, 39, 449 (1992).

8. P.C.Pandey, A.P.Mishra, Analyst, 113, 329, (1988).

9. M.Masoom, A.Townshend, Anal.Chim.Acta, 166, 111, (1984).

10. R.A.Bull, F.F.Fan, A.J.Bard, J.Electrochem.Soc., 129, 1009 (1982).

THERAPEUTIC POTENTIAL OF IMMOBILIZED ENZYMES

A. Kevser Pişkin
Hacettepe University
Faculty of Medicine
Department of Medical Biology
Sıhhiye 06100 Ankara
TURKEY

ABSTRACT

Enzymes are higly specific catalysts that have a large number of applications in diagnosis. However, their potential therapeutic usefulness is limited by liability, toxicity, poor cellular uptake and allergenicity. In view of these drawbacks, immobilized forms of enzymes are prepared using biological or synthetic materials to improve their efficiency. This paper reviews the immobilization studies carried out on enzymes with therapeutic potential.

INTRODUCTION

There are practical difficulties in using enzymes for therapy:

1. Enzymes rapidly loose their activity when administered into the body. This is largely due to the fact that most enzymes are unstable at 37°C. Attempts were made to overcome the problems of short enzyme lives by providing a source of enzyme by skin or by fibroblast transplantion which did not create a definite clinical improvement.

2. It is probable that the immune system of patients receiving such enzymes would produce antibodies to the "foreign" protein. This results in the rapid removal of the

G.G. Guilbault and M. Mascini (eds.), Uses of Immobilized Biological Compounds, 151–160.

enzyme by the reticuloendothelial system or more severe immunological reactions.

3. Enzymes introduced to the blood stream may not reach to the site where its action is needed. This situation may seem less difficult in diseases in which tissues rather than the central nervous system are involved. Modification or conjugation of the enzyme with a targetted molecule may facilitate enzyme therapy [1].

A recent approach in enzyme replacement is bone marrow transplantation. Bone marrow transplantation has been undertaken with encouraging results as a therapy for a variety of neurovisceral lysosomal storage disorders, such as the mucopolysaccharidoses and gaucher disease. Still to overcome, however, are problems of graft-versus-host disease and other complications of transplantation [2].

Enzymes immobilized to introduce them selective features remains to be a major prospect in enzyme therapy.

IMMOBILIZED ENZYMES

Enzymes can be immobilized by a large variety of procedures. However there are basicly four major techniques:

1. Covalent binding or adsorption
2. Crosslinking
3. Matrix entrapment
4. Microencapsulation

Immobilized forms of enzymes usually offer the following advantages over their free countparts:

1. Temperature and storage stability
2. Repeated use
3. Lack of immunogenicity
4. Possible use in column systems for extracorporeal perfusion
5. Possibility of being targetted to desired tissues.

The ultimate goal of immobilizing enzymes is the treatment of a number of clinical conditions ranging from suppurative wounds to cancer. An overall profile of work carried out on immobilized enzymes with therapeutic potential is as follows.

INBORN ERRORS OF METABOLISM

The underlying molecular defect in genetic metabolic diseases is the lack of a specific enzyme or the presence of its inactive form. Major forms include lipid storage diseases, mucopolysaccharidoses and erythrocyte enzyme deficiencies.

β-glucronidase: Mucopolysaccharidosis Type VI (Sly disease) is an incurable multisymptomatic disease that results from the deficiency of β-glucronidase [3]. We have immobilized this enzyme in alginate gels to improve its stability and obtain a delivery system. 71.1 % of β-glucronidase activity was recovered after immobilization. The substrate affinity was slightly lower (K_M values for free and immobilized enzyme was 1.43×10^{-4} M and 1.51×10^{-4} M respectively). This effect of immobilization may be due to diffusion limitations or a possible carrier-enzyme interaction. The stability of β-glucronidase was improved especially at 37°C as shown in Figure 1.

L-phenylalanine ammonia lyase: Phenylketonuna is known since 1947. However, low phenylalanine diet is the only treatment available. It is characterized by an increase in the level of phenylalanine which can be hydrolyzed by L-phenylalanine ammonia lyase. Bourget and Chang was able to lower the phenylalanine level in phenylketonuric rats by systemic administration of ultrathin cellulose nitrate microcapsules containing L-phenylalanine ammonia lyase [4]. In another study carried out by Ambros et.al., the enzyme was immobilized in a hollow fiber dialyzer and used in the treatment of phenylketonuric patients. The fiber enzyme reactor has shown remarkable efficiency in reducing the amount of circulating phenylalanine. In 5.5 hours of treatment total phenylalanine depletion was 1800 mg [5].

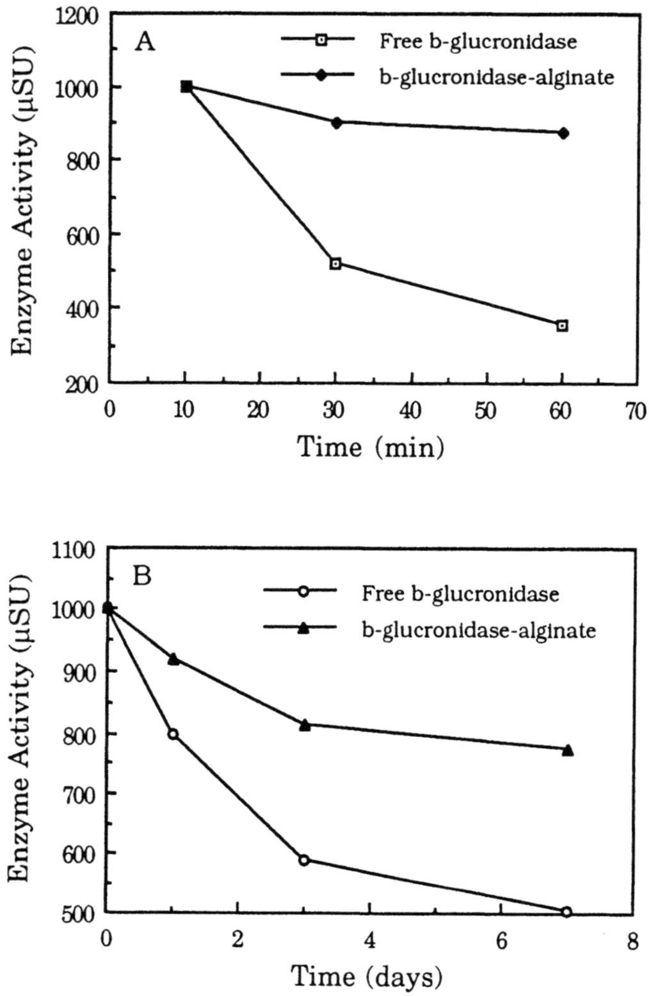

Figure 1. Stability of β-glucronidase (Enzyme Activity is Expressed in Sigma Units) .

<u>Gulonolactone Oxidase:</u> This is a microsomal enzyme which catalyzes the conversion of gulonolactone to ascorbic acid. Scurvy guinea pigs provide a convenient model to study the inborn errors of metabolism. Like humans, this species is unable to synthesize its own ascorbic acid. Hadley and Sato have tested the catalytic activity of polyethyleneglycol (PEG) conjugated gulonolactone oxidase in these animals. Intravenous injection of

the enzyme-PEG conjugate elicited ascorbic acid synthesis in a dose dependent manner. The modified of enzyme was stabilized to incubation at 37°C but was not protected from trypsin. Furthermore, its immunogenicity was eliminated [6].

Catalase: In acatalasemia, the low catalase activity leads to lack of oxygen. This creates great periodontological problems. Application of free catalase to oral lesions may create immune reactions. Microencapsulated catalase was successfully applied by Chang to the treatment of these lesions by topical administration [7].

Xanthine Oxidase: Lesch-Nyhan disease is an inborn error of metabolism due to hypoxanthine phosphoribosyltransferase deficiency. Increase in the level of hypoxanthine may lead to superoxide radical formation and thus toxicity. In a clinical trial by Palmaur et.al., microencapsulated enzyme was applied orally. It was concluded that within two weeks plasma and cerebrospinal fluid hypoxanthine concentrations were efficiently decreased without any toxic effect [8].

ORGAN FAILURE

In renal and liver failure certain toxic metabolites accumulate in the plasma. Removal of these molecules via enzymatic reactions is one of the most applicable approaches of therapy by immobilized enzymes.

Urease: In renal failure the accumulation of urea leads to toxic reactions. Urease is the first enzyme to show the feasibility of using immobilized enzymes in therapy. This was first demonstrated by Chang. In a study carried out using microencapsulated urease, blood urea could be hydrolyzed. Furthermore, gastrointestinal delivery of the immobilized form of urease was also able to remove blood urea when used in combination with an ammonia adsorbent [9,10]. Campbell and

Chang have formed a microencapsulated multienzyme system which efficiently converts urea into amino acids (Figure 2) [9].

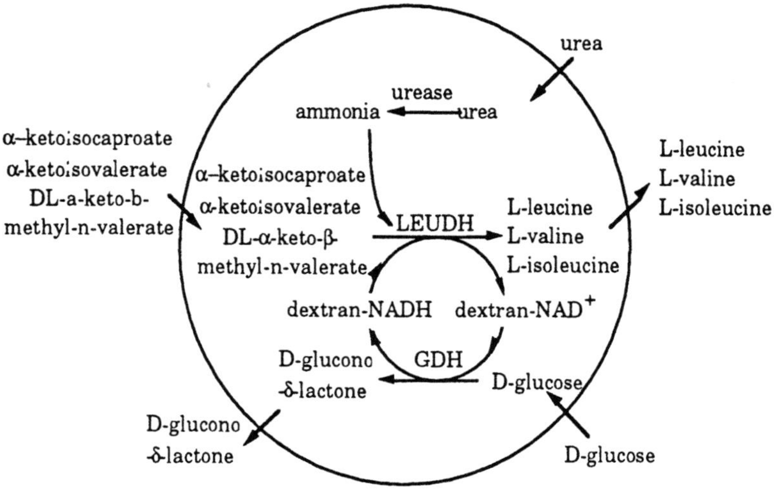

Figure 2. The Continuous Conversion of Urea into Amino Acids by a Multienzyme System (From Chang, 1990) [9].

To introduce urea removal into hemoperfusion columns, we have immobilized urease on activated carbon and encapsulated by cellulose nitrate membrane. Immobilization increased enzyme stability and was able to hydrolyze urea in hemoperfused rats as shown in Table 1 [11].

Table 1. Hydrolysis of Blood Urea.

Time (min)	% Conversion
5	94.8 ± 13.2
10	87.3 ± 17.1
30	87.8 ± 13.0
60	81.9 ± 18.6
90	74.1 ± 15.2

Urease have been immobilized by a large number of techniques to increase the recovered activity. For example, urease immobilized in hydrophilic copolymers by radiation retained 80 % of its activity [12,13].

Tyrosinase: In fulminant hepatic failure rats accumulated tyrosine and phenols were efficiently hydrolysed by microencapsulated tyrosinase as shown by Chang [9].

CANCER THERAPY

Immobilized asparaginase and arginase are used to deplete asparagine and arginine in substrate dependent tumors.

Asparaginase: Asparaginase-polyethyleneglycol conjugates was used to treat lymphoblastic canine leukemia with equal efficiency as canine native asparaginase. However, immobilized enzyme was less immunogenic than the free enzyme [14].

In another study asparaginase was attached to carboxymethyl cellulose. This form of enzyme had greater antileukemic activity and showed prolonged effect [15].

Arginase: Arginase causes growth inhibition of normal as well as tumor cell lines. In order to enhance the use of arginase in cancer therapy the enzyme is immobilized by various methods, including covalent trapment in a fibrin clot and attachment to polyethyleneglycol (PEG). Arginase bound PEG was injected to tumor-infected mice. The results were very positive in the case of the ascitic tumor but not so evident in murine Leukemia L 51784 [16,17].

THROMBOSIS AND DEVELOPMENT OF NONTHROMBOGENIC BIOMATERIALS

Enzymes with plasminogen activating potential such as urokinase and streptokinase are immobilized in order to increase their stability, eliminate their immunogenicity and also target them to the site of thrombus. For example, binding of urokinase to heparin increases their affinity towards thrombus material [18].

Some enzymes are also used to treat polymeric devices that are used in contact with blood. This improves the biocompatibility of the extracorporeal perfusion systems preventing the emboli formation 19].

Heparinase: This is another enzyme that is related to blood clotting. Extracorporeal devices require systemic heparin administration to prevent thrombus. However, heparin itself or the agents used in blood deheparinization may be harmful for the patient. Heparinase was immobilized on cellulose hollow fiber devices in order to remove heparin from the blood stream [20].

NEONATAL JAUNDICE

Over 6 % of all newborn infants are treated for jaundice, in premature infants and in β-glucronidase deficiency, the incidance is much higher. An immobilized enzyme reactor with bilirubin oxydase was designed and tested in Gunn rats which was able to produce 50 % reduction in serum bilirubin level [21].

SUPPURATIVE WOUNDS

Proteolytic enzymes in the treatment of purulent diseases of soft tissues of various locations have been employed to remove the necrotic debris. In order to increase the efficiency of these immobilization techniques are applied. The results of clinical applications of immobilzed trypsin, lysosyme, immozymase, however, are only available in Russian journals with English abstracts.

CONCLUSION

Although numerous enzymes have been defined and isolated there are more molecules with enzymatic activity to be determined. Especially in genetic process many unknown enzymes are thought to function as proposed by defined procaryotic counterparts. Discovery of novel enzymes may shed light upon a number of pathological conditions. Furthermore, the diverse nature of immobilization techniques and availability of a large selection of carriers offers the feasibility of designing efficient enzyme delivery strategies.

REFERENCES

1. Benson, F.F., Genetic Biochemical Disorders, p. 49, Oxford University Press, Oxford, (1985).
2. Bayever, E., Kamani, N., Ferrera, P., Machin, G.A., J. Inher. Metab. Dis., 15: 899, (1992).
3. Bou-Charios, G., Adams, G., Pace, P., Olsen, I., J. Inher. Metab. Dis., 15: 899, (1992).
4. Bourget, L., Chang, T.M.S., Biochim. Biophys. Acta, 883: 432, (1986).
5. Ambrus, C.M., Anthone, S., Horvath, C., Kalghetgi, K., et.al., Annals of Internal Medicine, 106: 531, (1987).
6. Hadley, K.B., Sato, P.H., Enzyme, 42: 225, (1989).
7. Chang, T.M.S., Biomat. Artif. Cells. Artif. Org., 15: 1, (1987).
8. Palmour, R., Goodyer, P., Reade, T., Chang, T.M.S., Lancet, 2: 687, (1989).
9. Chang, T.M.S., Ann. N.Y. Acad. Sci., 613: 109, (1990).
10. Gardner, D.L., Falb, R.D., Kim, B.C., Emmerkig, D.C., Trans. Amer. Soc. Artif. Organs, 17: 239, (1971).
11. Pişkin, K., Chang, T.M.S., Int. J. Artif. Organs., 3: 344, (1980).
12. Pişkin, K., Arca, E., Pişkin, E., Applied Biochem. Biotechnol., 10: 73, (1984).
13. Pişkin, A.K., Analytical Uses of Immobilized Biological Compounds for Detection, Medical and Industrial Uses, p. 17., G. Guilbault, Marco Mascini, eds., D. Reidel Publ. Co., Dordrecht, (1988).

14. MacEwen, E.G., Rosenthal, R.C., Fox, L.E., Loar, A.S., et.al., J. Vet. Intern. Med., 6(4) 230, (1992).
15. Wada, H., Imanura, I., Sako, M., Katagiri, S., Ann. N.Y. Acad. Sci., 618: 95, (1990).
16. Diez, A., Campo, M.L., Soler, G., Biotech. and Appl. Biochem., 12: 237, (1990).
17. Savoka, K.V., Davis, F.F., Vanes, T., McCoy, J.R., Palczuk, N.C., Cancer Biochem. Biophys., 7: 261, (1984).
18. Torchilin, V.P., Maksimenko, A.V., Mazaev, A.V., Methods in Enzymol., 137: 552, (1988).
19. Senatore, F.F., Bernath, F.R., Biotechnol. Bioeng., 28: 58, (1986).
20. Comfort, A.R., Albert, E.C., Langer, R., Biotechnol. Bioeng., 34: 1366, (1989).
21. Sung, C., Lawin, A., Keibanow, A., Langer, R., Trans. Am. Soc. Intern. Organs., 31: 264, (1985).

OPTIMIZATION OF MEMBRANES WITH GLUCOSE OXIDASE FOR GLUCOSE DETERMINATION USING OXYGEN ELECTRODE

MARIAN FILIPIAK[1], IGNACY ZAWICKI[2]
1. Department of Biochemistry and Microbiology, Academy of Economics, Poznan, Poland
2. Institute of Biocybernetics and Biomedical Engineering, Polish Academy of Sciences, Warsaw, Poland

Abstract. Several types of membranes were obtained by immobilization of glucose oxidase on different supports and used for glucose determination with an oxygen electrode. Some properties of these membranes were compared. The ways of extending a linear relationship between a glucose concentration and an electrode signal are discussed.

I INTRODUCTION

The measurement of sugars in foodstuffs and also biological fluids present special problems to the analyst. Specific biosensors enable to overcome most of these difficulties. There are numerous publications on using biosensors for sugar analyses [1,2]. Considerable attention has been given to the development of a glucose sensor. The essential parts of biosensors are enzyme membranes. The usefulness of biosensor and its quality is dependent mainly on enzyme membrane.

The aim of this work was to compare the properties of different membranes with glucose oxidase immobilized on solid supports and possibly to improve these properties mainly with respect to their application for glucose determination in different samples.

Special attention was given to the measuring range of constructed biosensors as one of the most important parameteres

G.G. Guilbault and M. Mascini (eds.), Uses of Immobilized Biological Compounds, 161–170.

characterizing these devices.

II EXPERIMENTAL

Several membranes were obtained by immobilization of glucose oxidase on different supports (table 1). The methods of immobilization were described elsewhere [3,4]. Obtained membranes

Table 1. Membranes used for glucose determination by means of an oxygen electrode

No	Support	Immobilization method
I	albumin	glutaraldehyde
II	gelatin	glutaraldehyde
III	glass fiber	silane/glutarald.
IV	polyamide	glutaraldehyde
V	enzyme entrapped by	dialysis membrane

were placed on the top of an oxygen electrode. Glucose determinations were performed in different ways, using specially constructed chambers for the enzyme electrode. One of them is schematically presented on figure 1. This is a device for single measurements. In some cases also a system of flow injection analysis (F.I.A) was used with a flow through chamber. Scheme of F.I. A. system is shown on figure 2.

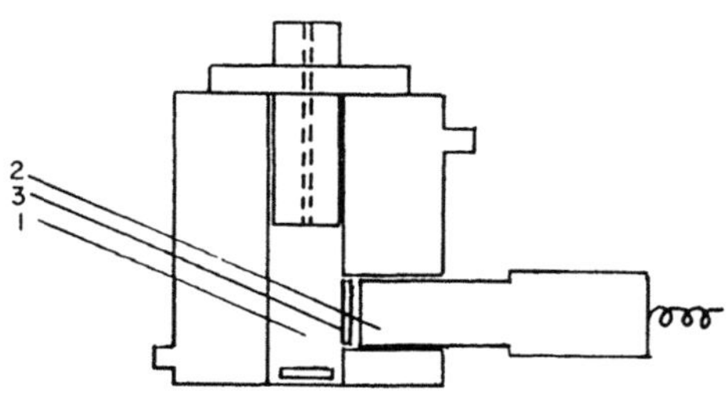

1 - sample
2 - oxygen electrode
3 - enzyme membrane

Fig. 1. Scheme of a chamber for single measurements

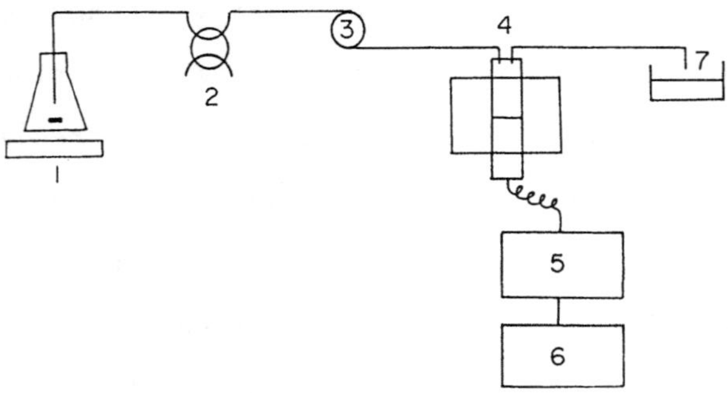

1 - water reservoir
2 - 4-way valve
3 - peristaltic pump
4 - flow-through chamber with oxygen electrode
5 - oxygen meter
6 - recorder
7 - waste

Fig. 2. Schematic diagram of a flow injection system for glucose determination

III RESULTS and DISCUSSION

Results of glucose determination using obtained membranes with an oxygen electrode in a chamber for single measurements are presented in table 2. For comparison glucose was also determined with an enzyme in solution. The experiments were performed using a commercial Radiometer oxygen electrode.

Table 2. Results of glucose determination using an oxygen electrode with different membranes and soluble enzyme

Membrane	Linear range up to... (mg/dl)	Correlation coefficient	Relative standard deviation	Response time (sec)
I	70	.9962	5.75	15
II	80	.9980	3.30	15
III	100	.9986	7.00	20
IV	100	.9956	11.50	60
V	150	.9941	10.00	10
enz. sol.	80	.9972	2.10	10

Linear relationship between a glucose concentration and an electrode signal was rather narrow and only membranes made of polyamide and glass fiber provided slightly extended range.

However for those membranes the value of relative standard deviation increased to 11.50 % (glass fiber). Use of a flow injection system made possible to extend the linear range significantly. At the flow rate of 1 cm^3/min the linear relationship reached glucose concentration of 400 mg/dl and increasing the flow rate to 1.21 cm^3/min enabled extension even up to 1000 mg/dl. Generally, a linear range depends on the type of enzymatic membrane to a certain degree only. It is more connected with the way of measurement. Detailed discussion on factors affecting linear range is included in the second part of this presentation.

Further experiments were performed with selected gelatin and albumin membranes. In figure 3 the relationship between the signal of the electrode equipped with an albumin membrane and amounts of glucose oxidase used for immobilization is presented. Application of more than 2.28 Units per 1 cm^2 of membrane did not generate any further increase of the signal. Therefore this amount of glucose oxidase was used in all experiments.

Temperature and pH profiles of glucose determination using an enzyme electrode equipped with albumin and gelatin membranes were estimated in the region of 20 - 70° C and pH 4 - 9 respectively. Results are presented on figure 4. Measurements of glucose using an enzyme electrode were very sensitive to temperature but almost independent on pH in the indicated region. An electrode signal increased on an average by 5.6% per 1° C. Soluble glucose oxidase started to inactivate already at the temperature of 60° C when enzyme membranes even at 70° C worked properly.

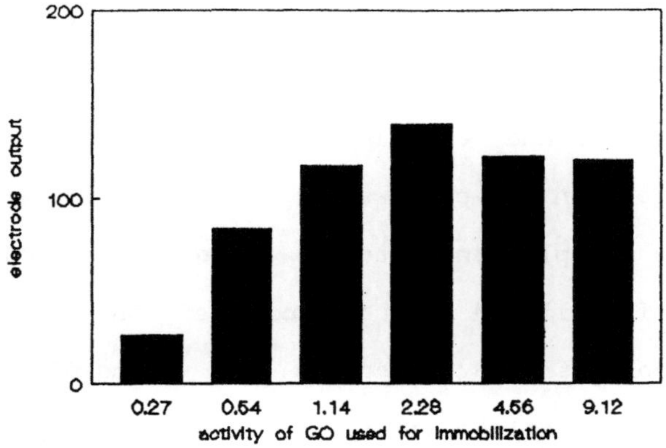

Fig. 3. Relationship between a signal of an electrode equipped with an albumin membrane and amounts of glucose oxidase used for immobilization

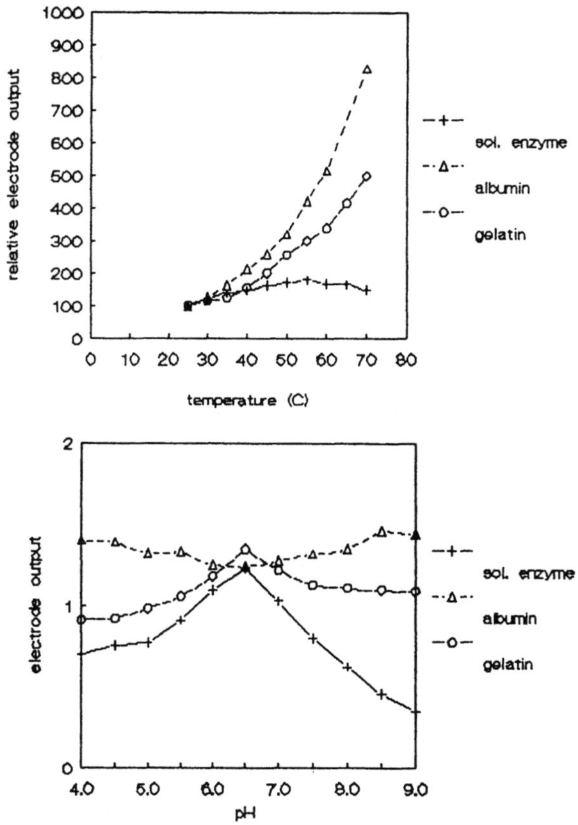

Fig. 4. Temperature and pH profiles of glucose measurements using an oxygen electrode with albumin and gelatin membranes

It is well known that one of the reasons for the relatively narrow measurement range of the amperometric glucose sensor based on oxygen detection and containing glucose oxidase membrane is low level of oxygen as compared to the glucose concentration present in the sample under test. Using for example gelatin membrane with immobilized glucose oxidase, one obtains linear range of the sensor from 0 to approx. 30 - 80 mg/dl (1.5 to 4.5 mM). For many practical applications it is not enough. Assuming that flow through measurement (including FIA) is not taken into account, one can consider the following approaches to extend the linear (measurement) range:

a) geometry of the sensor and type of materials can be selected in such a way as to limit the surface of glucose penetration to the enzyme electrode and to enlarge the surface available for oxygen diffusion [5]. This solution has not spread out however, perhaps for the reason of large response time of such sensor and its high nonlinearity.

b) electrolytic oxygen generation within the sensor [6]. Such solution has not widespreaded either.

c) the most common way of extending linear range is application of various polymer membranes which restrict the access of glucose to enzymatic membrane and at the same time do not influence much the permeability of oxygen. To this aim various polymers have been applied, including polyuretanes, polycarbons, cellulose acetates.

Application of the mechanicaly perforated gas-permeable membranes was also tried but the results were not reproducible [7].

Good results are reported with radiation track-etched polycarbon membranes modified with organosilanes [8].

The oxygen sensor designed for the experiments on extending linear range has been laboratory made. It contained Pt working electrode (diam. 0.5mm), Ag¦AgCl reference electrode and 0.1mM KCl inner electrolite. All this was covered by a set of membranes among which were oxygen permeable membrane (polyethylene) and the enzymatic one. Of the models among the sensors one was designed for measurements on small sample put on the top of the sensor as a drop. The results obtained with such sensor are reported in this presentation.

First of the sensors designed for measurements of glucose in a drop of blood required dilution of the sample in PBS in order to fall into the sensors range (aprox 0 -30 mg/dl). The results were good [9] as illustrated in fig 5., however the process of dilution was a little bit troublesome. Therefore some ways of extending the range were investigated and compared.

It has been confirmed that it is the deficiency of oxygen which esentially influences the range of the glucose sensor. To prove it the sensor and solutions of different concentrations of glucose were placed under a tent made of transparent plastic. Approximate size of the tent (L=60cm, W=40cm, H=30cm). The tent was supplied with pure oxygen for some time preceding the experiment in order to reach the saturation of the sensor and the samples with oxygen. Results are illustrated on fig. 6, curve 5.

It was expected that enzyme membranes containing glucose oxidase and catalase should assure wider range than membranes containing only glucose oxidase. This expectation has been

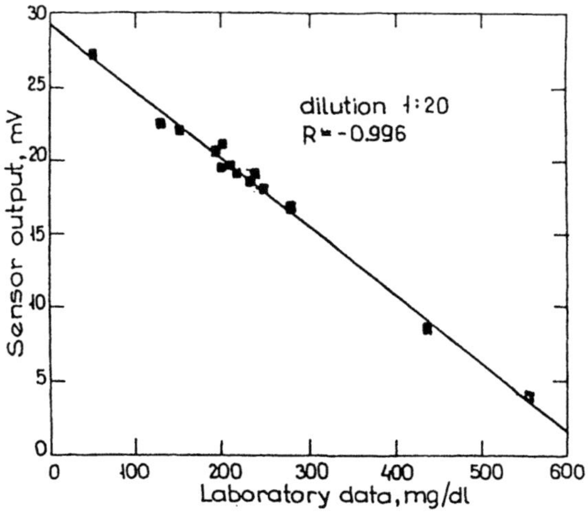

Fig. 5. Correlation between the sensor's output and the laboratory data for diluted blood

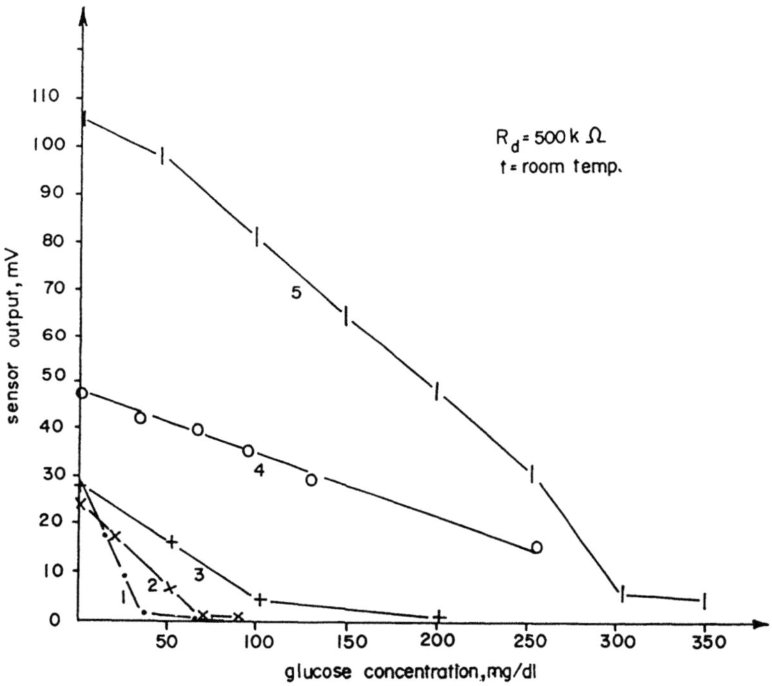

Fig. 6. Quasi steady-state (80 sec) output of the glucose sensor vs glucose concentration in PBS
1-glucose oxidase membrane, 2-glucose oxidase with catalase, 3-additional gelatin overcoat, 4- additional celulose acetate membrane, 5-pure oxygen

verified experimentally and illustrated in fig. 6, curve 2. The linear range has been extended here from 30 mg/dl to 60 mg/dl.
Further extention of the sensor range was obtained after applying additional coating of gelatin to the membrane containing immobilized enzymes. The overcoat was of approximately the same thickness as the enzymatic membrane (curve 3 on fig. 6).
The autors also tried to apply additional membranes over the enzymatic one in order to extend the range of the sensor.
The laboratory made membranes of celulose acetate of various proportions of components assured good results as far as the range of sensor was concerned (curve 4 on fig. 6.). However the long term stability was not satisfactory. This work is in progress. The track-etched polycarbonate, polyester and polypropylene membranes of various pore sizes and of porosity about 6 to 12% without any modification were not appropriate for this purpose. Permeability for oxygen had no preference over the permeability to glucose.
Surprisingly, positive preliminary results were obtained after application of an ordinary polypropylene membrane. For some samples of the material the range of the sensor could be extended as far as to 1000 mg/dl. However this results were not reproducible.

IVPROSPECTS

The membranes with immobilized glucose oxidase for use in glucose sensor based on oxygen electrode reveal different properties depending on the nature of a membrane support. One of the most important parameters of the glucose sensors is a range of linear relationship between a glucose concentration and an electrode output. It has been confirmed that the main reason of the narrow linear range is deficiency of oxygen in the sample. It is possible to extend this range by using an enzyme membrane with immobilized glucose oxidase and catalase, additional overcoating for example made of gelatin and some other membranes limiting access of glucose to the electrode.

ACKNOWLEDGMENT

This work was supported in part by the Polish Government's Committee for Scientific Research (KBN), Grant No.886529102.

REFERENCES

1. P. Wirth, Enzyme in der Lebensmitteltechnologie, 2. Symposium, 1988, GBF Monographs, 11, 271, 1991
2. E. Watanabe, M. Takagaki, S. Takei, M. Hoshi, C. Shu-gui, Biotechnol. Bioeng. 38, 99, 1991
3. M. Filipiak, J. Krauze, D. Kirstein, P. Mohr, P152808, 1981
4. J.J. Kulys, V.S. Laurinavicius, M.V. Peslakiene, Anal. Chim. Acta, 148, 13, 1983
5. D. A. Gough et al., Anal. Chem. 57, 2351, 1985
6. S. O. Enfors in: Biosensors: Fundamentals and Applications, (A.P.F. Turner, ed.), Oxford University Press, Oxford 1987, p.347
7. M. Mascini, D. Moscone, in: Advances in Biosensors, vol.1, (A. P. F. Turner ed.), JAI Press Ltd. London, England
8. W. H. Mullen et al., Anal. Chim. Acta 183, 59, 1986
9. I. Zawicki, M. Filipiak, Z. Adamaszek, J. Krzymien, R. Miarczynski, Z. Szczepaniak, M. Nalecz, in: Lecture notes of the ICB Seminar on Artificial Pancreas and Drug Delivery Systems, Madralin, April 1991, (V. Shumakov, M. Nalecz, eds.) ICB Warsaw, Poland 1992

MULTIENZYME BIOSENSORS - COUPLED ENZYME REACTIONS AND ENZYME ACTIVATION

FRIEDER SCHELLER, ULLA WOLLENBERGER[x], FLORIAN SCHUBERT[xx], DOROTHEA PFEIFFER, ALEXANDER MAKOWER AND C.McNEIL[xxx]
Max-Delbrück-Center of Molecular Medicine, O-1115 Berlin
[x]Fraunhofer Institut für Siliziumtechnologie, W-1000 Berlin 33
[xx]Physikalisch-Technische Bundesanstalt, W-1000 Berlin 10
[xxx]University of Newcastle, Newcastle upon Tyne, NE2 4HH, UK

Abstract Traditionally enzymes are used as recognition elements to measure substrate molecules by catalyzing the turnover of these species to detectable products. The signal generation at biosensors using immobilized biomolecules is based on the superposition of the following processes: (i) the mass transport of the analyte, (ii) biochemical recognition of the substance to be measured, and (iii) generation of the electrical output at the transducer interface. Optimal sensitivity for substrate measurement is reached at stoichometric analyte conversion within a highly permeable enzyme layer. On the other hand, substances which *effect* the enzyme activity get accessible to the determination by enzyme sensors provided that the overall signal is governed by the enzyme reaction. Despite the low sensitivity towards the substrate these *kinetically controlled* sensors allow the indication of submicromolar concentrations of inhibitors and activators.

I INTRODUCTION

Owing to their limited sensitivity the well established enzyme electrodes are applicable only to metabolites present in the micro- and millimolar concentration range. Highest sensitivity is reached when high enzyme activity within a thin layer is used

G.G. Guilbault and M. Mascini (eds.), Uses of Immobilized Biological Compounds, 171–179.

and effective external mass transfer is provided. Under these conditions, substrate measurement can be managed down to 1 μmol/l concentration with imprecisions below 2%.

The cell metabolism offers different principles to amplify weak chemical signals which have been used in the development of multi-enzyme sensors

- protein de novo synthesis is achieved by preincubating microbial sensors in appropiate measuring media,
- cyclic reactions produce a multiple of reaction products thus enhancing the sensor response,
- activation or inhibition of minute enzyme amounts lead to pronounced measuring signals.

II ENZYME SEQUENCES

Since not all enzyme-catalyzed reactions involve compounds detectable at an electrode, such as H^+, oxygen, or hydrogen peroxide, only a limited number of substances can be determined with one-enzyme sensors. In order to broaden the range of measurable substances, coupled enzyme reactions for analyte conversion in enzyme electrodes have been devised. In enzyme sequences the primary product of the analyte conversion is further converted enzymatically with the formation of a measurable secondary product.

In the late 70s Rechnitz' group[1] introduced enzyme sequences to enzyme electrodes. The scope of analytes has been expanded by *sequential* and *parallel* coupling of different enzymes with each other. Furthermore auxiliary enzymes have been combined with subcellular organelles and tissue slices to give *hybrid* sensors. Examples include ensembles of sensors based on glucose oxidase, lactate monooxygenase[2] and NAD(P)H oxidase[3]. Such enzyme sequence electrodes are known for sucrose, lactose, maltose, glucose itself, gluconate, glucosinolate, bilirubin, ATP, glucose-6-phosphate, cholesterol esters, fatty acid esters, acetylcholine, creatinine, hypoxanthine, glycerol, lactate, pyruvate and inosine.

III SEQUENCES WITH ACCUMULATION

An increase of sensitivity is gained if an intermediate of the analyte conversion is accumulated in front of the polarized indicator electrode. Stripping of the accumulated intermediate upon addition of a cosubstrate results in an increased reaction rate.

This principle has been applied for glycerol determination with a sensor using the sequence glycerol dehydrogenase (GlyDH)/lactic dehydrogenase (LDH)/lactate monooxygenase (LMO) [4].

The conversion of glycerol in the presence of an excess of NAD^+ generates the intermediate NADH. The accumulated NADH is stripped by addition of an excess of pyruvate. Due to the high concentration and diffusivity of pyruvate the reaction rate considerably exceeds that for the steady state current which is determined by the low concentration gradient of glycerol. Few seconds after the start with pyruvate the current reaches a peak. Since the NADH is consumed in the LDH-LMO reaction the current decreases and approaches a steady state identical to that without accumulation of NADH. The peak current rises with accumulation time. A limit is set by the equilibrium of the reaction. Therefore, the peak current levels off at long accumulation time. At an accumulation time of 6 min the sensitivity is increased by a factor of 64. Using the same procedure but with glucose dehydrogenase (GDH) inplace of Gly DH, the glucose signal has been amplified by a factor of 40.

IV LINEAR AMPLIFICATION BY ANALYTE RECYCLING

High signal amplification is obtained when the analyte ist shuttled between two enzymes. In each cycle *one* analyte molecule is regenerated and a detectable product is formed from the cosubstrate. For glucose amplification glucose oxidase (GOD) has been co-immobilized with GDH which reduces the gluconolactone in the presence of an excess of NADH. In this system an amplification of 10 is obtained. For the enzymatic recycling pair LOD/LDH the signal was amplified up to four orders of magnitude [5].

In such systems the analyte acts as a catalyst, being shuttled between both enzymes in the overall reaction of both cosubstrates. In this way, significantly more cosubstrate will be converted than the amount of analyte present in the enzyme membrane. According to theoretical considerations by Kulys et al[6], the amplification factor, G, is given by the following equation:

$$G = K_1 \times K_2 \times L^2/2(K_1 \times K_2) \times D$$

where K_1 and K_2 represent kinetic parameters, D the diffusion coefficient of the analyte, and L the membrane thickness.

The enzyme excess present in diffusion-controlled membranes is exploited in the analyte recycling. The amplification factor decreases with progressive enzyme inactivation during operation of the sensor. On the other hand the unamplified signal remains insensitive to the partial enzyme deactivation since diffusion control is still mantained for the simple process (FIG 1). For a given enzyme system membrane sensors show generally more effective amplification then reactor arrangement[7]. This behaviour is obviously based on the higher enzyme loading factor by applying the enzymes in a volume of a few microliters whilst the reactor volume is mostly in the order of 50-200 /ul. On the other hand the functional stability of amplification reactors by far exceeds that of the respective sensor systems.

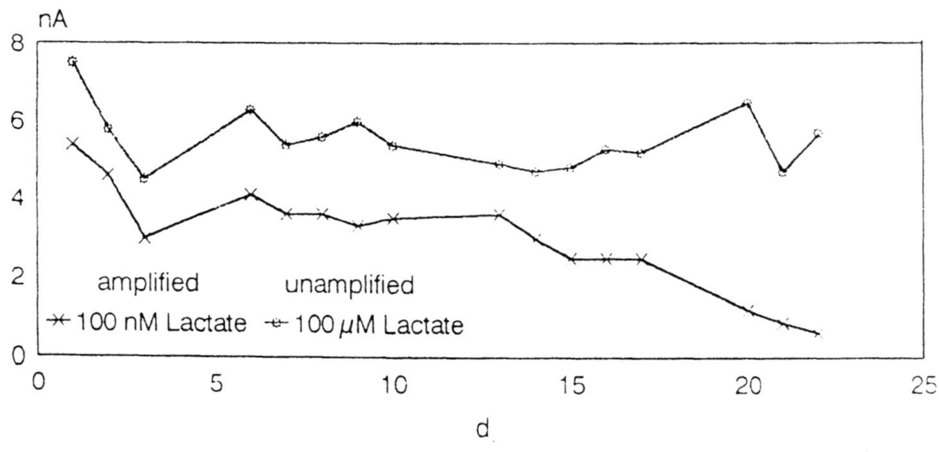

FIGURE 1 Functional stability of the LOD-LDH amplification system

V EXPONENTIAL AMPLIFICATION SYSTEMS

Tremendous signal amplification is expected if in the cycling reaction more than one analyte molecule is regenerated. This cascade-like process may be explained for the system myokinase (MK)/pyruvate kinase (PK).

Myokinase catalyses the formation of *two* ADP molecules from ATP and AMP. Each of the two ADP molecules is reconverted under the action of PK to give ATP and pyruvate. In this way in each cycle the amount of the catalyst is doubled. The reaction is driven by the excess of phosphoenol pyruvate (PEP) and AMP whereas the puruvate formation is indicated by the well known pyruvate oxidase reaction (FIG 2).

$$\text{ATP} + \text{AMP} \xrightarrow{\text{MYOKINASE}} 2\ \text{ADP}$$

$$\text{ADP} + \text{PEP} \xrightarrow{\text{PYRUVATE KINASE}} \text{ATP} + \text{pyruvate}$$

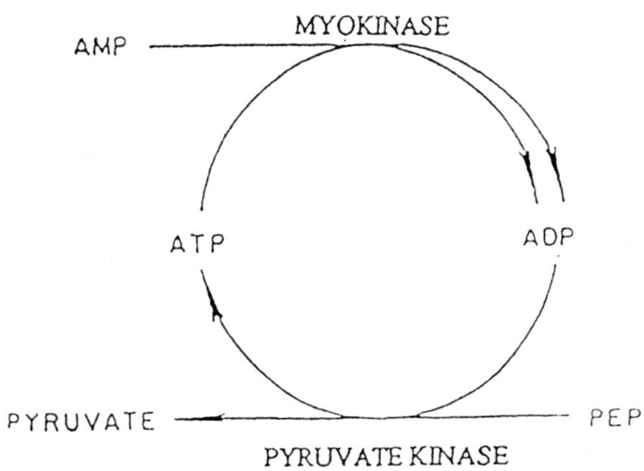

FIGURE 2 Scheme of the exponential cycle

Applying this exponential amplification system in homogeneous solution the expected exponential rise of pyruvate concentration with time has been established. However, even in the absence of added ADP an appreciable build-up of pyruvate was apparent, thus indicating that some component of the system contained either

ADP or ATP as a trace contaminant. This background rate could not be completely removed even after purification of AMP by ion-exchange chromatography. Under these conditions the limit of detection was around 1.5 nmol/l ADP.

Fixation of the two cycling kinases together with pyruvate oxidase in front of an oxygen electrode resulted in a different behaviour. The current-time curve of both the un- amplified and the amplified ADP measuring shows the normal sigmoid shape, reaching a steady state within 10 minutes. Obviously the coupling of mass tranfer with the enzymatic reactions causes different behaviours in solution and in the immobilized state. In the presence of an excess of AMP and PEP the signal of ADP is amplified by a factor of 800 resulting in a linear measuring range between 50 nmol/l and 2.000 nmol/l. This signal amplification is comparable with that obtained with linear cycling systems.

VI ELIMINATION OF INTERFERING SIGNALS

The capability of a diffusion-controlled enzyme membrane to completely convert the substrate permeating the membrane offers an interesting analytical application. Interfering substances are converted into non-disturbing products by 'filtering' the analyte flux of the indicator electrode.

To perform interference-free measurement of glucose the analyte-converting GOD may be co-immobilized with ascorbate oxidase. The potentials of this method were exemplified for glucose determination in mixtures of glucose and ascorbic acid. The measuring principle is as follows. The ascorbic acid interfering in anodic H_2O_2 oxidation is converted to inactive dehydroascorbic acid. Whith this method up to 2mM ascorbic acid in the sample was completely eliminated.

Elimination of electrochemically interfering signals of the sample is also achieved by starting the signal generation by adding an initiator substance. Whilst the basic current reflects all electrochemically active sample components, the current change is caused only by the intermediate and thus by the analyte of interest. This advantage has been used in the determination of triglycerides by using the aforamentioned enzyme electrode containing glycerol dehydrogenase, LDH and LMO. The triglycerides are hydrolyzed in alkaline methanol. After completion of this reaction the sample is injected into a buffered solution containing NAD^+. The addition of the methanolic solution causes a change of the basic oxygen sensor output resulting from the different oxygen contents in the sample and in the background solution. After establishing a stable baseline,which reflects the average oxygen concentration inside the bulk, pyruvate is added. The NADH formed in the oxidation of glycerol generates now an oxygen consumption via the LDH-LMO reaction sequence. In this way triglceride concentrations have been determined in serum samples with good correlation to the established method.

VII MEASUREMENT OF ALLOSTERIC EFFECTORS OF GLYCOGEN PHOSPHORYLASE

Glycogen phosphorylase b - the key enzyme of glycogenolysis - catalyzes the phosphorylation of the terminal glucose residue of glycogen to give free glucose-1-phosphate (G-1-P). Thus the rate of G-1-P formation represents the activity of the glycogen phosphorylase b. The G-1-P concentration is measured in the following three-enzyme sequence.

$$\text{G-1-P} \xrightarrow{\text{alkaline phosphatase}} \alpha\text{-Glucose} + \text{P}$$

$$\alpha\text{-Glucose} \xrightarrow{\text{mutarotase}} \beta\text{-Glucose}$$

$$\beta\text{-Glucose} + O_2 \xrightarrow{\text{glucose oxidase}} \text{Gluconolactone} + H_2O_2$$

At the compromise pH of 7.5 the sensitivity is sufficient to measure 1 μmol/l G-1-P with a CV < 5 percent.

The activity of glycogen phosphorylase b is controlled by the concentration of its allosteric effector AMP. By fixing the glycogen phosphorylase together with glycogen in front of the three-enzyme layer a new sensor for AMP has been developed[8]. The sensor responds repeatedly to micromolar concentrations of AMP with a response time of 20 s. Removal of AMP by washing the enzyme layer resulted in the reproducible establishment of the baseline. Based on the cascadic effect of enzyme activation by AMP the sensitivity of the electrode is higher than for the considerably smaller G-1-P.

Coentrapment of glycogen phosphorylase a with collagen in front of a phosphoglucomutase (PGM) and glucose-6-phosphat dehydrogenase (G6P-DH) containing layer allows to mimic the *glucose sensor of the liver*. In the presence of phosphate the formation of G-1-P leads to an oxygen consumption via G6P and

NADH. The reduced pyridine nucleotide is measured by the reaction with NMP^+ and its reoxidation under oxygen consumption. When 1 mmol/l phosphate was present in the background buffer containing 0.5 mmol/l NAD^+ at addition of 5 mmol/l glucose the G1P liberation was depressed by 15%. The presence of glucose did not affect the response on G1P injection, i.e. the PGM/G6PDH layer. Almost 50% inhibition of phosphorolysis was observed at a glucose concentration as high as 12 mmol/l. This correlates well with the inhibition constant of glycogen phosphorylase a of 7.9 mmol/l.

VIII INHIBITOR ANALYSIS

For the assay of organophosphorous and carbamate pesticides the cholinesterase enzymes can be considered the key enzymes. The detection is based on the inhibitory effect of these pesticides on cholinesterase.

A simple and reliable method of detection of organophosphorous pesticides is the measurement of cholinesterase inhibition in a membrane coupled directly to an amperometric electrode. Pesticide detection with an optimized bienzyme sequence electrode involves high choline oxidase activity, since this reaction is simply the indication of the acetylcholine conversion in the layer, and low esterase activity, allowing to affect the signal amplitude by influencing this enzyme. This can be interpreted also in terms of kinetically controlled acetylcholine response and diffusional limited choline detection (FIG 3). From the enzyme loading test 0.05 U/cm^2 was found to be the maximum esterase activity for sensitive indication of esterase inhibition. An amount of 1 U/cm^2 of choline oxidase was sufficient for diffusion limited choline response.

When the sensor was incubated for 3 min with 0,5 μmol/l dimethoate the inhibition was 6%. At 100 μmol/l a totally irreversible inhibition was observed. For the substances tested, the measuring ranges were 0.1 - 10 μmol/l for trichlorfon, 0.5 - 100 μmol/l for dimethoate, and 0.1 - 10 μmol/l for neostigmin.

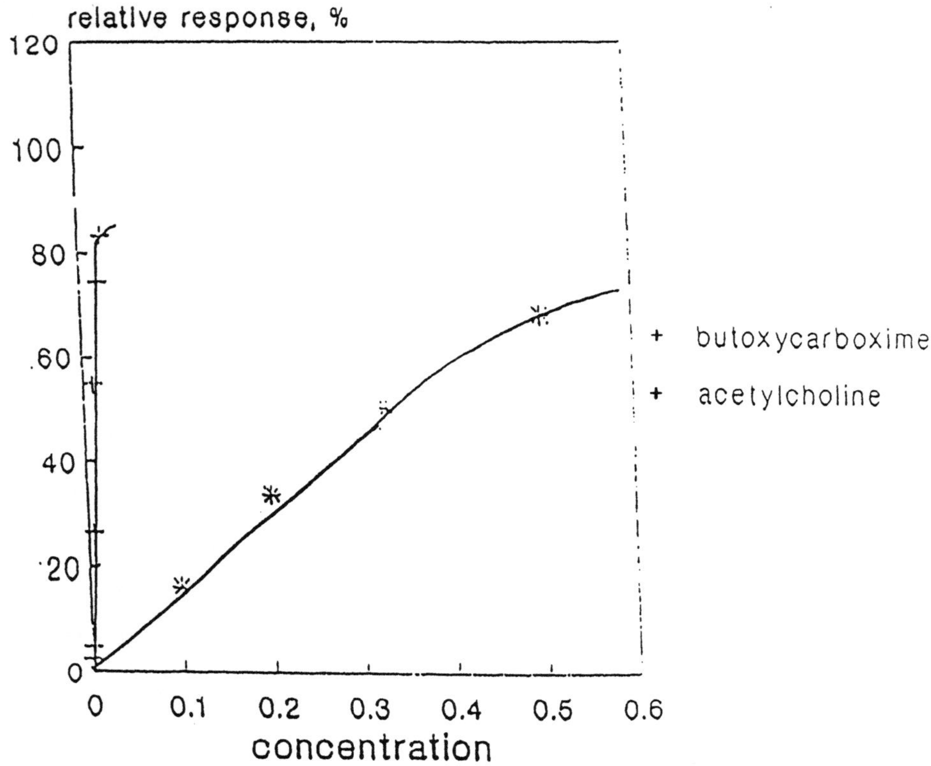

FIGURE 3 Concentration dependencies for the substrate and inhibitor

REFERENCES

1. M.A. Jensen, A. Rechnitz, J. Membr.Sci., 5, 117, (1979).
2. F. Schubert, U. Wollenberger, D. Pfeiffer, F. Scheller, Adv. Biosensors, 1, 77, (1991).
3. F.Scheller, F. Schubert, D. Weigelt, P. Mohr, U. Wollenberger, Makromol. Chem., 17, 429, (1988).
4. F. Scheller, A. Warsinke, R. Renneberg, F. Schubert, ACS Symp. Series, 487, 16, (1992).
5. F. Scheller, F. Schubert, B. Neumann, et al., Biosensors Bioelectr., 6, 245, (1991).
6. J.J. Kulys, V.V. Sorochinskii, R.E. Vidzinnaite, Biosensors, 2, 135, (1986).
7. E.H. Hansen, L. Norgaard, M. Pedersen, Talanta, 38, 275, (1991).
8. U. Wollenberger, F. Scheller, Biosensors Bioelectr., in press.

Three Different Approaches for Preparing GLOD and GLDH Reactors. Application for the Assay of Glutamates in Real Samples Using a FIA - System.

C.D. Stalikas, M.I. Karayannis and S.M. Tzouwara - Karayanni
Department of Chemistry, University of Ioannina, Ioannina, Greece.

Abstract

Glutamate Oxidase (GLOD) and glutamate dehydrogenase (GLDH) are immobilised on non-porous glass beads and Aminopropyl and Isothiocyanate Controlled Pore Glass (CPG) respectively. Three enzymatic methods are proposed for the determination of glutamic acid and glutamates, in food samples and pharmaceuticals. Various parameters are studied for the establishment of the optimum operating conditions for an in-house flow injection manifold. Many interfering compounds and several aminoacids were tested to verify the specificity of the systems. The accuracy of the proposed methods was tested by comparison of the results with those of official methods and the manafacturer's specifications for the analysed samples. Good agreement was attained.

Introduction

Glutamic acid is present in many foods as a flavour enhancer. Thus, the determination of glutamic acid is an important indicator of food quality. Glutamic acid is also an important parameter in clinical chemistry since it is the product of glutamic acid oxaloacetic transaminase and glutamic pyruvic transaminase. Glutamic acid and glutamine are the most concentrated amino acids occurring in the central nervous system.[1] Detailed studies revealed a possible role in the pathophysiology of mental disorders such as schizophrenia and depression.[2] As a result of this, an increasing trend is appearing for analytical methods for the reliable and accurate determination of low glutamic acid concentration in a small quantity of sample. In the past ten years many methods have been developed. For this purpose both glutamate oxidase (GLOD) and glutamate dehydrogenase (GLDH) immobilised in packed bed reactors (PBR) [3-5] and enzyme electrodes [6-8] performed with relative success.

In this work, GLOD is immobilised on non porous glass beads for the construction of Single Bead String Reactors (SBSR) and GLDH on aminopropyl - Controlled Pore Glass (CPG) as well as Isothiocyanate - CPG for the construction of packed bed reactors (PBRs).

In the case of GLOD the H_2O_2 generated by the enzymatic reaction is entrapped by the Trinder reagent [9] consisting of 4-Aminoantipyrine (AAP), 3,5-dichloro-2-hydroxybenzenesulfonic acid

G.G. Guilbault and M. Mascini (eds.), Uses of Immobilized Biological Compounds, 181–189.

(DCPS) and peroxidase (PO). The quinoneimine dye is then monitored at 510nm in a flow through cell. By using GLDH, the NADH produced by the enzymatic reaction is proportional to the original concentration of glutamic acid and measured fluorimetrically at an excitation wavelength of 340nm and an emission wavelength of 460nm.

The three enzymatic methods were employed for the determination of glutamic acid in food samples, pharmaceuticals and serum samples.

Experimental

Apparatus

Fig. 1 (a,b) includes schematic diagrams of the experimental flow injection manifolds. The inhouse flow injection systems consists of a four way pneumatically actuated injection valve, an eight channel peristaltic pump and either filter photometer or spectrofluorimeter (RF-551 Shimadzu) equipped with flow through cells.

The temperature of the reagents and the reactor was regulated by a thermostatically controlled water bath. Data processing and collection was performed with an IBM-compatible personal computer, software written in Quick-BASIC [10] and an interface unit RTL 800/815 multifunctional input/output board.

Reagents

All the chemicals were of analytical grade and the solutions were prepared in doubly distilled water (DDW). Glutamate oxidase (EC 1.4.3.11) from Streptomyces sp., 5U.mg^{-1} solid was a gift from Yamasa Shoyu (Chiba-ken, Japan). Glutamate dehydrogenase (EC 1.4.1.3) 120 U.mg^{-1} enzyme protein from beef livers, as well as Nicotinamide Adenosine Dinucleotide (NAD grad I, 100%) were purchased from Sigma (St. Louis, MO, USA).

Sample Preparation

For soups and pharmaceuticals a 0.2 g sample was accurately weighed and homogenised in a mortar. For meat products 3 g of sample were mixed with sand and homogenised in a mortar. In both cases the samples were transferred into a 100ml beaker with 30ml of water at room temperature and sonication was applied until all the water soluble substances were dissolved. For cheese samples the Water Soluble Fraction (WSF) was extracted according to a method described elsewhere.[11] The samples were vacuum filtered, the residue on the filter was washed with DDW and the filtrate and the washings were collected in a 50ml volumetric flask. In the case of serum samples, depending on the method, 1 in 20 or 1 in 7 dilutions were made with the working buffer. For the removal of proteins, ultrafiltration is applied, through membranes MWCO 10000 (Millipore) or the sample is chemically treated with $Ba(OH)_2$, $ZnSO_4$ and finally centrifuged. In both treatments the time taken to prepare the protein free solutions does not exceed 15 minutes.

Analytical Procedure

In the case of photometric detection the procedure entails injection of the sample into the carrier stream and the production of H_2O_2 on the SBSR composed of non-porous glass beads containing the immobilised GLOD. The reaction mixture meets the Trinder reagent (R) at the output of the

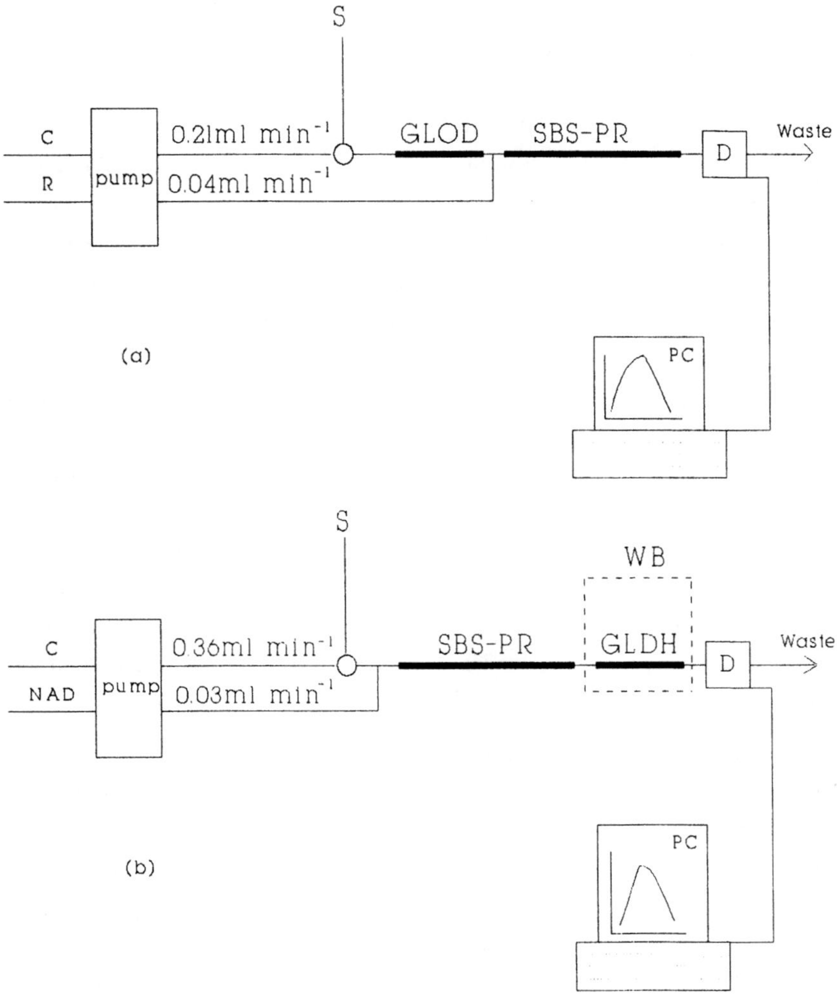

Fig.1 F.I.A. set up for glutamic acid determination:
(a) Photometric, (b) Fluorimetric detection.
SBS-PR: Single Bead String-Plain Reactor
C: Carrier, R: Trinder Reagent, S: Sample, WB: Water Bath
D: Detector, PC: Personal Computer.

enzymatic reactor and flows through a second SBS reactor (PR) constructed by untreated glass beads of a length of 40cm, giving sufficient time for the thorough mixing and the complexation of the reaction:

$$H_2O_2 + AAP + DCPS \rightarrow \text{Red Dye}$$

The maximum of the absorbance-time curve, A_{max}, is measured automatically at 510nm. When fluorescence detection is used, the solution for circulation with the sample is driven to a 15cm mixing SBSR positioned after the merging with NAD, and passes through the immobilised GLDH reactor. The peak height of the fluorescence is used for the quantification of glutamic acid.

Procedure for Immobilisation of the Enzymes

The immobilisation of the GLOD on glass bead 0.6mm d. is described elsewhere and consists of the following steps. [12,13] (i) Surface Cleaning of the Beads, (ii) Whisker formation, (iii) Silanisation, (iv) Glutaraldehyde Attachment and (v) Enzyme Attachment using a teflon tubing of i.d. 0.8 mm and composition of SBSR. The immobilisation of GLDH involves quite a different procedure. Specifically, reactors were prepared by either packing Aminopropyl CPG or Isothiocyanate CPG in glass columns (1.7mm x 4cm). Glutaraldehyde solution (5%v/v) in 0.1M $NaHCO_3$ is pumped only through the column of Aminopropyl CPG at 0.32ml/min to activate the support and the column is washed with DDW for 30 minutes. GLDH is loaded onto the supporting materials by circulating the enzyme in phosphate buffer with the appropriate pH for 8 hours at 0.32ml/min, at room temperature. The reactors are stored at 4^oC filled with the buffer of immobilisation.

RESULTS AND DISCUSSION

The optimum efficiency for immobilisation of GLOD is obtained at around pH 7.0 with the use of 0.1M phosphate buffer, while 1.0mg/ml is adequate to saturate 0.12g pretreated non-porous glass beads. 0.12g glass beads is sufficient for fabricating three 5cm reactors. It has been found that 82% of the initial enzyme protein attached itself to the beads. The efficiency of immobilisation on the glass derivatives, for constant mesh 200-400 gave the following results. The greater the number of amine and isothiocyanate groups and the smaller the pore sizes, the lower the efficiency of immobilisation of the reactor and therefore the lower the response of the system. This point is crucial because the differences to the resulting reactors, ultimately, are very large. The optimum pHs for the immobilisation of GLDH on Aminopropyl and Isothiocyanate CPG are 7.0 and 6.5 respectively. GLDH was loaded on the surface of the above derivatives with 76% and 80.0% yields respectively.

Conditions for the determination of glutamic acid were optimised by studying the effect of several parameters. For the optimisation of the analytical systems the univariate approach was applied in which one parameter was varied while all the others were kept constant.

The composition of the buffers and the pHs of the systems were primarily investigated. Figs 2 and 3 (a,b) show that 0.1M phosphate buffer is the most appropriate for the operation of the systems, in particular pH = 7.8 and 8.0 are relevant as working pHs for SBSR-GLOD and Aminopropyl, Isothiocyanate CPG respectively.

Figure 2

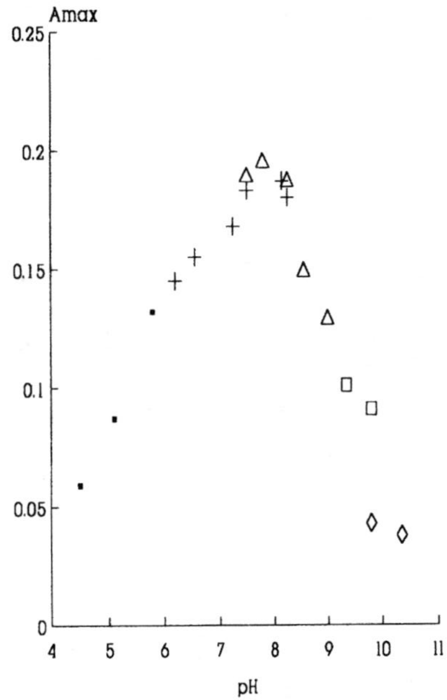

Effect of pH on the response of the system SBSR-GLOD(glass beads)
Buffers: ■Acetate,+ Phosphate,□Carbonate,△Trizma,◇Borate

Figure 3(a)

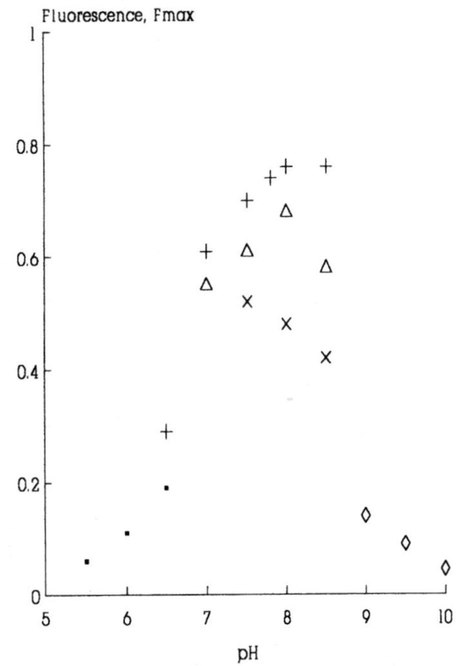

Effect of pH on the response of the system PB-Aminopropyl- GLDH
Buffers:■ Acetate,+ Phosphate,△Trizma,×Triethanolamine,◇Borate

Figure 3(b)

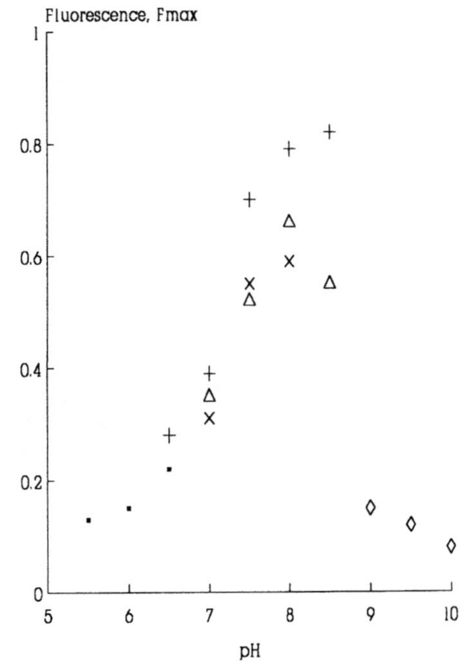

Effect of pH on the measuring system PB-Isothiocyanate-GLDH
Buffers:■Acetate,+ Phosphate,△Trizma,×Triethanolamine,◇Borate.

A 5cm reactor length for SBSR-GLOD and 4cm for Aminopropyl and Isothiocyanate CPG yield acceptable peak heights, while minimising the back pressure in the flowing stream. The length of the reactors (PR) filled with plain beads were 40cm and 15cm, as discussed before.

A sample volume of 52μL was chosen for the measurements as no significant increase was observed by further increasing the volume while this volume reduces the necessary sample volume as much as possible. Several flow ratios were studied keeping the overall flow rate constant. A 16:3 ratio (Carrier Buffer:Trinder Reagent) and 12:1 (Carrier Buffer:NAD solution) were selected for each system, offering low consumption of the reagents and dilution of the sample with satisfactory peak height. The overall flow rates of 0.250 $mlmin^{-1}$ and 0.390 $mlmin^{-1}$ were selected for photometric and fluorimetric methods, respectively, which reconciles fairly high sensitivity and satisfactory sampling throughput (40 samples/hour). Higher flow rates result in a decrease of peak height while lower minimise sampling throughput.

Concentrations of AAP, DCPS and PO in the Trinder reagent were adopted from Stults ***et al*** (12) with a minor modification in the concentration of PO which was 0.4 instead of 0.8mg/ml which did not affect the response. Concerning the NAD concentration the peak height increased with increasing concentration, rapidly at first and then gradually until a plateau is reached at 6×10^{-3}M. The final concentration of 2×10^{-3}M was chosen for subsequent experiments due to the high cost of the chemical. The enzyme was not activated by ADP in contrast with that for soluble enzyme [14]and therefore it was judged to be unpreferable. Also, ADP did not show any stabilising effect on GLDH, when added to the storage buffer. The final variable studied was the temperature of the reactors. The sensitivity increased sharply with temperature, reaching a plateau between 35° and 52°C, in the case of SBSR-GLOD. When using Aminopropyl CPG reactor, the maximum response appears at 30° to 37°C, while in the case of the Isothiocyanate reactor it increases until 45°C relatively rapidly.

The specificity of the enzymatic reactors was also studied. Various aminoacids were tested at concentrations of 1mM as was a mixture of them. Although it was reported that GLDH oxidises a number of aminoacids, only glutamine, leucine, valine interfered at 1.8%, 0.12% and 0.16% in Aminopropyl CPG reactor and 2.0%, 0.15% and 0.17% in the Isothiocyanate CPG reactor. In the case of SBSR-GLOD only glutamine interfered at 0.08%. Many inhibitors were also tested for their influence on the measurements. The results are presented in Table I.

Glutathione, ascorbic acid and uric acid are sources of serious inhibition because of their reducing action on participating in the Trinder reaction. GLDH systems are susceptible to metallic cations and activated by EDTA.

Standard solutions of glutamic acid were prepared in phosphate buffer. Under optimum working conditions fairly good linearity was obtained between A_{max} or F_{max} and glutamic acid concentration:

System	**Linear Range** (μM)	**Correlation Coefficient**
SBSR-GLOD	10 -500	0.999
PB-Aminopropyl-GLDH	2 - 100	0.998

PB-Isothiocyanate-GLDH	1 - 100	0.998

Table I: Interference Effects of Various Compounds in Concentrations of 1mM in the Assay of Glutamic Acid.

Interfering Compound	Relative Activity		
	SBSR-GLOD	PB-Aminopropyl	PB-Isothiocyanate
None	100.0	100.0	100.0
EDTA	100.7	102.8	115.3
2,2-Bipyridyl	100.3	97.50	96.70
Glucose	95.00	98.00	98.20
Glutathione	46.10	104.0	105.0
Ascorbic Acid	25.00	99.00	98.30
Starch	90.00	95.20	94.10
KCl	104.8	95.80	96.40
NaCl	101.3	100.0	100.0
$BaCl_2$	96.50	88.00	88.30
Uric Acid	71.00	99.00	98.60
Urea	98.00	99.50	99.00

The reactors showed very good operational and storage stability when washed with the buffer of immobilisation and stored at 4°C when not in use. They require more frequent calibration during the first two days due to their partial loss of superficial activity. SBS-GLOD kept its initial activity after a working period of two months while the PB-GLDH maintains its activity for even longer and is usable for about 2000 runs.

The methods were applied to food samples, pharmaceuticals and serum samples. The results are summarised in table II and III.

Comparison for glutamate content was made with the AOAC method [15] and colorimetrically with Boehringer Kit according to the manufacturer's instructions. [16] Good correlation was obtained. All assays were performed after appropriate dilutions of samples with buffer. In the case of serum samples deproteinisation was made either by ultrafiltration (PB-GLDH reactors) or $Ba(OH)_2/ZnSO_4$ (SBS-GLOD reactors) followed by centrifugation. The accuracy and matrix effect was also studied by performing recovery experiments which showed 97-105% recovery.

Table III: Determination of Glutamate in serum samples.

SAMPLE No.	Glutamate, mM		
	SBSR	Isothiocyanate	Aminopropyl
1	0.21	0.22	0.21
2	0.18	0.19	0.18
3	0.17	0.17	0.17
4	0.17	0.18	0.18
5	0.26	0.27	0.28
6	0.18	0.18	0.19

Table III: Determination of glutamate in food samples and pharmaceuticals. Values are compared with those found applying reference methods and Relative Errors (%RE) are given for each proposed method.

Sample	Glutamate % m/m *						
	Reference Method**	Isothio-cyanate	%RE	Amino-propyl	% RE	SBSR	%RE
Guda Cheese	0.64	0.63	1.50	0.63	1.50	0.62	3.10
Cheddar Cheese	0.78	0.79	1.30	0.78	0.00	0.78	0.00
Tomato Cream	5.43	5.38	0.90	5.32	2.00	5.46	0.50
Beef Cube	6.82	6.71	1.60	6.79	0.40	6.65	2.50
Beef cream	1.48	1.58	6.70	1.52	2.70	1.55	4.70
Chicken Cube	8.68	8.75	0.80	8.78	1.10	8.54	1.60
Mushroom Sauce	3.36	3.40	1.20	3.51	4.50	3.23	3.90
Zwan (canned meat)	0.04	0.04	0.00	0.04	0.00	0.04	0.00
Your Life Tablet	8.20	8.50	3.60	8.30	1.20	8.60	4.80
Brain Tablet (mg/tablet)	75.0	74.5	3.30	74.2	1.10	77.0	2.70

* Average of three runs

** For cheese samples the Boehringer Mannheim Kit, for tablets the manafacturer's specifications and for the rest of the samples the AOAC method was applied as the reference method.

In conclusion both systems are successfully applicable to glutamic acid determination however with some shortcomings. The SBS-GLOD reactor has some advantages such as limited dispersion, longer residence time, low consumption of the extremely expensive GLOD and excellent matching to the conditions of a Flow Injection Systems. The soluble PO used in the system slightly reduces the economical features of the method owing to its low consumption. The high detection limit restricts its use on samples with relatively high content of glutamic acid while the interference of reducing compounds requires a special treatment in some instances. On the other hand, PB-GLDH reactors have proved to be successful since CPG derivatives seem to be very positive functionalised supports for further activation and binding of the relatively cheap GLDH. Owing to the limited amount of loaded enzyme on SBS reactors, the PB-GLDH offers an important advantage against the former with a subsequent higher sensitivity, but higher dispersion and back pressure. The procedure for loading the enzyme is simpler and more efficient compared to the SBS Reactors.

REFERENCES

1. T.L. Perry, S. Hansen, and J. Kennedy, J.Neurochem., 24, 587, (1975).
2. B. Engelsen, Acta.Neurol.Scand., 74, 337, (1986)
3. H. Kusakabe, Y. Midorikawa, T. Fujishama, A. Kuninaka, and H. Yoshino, Agric.Biol.Chem., 47, 1323, (1983)
4. N. Kiba, T. Moriya, M. Furusawa, Anal.Chim.Acta., 256, 221, (1992)
5. R. Puchades, L. Lemieux, R.E. Simard, J.Food.Sci., 54, 423, (1989)
6. R. Villarta, D. Cunningham, G. Guilbault, Talanta, 38, 49, (1991)
7. H. Yamauchi, H. Kusakabe, Y. Mio Kawa, T. Fujishama, A. Kuniinaka, paper presented at the Third European Congress of Biochemistry, Vol. I, p.705, (1984)
8. W. Vahjen, J. Bradley, U. Bilitewski, R. Schmid, Anal.Lett, 24, 1445, (1991)
9. P. Trinder, Ann.Clin.Biochem., 6, 24, (1969)
10. C. Patton, and S. Crouch, Anal.Chim.Acta., 179, 189, (1986)
11. C.N. Kuchroo, and P.F. Fox, Milcchwissenschaft, 38, 76, (1983)
12. C. Stults, A. Wade, and S. Crouch, Anal.Chim.Acta., 192, 155, (1987)
13. E. Kiranas, S. Karayanni-Tzouwara, and M.I. Karayannis, Acta.Chim.Hung. - Models in Chemistry, 129, 461, (1992)
14. P. Lund, in H.U. Bergmeyer (Ed.), Methods of Enzymztic Analysis, Vol. VI, Verlag Chemie, Weinheim, p.357, (1984)
15. Official Methods of Chemical Analysis of the Association of Official Analytical Chemists, AOAC, Washington D.C., 12th Edn., Sections 20.149 - 20.151, (1975)
16. H.O. Beutler, in H.U. Bergmeyer (Ed.), Methods of Enzymatic Analysis, Vol. VI, Verlag Chemie, Weinheim, p.369, (1984)

ENZYME IMMOBILIZATION ON CHITIN AND CHITOSAN FOR CONSTRUCTION OF ENZYMATIC SENSORS

JÚLIA M.C.S.MAGALHÃES AND ADÉLIO A.S.C.MACHADO*
Departamento de Química, Faculdade de Ciências, P4000 Porto, Portugal

Abstract: The literature on enzyme immobilization on chitin and chitosan is briefly reviewed. Preliminary results on the preparation of a potentiometric electrode composed of a membrane of urease/chitosan applied on a nonactin/PVC ammonium selective electrode are presented and discussed.

I. INTRODUCTION

Chitin [1-5] is the second most common natural polysaccharide after cellulose, produced mainly as the skeletal integument of crustaceous as well as by insects and fungi. The shells of crustaceous, constituted by 15-20% of chitin and 25-40% of proteins (on a dry basis) [4] besides calcium carbonate, are the main industrial source of the substance, which is obtained in industrial amounts from waste products of the crab and shrimp processing industries by acid treatment for demineralization followed by diluted alkali treatment for deproteination.

Chitin is a linear polymer composed of near straight-chains of β(1->4)2-acetamido-2-deoxy-D-glucopyranose units (Fig.1.A), kept together by strong interchain hydrogen bonding. This structure is similar to that of celullose except that the 2-OH group of each glucose unit is replaced by an acetylated amino group ($-NHCOCH_3$). The extraction procedures cause some degradation of the polymeric chains and therefore the molecular

*) Author for correspondence

G.G. Guilbault and M. Mascini (eds.), Uses of Immobilized Biological Compounds, 191–200.

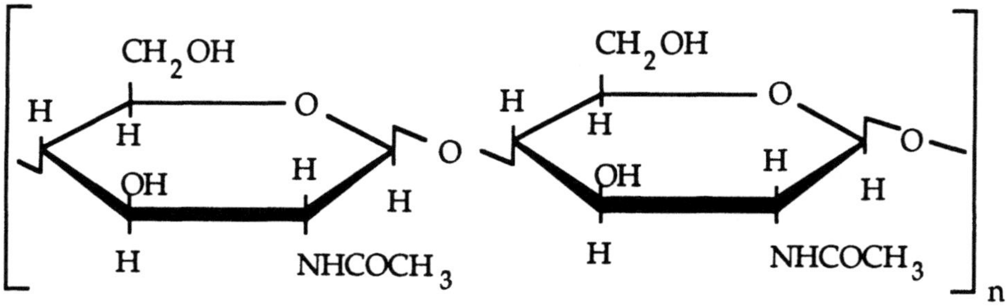

(A)

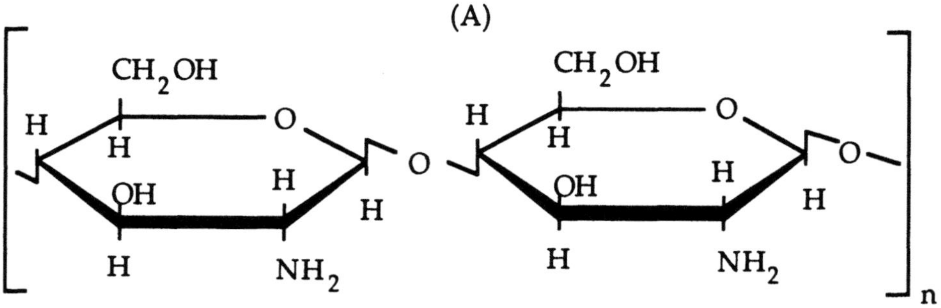

(B)

Fig. 1 - The chitin (A) and chitosan (B) units

weight of the commercial product is much lower than for the native substance (1-5 x 10^5 *vs.* 10^6 D) [4].

Upon deacetylation by further reaction with hot concentrated alkali, a related substance, chitosan (Fig.1.B), is obtained (the commercial material is only 70-90% deacetylated). This material is much easier to handle in the laboratory than chitin. While chitin is insoluble in water, dilute acids and alkalis, and ordinary solvents, chitosan, although insoluble in water, dissolves in aqueous organic acids (e.g., acetic acid), as well as in inorganic acids, due to protonation of the amino groups (upon positive charging of these, the polymer chains fall apart). From these viscous solutions, the solid can be reprecipitated by heating or neutralization.

Both chitin and chitosan found a large number of applications in a variety of industries [4,6], including pharmaceutical [6,7], cosmetical [6,8], and surgical and biomedical uses [6,9-11]. These include contact lenses [9,12], surgical sutures [13], burn therapy [9], etc. These are possible because the substances show no toxicity and are biocompatible (see ref.13-14) and show bacteriostatic action [9].

The occurrence of biocompatibity makes chitin and chitosan attractive as sensor materials for *in vivo* measurements. On the other hand, as discussed in detail in the next section, these materials have been used as supports for enzyme immobilization, and recently the preparation of a chitosan membrane with urease immobilized was reported [15]. These facts prompted us to attempt the use of chitin and chitosan as supports for enzyme immobilization for the construction of enzymatic sensors. This paper reports work on a potentiometric sensor for urea based on urease immobilized on a chitosan membrane applied to an ammonium selective electrode. Very recently a glucose amperometric sensor based on glucose oxidase immobilized on chitin was reported [16].

II. LITERATURE ON ENZYME IMMOBILIZATION ON CHITIN AND CHITOSAN

Chitin and chitosan have been studied as supports for several enzymes for industrial applications since the seventies (most of the work has been published in technical literature and patents, and little fundamental work appears reported in scientific chemical literature until recent years). Recent literature on enzyme immobilization on chitin and chitosan is shown in Table 1 (for earlier references, see ref.17) and includes a few papers of Japanese origin on analytical uses, as noted in the column "Remarks" of the Table. For instance, besides the above mentioned glucose oxidase/chitin amperometric sensor [16], a system for determination of fish freshness based on use of two reactors with, respectively, nucleoside phosphorylase (NP)/xanthine oxidase (XOD) and NP/XOD/ 5-nucleotidase(NT) immobilized in chitosan beads, was reported recently by Okuma *et.al.* [18].

Table 1 - Recent works on immobilization of enzymes on chitin and chitosan

Enzyme	Technique	State	Remarks	Ref.
		CHITIN		
Glucose oxidase	Adsorption	Membrane	**Enzymatic electrode**	16
Glucoamylase	Adsorption	Plates	Industrial application	19
		CHITOSAN		
Urease	Glutaraldehyde	Membrane	Artificial kidney	15,20
NP/XOD * NT/ND/XOD	Glutaraldehyde	Beads	**Fish freshness detector**	18
Glucose oxidase(?)			**Enzyme electrodes**	21
Urease			**Column/urea in blood**	22
Penicillin G Acylase	Glutaraldehyde	Powder/beads	Industrial application	23
α-Galactosidase Glucoamilases	Adsoption	Beads	Industrial application	24
Glucoamylases	Adsorption	Powder	Industrial application	25

*) NP, Nucleoside Phosphorylase
XOD, Xanthine Oxidase
NT, 5-Nucleotidase

III. IMMOBILIZATION OF UREASE ON A CHITOSAN MEMBRANE FOR CONSTRUCTION OF A POTENTIOMETRIC SENSOR

A procedure similar to that described by Krajewska *et. al.* [15] was followed for preparation of chitosan membranes and immobilization of urease on them. However, attempts were also made to prepare chitosan membranes without any mesh for support.

III.1. Experimental

Preparation of chitosan membranes. A solution was prepared by stirring overnight 1 g of chitosan (from crab shells, SIGMA C 0-0792) in 100 mL of 0.8 % (w/v) acetic acid. The membranes were cast on polyethylene plates on a nylon mesh (or without it in some experiments, see below) from a measured volume of solution per surface area (the electrode described below was prepared with 0.34 mL/cm^2, which, in terms of mass of chitosan, corresponds to 3.4 mg /cm^2). The membranes formed upon drying at 60°C overnight. The next day they were neutralized with 1% NaOH solution during 30 min and washed with water. The membranes were kept under water until use for enzyme immobilization.

Immobilization of urease on chitosan membranes. A 0.01% (w/v) glutaraldehyde solution was prepared (from a 25% solution, BDH). The wet chitosan membranes were treated with this solution for 1 1/2 hr and then thoroughly washed with water for removal of the glutaraldehyde. The membranes were then placed in an urease (from jack beans, Sigma, type III, EC 3.5.1.5) solution of concentration 0.5 mg/mL or larger, prepared in pH=5.6 phosphate buffer, for 1 hr at room temperature and left in it overnight at 5°C. After this procedure, the membranes were kept at 5°C under a 0.1 **M** (pH =7) TRIS buffer.

Assembly of the electrode. The membranes were applied to the tip of the ammonium electrodes (ϕ = 10 mm) described below and fixed with a silicone ring.

Evaluation of the electrode. Calibration curves were obtained by titration of 50 mL of a in 0.1 **M** (pH =7) TRIS buffer with a 0.1 **M** urea solution prepared with the same buffer as solvent.

Determination of enzyme activity. Measurements were made with an ammonium electrode following a conventional procedure (see ref.26). 50 mL of 0.1 **M**

urea solution in 0.1 **M** (pH =7) TRIS buffer were placed in a cell and, after stabilization of the potential difference, a piece of membrane whose surface area had been previously mesured (2-4 cm^2) was introduced in the solution and the variation of potential with time was recorded. The amount of ammonium ion produced was calculated from a previously obtained calibration curve. The membrane used in the electrode described below had an activity of 1.5 µmol/min/cm^2.

Construction of the ammonium electrode. All solid state ammonium ion selective electrodes with a nonactin cocktail dispersed in a PVC membrane [27] applied on a conductive epoxy support [28] were used as base electrodes. Graphite powder (≤ 50 µm, Merck) was used to make the epoxy (Epoxy Technology Inc. H54-UNF) conductive [29,30]. This type of ammonium electrode has shown good performance both in enzymatic membranes [30-32] and in other situations [33].

The membranes were constituted by evaporation of the solvent from a THF (Merck p.a.) solution of nonactin (FLUKA Selectophore®, Ionophore I) (2%), bis(2-ethylhexyl)adipate (FLUKA Selectophore®) (68%) and PVC (high molecular weight, FLUKA Selectophore®) (30%). The electrode was calibrated with ammonium chloride in 0.1 and 0.01 **M** (both pH =7) TRIS (Merck p.a.) buffers, showing Lower Limit of Linear Range (LLLR) of respectively ca. 5×10^{-5} and 2×10^{-5} **M** with slopes of *ca.* 60 mV/dec. Calibrations in the presence of urea (in 3-10 x 10^{-3} **M** concentration range) showed no significative alteration of these calibration parameters.

Other equipment. An Orion 90.02 double junction reference electrode (with TRIS buffer in the external compartment) and a Crison 2000 mVmeter (±0.1 mV) with a PM8222 Philips recorder were used. Calibrations were made in thermostatised cells at 25±0.2°C.

III.2. Results

Evaluation of the electrodes. For membranes prepared as described above with 3.4 mg/cm^2 chitosan and urease activity of 1.5 µmol/min/cm^2, the calibration of the electrode against urea as described above (Fig. 2) showed the following values for the parameters: LLLR, *ca.* 5×10^{-5} **M**, slope *ca.* 50 mV/dec, upper limit of linear response

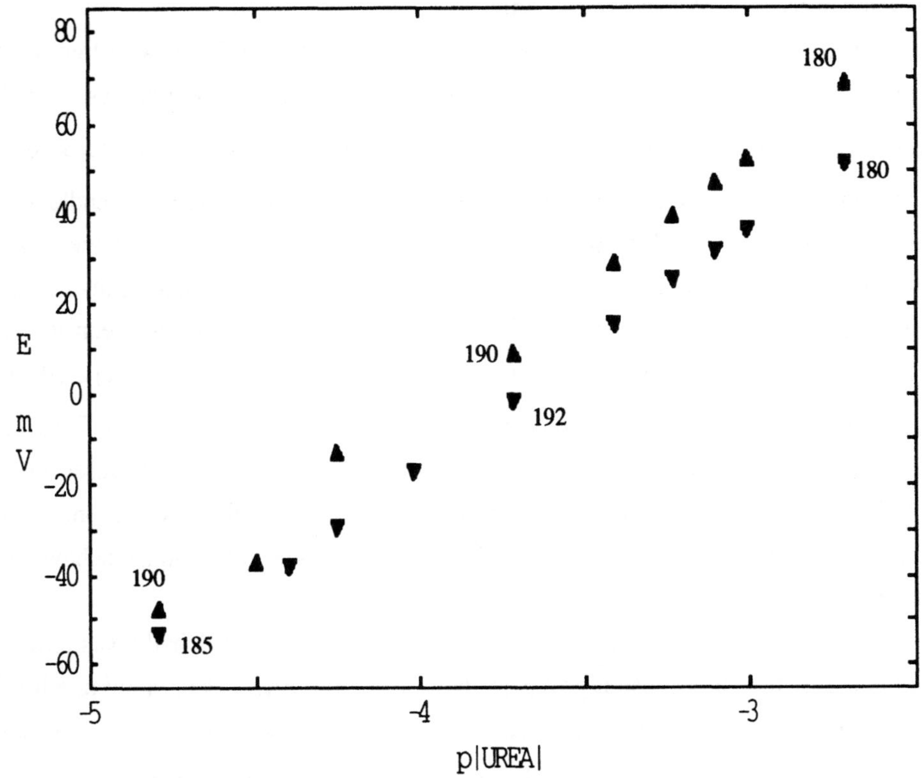

Fig.2 - Calibration curves for the urea sensor showing the diferences between the second (triangles up) and the third (triangles down) day. Numbers are response times (in s).

(ULLR), > 2 x 10^{-3} **M**. When the electrode was calibrated again 24 and 48 hrs later, after being kept in 0.1 **M** (pH =7) TRIS, about the same values were obtained.However, after the third day, the slope started decreasing (Fig.2) and the membrane showed a increasing loss of activity.

When an activated membrane was kept in a 0.5 mg/mL urease solution in 0.1 **M** (pH =7) TRIS buffer at 5°C and used about 45 days later, the electrode showed the same calibration parameters as for a recently activated membrane. This behaviour agrees with the reported in ref. [15].

Response times. Values of response times (typical values are included in Fig.2

for some of the points), were about 3 min.

Optimization of the preparation of chitosan membranes. Attempts to prepare chitosan membranes without nylon mesh were made. Membranes with 3.4 mg/cm^2 chitosan or thiner (2.9, 2.4, 1.8, 1.3, 0.9 and 0.6 mg/cm^2) were prepared but those with less than 1.8 mg/cm^2 bent very easily and were very fragile to handle. Even for larger surface weights, when membranes of this type (after treatment with glutaraldehyde) were applied to ammonium sensitive PVC membranes constituted in sandwich cells for use in flow-through measurements and squeezed between the two pieces of the sandwich, they tore very easily. When membranes were prepared over nylon mesh, thiner membranes could be prepared (2.1, 1.8, 1.4, 1.1 and 0.7 mg/cm^2 chitosan) which showed more resistence during handling and could be placed in and withdrawn from flow-through cells repeatedly. If and how these variations in the preparation of the chitosan base affect the immobilization procedure and the behaviour is currently under investigation.

Immobilization of urease. A low concentration of glutaraldehyde was used for membrane treatment because it provokes the reticulation of chitosan, yielding a britlle material (for instance, when glutaraldehyde is added to a solution of chitosan). When a 2.0 mg/cm^2 chitosan membrane was divided in two halves, and only one was treated with glutaraldehyde before enzyme immobilization, the only difference was that a more rigid membrane was obtained, no significant difference being observed in the extent of immobilized urease.

IV. PROSPECTS

The work developed so far on the use of chitosan membranes for the immobilization of urease for preparation of an urea potentiometric sensor suggests that this material is suitable for obtaining membranes with immobilized enzymes to be used in the construction of sensors. A more detailed study of the techniques for membrane constitution and enzyme immobilization is however required to obtain sensing membranes with adequate mechanical, sensing and stability characteristicis.

Besides the points discussed above, also the following deserve attention and are being or will be considered in further work:

Simultaneous constitution of the membrane and enzyme

immobilization. It is worth exploring procedures for performing simultaneously the composition of the membrane and the immobilization of the enzyme [16]. In principle, such operations yield producects with large amounts of fixed enzyme.

Nature of the chitosan with respect to the degree of N-acetylation. In a detailed recent study [34] on the adsorption of chitosan on cellulose, it was found that the degree of N-acetylation of the chitosan samples affects markedly the adsorption properties of the substance. More precisely, the extent of adsorption decresased when the amine group content increased, *i.e.* , when the degree of N-acetylation decreased, which was explained in terms of increased solubility, increased hydrodynamic volume and increased charge repulsion (upon protonation). It is then expected that smaller extents of the N-deacetylaytion of natural chitin in the preparation of chitosan produce better substances with respect to their aggregation in membranes.

Resistence to interferences. It has been found that when urease is immobilized on chitosan it shows extra resistence to inactivation by certain metal ions [20]. This effect is probably due to the chelating effect of chitosan and may have favourable consequences with respect to interferences on the sensing membranes.

ACKNOWLEDGEMENTS

This work has been supported by INIC (Instituto Nacional de Investigação Científica) and JNICT (Junta Nacional de Investigação Científica), Lisbon, Portugal.

REFERENCES

1. G.Skja-Braek, T.Anthonsen and P.Sandford, Chitin and Chitosan: Sources, Chemistry, Biochemistry, Physical Properties and Applications , Elsevier, 1989
2. J.P.Zikakis (ed), Chitin, Chitosan and Related Enzymes , Academic, 1984
3. R.A.A.Muzzarelli, Chitin , Pergamon, 1977
4. N.K.Mathur and C.K.Narang, J.Chem.Ed. , 67 , 938 (1990)
5. P.R.Austin, C.J.Brine, J.E.Castle and J.P.Zikakis, Science , 212 , 749 (1981)
6. P.A.Sandford, "Chitosan: Commercial Uses and Potential Applications" , *in* ref.1, p. 51-69
7. T.Nagai, Y.Sawayanagi and N.Nambu, "Application of Chiin and Chitosan to Pharmaceutical Preparations" , *in* ref.2, p.21-39
8. G.Lang e T.Clausen, "The Use of Chitosan in Cosmetics" , *in* ref.1, p.139-147
9. G.G.Allan, L.C.Altman, R.E.Bensinger, D.K.Ghosh, Y.Hirabayashi, A.N.Neogi and S.Neogi, "Biomedical Applications of Chitin and Chitosan" , *in* ref.2, p.119 -133
10. J.Dutkiewicz, L.Dutkiewicz, A.J.Papiewski, M.Kucharska and R.Ciszewski,

"Some Uses of Krill Chitosan as Biomaterial", *in* ref.1, p. 719-730
11. R.Olsen, D.Scwartzmiller, W.Weppner and R.Winandy, "Biomedical Application of Chitin and Its Derivatives, *in* ref.1, p.813-828
12. M.L.Markey, L.M.Bowman and M.V.N.Bergamini, "Contact Lenses Made of Chiosan", *in* ref.1, p.713-717
13. M.Nakajima, K.Atsumi and K.Kifune, "Development of Absorbable Sutures from Chitin", *in* ref.2, p. 407-410
14. M.Maeda, Y.Inoue, H.Iwase and K.Kifune, "Characteristics of Chitin for Orhopedic Use", *in* ref.2, p. 411-415
15. B.Krajewska, M.Leszko and W.Zaborska, J.Chem.Tech.Biotechnology, 48, 337 (1990)
16. E.Ohashi and T.Koriyama, Anal.Chim.Acta, 262, 19 (1992)
17. A.Rosevear, J.F.Kennedey and J.M.S.Cabral, Immobilised Enzymes and Cells, Hilger, 1987, p.89 and 118
18. H.Okuma, H.Takahashi, S.Yazawa and S.Sekimukai, Anal. Chim. Acta, 260, 93 (1992)
19 M.A.Sanroman, M.J.Nunez and J.M.Lema, Afinidad, 40(440), 231 (1992)
20. B.Krajewska, J.Chem.Tech.Biotechnology, 52, 157 (1991)
21. H.Katayama, JP 63 78,062 [CA 110:228160c (1989)]
22. H.L.Tang, C.T.Chang, Y.H.Sung and J.C.Su, Chung-kuo Nung Yeh Hun Hsueh Hui.Chih, 27(2), 257 (1989) [CA 111:190299k (1989)]
23. J. Braun, P. Le Chanu and F. Le Goffic, Biotechnol. Bioeng., 33, 242 (1989)
24. A.Ohtakara, G.Mukerjee and M.Mitsutomi, "Immobilization of α-Galactosidase and Glucoamilases on Crosslinked Chitosan Beads", in ref.1, p.643-651
25. T.Takahashi and N.Kayama, Chem.Pharm. Bull., 40(10), 2775 (1992)
26. M.Mascini and P.Palleshi, Anal.Chim.Acta, 145, 213 (1983)
27. O.G.Davies, G.J.Moody and J.D.RThomas, Analyst, 113, 498 (1988)
28. J.L.F.C.Lima and A.A.S.C.Machado, Analyst, 111, 799 (1986)
29. S.Alegret, J.Alonso, J.Bartroli, J.L.F.C.Lima and A.A.S.C.Machado, "All-solid-state PVC Membrane Flow-through Potassium Electrode Based on Valynomicin", *in* J.L.Aucoutorier *et.al.* (eds), Proceedings of the Second International Meeting on Chemical Sensors, Bordeaux Chemical Sensors, Talance, 1986, p.751-4
30. S.Alegret and A.A.S.C.Machado, "Extension to Enzymatic Electrodes of a General Procedure for the Construction of Flow-trough ISEs Based on the Use of Conductive Epoxy as Support of the Sensor", *in* G.G.Guilbault and M.Mascini (eds), Analytical Uses of Immobilized Biological Compounds for Detection, Medical and Industrial Uses, NATO ASI Series C226, Reidel, 1988, p.309-17
31. G.Palleschi, M.Mascini, E.Martínez-Fàbregas and S.Alegret, Anal.Lett., 21, 1115 (1988)
32. S.Alegret and E.Martínez-Fàbregas, Biosensors, 4, 287 (1989)
33. S.Alegret, J.Alonso, J.Bartroli and E.Martínez-Fàbregas, Analyst, 114, 1443 (1989)
34. J.G.Domszy, G.K.Moore and G.A.F.Roberts, "The Adsorption of Chitosan on Cellulose", in J.F.Keneddey *et.al.* (eds), Cellulose and its Derivatives: Chemistry, Biochemistry and Applications, Ellis Horwood, 1985, Chapter 42, p.463-73

ELECTROCHEMICAL BIOSENSORS

ANA MARIA OLIVEIRA BRETT
Departamento de Química, Universidade de Coimbra,
3049 Coimbra, Portugal

Abstract Electrochemical biosensors have been developed to be used as a stationary amperometric enzyme sensor for hydrogen peroxide and cholesterol and as a hydrodynamic amperometric sensor for glucose with flow injection analysis. The importance of a biocompatible electrode surface has been evaluated.

I INTRODUCTION

Electrochemical biosensors have been developed to be used to quantify analytical species in biological fluids, usually in blood serum and urine, in food, in environmental analysis and to detect biological compounds[1]They combine the transducer, the electrochemical sensor, with a biologically active substance: enzyme, bacterial cell, antibody, receptor, tissue, etc. The biosensor should be specific, sensitive, have a good response time and give reliable and reproducible results during its lifetime. Besides this, if it is going to be used *in vivo*, both blood and tissue compatibility are problems to be solved before an invasive biosensor will be used on humans.

The biosensors studied in the work described here are a stationary amperometric enzyme sensor for hydrogen peroxide and cholesterol, and a hydrodynamic amperometric glucose sensor using a mini-reactor containing immobilized glucose oxidase in a flow through system.

In order to protect the electrode from serum constituents the electrode was also pretreated with a terpolymer[2] of acrylonitrile poly(ethyleneoxide) methacrylate and trimethylaminoethyl chloride methacrylate, by physisorption, which leads to a diffuse layer of polymer segments covering the surface of the electrode whilst enabling small electroactive species to react at the electrode surface.

G.G. Guilbault and M. Mascini (eds.), Uses of Immobilized Biological Compounds, 201–210.

II STATIONARY AMPEROMETRIC ENZYME SENSOR FOR HYDROGEN PEROXIDE AND CHOLESTEROL

The importance of a quick, reliable and easily applied method for a laboratory assay either for hydrogen peroxide or cholesterol does not need to be stressed.

An electrochemical bienzyme membrane system was designed to function using a graft co-polymer as membrane support for the binding through covalent links of the enzymes catalase and cholesterol oxidase. The membrane was placed over a Clark-type oxygen electrode.

The consecutive reactions of the two enzymes, cholesterol oxidase and catalase, occur on the Teflon membrane and the oxygen released is measured at the oxygen electrode:

- on the membrane

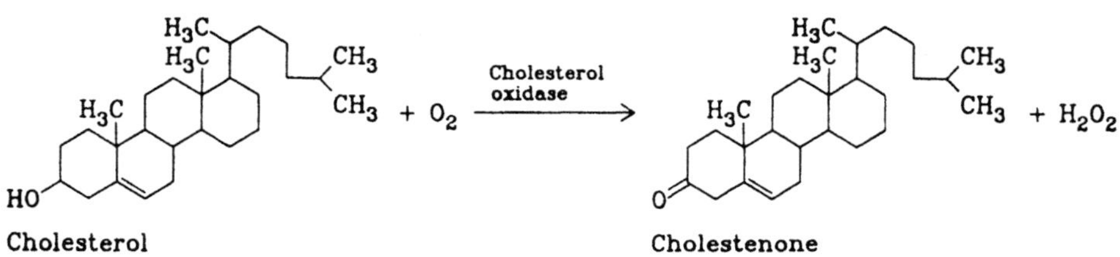

$$H_2O_2 \xrightarrow{\text{Catalase}} H_2O + \frac{1}{2}O_2$$

- at the electrode

$$O_2 + 4\,H^+ + 4\,e^- \longrightarrow 2\,H_2O$$

The electrode measures the oxygen produced in the oxidation of the hydrogen peroxide by catalase on the inner side of the membrane; the oxygen reacting with cholesterol comes from the sample solution, as there is no cholesterol in the electrolyte solution enclosed between the gold electrode and the Teflon membrane.

The grafted Teflon membrane was prepared using gamma radiation at 400 Gy h^{-1} for different intervals of time [3] .Through this procedure carboxylic acid groups were introduced onto a hydrophobic support creating a good environment for enzyme immobilization.

The enzymes, catalase and cholesterol oxidase, were consecutively coupled in phosphate buffer, using carbodiimide metho-*p*-toluenosulphonate as activating agent, in order to couple the enzymes to the carboxylic groups of the grafted poly(acrylic acid)[4] The percentage of grafting of 22% was determined by weight and the presence of poly(acrylic acid) was confirmed by infrared spectroscopy.

The response time for this bienzyme sensor is very fast for hydrogen peroxide and is slower, as predicted, for cholesterol in agreement with the characteristics of each enzyme, catalase and cholesterol oxidase, as shown in Figure 1.

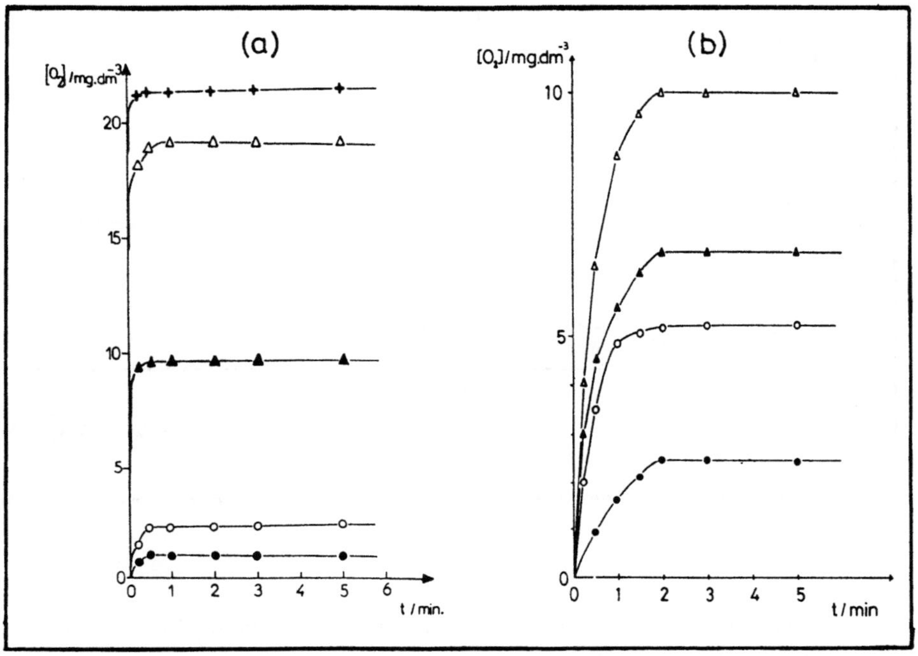

FIGURE 1 Biosensor response time, in pH = 7.0 and 0.05 mM phosphate buffer (1 mM O_2 = 32 mg dm^{-3}), increasing concentrations of:

(a) H_2O_2 (0.53 mM; 1.54 mM; 3.34 mM; 6.64 mM; 9.74 mM)

(b) Cholesterol (0.18 mM; 0.28 mM; 0.33 mM, 0.55 mM).

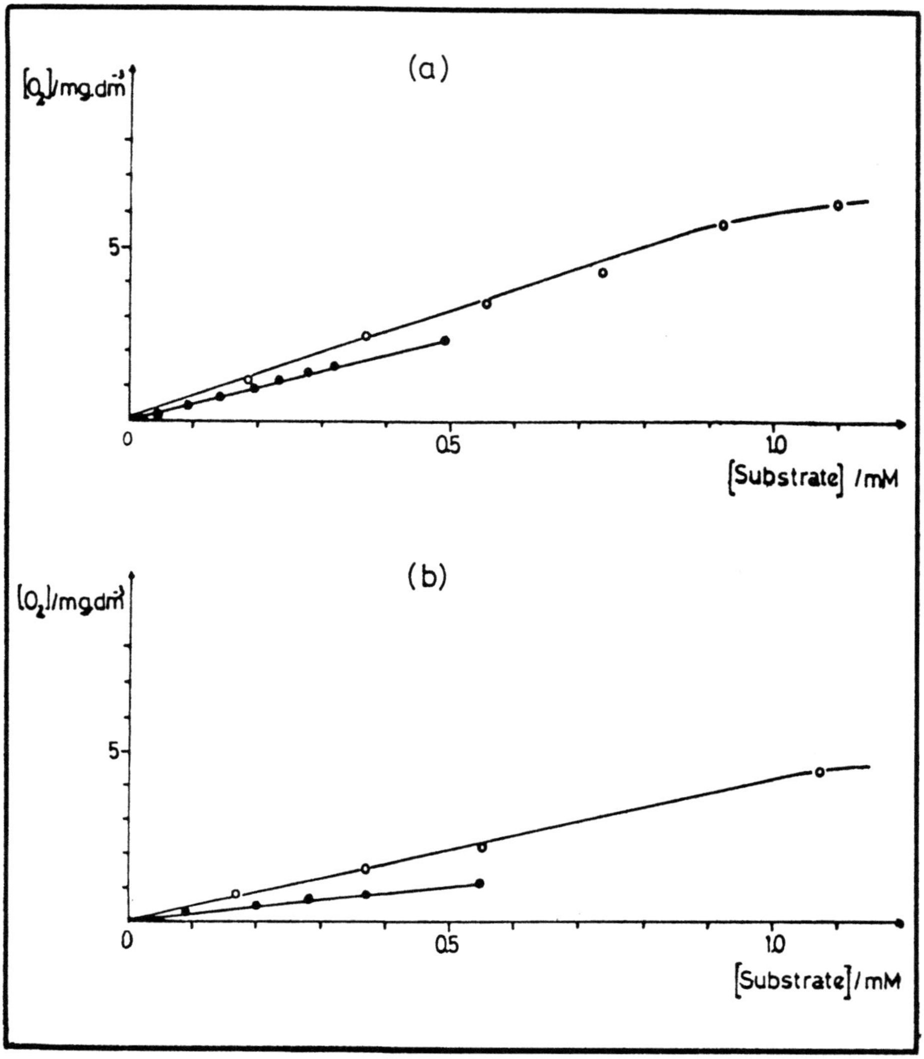

FIGURE 2 Calibration curve for the biosensor on (a) the 15th day and (b) the 35th day: o H_2O_2; ● cholesterol.

Calibration graphs for hydrogen peroxide and cholesterol, Fig.2, showed a lifetime of one month, the membrane being left on the electrode at room temperature in distilled water. However when only immobilized catalase was used, the sensor showed an excellent lifetime of more than four months.

The linear range obtained for measurement of free cholesterol was from 90 μM to 0.55 mM and the detection limit was 50 μM. The linear range obtained for hydrogen peroxide was from 10 μM to 6 mM and the detection limit was 4.7 μM.

These experimental results will enable the sensor to be mounted in on-line continuous flow and be easily automated to analyse samples injected into the flow stream.

It was shown that the use of a single membrane with two immobilized enzymes, catalase and cholesterol oxidase, placed onto the standard oxygen electrode, enables quick determination of hydrogen peroxide or free cholesterol using already existing commercial, easy to operate and portable equipment.

III HYDRODYNAMIC AMPEROMETRIC SENSOR FOR GLUCOSE

Hydrodynamic electrochemical systems have the advantage of an increased transport of electroactive species to the electrode, leading to higher currents and thence a greater sensitivity and reproducibility. Their use in flow injection analysis is obviously of wide application.

In the wall-jet electrode system, a jet of fluid impinges on the centre of a disc electrode and spreads out radially. Only solution originating from the jet can reach the electrode surface, so that the electrode always receives fresh solution. A typical wall-jet cell is shown in Fig.3.

A flow through mini-reactor containing a membrane with immobilized glucose oxidase was attached to the inlet of a wall-jet electrochemical detector, Fig.3. It was applied to the determination of glucose in serum, thus combining the advantages of an electrochemical hydrodynamic system with the specificity of the enzyme used.

The hydrogen peroxide produced in the reaction of glucose with glucose oxidase is detected by electrochemical oxidation at the platinum wall-jet electrode.

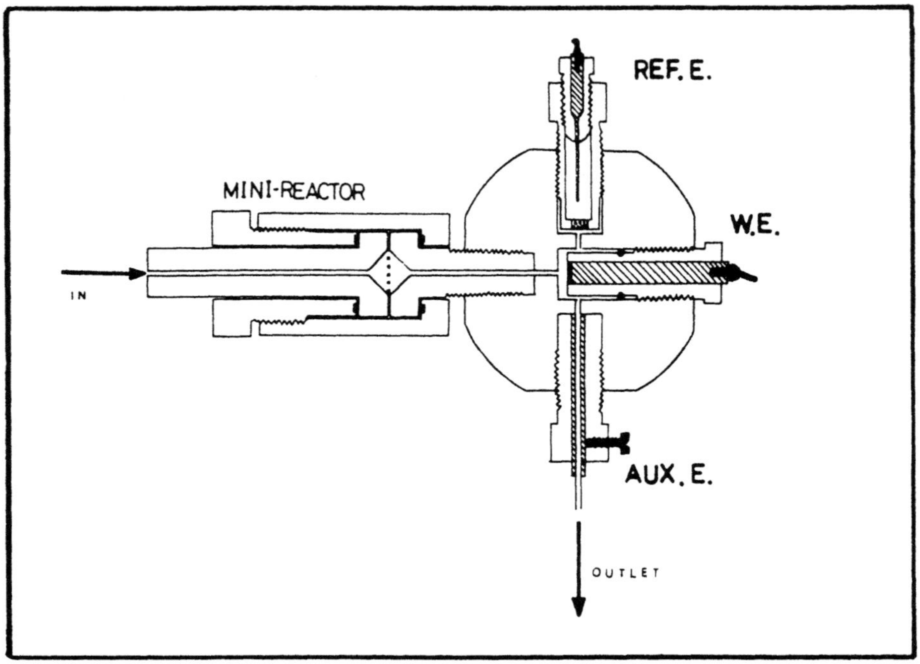

FIGURE 3 Mini-reactor attached to the wall-jet electrochemical detector.

The membrane used was a porous Nylon-66 fabric on which glucose oxidase was covalently bound using a multi-step procedure [5]. This procedure consists in first activating the membrane with glutaraldehyde followed by coupling with bovine serum albumin denatured by heat, activating a second time with glutaraldehyde and coupling with hexamethylenediamine, and after activating a third time with glutaraldehyde, finally coupling with glucose oxidase.

This long spacer arm gives the membrane a simulated biological environment which favours the covalent immobilization of glucose oxidase. Membranes prepared in this way had good immobilization yields and a good enzyme retention, maximum enzyme activity and a mimimum K_M', i.e. maximum enzyme affinity.

The membrane was put in the mini-reactor and the system: flow-through glucose mini-reactor with wall-jet electrochemical detector, was used for flow injection analysis of blood serum samples [6].

The flow injection system was calibrated with glucose standards. Results were verified using bio-spectrophotometric and colorimetric methods and were in good agreement.

However a decrease in the electrochemical signal during the first 10 injections was observed, after which it remained at a constant level. This diminution in current was due to poisoning of the electrode by adsorption of substances present in serum, fats, triglycerides, esterified cholesterol and others.

Pretreatment of the electrode with biocompatible polymers can prevent its surface being blocked by biological constituents of the contacting medium, and electrode rejection and formation of connective tissue onto the electrode surface when used *in vivo.*

The platinum working electrode was pretreated by physisorption with a terpolymer of acrylonitrile poly(ethyleneoxide) methacrylate and trimethylaminoethyl chloride methacrylate synthesized with the aim of creating hemocompatibility of the blood lines leading to and from a dialysis module.

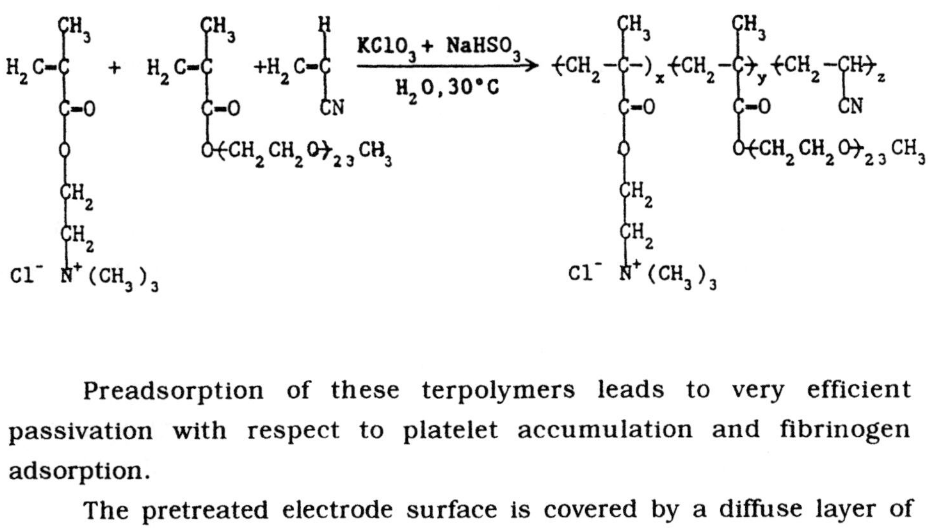

Preadsorption of these terpolymers leads to very efficient passivation with respect to platelet accumulation and fibrinogen adsorption.

The pretreated electrode surface is covered by a diffuse layer of the terpolymer segments near the interface, which has a thickness of the order of a few angströms. The diffuse layer forms a non-conducting barrier. However the polymers are never in equilibrium, and the surface is dynamic, shows mobility and surface rearrangement, responding to external environmental perturbations, and allows small electroactive species such as hydrogen peroxide to reach the electrode surface, Fig.4. The fluid velocity within the polymer diffuse layer is nearly zero and the transport to the electrode surface is purely diffusive.

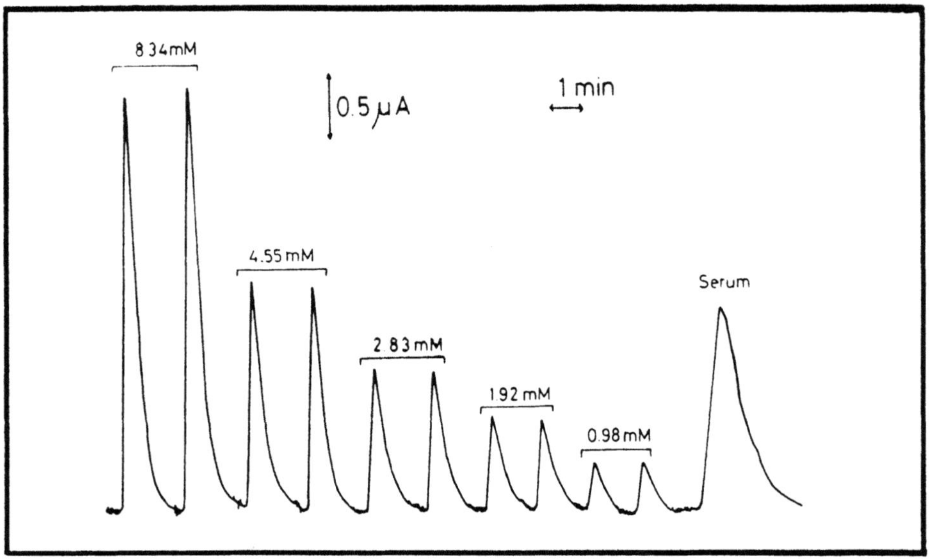

FIGURE 4 Flow injection analysis of glucose samples in 0.1 M acetate buffer using a platinum working electrode coated with the terpolymer hemocompatible film.

Comparison of the results obtained using an uncoated and a coated electrode, Fig.5, shows that, although the signal at the coated electrode was not as high to start with, the decrease in signal size with increasing number of injections was smaller, whereas for the uncoated electrode the signal was continuously decreasing.

This flow-through mini-reactor wall-jet electrochemical detector will measure glucose in the range of 10 μM to 15 mM and the detection limit is 5 μM.

The response time of the sensor is very rapid allowing flow injection analysis of ~60 samples per hour and the membrane lifetime is of one month when stored at room temperature.

The sensor can be automated and run continuously with one or more enzymes immobilized in porous supports.

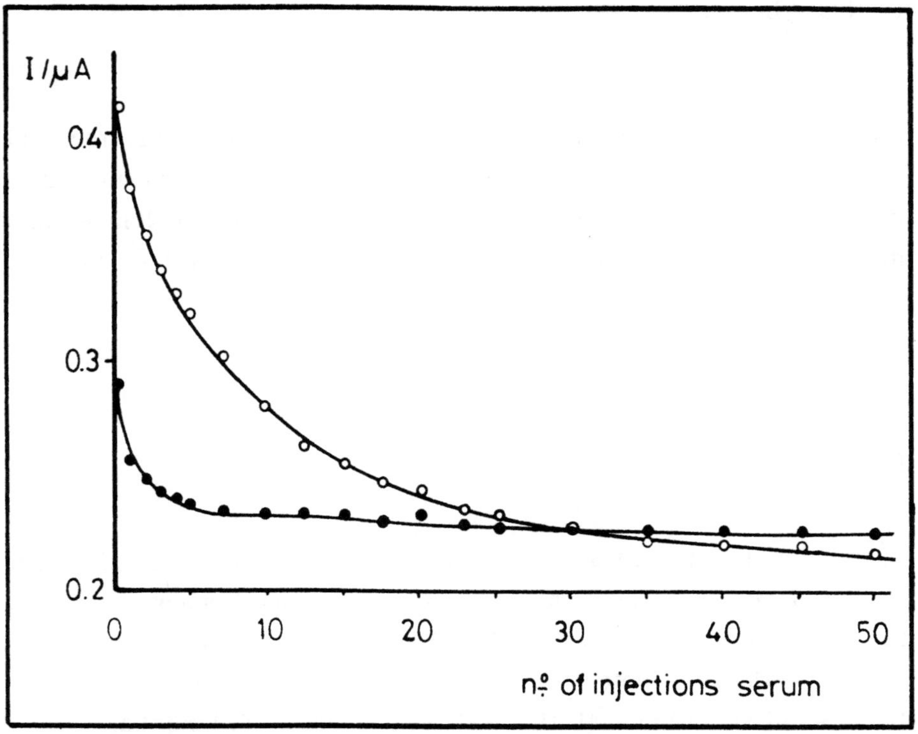

FIGURE 5 Plot of flow injection of glucose standard samples, 8.34 mM, in 0.1 M acetate buffer, followed by injections of serum using a platinum working electrode (O) uncoated; (●) coated with the terpolymer hemocompatible film.

Thus, electrodes modified physically, through the adsorption of synthetic polymers that have been developed with the aim of creating a hemocompatible surface, seem a promising way of reducing the problem of electrode fouling for *in vivo* aplications and determinations in blood serum. The polymer covering is permeable to the small molecules which are the products of the enzyme reaction with the substrates, and which react at the electrode.

However, materials to modify the surface of the electrodes have to be chosen with great care so as not to cause toxication, thrombosis or inflammatory effects due to irritation of blood vessels or tissue. Physically modified electrodes using a hemocompatible surface modifier are a promising way of reducing the problem of electrode fouling for *in vivo* applications.

IV CONCLUSIONS

Electrochemical biosensors is an interdisciplinary research area that needs to use new technologies as well as established ones, having in mind the problems that need to be solved and what can be done so that miniaturization of the working systems will enable *in vivo* measurements.

REFERENCES

1. J. M. Kauffmann and G. G. Guilbault, Bioanalytical Applications of Enzymes, 36,63,(1992), Ed. John Wiley and Sons,Inc.
2. F. Yan, Ph. Déjardin and A. Schmitt, Hemocompatible composite material, Eur. Patent Appl. 91420199.1-, 18 June 1991.
3. M. Alves da Silva, M. H. Gil, A. P. Piedade, J. S. Redinha, A. M. Oliveira Brett and J. M. Caridade Costa, J. Polym. Sci., Part A, 29. 269, (1991).
4. A. M. Oliveira Brett, M. H. Gil and A. P. Piedade, Bioelectrochem. Bioenerg., 28, 105, (1992).
5. M. Alves da Silva, M. H. Gil, J. S. Redinha, A. M. Oliveira Brett and J. L. C. Pereira, J. Polym. Sci., Part A, 29. 275, (1991).
6. J. S. Redinha, A. M. Oliveira Brett, J. L. C. Pereira and M. H. Gil, Bioelectrochem. Bioenerg., 22, 379, (1989).
7. A. M. Oliveira Brett, J. L. C. Pereira and Ph. Déjardin, Bioelectrochem. Bioenerg., 28, (1993), in press.

NOVEL APPROACHES TO THE DEVELOPMENT OF MEDIATED BIOSENSORS AND ENZYME ASSAY

JOHN H.T. LUONG, KEITH B. MALE, CARL A GROOM
and SHISHAN ZHAO

Biotechnology Research Institute
National Research Council Canada
Montreal, Quebec, Canada H4P 2R2

ABSTRACT

2-Hydroxypropyl-β-cyclodextrin (hp-β-CyD), a cyclic and non-reducing oligosaccharide was used to enclose the hydrophobic guest molecules, ferrocene ($FeCp_2$) and 1,1'-dimethylferrocene ($DMFeCp_2$) to form a water-soluble yellow complex. At 300 mM, hp-β-CyD enclosed up to 100 mM $FeCp_2$ or $DMFeCp_2$. The cyclic voltammogram of the $DMFeCp_2$-hp-β-CyD complex exhibited one single electron transfer process with a half-wave potential of 170 mV. The oxidation potential of the $DMFeCp_2$-β-CyD complex was not affected by pH ranging from pH 4.7 to 8.8 and the difference between the anodic and cathodic peak potentials was observed to be 95 mV.

The yellow complexes were electrochemically oxidized (platinum vs Ag/AgCl poised at +450 mV) to form the blue dyes, ferricinium ($FeCp_2+$) and 1,1'-dimethylferricinium ($DMFeCp_2+$). This is a one-electron transfer process and the ferricinium cations formed exhibited absorption peaks at 620 and 650 nm, respectively. The concentrated $DMFeCp_2+$ was much more stable (4 months) than the $FeCp_2+$ (2 days) and both oxidized dyes were insensitive to a wide pH variation (pH 2-11). The $DMFeCp_2+$ was reduced easily by various reducing agents such as ascorbic acid, uric acid, and sulfite and exhibited an absorption coefficient of 325 $cm^{-1}M^{-1}$. Application of the novel $DMFeCp_2+$ complex as a colorimetric dye for the enzymatic oxidation of glucose, glutamate, lactate, phenylalanine, xanthine and hypoxanthine was successfully demonstrated. The $DMFeCp_2+$ complex could also be utilized to monitor activity of oxidases under saturating substrate concentrations.

The $DMFeCp_2$-hp-β-CyD complex was used together with immobilized glucose oxidase to construct a mediated biosensor for the determination of β-D-glucose. The enzyme was either covalently immobilized on a preactivated nylon membrane or retained onto the sensing area of the electrode by a dialysis membrane. In the latter preparation, the enzyme electrode exhibited a detection limit of 0.02 mM and a linear range of 0.02 to 0.5 mM. About 2 min was required to attain 90% of the full response. The biosensor was effective for the determination of glucose in food and beverage samples.

G.G. Guilbault and M. Mascini (eds.), Uses of Immobilized Biological Compounds, 211–230.

INTRODUCTION

Ferrocene [*bis* (η^5-cyclopentandienyl) iron] and their derivatives have been studied extensively to develop mediated biosensors because of their electron-exchange properties with various enzymes [1,2]. Among several known ferrocene-related compounds, ferrocene and 1,1'-dimethylferrocene possess the lowest redox potentials [3]. To function as a mediator, ferrocene is anodically oxidized to ferricinium through the loss of one electron from the metal atom. This cation species is water-soluble and able to react with several reduced oxidoreductases as follows:

$$\text{Substrate} + \text{Enzyme}_{ox} \longrightarrow \text{Enzyme}_{red} + \text{Product} \quad (1)$$

$$2\ \text{Ferricinium} + \text{Enzyme}_{red} \longrightarrow 2\ \text{Ferrocene} + \text{Enzyme}_{ox} + 2H^{+} \quad (2)$$

One of the most interesting features of the ferrocene-ferricinium redox couple is their difference in UV-VIS absorption spectra. Ferrocene in 95 % ethanol is yellow and exhibits two wavelength bands at 325 and 440 nm [4]. On the other hand, ferricinium solutions exhibit a characteristic dichroism with an absorption peak at 620 nm, i.e, dilute solutions appear blue or green, while more concentrated solutions are blood red 4 .

Apparently, a water-soluble form of ferrocene for homogeneous reactions is desired and of great interest in bioelectrocatalysis. In addition, ferricinium or its derivatives could be used in spectrophotometric assays involving oxidoreductases if they can be prepared from water-soluble forms of ferrocene-related compounds. Unfortunately, ferrocene and its derivatives (except for carboxyl ferrocene) are virtually water-insoluble [3].

The hydrophobic non-charged ferrocene and its derivatives may form an inclusion (host/guest) complex with water-soluble cyclodextrins; known as cyclic and non-reducing oligosaccharides consisting of glucose units linked by α-1, 4 bonds. Cyclodextrins can form inclusion complexes with various organic chemicals by incorporating such guest molecules into their hydrophobic cavities [5]. Hp-β-CyD possesses a very high solubility (50% w/v in aqueous media) and a cavity size of ca. 7.8 Å in diameter and 7.8 Å in depth. Hp β-CyD should also be able to enclose ferrocene since this mediator exhibits a pentagonal antiprismatic structure in the crystal state with the mean iron-carbonyl distance of 2.04 Å and the carbon-carbon distance of 1.40 Å [4].

In this study, the formation of the water-soluble 1, 1'-dimethylferrocene/hp-β-CyD complex and its electrochemical and bioelectrocatalytic properties were investigated for biosensor applications. The electrochemical oxidation of ferrocene and 1,1'-dimethylferrocene was also investigated to develop a spectrophotometric assay for various important analytes.

EXPERIMENTAL PROCEDURES

Dissolution of Ferrocene and 1,1'-Dimethylferrocene in hp-β-CyD. Solutions of α-cyclodextrin and hp-β-CD were prepared in 20 mM acetate, pH 5.2 buffer to maximum concentrations of 80 mM and 300 mM, respectively. Ferrocene and 1, 1'-dimethylferrocene were solubilized in various concentrations of the two cyclodextrins by constant stirring for 2 h at room temperature. The concentration at maximum ability was determined spectrophotometrically (Beckman DU-7) using the absorption coefficients of 100 $cm^{-1}M^{-1}$ for ferrocene and 1, 1'-dimethylferrocene at 440 nm and 435 nm, respectively.

Preparation of Ferricinium and 1,1'-Dimethylferricinium. Solutions (25 mL) of ferrocene (ca. 80 mM) or 1, 1'-dimethylferrocene (ca. 90 mM) solubilized in hp-β-CyD (250 mM in 100 mM KCl, 100 mM acetate, pH 5.2) were electrochemically oxidized at a platinum foil electrode poised at + 450 mV vs an Ag/AgCl reference electrode in a three-electrode system. The counter and working electrodes were separated by a 2M KCl bridge and the electrochemical oxidation was performed using a polarographic analyzer/stripping voltammeter (Princeton Applied Research, Princeton, NJ). During the course of oxidation, the current was recorded and the absorbencies of ferricinium and 1,1'-dimethylferricinium formed were monitored at 620 nm and 650 nm, respectively. The absorption coefficients for these two cations were determined by adding subsaturating known concentrations of ascorbic acid or uric acid.

Amperometric Electrodes with Hp-β-CyD. Glucose oxidase was covalently immobilized onto a preactivated Immunodyne nylon using a variation of the method employed by Luong and Male for immobilization of xanthine oxidase [6]. All membranes were stored at 4 °C in 0.1 M phosphate buffer pH 6 and under this condition were stable for several months. The Immunodyne enzymic membrane was attached to the surface of the working platinum electrode using an O-ring. Electrodes were alternatively constructed by adding 80 units of concentrated enzyme in buffer (5,000 U/ml) to the electrode's surface and then covering it with a 8, 000 MW cut-of dialysis membrane. An operating potential of 100 mV versus Ag/AgCl was selected for the working electrode.

Solutions in the reaction chamber (containing 5 mM $DMFeCp_2$ in hp-β-CyD) were continuously mixed and sparged with nitrogen to remove dissolved oxygen. After 30 min the background current due to the electrochemical oxidation of 1, 1'- dimethylferrocene was offset to zero. Standard solutions of glucose (1 to 100 mM equilibrated for 4 h at 25 °C to give stable optically active α and β forms of D- glucose) were added to the reaction chamber and the resulting change in current was recorded. The reaction chamber was used together with an EVA-pump and EVA-injector valve (Eppendorf North America, Madison, WI) for continuous operation. An injection of 100 mM glucose was delivered every 20 min and the system was allowed to continue for 10-24 h. The BAS electrodes and cell stand were used in conjunction with an amperometric biosensor detector (Universal Sensors, New Orleans, La). Urine samples collected from several healthy males were added undiluted to the stirred cell as in the case for glucose solutions. Immunodyne and dialysis membranes without enzyme were used to measure the high nonspecific response of the mediated electrode to the urine samples. Solid food samples were prepared as described by Male and Luong [7]. Juice samples were diluted 100-fold and then directly added to the reaction chamber. For comparison, the reference hexokinase method was used for the measurement of glucose in food, juice and urine samples [8].

RESULTS AND DISCUSSION

1.1 Characterization of $FeCp_2$ and $DMFeCp_2$ Complexes with Cyclodextrins. Both $FeCp_2$ and $DMFeCp_2$ were dissolved in methanol or chloroform and exhibited an absorption coefficient of 100 $cm^{-1}M^{-1}$ at 440 nm and 435 nm, respectively. Such a value was then used to estimate the solubility of $FeCp_2$ or $DMFeCp_2$ in cyclodextrin. $FeCp_2$ and $DMFeCp_2$ were only slightly soluble in aqueous solutions (< 0.2 mM), but their solubilities were greatly increased by the formation of inclusion complexes with the water-soluble α-CyD or hp-β-CyD. If a known subsaturating concentration of either $FeCp_2$ or $DMFeCp_2$ was solubilized in hp-β-CyD (167 mM), the absorption coefficient obtained was still 100 $cm^{-1}M^{-1}$. Hp-β-CyD exhibits a higher solubility (50 % w/v) than the α-CyD (12-16 % w/v) and as a result 300 mM and 80 mM stocks could be prepared, respectively. The inclusion of either $FeCp_2$ or $DMFeCp_2$ by α-CyD was not very effective since their addition into α-cyclodextrin solutions reduced the solubility of the complexing agent and caused precipitation. The maximum solubility of $FeCp_2$ and $DMFeCp_2$ was about 3-4 mM before precipitation occurred, regardless of the concentration of the α-CyD tested (26-80 mM). Both β-CyD and γ-CyD were not tested since the former exhibits a very low solubility in aqueous solutions (1.8 % w/v) while the latter is exorbitantly expensive.

$FeCp_2$ and $DMFeCp_2$ were both readily soluble in hp-β-CyD. The increased solubility was likely due to the larger cavity (7.8 Å in diameter and 7.8 Å in depth) of the complexing agent when compared to the α-cyclodextrin cavity (4.7 Å-6.0 Å). As shown in Figure 1 the molar ratio of hp-β-CyD to $FeCp_2$ in $FeCp_2$ saturated solutions was 4.2 at the lowest concentration of hp-β-CyD and the ratio decreased slowly to 3.1 at the highest concentration.

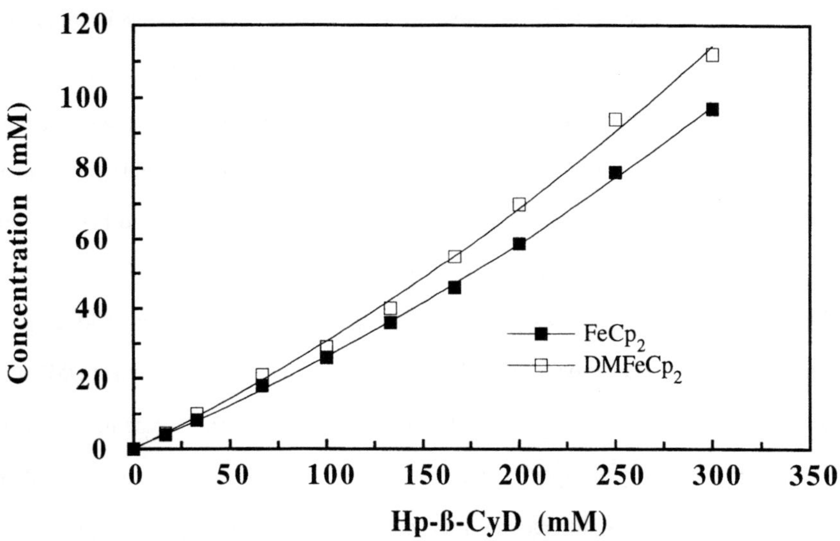

Figure 1. Dependence of the dissolution of ferrocene ($FeCp_2$) and1,1'-dimethyferrocene ($DMFeCp_2$) on the concentration of hp-ß-CyD in 20 mM acetate buffer pH 5.2

Similarly, the molar ratio of hp-β-CyD to $DMFeCp_2$ was 3.7 at the lowest concentration and 2.7 at the highest concentration. Therefore it was possible to prepare solutions of $FeCp_2$ or $DMFeCp_2$ as high as 100 mM.

1.2 Electrochemical Preparation of $FeCp_2^+$ and $DMFeCp_2^+$. During the time course of oxidation, the current dropped and the solution turned from yellow to green and finally to dark blue. The development of the blue color was monitored at the absorption maximum for $FeCp_2^+$ at 620 nm or $DMFeCp_2^+$ at 650 nm (Figure 2 A, B). After about 12-16 h, the absorbance reached a plateau while the current continued to decrease. For the preparation of $FeCp_2^+$, if the reaction was allowed to proceed further, an absorbance decrease was then noted, indicating the decomposition of the oxidized product. The $DMFeCp_2^+$ was more stable and did not show the same phenomenon until another 6-8 h had passed. The $FeCp_2^+$ preparation if terminated in time, still began to degrade after 2-3 days if left in the concentrated form although it was more stable in diluted form (4 months). On the contrary, the concentrated $DMFeCp_2^+$ was stable for at least 4 months when stored at 4°C. It was also noted that the $DMFeCp_2$-hp-β-CyD complex stored at room temperature began to turn green, indicating a preference toward the oxidized state. As also reported, the redox potential of $DMFeCp_2$ is lowest (100 mV) among several known ferrocene-related compounds [3] and such behavior may contribute to the stabilility of $DMFeCp_2^+$ in comparison to $FeCp_2^+$. Both ferriciniums were insensitive to pH variation from 2-11. This characteristic is of advantage since the assay can be performed regardless of some fluctuation in pH and is applicable for a variety of oxidases with different pH optima. Due to the instability of the concentrated $FeCp_2^+$, further studies focused on the electrochemically produced $DMFeCp_2^+$.

1.3 Chemical Oxidation of $FeCp_2^+$ and $DMFeCp_2^+$. Both $FeCp_2$ and $DMFeCp_2$ were also oxidized by ceric sulfate or iodine/KI to $FeCp_2H^+$, known as ferrocenonium and $DMFeCp_2H^+$ (1, 1'-dimethylferrocenonium), respectively [9]. The reaction was almost instantaneous and the spectra of these two cation species were identical to those of ferricinium and 1,1'-dimethylferricinium. However, the resulting products were not stable to changes in pH and often precipitation occurred with time. Such results were not completely unexpected since in strong acid media ferrocene was reported to undergo extensive protonation on the metal atom [4]. The NMR

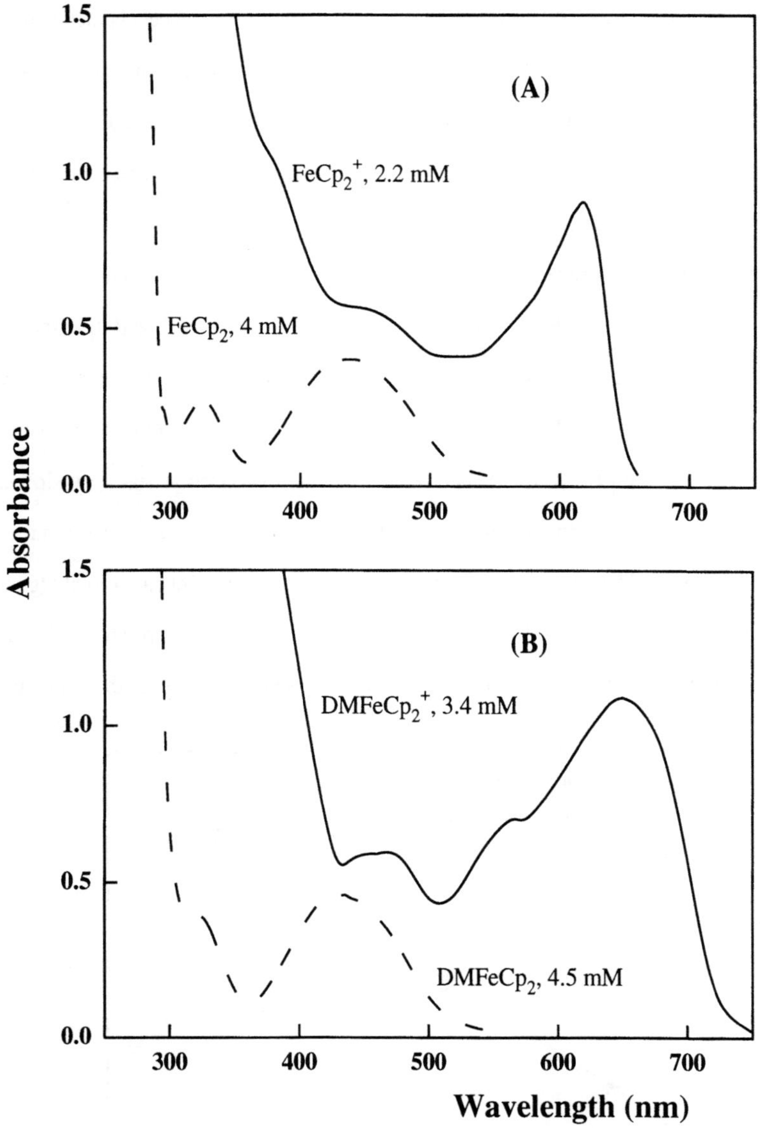

Figure 2. Spectrum of ferricinium dyes in 100 mM acetate, 100 mM KCl buffer, pH 5.2

spectra of the cation in boron trifluoride hydrate exhibited high field resonance characteristics of metal bound protons [4]. As well, in the protonated ferrocene the cyclopentadienyl rings are no longer parallel but are somewhat tilted so that the angle ω, defined by perpendiculars to each ring from the metal atom, lies between 135 and 180°. As a result of this finding, the chemical oxidation pathways were not pursued further.

1.4 Effect of Reducing Agents on the $DMFeCp_2^+$. Ascorbic acid reduced the dye instantly over a wide pH range of 5-9. As shown in Figure 3 the slope of the line (0.65 $OD_{650\,nm}$/ mM) results in an absorption coefficient of 325 $cm^{-1}M^{-1}$ for the $DMFeCp_2^+$. The same value was obtained when sulfite and uric acid were used as reducing agents. Uric acid only reduced the $DMFeCp_2^+$ rapidly at alkaline pHs (> 8). At neutral pH, the rate of reduction was very slow and below pH 6, uric acid exhibited no reducing capability. Similarly, sulfite was able to reduce the dye rapidly at pHs above 8. However, at pH 7 and 5.2, the reduction was much more sluggish (5 and 10 min for a complete reduction). The pH effect on the reduction of the dye should be taken into account if sulfite is used for oxygen removal as discussed later. From the absorption

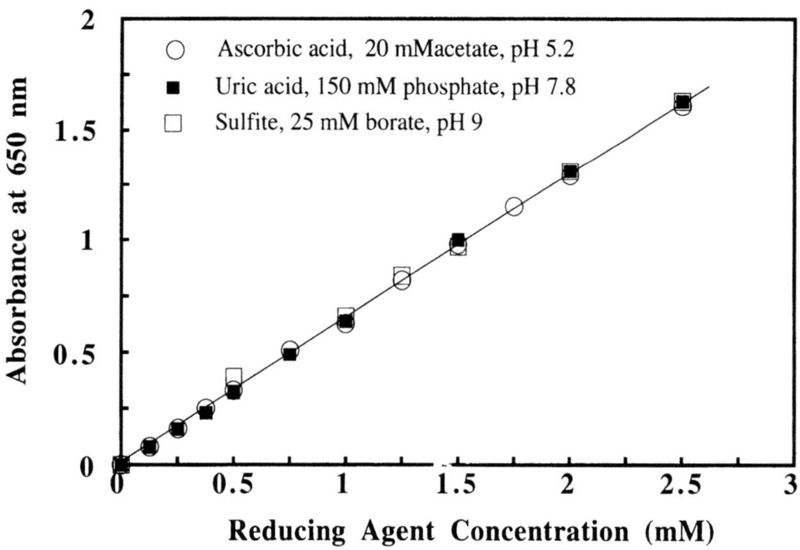

Figure 3. Effect of reducing Agents on $DMFeCp_2^+$

coefficient of 325 $cm^{-1}M^{-1}$, the concentration of the stock $DMFeCp_2^+$ was determined to be 68 mM, corresponding to 76% conversion of the starting $DMFeCp_2$/hp-β-CyD complex. In comparison, the conversion of $FeCp_2$ to $FeCp_2^+$ was 56 % with a final concentration of 45 mM ferricinium. The absorption coefficient obtained for $FeCp_2^+$ was 425 $cm^{-1}M^{-1}$ which agreed favorably with the reported value [10].

1.5 Response of $DMFeCp_2^+$ to Various Metabolites. $DMFeCp_2^+$ had a calibration linearity up to 9 mM (OD_{650} = 3.0) when using plastic cuvettes (1cm Sarstedt). However, the linearity would be less at about 5 mM (OD_{650} = 1.7) with other cuvettes such as quartz or glass. The $DMFeCp_2^+$ dye was investigated with various concentrations of glucose and the reduction was initiated by adding glucose oxidase. The response to glucose was nonlinear rather than a straight line with the anticipated slope of 0.65 OD/mM. Such a result would not be totally unexpected since there was a competition for the reduced enzyme between the $DMFeCp_2^+$ mediator and oxygen. Oxygen (normally 0.2-0.3 mM) in the buffer or glucose containing samples, could be removed by nitrogen sparging or by adding excess sodium sulfite (0.4 mM) for 5-10 min. Since the glucose oxidase reaction was carried out at pH 5.2, when sulfite was added to remove oxygen, 10 min was allowed for a complete reaction between the residual sulfite and the dye (50 μL), before enzyme was added. As expected, the reduction of the dye before enzyme addition was equivalent to 0.1-0.2 mM and the removal of oxygen by either nitrogen sparging or the sulfite method resulted in a straight line for glucose calibration with a slope of 0.67 OD/mM.

Similar to glucose measurement, for the assay of glutamate by glutamate oxidase, the removal of oxygen was necessary to generate a linear relationship. The residual sulfite was expected to react instantly with the dye since the assay was performed at pH 7.8. The slope of the calibration was 0.65 OD/mM. Although in both cases the mediator concentration was in excess compared to oxygen (6.8 mM vs 0.2-0.3 mM) the competition could still arise if the K_m of the oxygen for the reactions is much lower than that of the mediator. Using lactate oxidase and amino acid oxidase, both lactate and phenylalanine monitored with $DMFeCp_2^+$ resulted in slopes of 0.66 and 0.65 OD/mM, respectively. In both cases the removal of oxygen did not alter the results indicating that there was no competition between oxygen and the mediator dye for the reduced enzyme concerned.

For xanthine and hypoxanthine the reaction was first conducted at pH 5.2. There are two oxidation steps from hypoxanthine to uric acid in the presence of xanthine oxidase and only one oxidation step from xanthine to uric acid. It should be recalled that each reduced enzyme will require two ferriciniums (Eq.2).

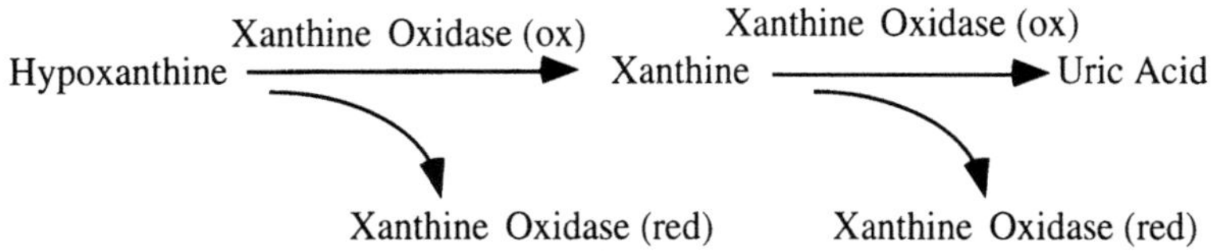

(3)

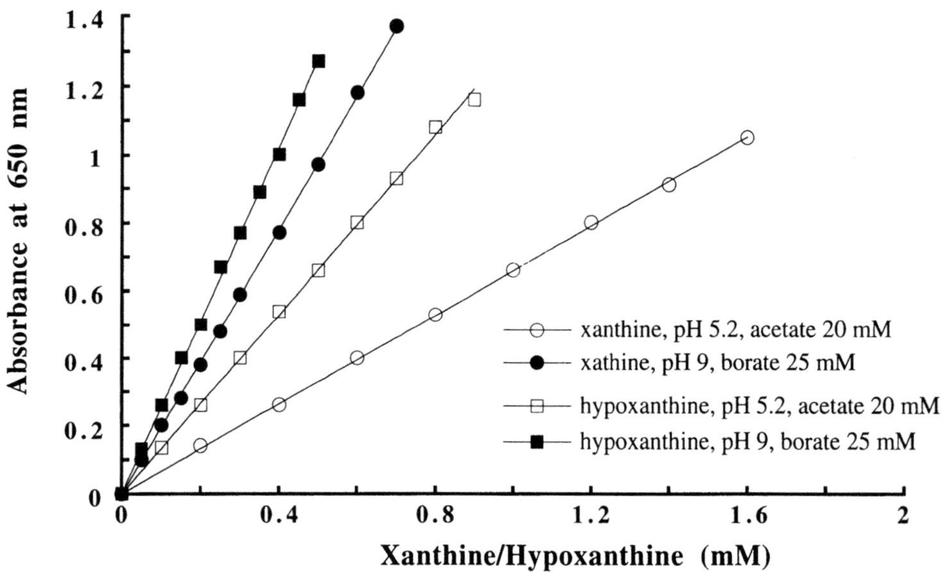

Figure 4. Reduction og $DMFeCp_2^+$ by hypoxanthine/xanthine in the presence of xanthine oxidase at different pH conditions.

As expected, the results indicated two reducing equivalents (slope = 0.66 OD/mM) for xanthine and four reducing equivalents (slope =1.32 OD/mM) for hypoxanthine (Figure 4) since the uric acid produced could not reduce the $DMFeCp_2^+$ at this pH. However, when the pH was raised to 9, the results of Figure 4 indicated six reducing equivalents for xanthine (slope = 1.95 OD/mM) and eight reducing equivalents for hypoxanthine (slope = 2.56 OD/mM). Results indicated that at this pH, the uric acid produced from the xanthine oxidase was further oxidized by the crude enzyme source and reduced the $DMFeCp_2^+$ by four reducing equivalents. If uric acid was allowed first to react completely with the $DMFeCp_2^+$ before adding xanthine oxidase no further reduction of the mediator was noted (only two reducing equivalents). Adding uric acid and the enzyme simultaneously resulted in four reducing equivalents which may imply that the oxidation process for uric acid was somewhat different in the two cases. Allantoin, one of the known possible products of uric acid oxidation, may interact with the dye to produce two reducing equivalents.

Experimental data, however, revealed that at this pH the $DMFeCp_2^+$ dye did not react with allantoin alone or a mixture of enzyme and allantoin. Due to these additional reducing equivalents the sensitivity of the dye system for detecting hypoxanthine was four-fold greater than any of the other metabolites tested.

1.6 Determination of Oxidase Activities. The five enzymes monitored at saturating concentrations of their substrates resulted in linear relationships between ΔOD_{650}/min and number of enzyme units. Oxygen was not removed by sulfite in the assays for glucose and glutamate as no increase in rate was observed. The slope of the lines in ΔOD/min /unit vary due to the differences in the definition of the unit by the manufacturers for each enzyme. In addition, the assay conditions specified by the manufacturer with respect to pH and temperature are not necessarily the same conditions used for the $DMFeCp_2^+$ assay. Also the unit definitions of the manufacturers are using oxygen as the co-substrate of the metabolites rather than $DMFeCp_2^+$. However, from the slopes of these lines a correlation could be made between an unknown source of an oxidase being purified and the conventional definition given for the same commercially available enzyme. The slope values obtained for xanthine oxidase with xanthine and hypoxanthine were close to the predicted ratio of 3 to 4 based on the reducing equivalents (Figure 5). Another alternative would be to define the unit for any particular oxidase activity based simply on the dye assay itself; such that one unit of enzyme activity corresponds to the reduction of 1 μmole of $DMFeCp_2^+$ per min at 25°C.

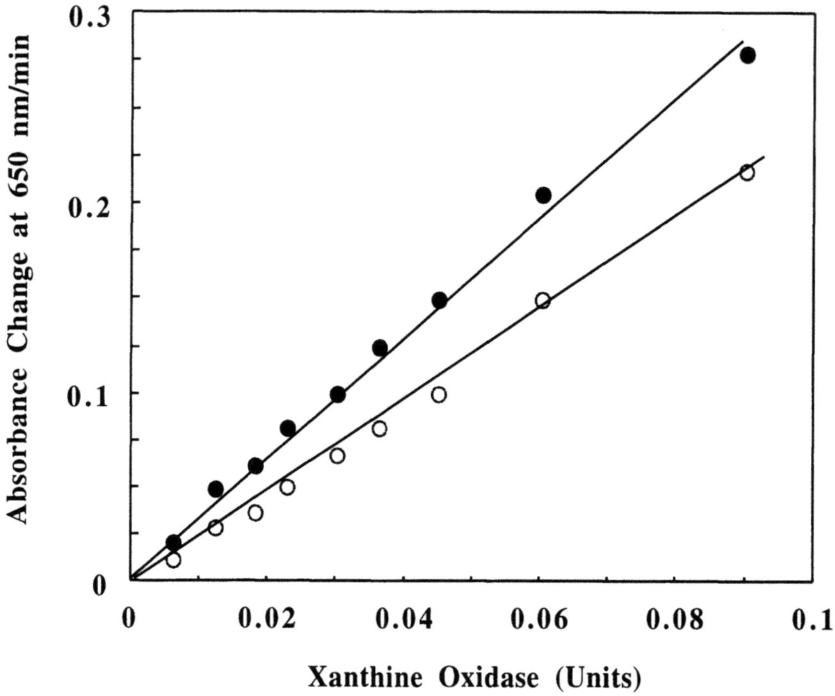

Figure 5. Rate of reduction of $DMFeCp_2^+$ by xanthine oxidase at pH 9.0 (slope is determined at 95 % confidence interval).

(O) xanthine (2.39 OD_{650} nm / unit); slope = 2.39 ± 0.10; R^2 = 0.999
(●) hypoxanthine (3.28 OD_{650} nm / unit); slope = 3.28 ± 0.10; R^2 = 0.999

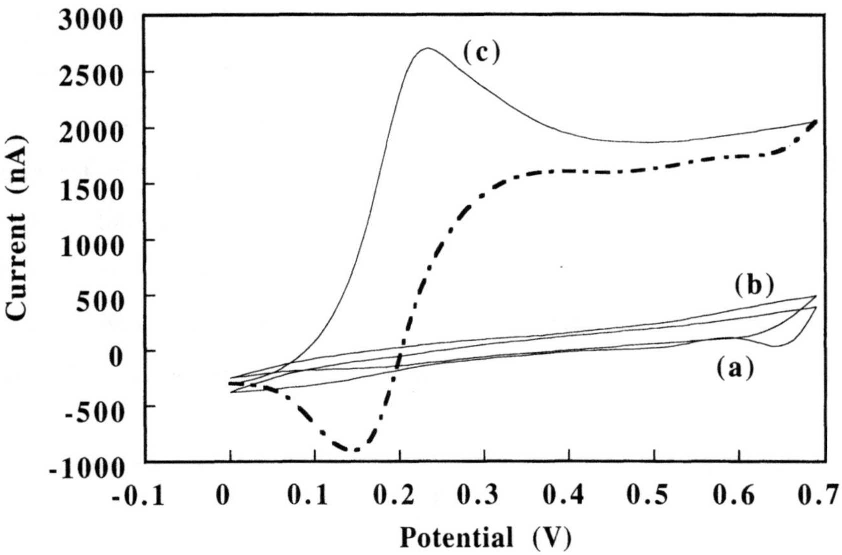

Figure 6. Electrovoltammetry of the $DMFeCp_2$-hp-ß-CyD complex

2.1 Electrovoltammetry of the $DMFeCp_2$-hp-β-CyD Complex. The cyclic voltammogram of either phosphate or 333 mM hp-β-CyD (Figures 6a &b) was essentially featureless over the potential range and operating conditions applied. As expected, the $DMFeCp_2$-hp-β-CyD complex exhibited an almost reversible one-electron transfer voltammogram with a half-wave potential of 170 mV (Figure 6c). For this system, the peak separation (difference between the anodic and cathodic peak potentials) is about 95 mV (250 mV-145 mV) and thus differs from the peak separation for an ideally reversible one electron transfer (59 mM at 298 °K). The oxidation potential of the $DMFeCp_2$-hp-β-CyD complex was not affected by pH ranging from pH 4.7 to 8.8 (figure not shown). Immobilized glucose oxidase was reported to respond maximally to glucose over the pH range 4.5-7 [7]. In view of this, a strict control of pH during the measurement is no longer necessary and one can carry out the mediated reaction regardless of some fluctuations in pH. The solubilized $DMFeCp_2$ in hp-β-CyD also displayed a positive shift in oxidation potential of 70 mV in comparison to the half-wave potential of a $DMFeCp_2$ saturated in 0.1 M phosphate buffer, pH 6. The cause of this phenomenon has yet to be identified definitively. Ferricinium is water-soluble and may not reside in the hydrophobic cavity of hydroxypropyl β-cyclodextrin. If this prevails, an increased oxidation potential would then be necessary to counter energies required for movement of the reduced $DMFeCp_2$ within the hp-β-CyD hydrophobic cavity.

Without glucose oxidase, a typical cyclic voltammogram with a diffusion limited anodic peak current at 250 mV was observed for the solution containing glucose and the $DMFeCp_2$-hp-β-CyD complex (Figure 7a). The change in the voltammogram upon the addition of glucose oxidase can be clearly seen in Figure 7b. The peaks completely disappeared and instead a large catalytic current flowed at this oxidizing potential, indicating a reaction limited by enzyme kinetics. Such behavior

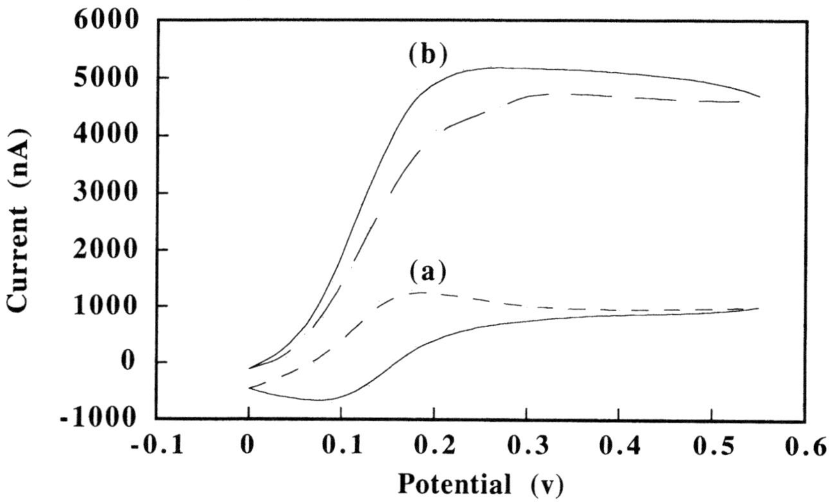

Figure 7. Cyclic voltammogram in the presence and absence of glucose oxidase

was expected from the reduction of 1,1'- dimethylferricinium by the reduced enzyme. Most of the ferricinium will react with the reduced enzyme to regenerate ferrocene as confirmed by voltammograms. However, some of the 1,1'-dimethylferricinium may also escape from the electrode surface and remain sufficiently stable in solution, at least during the course of measurement. This is the principal cause of mediator depletion for most ferrocene mediated electrodes prepared by depositing ferrocene or its derivatives onto the sensing area of the electrode {3]. In addition, some chelations are anticipated between ferricinium and ion species such as oxalate, phosphate, iodine, etc. depending on the nature of the bulk solution [4]. When considering the reduction of mediator by enzyme, since the oxidized form of ferrocene is no longer included by the hp-β-CyD host, one would expect that the kinetic constant for the enzymatic reduction of $DMFeCp_2$ to be identical in the presence or absence of hp-β-CyD.

2.2 Mediated-amperometric electrode for glucose. The electrode response increased with an increase in the amount of glucose oxidase immobilized onto the membrane up to 80 U. In this experiment, the biocatalysis was followed by monitoring hydrogen peroxide at the platinum electrode poised at + 0.7 V vs Ag/AgCl. Cyclic voltammograms obtained using a platinum working electrode with an enzymic membrane were identical with or without glucose (data not shown). This phenomenon can be attributed to the slow diffusion of $DMFeCp_2$-hp-β-CyD and glucose across the membrane relative to the scan rate (5-10 mV/sec) performed.

The effect of $DMFeCp_2$ concentration was examined for electrodes using Immunodyne membranes and for soluble enzyme (80 units) in oxygen free buffer. The response to excess glucose (1.8 mM) increased with $DMFeCp_2$ to form a broad maximum ranging from 5 to 30 mM $DMFeCp_2$. Subsequent preparation of standard curves for glucose at $DMFECP_2$ concentrations of 5 and 20 mM $DMFeCp_2$ revealed that the linear range of the electrodes decreased with $DMFeCp_2$ concentration and the electrodes were therefore operated at 5mM $DMFeCp_2$. It is worth noting that at a higher $DMFeCp_2$ concentration, the bulk liquid became more viscous and increased the response time of the electrode. In view of this, a balance must be compromised to account for response time, linear range and magnitude of steady state response.

The presence of oxygen in the reaction fluid was shown to significantly reduce the current produced by a given amount of glucose (Figure 8). A decreased current in the presence of oxygen was also observed for bare platinum electrodes and for electrodes covered with a blank Immunodyne or dialysis membrane. It is likely that dissolved oxygen affects the equilibrium of reduced to oxidized $DMFeCp_2$ at the electrode surface and also competes with ferricinium at the active site of the enzyme. The use of solubilized $DMFeCp_2$ at a concentration of 50 mM appeared to reduce the effect of oxygen but also decreased the response of the biosensor (figure not shown).

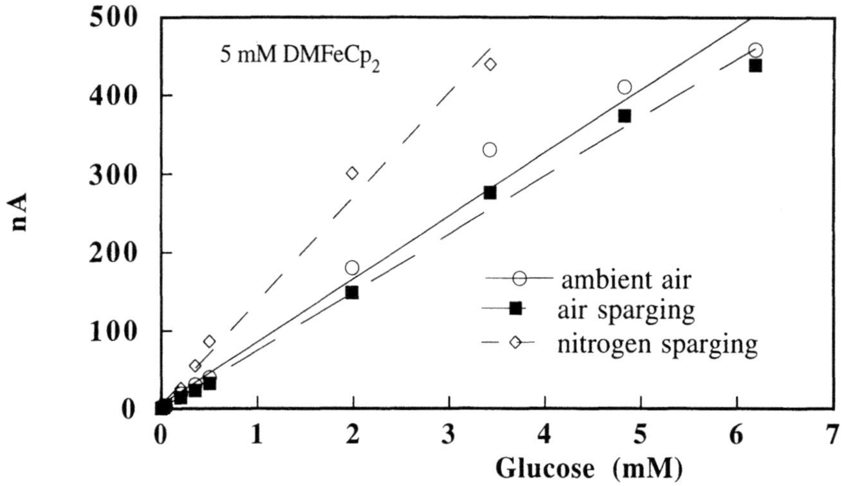

Figure 8. The effect of oxygen on the response of the mediated biosensor

One of the principal reasons for using mediated biosensors is to alleviate the interferences of electroactive compounds such as ascorbic acid, uric acid and acetaminophen. Figure 9 demonstrates the performance of the solubilized $DMFeCp_2$ system relative to an identical system using aeration and the detection of hydrogen peroxide at + 0.7 V vs Ag/AgCl. Using 1 mM glucose the mediated system was shown to produce a current approximately 60 % of that of the conventional sensor, but reduced the interference due to ascorbic acid by half and eliminated the interference due to the other two compounds. Up to 4.5 mM uric acid and up to 0.33 mM ascorbic acid can be normally found in urine samples [11].

The immobilized enzyme membrane (80 U) with the solubilized $DMFeCp_2$ (5 mM) was then used for the measurement of glucose in solution. The detection limit and the discriminative capability of the system were 0.02 mM glucose. About 8 min was required to achieve 90 % of the full response and the linear range was typically 0.02 to 5 mM. The operational stability of the sensor was tested by using the stirred cell with a flow injection analysis (FIA) system for a 24 h continuous operation. After 100 injections (final glucose concentration = 4 mM) no decrease in activity was evident. For the enzyme electrode prepared by entrapping glucose oxidase by a 8, 000 MW cut-off dialysis membrane (80 U enzyme and 5 mM $DMFeCp_2$), the cyclic voltammogram in

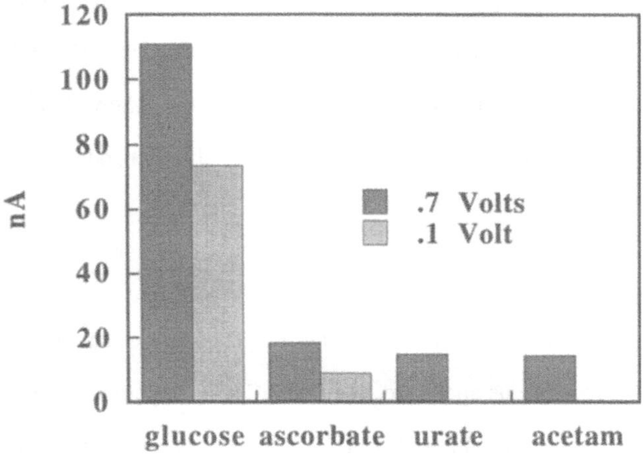

Figure 9. Response of the mediated electrode to interferences relative to conventional amperometric electrode operated with aeration at + 0.7 V vs Ag/AgCl reference.

the presence of glucose depicted only a small diffusion current. Again, this phenomenon can be attributed to the slow diffusion of DMFeCp$_2$-hp-β-CyD and glucose across the membrane relative to the scan rate (5-10 mV/sec) performed. However, the diffusional resistance of the dialysis membrane was lower than that of the enzymic Immunodyne membrane, as demonstrated by a response time of 2 min to achieve 90 % of the full response to an injection resulting in 0.2 mM glucose. The electrode exhibited a detection limit of 0.02 mM and a linear range of 0.02 to 0.5 mM. When tested using a continuous flow system with repeated injections, the electrode lost 25 % of its initial response in the first two hours of operation, and subsequently lost no activity for the remainder of the test (10 h, 57 injections, 4 mM final glucose concentration). The high level of stability of amperometric electrodes using glucose oxidase is a well established phenomenon [12].

The two types of enzyme electrodes were tested for the detection of glucose in food, juice and urine samples. The results were compared with that of the reference hexokinase method. The results demonstrated good agreement for the determination of glucose in food and juice samples, but a nonspecific interference was noted in the case of urine samples (Table 1). Addition of the urine samples to electrodes covered only with a blank membrane followed by subtraction of the blank current from the biosensor current resulted in glucose values generally lower than that of the reference method.

CONCLUSIONS

The formation of water-soluble $DMFeCp_2$-hp-β-CyD complex and its bioelectrocatalytic efficacy opens up homogeneous applications of $DMFeCp_2$ in amperometric biosensor systems. The complexation of $DMFeCp_2$ with hp-β-CyD exhibits a high solubility and this complex has shown to be a chemically stable and efficient mediator for bioelectrocatalysis. In addition, a new dye prepared from the water-soluble $DMFeCp_2$-hp-β-CyD has proven to be very useful in monitoring the activity of various oxidases under saturated substrate condition. Therefore, it may serve as a new methodology for screening, isolation and purification of these enzymes. Alternatively, together with oxidases the new dye could be applicable for determination of several important target analytes. The advantages of this new dye are two-fold: (i) the new dye should be more applicable to various colored samples since the absorbance can be followed at 650 nm and (ii) it requires no other coupling enzymes and reagents.

TABLE 1

Comparison of Samples Measured using Hexokinase and Solubilized 1,1'-Dimethylferrocene Mediated Electrode Methods.

Immunodyne Membrane Electrode

	Hexokinase (mM)	Electrode (mM)	% Difference
Apple juice	145 ± 4.7	155 ± 8.68	4.6
Orange juice	135 ± 2.7	143 ± 7.18	6.4
Peach juice	233 ± 4.7	243 ±17.1	7.2
fresh corn	4.31 ± 0.086	4.15 ± 0.186	3.7
green beans	5.78 ± 0.116	5.64±0.248	2.4
snow peas	29.59 ± 0.592	30.63 ± 1.41	1.4
Urine			
#1	0.340 ± 0.007	0.518 ± 0.026	52.5
#2	0.326 ± 0.006	0.498 ± 0.025	52.7
#3	0.313 ± 0.006	0.788 ± 0.039	152
#4	1.29± 0.024	1.39 ± 0.063	14
#5	0.193 ± 0.004	0.335 ± 0.017	73.6

Dialysis Membrane Electrode

	Hexokinase (mM)	Electrode (mM)	% Difference
Apple juice	107.4 ± 2.05	109.8± 2.4	2.3
Orange juice	136.5 ± 1.06	139.7± 0.14	2.3
Peach juice	250.1 ± 0.3	250.8 ±8.07	0.3
fresh corn	3.22 ± 0.06	3.18 ± 0.10	1.2
green beans	8.14 ± 0.109	8.04±.044	1.2
snow peas	24.77 ± 0.677	24.63 ± 0.402	0.57
Urine			
#1	0.388 ± 0.004	2.08 ± 0.339	436
#2	0.077 ± 0.006	6.795 ± 0.078	8725

REFERENCES

1. Cass, A.E.G., Francis, D. G., Hill, H.A.O., Aston, W.J., Higgins, I.J.; Plotkin, E.V., Scott, L.D.L., and Turner, A.P.F. (1984). *Anal. Chem.* **56**, 667-671.

2. Cass, A. E. G., Davis, G., Green, M. J., and Hill, H. A. O. (1985). *J. Electroanal. Chem.* **190**, 117-127.

3. Hill, H.A.O., and Sanghera, S. (1990). Mediated amperometric enzyme electrodes. In *Biosensors, a Practical Approach* , ed. A.E.G. Cass, IRL Press, Oxford, pp. 19-46.

4. Roseblum, M. (1965). Chemistry of the Iron Group metallocenes: Ferrocene, Ruthencene, Osmocene: Part One, pp.29-61, John Wiley & Sons, New York.

5. Thoma, J.A., and Stewart, L. (1965). Cycloamyloses. In *Starch: Chemistry and Technology, Volume I, Fundamental Aspects*, eds. R.L. Whistler, E.F. Paschall, J.N. Bemiller, and H.J. Roberts, Academic Press, New York, pp. 209-249.

6. Luong, J.H.T., and Male, K.B. (1992). Development of a new biosensor system for the determination of the hypoxanthine ratio, an indicator of fish freshness. *Enzyme Microb. Technol.* 1992 **14**, 125-30.

7. Male, K.B., and Luong, J.H.T. (1993). Improvement of the selectivity of a FIA Amperometric biosensor system for glucose. *Biosensors Bioelectronics* (In Press).

8. Lowry, O. H., and Passonneau, J.V. (1972). *A Flexible System of Enzymatic Analysis*. Academic Press, New York, pp.174-77.

9. Cramer, F., Saenger, W., and Spatz, H.-ch. (1967). *J. Am. Chem. Soc.* **89**, 14-20.

10. Holecek, J., Handlir, K., Klikorka, J., and Dinh-Bang, N. (1979) *Collection Czechoslov. Chem. Commun.* **44**, 1379-1387.

11. Tietz, N.W., and Logan, N.M. (1987). Reference ranges. In *Fundamentals of Clinical Chemistry* , ed. N.W. Tietz, 3 rd edition, W. B. Saunders Co., Philadelphia, p.944.

12. Wilson, R., and Turner, A.P.F. (1992). *Biosensors Bioelectronics*, **7**, 165-85.

POLY(AMPHIPHILIC PYRROLE)-ENZYME ELECTRODE : A NEW APPROACH FOR BIOSENSOR CONSTRUCTION

S. COSNIER AND P. LABBE

Laboratoire d'Electrochimie Organique et de Photochimie Rédox, (URA CNRS 1210), Université Joseph Fourier Grenoble 1, BP 53X, 38041 Grenoble-Cédex, France.

ABSTRACT. A novel immobilization method allows construction of biosensors exhibiting exceptional performance both in water and in organic solvents. Microelectrodes and complex biostructures are also feasible.

INTRODUCTON

The entrapment of enzymes in polymer films such as polypyrrole by electrochemical polymerization of a solution containing monomer and enzyme provides a simple method of enzyme immobilization. However, this procedure suffers from severe limitations. High amounts of monomer but also of enzyme must be used. Moreover, the quantity and the real activity of the enzyme embedded in the polymer cannot be easily determined without enzyme denaturation.

Here we report a new approach for biosensor construction based on the unique capacities of the following amphiphilic pyrroles for retention and adsorption.

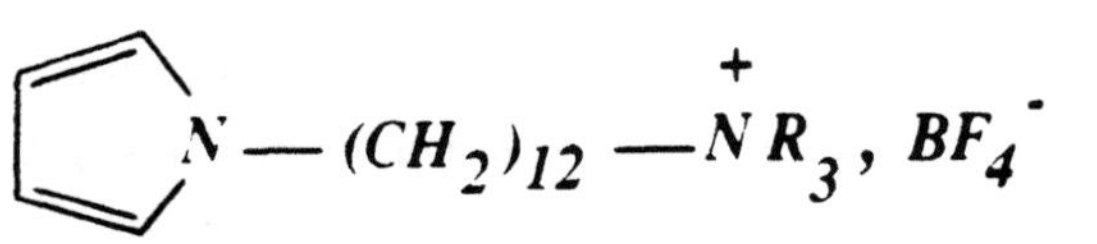

$R = C_2H_5 \longrightarrow A_2$
$R = C_3H_7 \longrightarrow A_3$
$R = C_4H_9 \longrightarrow A_4$
$R = C_6H_{13} \longrightarrow A_6$

This method consists of the adsorption on the electrode surface of a pyrrole amphiphilic monomer-enzyme mixture followed by the electropolymerization of the adsorbed layer (Scheme 1).

Following this procedure several inexpensive, mechanically robust, sensitive and fast responsive bioelectrodes have been obtained [1-3].

The characteristics (amount of immobilized enzyme, response time and sensitivity) of this type of biosensor have been examinated by entrapment of a polyphenol oxidase (tyrosinase) as enzyme model in the polypyrrolic matrix. The effect of the enzyme loading on the biosensor response as well as its kinetic parameters have been also studied, indicating that the enzymatic reaction is the rate limiting step in this sytem, owing to the good permeability of the polymeric films.

G.G. Guilbault and M. Mascini (eds.), Uses of Immobilized Biological Compounds, 231–244.

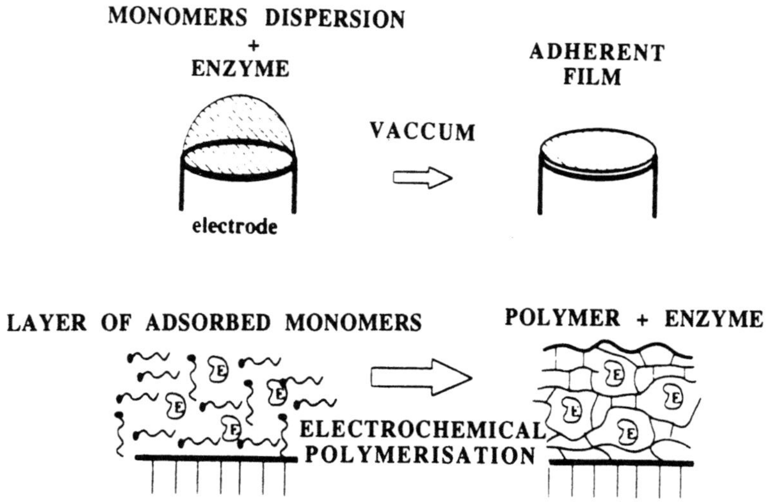

Scheme 1 : Principe of biosensor construction

The properties of different polypyrrolic matrixes in terms of enzyme retention, enzymatic and electrochemical activities were compared through the immobilization of several enzymes exhibiting different molecular weights (40 kDa to 279 kDa) and isoelectric points (4.2 to 12).

In addition, the potentialities of this technology for the elaboration of composite layer structures (multi-enzyme entrapment or incorporation of hydrophilic clay particles) and for the transfer of an enzymatic activity in organic solvents have been investigated.

RESULTS AND DISCUSSION

A) Characteristics of the biosensor

*** *Composition of the biomaterial***

As reported previously [4] for the pyrrole monomer A_2, these amphiphilic pyrroles are only weakly soluble in pure water but they can be dispersed into a stable emulsion. The adsorption of the resulting monomer aggregats on the electrode surface leads to an electroactive layer which gives rise upon oxidative electropolymerization in aqueous electrolyte to a stable poly(pyrrole-ammonium) film.

The oxidative electropolymerization of these monomers can be accomplished either by repetitive scanning over the - 0.2 to 0.8 V vs standard calomel electrode (SCE) range (Figure 1), or by controlled-potential oxidation at 0.76 V. The modified electrodes exhibited in aqueous electrolyte the stable electrochemical response of N-substituted polypyrroles, with a quasi-reversible redox peak system around 0.5 V.

This method of biosensor construction has been employed for the immobilization of a polyphenol oxidase (tyrosinase) as enzyme model to illustrate the enzymatic and electrochemical characteristics of this kind of biosensor.

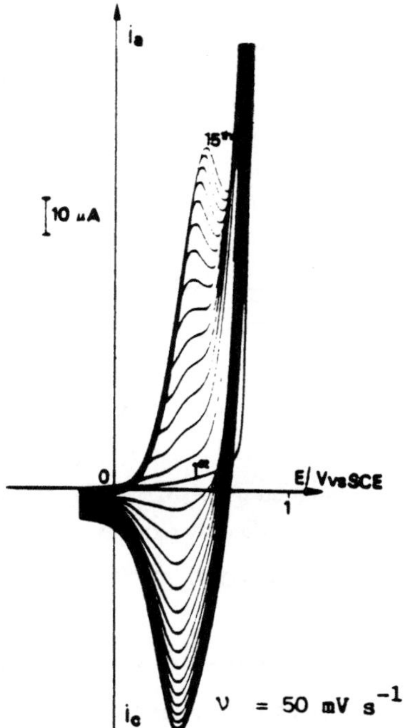

Figure 1 : Electropolymerization of A_2 (5.1×10^{-8} mol adsorbed on a 5 mm diameter Pt electrode) by successive potential scanning in 0.1 M $LiClO_4$ aqueous solution ; $v = 50$ mV s^{-1}.

Since tyrosinase has an isoelectric point of 4.7 [5], it carries negative charges at the pH 7 of the dispersion of A_2 in water and has attractive electrostatic interactions with A_2. Moreover, hydrophobic interactions presumably occur between the long hydrocarbon tail of A_2 and hydrophobic regions of the protein. Owing to these interactions, an enzyme-monomer A_2 association can be adsorbed onto the electrode surface and electropolymerized by controlled potential electrolysis. This novel immobilized-enzyme technology procedure is monomer and enzyme-saving.

The voltammogram of the resulting modified electrode prepared with 0.3 mg of tyrosinase and 0.04 mg of A_2 depicts in the anodic area the reversible oxidation of the polypyrrole matrix (Figure 2). The apparent surface coverage Γ_1 can be estimated from the charge recorded under the polypyrrole oxidation wave, assuming that one in three pyrrole units is oxidized [6]. Since this biosensor construction is based on electropolymerization of an adsorbed enzyme-monomer A_2 mixture, the leakage of tyrosinase during the polymerization step can be determined from the enzymatic activity of the electrolyte solution. Thus, the real quantity of tyrosinase trapped in the poly-A_2 film was evaluated at 70-75 % of the initially adsorbed amount. Moreover, as the activity of the tyrosinase electrode has been estimated by spectrophotometric measurements we can determine the specific activity of the immobilized enzyme which has been evaluated at 9 % of the expected value corresponding to the free enzyme.

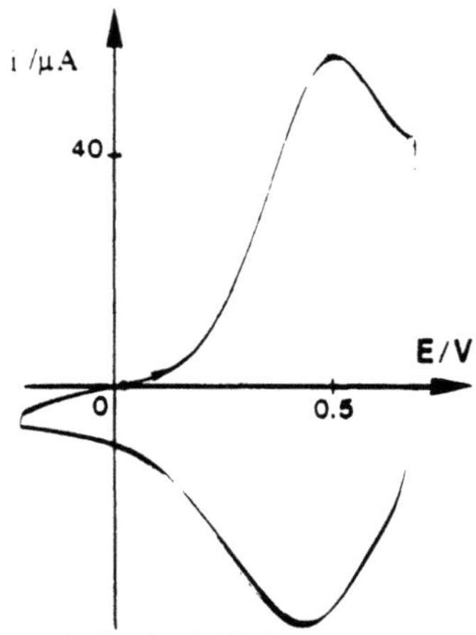

Figure 2 : Cyclic voltammogram in 0.1 M $LiClO_4$ aqueous solution (pH = 7) for glassy carbon disk electrode (diam. 5 mm) modified by oxidative electro-polymerization of an adsorbed A_2 (90 nmol) - tyrosinase (2 nmol, 0.3 mg) layer) ; v = 0.1 V s^{-1}.

*** *Current response of the poly-A_2 tyrosinase electrode***

Since tyrosinase catalyses the oxidation of several monophenols and *o*-diphenols to *o*-quinones, the analytical capabilities of this biosensor have been investigated in "air-saturated" 0.1 M phosphate buffer by applying a potential of - 0.2 V vs SCE to detect amperometrically the biocatalytically generated quinoid products. Specific substrates of tyrosinase were added to this stirred solution and the current corresponding to the reduction of the *o*-quinones was recorded (Scheme 2).

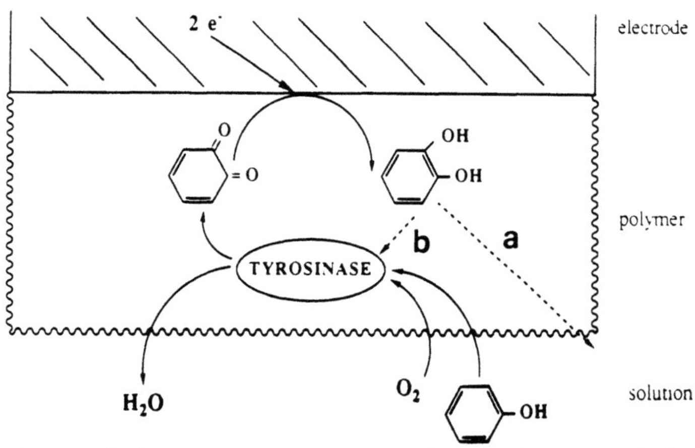

Scheme 2 : Description of the electroenzymatic cycle for the detection of phenol as substrate ; (a) diffusion to the solution ; (b) enzymatic reoxidation.

Figure 3 presents the steady-state current response of the modified electrode upon successive additions of identical amounts of catechol, illustrating the reproducibility and the fast response time of the electrode (3 s), owing to the proximity at the molecular level between the catalytic sites and the electrode surface. The calibration curve obtained for catechol with the poly A_2-tyrosinase electrode shows a linear response over the concentration range 5.0 x 10^{-9} M - 5.0 x 10^{-5} M with a detection limit of 2.0 x 10^{-9} M (based on a signal-to-noise ratio of 3). This detection limit is, to our knowledge, the lowest concentration value that can be detected using sensors based on tyrosinase.

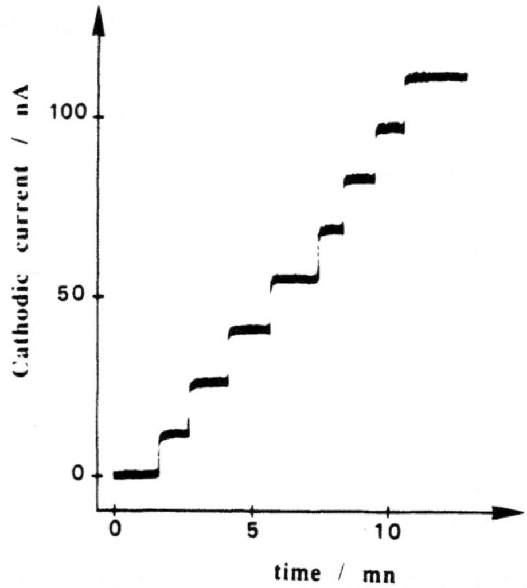

Figure 3 : Steady-state current-time responses of the C/poly A_2-tyrosinase electrode (diam. 5 mm) with increasing concentration of catechol in 50 nM steps. Applied potential, - O.2 V ; air saturated 0.1 M phosphate buffer (pH 6.5) kept under vigorous stirring.

Investigation of the response of the poly A_2-tyrosinase electrode to various phenols and catecholamine neurotransmitters yields linear calibration curves (Figure 4).

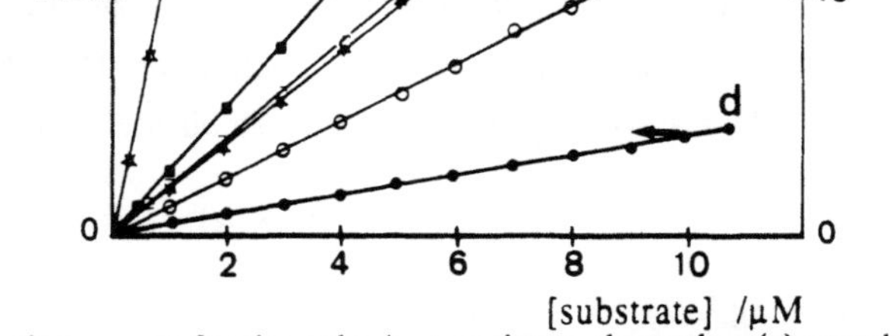

Figure 4: Calibration curves for the poly A_2-tyrosinase electrode ; (a) catechol, (b) phenol, (c) p-cresol, (d) dopamine, (e) L DOPA, (f) L tyrosine ; other experimental conditions as in Figure 3.

The results are shown in table 1. The sensitivity of the enzyme electrode varies over the range 1-300 mA M^{-1}, depending on the substrate and illustrating the selectivity of the biosensor afforded by enzyme catalysis. Furthermore, this biosensor provides very low limits of detection with very fast response times whatever the tested substrate. In particular with dopamine whose determination is of biological importance, the detection limit is 5×10^{-8} M. This value is markedly lower than those usually reported ($\geq 10^{-6}$ M) for tyrosinase-based sensors pre-viously described in the literature [7, 8].

Table 1 : Bioanalytical performance of the poly A_2-tyrosinase electrode towards various substrates.

Substrate	Sensitivity (mA M^{-1})	Detection limit (M)
catechol	300	2×10^{-9}
phenol	70	5×10^{-9}
p-cresol	50	5×10^{-9}
dopamine	11	5×10^{-8}
noradrenaline	7	5×10^{-7}
L DOPA	2	8×10^{-7}
L-tyrosine	1	8×10^{-7}

*** *Kinetic parameters***

Since the rate of the enzymatic reaction may either be mass transport or catalysis controlled, the amperometric response of the enzyme electrode was recorded in hydrodynamic conditions to circumvent any mass transport (or diffusion) limitations. Further evidence supporting this hypothesis was gathered by rotating-disk electrode (RDE) experiments. The effect of the electrode rotation rate on the current response at several concentrations of catechol was measured and plotted as a Levich plot. At rotation rates, ω, above 200 rpm (up to 2000 rpm) the steady- state current is independent of ω while below 100-200 rpm, the response decreases on decreasing the rotation rate.

Thus for forced-convective systems, the absence of a rotation rate dependence in this enzyme electrode indicates that the kinetic process limiting the current is not substrate mass transport. It could be therefore concluded that the enzymatic reaction is the rate limiting step. Thus, the apparent kinetic parameters of the immobilized tyrosinase may be estimated from a Michaelis-Menten analysis of the subtrate calibration curve.

*** *Effect of enzyme loading***

One of the advantages of this procedure of biosensor construction is the possibility to control the composition of the enzyme-polymer layer. Therefore, the effect of varying the enzyme loading on the electrode performance has been evaluated. Figure 5 shows the calibration curves for catechol obtained at poly A_2-tyrosinase electrodes prepared with different amounts of enzyme, the amount of monomer A_2 remaining constant. The sensitivity in the linear region of these calibration graphs increases quasi-linearly with increasing enzyme loading. Furthermore, the K_M^{app} values are independent of the enzyme loading whereas i_{max} increases with increasing enzyme concentration. Thus, the biosensor response appears to be limited by the amount of enzyme in the polymer. The enzyme turnover capacity and not internal diffusional limitations seems to control the electrode kinetics. Although the enzyme molecules are entrapped within the poly A_2 film, the surrounding matrix seems not to behave as a diffusion barrier for substrate.

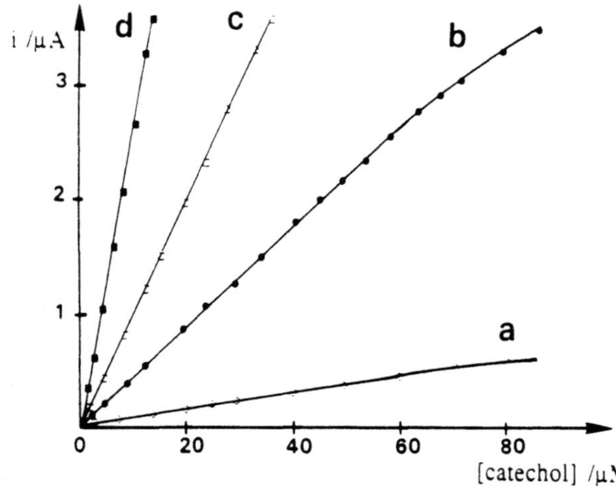

Figure 5 : Calibration curves for catechol obtained at poly A_2-tyrosinase electrode ($\Gamma_1 = 3.9 \times 10^{-8}$ mol cm^{-2}) prepared with 0.05 (a), 0.1 (b), 0.2 (c) and 0.3 (d) mg tyrosinase. Other detection conditions as in Figure 3.

*** *Capacities of this immobilization method for biosensor construction***

The properties of different polypyrrolic matrixes based on A_2, A_3, A_4 and A_6, were compared in terms of enzyme retention, enzymatic and electrochemical activities through the immobilization of the following enzymes : horseradish peroxidase (HRP), galactose oxidase (GAO), tyrosinase (PPO), glucose oxidase (GOD) and xanthine oxidase (XOD).

After the electrochemical polymerization of the adsorbed enzyme-monomer mixtures, the amounts of enzymes immobilized in the different resulting matrixes have been determined (Table 2).

All matrixes present the same behaviour exhibiting good retention properties for enzymes of low molecular weights and a strong decrease of the amount of immobilized enzyme for enzymes above 130 kDa.

Table 2 : Enzyme retention* by the electropolymerization of different amphiphilic pyrroles.

Enzyme (molecular weight) \ Matrix	A_2	A_3	A_4	A_6
HRP (40 000)	60	66	66	68
GAO (68 000)	87	92	90	80
PPO (128 000)	70	55	45	35
GOD (160 000)	20	10	8	20
XOD (275 000)	10	10	5	5

* calculated in percentage of the initially adsorbed amount of enzyme

The efficient retention of GAO, a highly positively charged enzyme (isoelectric point of 12) and of PPO which is negatively charged (pI 4.7) at pH 7 used for the adsorption of enzyme-monomer mixture, suggests that this procedure of enzyme immobilization is generally applicable to enzymes whatever their isoelectric point. On the other hand, the comparison of the electrochemical activities of the resulting bioelectrodes indicates that the poly A_2 film is the best host matrix (Table 3).

Table 3 : Electrochemical characteristics of the different biosensors containing GAO and GOD.

Enzyme	Matrix	A_2	A_3	A_4	A_6
GAO	Sensitivity ($mA\ M^{-1}\ cm^{-2}$)	14.2	4.7	5.0	2.8
	I max ($mA\ cm^{-2}$)	30.5	10.5	13.0	11.0
GOD	Sensitivity ($mA\ M^{-1}\ cm^{-2}$)	21.5	6.5	6.5	6.5
	I max ($mA\ cm^{-2}$)	100.0	33.5	35.1	40.5

Moreover the determination of the enzymatic activities of the different biosensors confirms this trend, the bioelectrodes based on poly A_2 film exhibiting the highest values.

This may be attributed to a permeability effect, assuming that matrixes formed by electro-polymerization of A_3, A_4 and A_6 are more hydrophobic than those made with A_2 and therefore are less permeable than the poly A_2 films.

B) Composite layer structures

One of the advantages of this procedure of biosensor construction is the possibility to control the composition of the enzyme-polymer layer.

*** *Heterogeneous enzyme location***

The effect of an additional polymer coating on the electrode performance has been evaluated in the case of bioelectrodes containing GAO.

The poly A_2-GAO electrode prepared as previously described presents relative rates of oxidation of several galactosides which are similar to those published for free galactose oxidase (Table 4). A poly A_2/poly A_2-GAO electrode is realized by the additional adsorption and electropolymerization of A_2 on a poly A_2-GAO film. The substrate specificity of such "multilayer" biosensor differs markedly from the substrate specificity of the preceding "monolayer" biosensor (Table 4). The presence of an overlying poly A_2 film provides an additional resistance for diffusion of galactosides to the active enzyme layer, the overall kinetics being controlled by this diffusion process contrarily to the poly A_2-GAO electrode.

Table 4 : Substrate specificity of free GAO and of GAO based biosensors.

Galactoside (molecular weight) / Enzyme system	Raffinose (504)	Galactose (180)	Lactose (342)	Maltose (342)
free GAO	100	83-70	8-4	4-1
poly A_2-GAO electrode "monolayer"	100	75	7	0.5
poly A_2/poly A_2-GAO electrode "multilayer"	100	150	134	20

Thus, the biosensor selectivity results from the substrate specificity of the enzyme and from the steric hindrances afforded by the permeability of the outer poly A_2 layer.

*** *Multi-enzyme entrapment***

The sequential adsorption and electropolymerization of different enzyme-monomer mixtures allows the elaboration of multi-enzymatic materials with ordered enzyme-location.

Following this procedure, bienzymatic sensors for glucose determination in presence of interfering agents have been elaborated :

poly A_2-GOD/poly A_2-AO electrode I

poly A_2-GOD/poly A_2-PPO electrode II

The immobilization of ascorbate oxidase (AO) or Tyrosinase (PPO) in the inner layer and glucose oxidase (GOD) in the outer layer was used to prevent interferences due to the oxidation of ascorbate or acetaminophen on a platinum electrode at + 500 mV vs SCE.

The enzymes AO and PPO induce a protection against these interfering effects by an enzymatic oxidation of ascorbate and acetaminophen.

$$\text{4-acetamidophenol} + O_2 \xrightarrow{\text{PPO}} \text{4-acetamido-1,2 benzoquinone} + H_2O$$
$$\text{ascorbate} + O_2 \xrightarrow{\text{AO}} \text{dehydroascorbate} + H_2O$$

The magnitude of the bias introduced by the presence of 5 μM ascorbate or 5 μM acetaminophen on the measurement of a 0.2 mM glucose concentration by a poly A_2-GOD electrode represents 75 % or 40 % respectively of the biosensor response. For example, a 200 μM concentration of ascorbate produces a current which would be generated by a glucose concentration of 8 mM. Figure 6 shows how ascorbate interferes with the sensor measurement of glucose. As

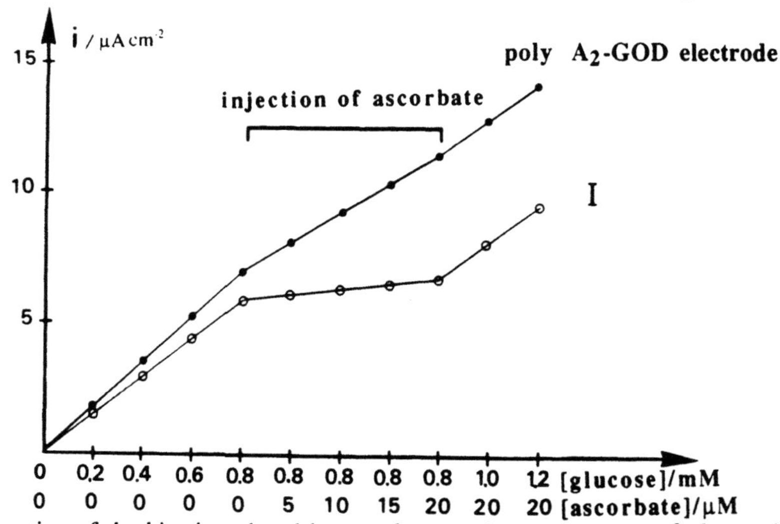

Figure 6 : Estimation of the bias introduced by ascorbate on the measurement of glucose by biosensors containing the same amount of GOD.

demonstrated by the calibration curve obtained with the sensor I, the interfering effects can be drastically reduced by appropriate adjustement of the amount and location of ascorbate oxidase in the biocatalyst layer without decrease of the biosensor sensitivity for glucose. A similar phenomenon was observed with sensor II which minimizes the bias introduced by acetaminophen. The elaboration of a trienzymatic GOD-PPO-AO sensor is now under way.

C) Inorganic and organo-inorganic biomatrixes

Some preliminary results concerning the use of organo-inorganic composite films as a way to immobilize enzymes on an electrode surface are now presented : poly A_2 films can be doped by hydrophilic colloidal inorganic particles such as clay. This inorganic lamellar material can be delaminated into colloidal particles when suspended in water [9]. Under ultrasonic agitation, mixed dispersion of laponite and amphiphilic monomer A_2 were prepared. This dispersion has been used to immobilize tyrosinase following the same procedure as described previously. Figure 7 demonstrates that the sensitivity of the phenol sensor is threefold higher when the poly A_2 film has been doped by laponite clay particles. Since the amount of effectively entrapped enzyme within the film is the same (as determined by measuring the activity of the polymerisation electrolyte), it can be deduced that the greater sensitivity of the biosensor results from a higher activity of the tyrosinase in the laponite-doped film. This electrochemical observation is confirmed by measuring the specific activity of the immobilized tyrosinase which is nearly threefold higher in the presence of laponite particles. This result suggests that the presence of hydrophilic laponite modifies the microenvironment of the enzyme which allows a better flexibility of the protein and then a better activity.

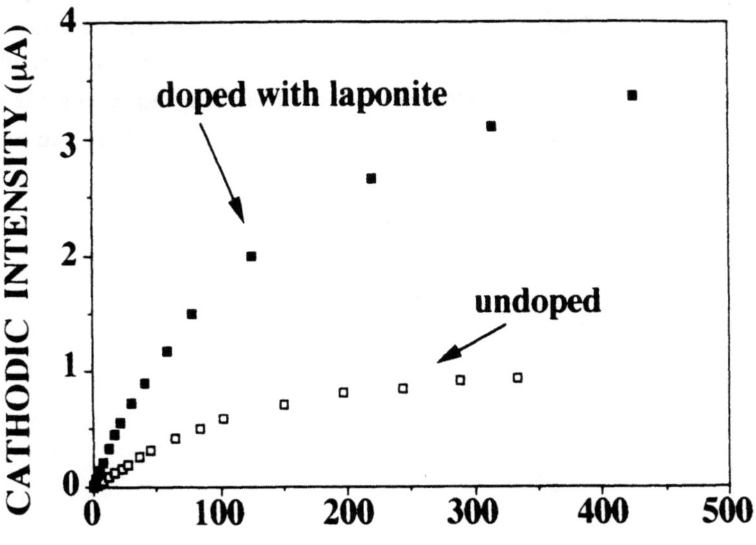

Figure 7 : Comparison of the phenol calibration curves obtained with a poly A_2-tyrosinase electrode doped or undoped with laponite clay particles.

Recently elaboration of electrodes modified by inorganic particles such as clay [10], zeolite [11], hydrotalcite [12] or aluminium oxide [13] has received a growing interest. This approach appears very promising because spatial contraints and surface chemical effects induced by these materials could lead to new patterns of reactivity and selectivity. In particular, clay modified electrodes can be elaborated under very mild conditions since they are easily formed by air-drying an aqueous colloidal clay dispersion onto an electrode surface [9]. Thus enzymic layers can be fixed at an electrode surface in a simple way by air-drying or spin-coating an aqueous colloidal mixture of enzyme and clay. This new method has been successfully applied for the detection of glucose and various phenolic compounds with a glucose oxydase and a polyphenol oxydase clay modified electrode [14]. Work is in progress in the characterization and the uses of this new kind of biomaterial.

D) Organic phase enzyme electrodes

For many years, enzymes have been used in organic solvents for synthesis. Recently, this ability of enzymes to function in nonaqueous media has led to the development of organic-phase biosensors [15-16]. Indeed, such enzyme electrodes offer opportunities for the detection of unusual (hydrophobic) substrates of enzymes or for bioelectrocatalysis in organic media.

Since the different poly A_2-A_6 matrixes are stable in a number of nonaqueous solvents, the entrapment of enzymes in such polymers should produce sensors able to operate in organic media.

This possibility has been investigated with tyrosinase as an enzyme model. The resulting biosensors have been tested towards the amperometric detection of phenolic and *o*-diphenolic compounds in anhydrous chloroform. As demonstrated by comparing the calibration curves obtained with these sensors for catechol (Figure 8), only the poly A_2-PPO electrode exhibits an electroenzymatic activity. Therefore, it seems that only the amphiphilic monomer A_2 might be able to retain simultaneously the enzyme and the water molecules which are necessary to its activation.

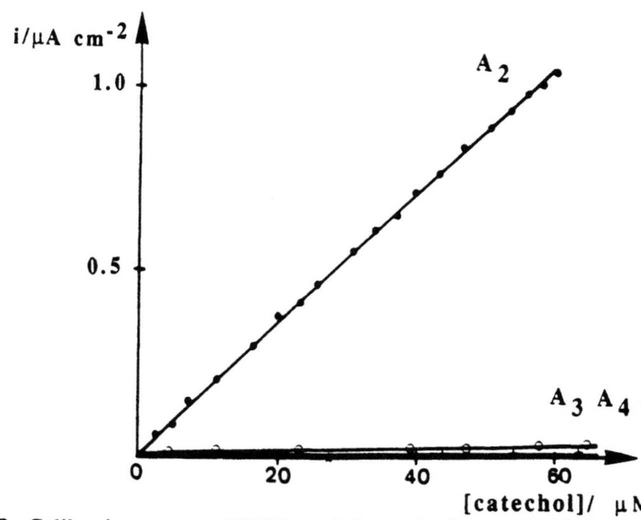

Figure 8 : Calibration curves of PPO based electrodes for catechol in anhydrous $CHCl_3$.

On the other hand, the saturation of chloroform solution with 0.1 % of aqueous phosphate buffer (0.1 M, pH 6.5) induces a marked increase of the sensitivity of the poly A_2-PPO electrode (20 → 250 mAM^{-1} cm^{-2}) and the strong emergence of electroenzymatic responses for the biosensors based on the other amphiphilic pyrroles (Figure 9). These biosensors display higher sensitivities than the poly A_2-PPO electrode's one although they contain less enzyme than the A_2 based biosensor (Table 2).

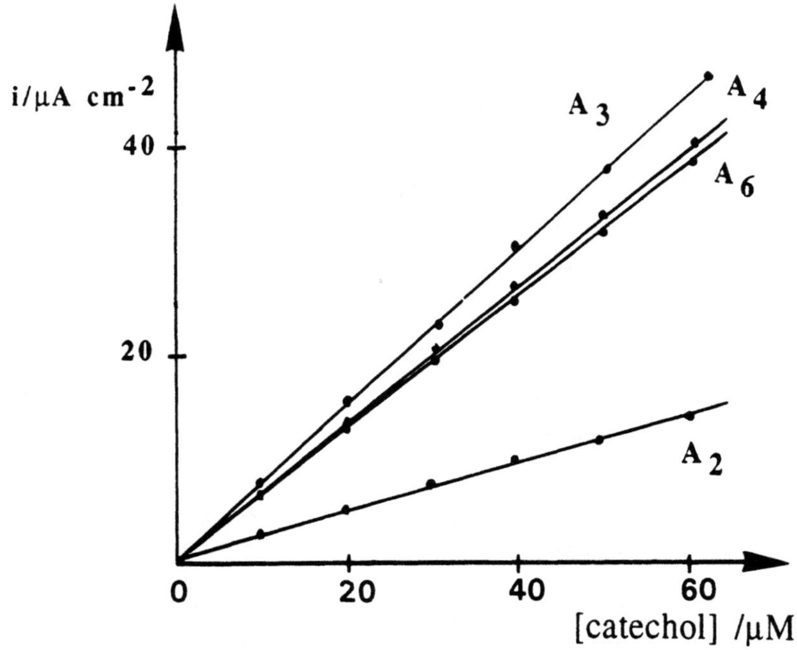

Figure 9 : Calibration curves of PPO based electrodes for catechol in $CHCl_3$ saturated with 0.1 % of phosphate buffer (0.1 M, pH 6.5)

This phenomenon could be due to an increase of the permeability of the host matrixes from poly A_2 to poly A_6 which should facilitate the substrates and products diffusion and also a better hydratation of the surroundings of this enzyme. This permeability increase results from the increase of the monomer volume which induces the formation of expanded polymeric structures. In addition, the hydrophobic character of the monomer increases from A_2 to A_6 involving a better swelling of poly A_6 than poly A_2 in hydrophobic solvents.

CONCLUSION

In conclusion, this novel enzyme immobilization allows the elaboration of biosensors exhibiting exceptional performances in water but also in organic solvents. Moreover, this procedure is a powerful tool for the functionalization of microelectrodes and the build-up of complexes biostructures.

Aknowledgments : Part of this work is supported by the Commission of the European Communities under the BRITE-EURAM Programme by the way of the Concerted European Action : Pillared Layered Structures (CEA-PLS).

REFERENCES

[1] S. Cosnier and C. Innocent, J. *Electroanal. Chem.*, **328**, 361 (1992).

[2] L. Coche-Guérente, S. Cosnier, C. Innocent, P. Mailley, J.-C. Moutet, R. Morelis, B. Leca and P. Coulet, *Electroanalysis*, in press.

[3] S. Cosnier and C. Innocent, *Bioelectrochem. Bioenerg.*, in press.

[4] L. Coche-Guérente, A. Deronzier, B. Galland, P. Labbé, J.-C. Moutet and G. Reverdy, *J. Chem. Soc., Chem. Commun.*, 386 (1991).

[5] D. Kertesz and R. Zito, *Biochim. Biophys. Acta*, **96**, 447 (1965).

[6] G.K. Chandler and D. Pletcher, *Spec. Period. Rep. Electrochem.*, **10**, 117 (1985).

[7] J. Wang and M.S. Lin, *Anal. Chem.*, **60**, 1545 (1988).

[8] A. Navaratne, M.S. Lin and G.A. Rechnitz, *Anal. Chim. Acta*, **237**, 107 (1990).

[9] a) B. Brahimi, P. Labbé and G. Reverdy, *Langmuir*, **8**, 1908 (1992).
b) B. Brahimi, P. Labbé and G. Reverdy, *J. Electroanal. Chem.*, **267**, 343 (1989).

[10] P.K. Ghosh and A.J. Bard, *J. Am. Chem. Soc.*, **105**, 5591 (1983).

[11] C.G. Murray, R.J. Nowak, D.R. Rolison, *J. Electroanal. Chem.*, **164**, 205 (1984).

[12] K. Itaya, H. -C. Chang AND I. Uchida, *Inorg. Chem.*, **26**, 624 (1987).

[13] C. A. Goss, C. J. Miller and M. Majda, *J. Phys. Chem.*, **92**, 1937 (1988).

[14] J.-L. Besombes, G. Reverdy, P. Labbé, C. Mousty (Oral communication)
CEA-PLS Topical Meeting on sensors, Swansea 16-17 april 1993.
Concerted European Action on Pillared Layered Structures (CEA-PLS).
Brite Euram Program.

[15] G.-F. Hall and A.P. Turner, *Anal. Lett.*, **24**, 1375 (1991).

[16] J. Wang and A.J. Reviejo, *Anal. Chem.*, **65**, 845 (1993) and references therein.

EFFECT OF ORGANIC SOLVENTS ON THE BEHAVIOUR OF A GLUCOSE OXIDASE-BASED BIOSENSOR

EMMANUEL I IWUOHA AND MALCOLM R SMYTH
School of Chemical Sciences Dublin City University,
Dublin 9, Ireland.

Abstract Organic phase application of a glucose oxidase (GOx) electrode has shown that the catalytic efficiency of the biosensor in acetonitrile, and its steady-state response in 2-butanol, are comparable to those obtained in phosphate buffer. The behaviour of the sensor in acetonitrile is characterized by enhanced catalytic and steady state currents. In 2-butanol, the sensor exhibits a decrease in catalytic efficiency and an increase in maximum steady state current (I_{max}). These enhancements have been attributed mainly to solvent-induced high diffusion coefficients of the substrate and the mediator in the immobilized enzyme layer. The diffusion coefficients are reflected in improved response of the sensor only when the solvent system contains some amount of water. These solvent-related factors culminate in an increase in the heterogeneous rate constant of the enzyme electrode, which translates into faster responses to glucose.

I INTRODUCTION

The dynamics of enzyme biosensors are affected by the diffusion and partition coefficients of the substrate and other analytes in the enzyme layer, which are normally determined by the solvent characteritics and immobilisation technique. Some immobilisation methods, such as cross-linking with glutaraldehyde, simply result in the adsorption of the enzyme on the electrode surface, without any major alteration in the enzyme structure. Other methods involve both physical and chemical perturbation of the protein. In the latter category of modified enzyme electrodes, the enzyme active site is exposed to ensure fast electron-transfer rates at low over-potentials. Electron transfer mediators are often employed in solution or incorporated

G.G. Guilbault and M. Mascini (eds.), Uses of Immobilized Biological Compounds, 245–254.

in the biosensor to relay electrons between the enzyme active site and the electrode surface.

So much emphasis has been laid on immobilisation strategies for biosensor construction, that virtually no interest has been shown in exploring the potentialities of improving sensor response by the use of organic solvents. Yet organic solvents have the advantage of not only rendering the substrate and mediator more soluble, but due to viscosity effects, may increase the mobility of reacting species across and within the enzyme layer. In addition, interferences from side reactions can be effectively controlled by the use of appropriate solvents. The use of organic solvents in improving sensor performance becomes more important when the enzyme involved has its active site embedded deep in the protein matrix. GOx is an example of this class of enzyme. The redox-active flavin adenine dinucleotide of GOx is shielded by a carbohydrate shell[1]. Instead of reducing the tunneling distance required for electron transfer by modifying the protein, organic solvents can be used in combination with simple immobilisation methods.

In this paper, GOx was immobilized on a glassy carbon electrode by cross-linking with glutaraldehyde. The biosensor amperometric responses to glucose were investigated in acetonitrile and 2-butanol, and in solvent-water (90 + 10 % v/v water) mixture. Details of the kinetic studies of the sensor in these solvent systems have been reported in our other communications[2,3].

II BIOSENSOR CATALYTIC EFFICIENCY IN ORGANIC SOLVENTS

Where an amperometric sensor requires anodic detection, the ratio of the anodic peak current in the presence to that in absence of enzymatic activity (I_K/I_D) obtained from the cyclic voltamograms (CVs) of the sensor, can be used as a measure of its catalytic efficiency[4]. Analyses of the CVs in Fig. 1, show that the half-wave potential when glucose is present in the solution is 0.27 V, which corresponds to the standard oxidation potential of FMCA.

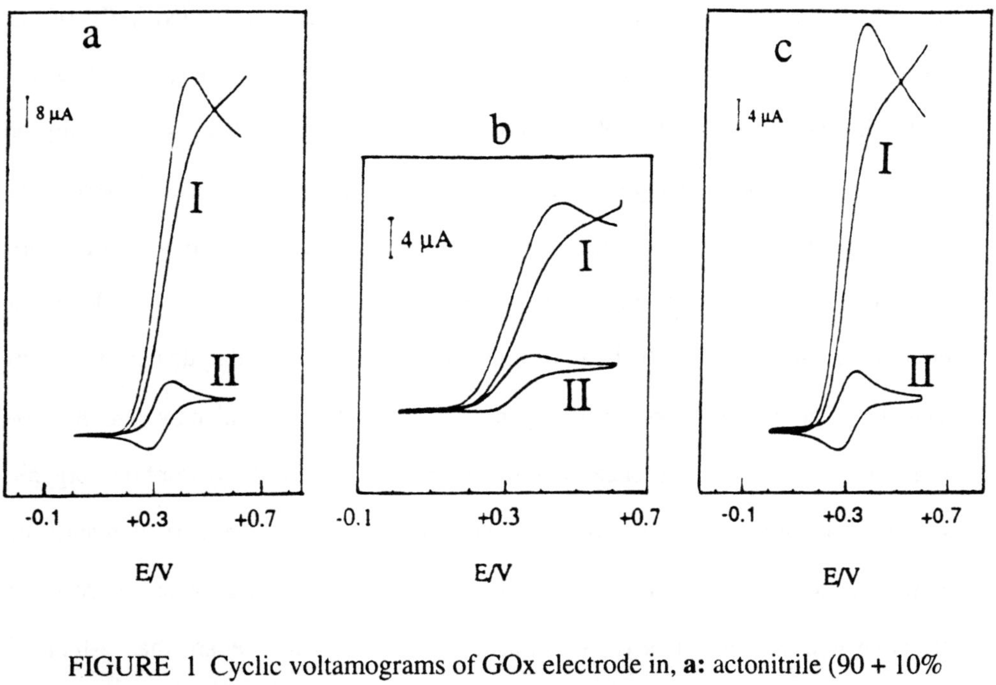

FIGURE 1 Cyclic voltamograms of GOx electrode in, **a:** actonitrile (90 + 10% v/v water); **b:** 2-butanol (90 + 10% v/v water); **c:** 0.1 M phosphate buffer, pH 7.0; with 2.5 mM FMCA, 0.1 M KCl, in the presence, **I;** and absence, **II;** of 20 mM glucose. $E_i = 0.00$ V; $E_\lambda = 0.60$ V; at 20 mV/s.

This means that the current observed in the presence of glucose is catalytic current[5] I_K. The ratio of the anodic peak current to the square root of the scan rate, for all the solvents, was found to be independent of scan rate. Therefore, the peak current is diffusion controlled current, I_D. The I_K/I_D value of the sensor in acetonitrile (7.0) is in good agreement with that in phosphate buffer (6.9). However, 2-butanol decreased the catalytic efficiency of the sensor ($I_K/I_D = 4.8$).

Compared with the I_K of the sensor in 0.1 M phosphate buffer, ($I_K = 52\ \mu A$), the I_K increased by a factor of 2 in acetonitrile and decreased by more than the same factor in 2-butanol (see Table 1). The results show that glucose responds faster in acetonitrile than in 2-butanol or phosphate buffer. The decreased I_K and I_K/I_D of the sensor in butanol can be explained on the basis of the stripping of the active site water by the hydrophilic solvent (dielectric constant, $\varepsilon = 15.8$ at 25 °C) . It is worthy of note that in completely anhydrous 2-butanol, the biosensor is stripped of most of its active site water within 1 hour, and the immobilized enzyme layer hardens up, bringing about drastic reduction in the CV peak currents. For acetonitrile, which is a more polar solvent ($\varepsilon = 37.5$ at 25 °C) than butanol, decreased catalytic current would have been predicted. However it has been pointed out by Klibanov and his co-workers[6], that in the presence of correct amount of water, hydrophilic organic solvents increase the rate of enzyme reaction by causing the enzyme to release its active site water, as substrate binds. Since this phenomenon is expected to be more effective with solvent polarity, acetonitrile acts better in this regard than 2-butanol (compare CVs in Figs. 1 and 2 for acetonitrile and 2-butanol).

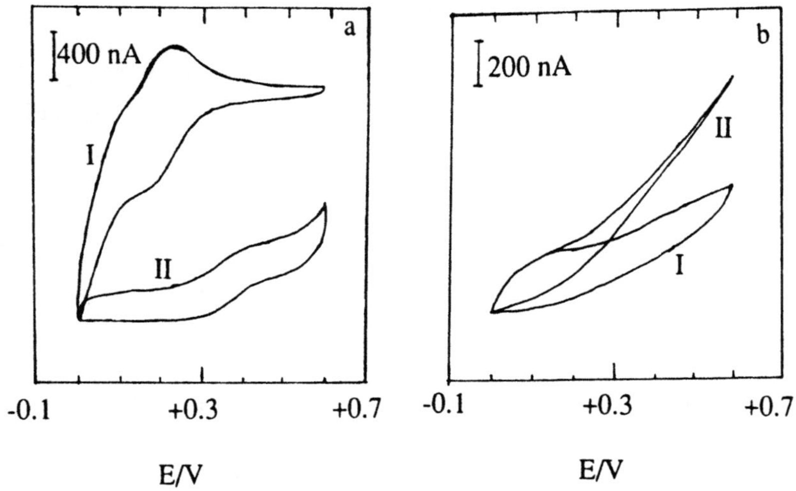

FIGURE 2 Cyclic voltamograms of GOx electrode in, **a:** 100% acetonitrile; and **b:** 100% 2-butanol ; in the presence, **I;** and absence, **II;** of 20 mM glucose. Added electrolyte 0.05 M tetraethylammonium *p*-toluenesulphonate. Other conditions are as in Fig. 1.

III EFFECT OF SOLVENT CHARACTERISTICS ON BIOSENSOR RESPONSE

Diffusion coefficients of glucose, D_G, and the electron transfer mediator (FMCA), D_F, were evaluated from the slopes of the Koutecky-Levich plots shown in Fig. 3. These values are higher in the two organic phases than in the aqueous phosphate buffer (see Table 1).

TABLE 1

Kinetic parameters of GOx electrode **A:** acetonitrile; **B:** 2-butanol and **C:** 0.1 M phosphate buffer pH 7.0. Other conditions as in Figs. 1 and 3.

Parameters	A	B	C
$I_K/\mu A$	96.0	20.4	55
I_K/I_D	7.0	4.8	7.1
$10^9\ D_G/cm^2\ s^{-1}$	117.8	0.63	0.48
$10^9\ D_F/cm^2\ s^{-1}$	465.0	2.48	1.90
$10^5\ k'_{MF}/cm\ s^{-1}$	645	3.86	2.01

In these polar organic solvents, the effect of active site water desorption is opposed by the increased diffusion of solutes into the enzyme layer. The diffusion factor outweighs the water stripping effect, with the result that in 2-butanol the biosensor does not just have the same type of steady-state response as is observed in phosphate buffer, but in fact has a higher I_{max} (1.54 μA and 1.0 μA for 2-butanol and phosphate buffer, respectively). On the other hand, in acetonitrile, the reacting species exhibit diffusion coefficients which are by two orders of magnitude higher than the corresponding values in phosphate buffer. This culminates in a six-fold increase in steady-state currents compared with the responses in phosphate buffer (compare values in Fig. 4).

IV RATE COEFFICIENT OF THE GOx ELECTRODE IN ORGANIC SOLVENTS

The heterogeneous rate constants, k'_{ME}, of the glucose sensor reactions in the various solvent media, which were calculated from the intercept of the Koutecky-Levich plots, show that the electrode reactions are faster in the organic phases than in the aqueous phase (k'_{ME} have values 6.45 x 10^{-3} cm s^{-1}, 3.86 x 10^{-5} cm s^{-1}, and 2.01 x 10^{-5} cm s^{-1}; for acetonitrile, 2-butanol, and phosphate buffer, respectively).

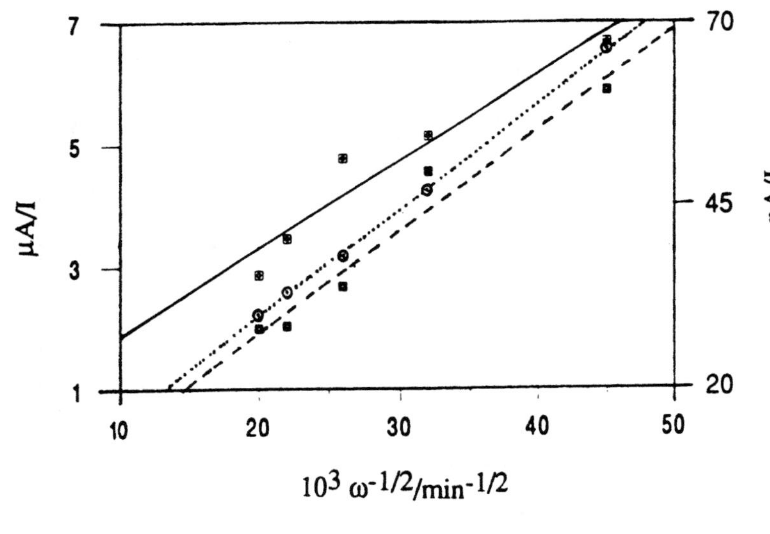

FIGURE 3 Koutecky-Levich plots for GOx electrode. —⊠: 0.1 M phosphate buffer pH 7.0; and ---■: 2-butanol (left ordinate); ···⊙: acetonitrile (right ordinate). 10 mM glucose; 4 mM FMCA; 0.1 M KCl and 0.3 V.

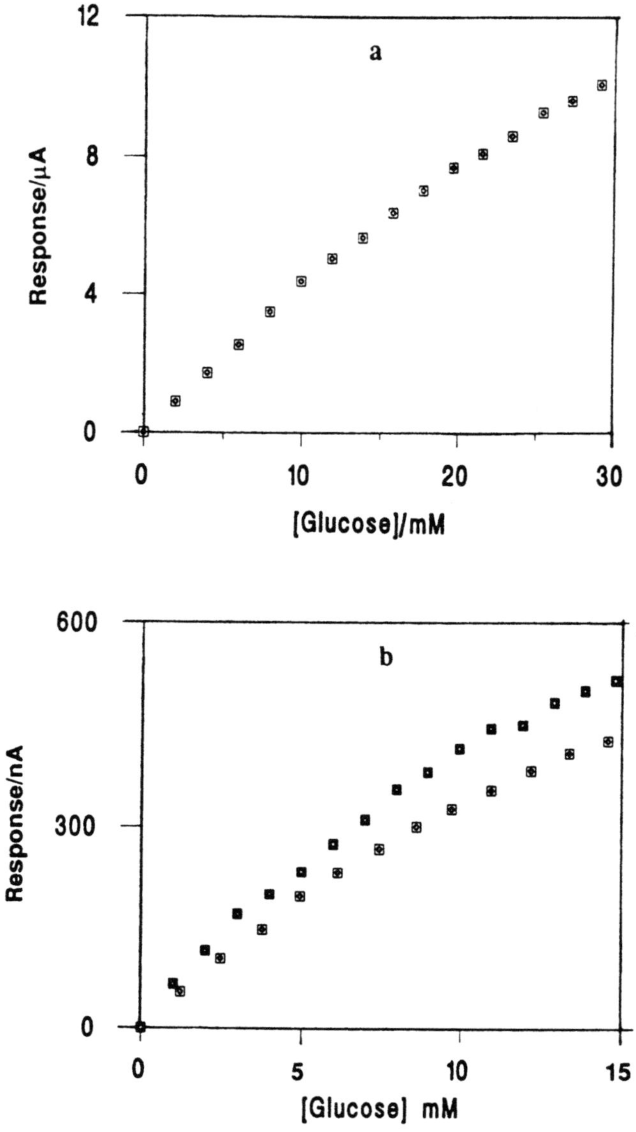

FIGURE 4 Comparison of biosensor responses in organic and aqueous phases. **a:** acetonitrile; **b:** 0.1 M phosphate buffer pH 7.0 ■; and 2-butanol ▣. E = 0.3 V; 1000 rpm; 0.1 M KCl; 4 mM FMCA.

Again we see that the rate of GOx catalyzed oxidation of glucose in acetonitrile is by more than 2 orders of magnitude faster than in aqueous buffer. This is attributed to greater diffusion of the solutes in acetonitrile, brought about by low kinematic viscosity,ν, (ν_{CH_3CN} = 0.0046 cm^2 s^{-1} at 20 °C) which allows greater mobility of both glucose and FMCA in the enzyme layer. The corresponding value for 2-butanol ($\nu_{2\text{-}C_4H_8OH}$ = 0.0522 at 20 °C) is 10 times the viscosity of acetonitrile and 5 times that of aqueous phase, which is 0.01 at 20 °C. This suggests a relationship between the mobility of reacting species and the sensor response. In fact the combination of low viscosity and high dielectric constant makes acetonitrile an efficient organic phase for charge transfer processes: a condition which is favourable for the electrochemical reactions taking place at the enzyme electrodes that produce charged species.

V PROSPECTS

The acetonitrile induced responses of the glucose sensor are better than those reported[6,9] for the modification of the enzyme electrode by special techniques to perturb the carbohydrate shield of the enzyme redox centre. This suggests that simple immobilisation strategies, which do not have a drastic effect on the form and functions of the enzyme, can be combined with proper choice of solvent to construct new generation organic phase enzyme biosensors. As most enzymes are not soluble in organic solvents, their retention on the electrode surface is ensured. This promises wider application of biosensors to analyte matrices that were, hitherto, inaccessible.

REFERENCES

1. Y. Degani and A. Heller; J. Am. Chem. Soc., 110, 2615, 1988.
2. E. Iwuoha and M. Smyth, J. Roy. Soc. Chem., (submitted, 1993).
3. E. Iwuoha and M. Smyth, Analyst, (submitted, 1993)
4. N. Foulds and C. Lowe, Anal. Chem., 60, 2473, 1992.
5. S. Sakura and R. Buck; Bioelectrochem. Bioenerg., 28, 387, 1992
6. A. Klibanov, Acc. Chem. Res., 23, 114, 1990.
7. J. Sakurai, A. Margolin, A. Russell and A. Klibanov, J. Am. Chem. Soc., 110, 7236, 1988.
8. P. Bartlett, D.Dawson and J. Farrington, J. Chem. Soc. Faraday Trans., 88, 2685, 1992.
9. J. Dics, M. Cardosi, A. Turner and I. Karube, Electroanalysis, 5, 1, 1993.

ORGANIC-PHASE BIOSENSORS

JOSEPH WANG
Department of Chemistry and Biochemistry
New Mexico State University, Las Cruces, NM 88003, U.S.A.

Abstract The realization that enzymes can function in nonaqueous media has led to the development of organic-phase enzyme electrodes. Such biosensing probes greatly expand the possibilities of biosensors toward previously inaccessible samples or analytes, and offer many other improvements and advantages. Such unique analytical opportunities, as well as fundamental aspects of organic-phase biosensors are discussed.

1. INTRODUCTION

Enzymatic conversions and accordingly biocatalytic sensing, have been traditionally performed in aqueous environments. However, recent research - primarily in Klibanov's laboratory [1] - has resulted in the remarkable finding that many enzymes can function in organic solvents, containing little or no added water. While the main interest in nonaqueous enzymology is for preparative/synthetic organic work, the field of biosensors can greatly benefit from this important finding. In particular, organic-phase enzyme electrodes can be applied to assays of new challenging environments or towards additional (hydrophobic) substrates, can offer extended operational stability, simplified immobilization schemes, reduced side reactions and no microbial inactivation. These immense sensing opportunities were first realized by Prof. Turner from Cranfield Institute of Technology [2, 3] Since then, several useful organic-enzyme electrodes have been developed (mainly in Cranfield and Las Cruces) utilizing enzymes such as horseradish peroxidase, tyrosinase, cholesterol oxidase or alcohol dehydrogenase (Table 1). The number of papers on this topic has surged over the past three years. The present chapter examines the behavior of organic-phase enzyme electrodes and discuss the analytical opportunities resulted from the development of these novel devices.

G.G. Guilbault and M. Mascini (eds.), Uses of Immobilized Biological Compounds, 255–262.

Table 1. Organic-phase enzyme electrodes.

Biocomponent	Analyte	Transducer	Reference
Tyrosinase	Phenolic compounds	Graphite foil	2
Horseradish peroxidase	Peroxide species	Carbon electrode	4,5
Ta. Brockii alcohol dehydrogenase	Secondary alcohols	Glassy carbon electrode	6
Cholesterol oxidase	Cholesterol	Oxygen electrode	7
Laccase	Catechols	Glassy carbon electrode	8
Chymotrypsin	Amino acids	pH electrode	9
Plant tissues	Phenol, peroxide	Graphite electrode	10

II BIOCATALYSIS IN NONAQUEOUS MEDIA

In order to take full advantage of the unique sensing opportunities accrued from organic-phase enzymology, it is essential to understand the behavior of biocatalyst in nonaqueous environments. In particular, knowledge of the solvent effect on the enzyme activity and stability, and of the role of water, are crucial for optimizing the enzymatic reaction, and hence the sensor performance.

Although our understanding of the solvent effect on the activity and stability of enzymes is growing rapidly, it is not yet at a stage where it can be used predictively to select a solvent for a given sensing application. Many factors related to the solvent effect on the biocatalytic activity are not yet well understood. In general (but not always), the enzymatic activity is correlated to the polarity of the solvent, as expressed by its log P value [11, 12]. (P is the partition coefficient of the solvent in the two-phase octanol/water system.) Hydrophobic solvents (with log P values higher than 4) are preferred because they do not strip the essential water layer surrounding the enzyme. Other factors, such as the thermostability and solubility of the substrate and product (and their partition to and from the aqueous environment of the enzyme) should also play a role in the solvent selection. Hydrophobic solvents also offer increased structural rigidity and hence thermostability. For example, porcine pancreatic lipase has been shown to tolerate heating at 100°C, and to exhibit high catalytic activity at this temperature [13]. This behavior is attributed to the structural rigidity of proteins in organic media. Such minimization of the enzyme unfolding (as compared to the conformational flexibility in aqueous media) may greatly extend the lifetime of biosensors. The improved thermostability may be useful for in-situ monitoring of many industrial processes occurring at high temperatures.

Related to the media is the amount of water. While some water is needed for the biocatalytic activity this essential water can be localized around the enzyme molecules. The essential water content differs from enzyme-to-enzyme and from solvent-to-solvent. For example, while chymotrypsin needs only 50 molecules of water per enzyme molecule for activity in octane, the activity of tyrosinase in chloroform requires about $3.5x10^7$ molecules of water per enzyme molecule [14]. The thermostability, however, decreases upon increasing the water content.

Solvent-induced kinetic changes in the biocatalytic activity may offer advantages for biosensing work. For example, extended linearity may result from the change in the apparent Michaelis-Menten constant in organic media, enhanced sensitivity may be offered by changes in the reaction velocity, while substrate specificity can be regulated by the solvent.

As for the pH, while organic solvents have no distinct measurable pH, the enzyme "remembers" the pH of the last aqueous solution to which it has been exposed. Hence, it is often desired to employ enzymes that have been lyophilized from an aqueous solution of optimal pH for the enzymatic conversion.

III ENZYME IMMOBILIZATION FOR ORGANIC-PHASE OPERATIONS

Enzyme immobilization schemes, originally developed for work in aqueous systems, are not always compatible with the nonaqueous medium. The organic-phase operation may actually simplify the immobilization procedure. One simple, rapid, and effective immobilization scheme for organic-phase operation is direct adsorption onto the transducer surface [2]. Other useful enzyme immobilization schemes have been explored in our laboratory, including entrapment within Eastman AQ coatings [15] or incorporation within the bulk of composite carbon materials [16]. Whenever needed, the mediator should be insoluble in the organic media (so it can be stably coimmobilized by adsorption or entrapment). For example, the coimmobilization (coadsorption) of ferrocyanide and peroxidase onto a graphite foil surface yielded a reagentless sensor for hydrogen peroxide in organic media [4].

Unlike most polymeric films, the Eastman AQ ionomer is highly stable in different organic environments. This poly(ester-sulfonic acid) is thus very attractive for entraping enzymes for organic-phase biosensing [15]. Such an immobilization scheme offers additional advantages of permselectivity, speed of response and high enzyme loadings. The incorporation of enzymes within the bulk of a rigid graphite-Teflon matrix is another useful immobilization scheme [16]. Such use of Teflon as binder, instead of mineral oil, eliminates dissolution problems characterizing the operation of conventional carbon paste biosensors in organic solvents. This type of bioelectrode can also be renewed by polishing, and offers a fast and sensitive response (due to the absence of membrane barriers).

IV ANALYTICAL UTILITY OF ORGANIC-PHASE ENZYME ELECTRODES

There are numerous analytical opportunities in employing enzymes in organic (as opposed to aqueous) media. Most organic compounds of commercial interest are very sparingly soluble in aqueous media. In addition, many important samples from the food, pharmaceutical, cosmetic and petrochemical industries are highly hydrophobic and require time-consuming sample manipulations prior to their biosensing in aqueous solutions. Organic-phase biosensors, in contrast, obviate the need for such sample pretreatment procedures and allow rapid assays of challenging sample matrices. Such direct and fast bioassays should be extremely attractive for quality control testings and industrial process control. Recent advances in organic-phase biosensors can greatly facilitate these industrial applications. These include the use of ultramicroelectrodes for minimizing ohmic drop distortions in non-polar media [17] and the development of fast-responding organic-phase biosensing flow detectors [18]. The fast, sensitive and reproducible response indicates that the excellent performance inherent to on-line amperometric biosensors is not compromised by the use of organic media. The recent introduction of noncovalently immobilized enzyme reactors operating in nonaqueous media [19] should also facilitate the development of analytical flow systems for monitoring organic streams. On-line separations (e.g. liquid chromatography or solvent extraction) may also benefit from these organic-phase flow detectors and reactors.

The food industry, with its many challenging hydrophobic products, represents a prospective market for organic-phase biosensors. Such devices offer direct assays of many previously inaccessible food samples. For example, the phenol content of olive oil is an important factor in the quality and stability of this product. Accordingly, we developed recently a rapid flow injection procedure, based on a tyrosinase electrode, for determining the phenol content of olive oil samples [20]. By adsorbing the tyrosinase onto the surface of the graphite wall-jet flow detector, and by using chloroform as a carrier solution, reliable quantitation of the phenol content of various Italian and Spanish olive oils was achieved with sample frequency of 60 h^{-1}.

Butter and margarine represent other challenging food samples. Hall and Turner [7] demonstrated recently the utility of a cholesterol-oxidase/Clark electrode sensor for rapid measurements of cholesterol in butter and margarine samples (after dissolution in chloroform/hexane). Organic-phase biosensors should benefit assays of other fats and oils. For example, the organic-phase activity of lipase and peroxidase could be exploited for measuring triglycerides and peroxide species in these matrices.

Other challenging samples come from the pharmaceutical and cosmetic industries. The tyrosinase- and peroxidase organic-phase enzyme electrodes were applied recently for reliable quantitation of phenol and peroxide antiseptics in different anti-infective formulations [21]. The phenol values (resulted from such standard additions experiments) were in excellent agreement with the labelled values. A similar performance was obtained for measurements of peroxide bleaching agents in cosmetic products.

Assays of petrochemical samples can also benefit from the organic-phase biosensor operation. A new organic-phase alcohol sensor, based on the entrapment of alcohol dehydrogenase (from the thermophilic bacteria *Ta. Brockii* and its $NADP^+$ cofactor with an Eastman AQ polymeric coating was applied to measurements of secondary alcohols in untreated gasoline samples [6]. Clinical, environmental and military applications of organic-phase biosensors were discussed by Turner and coworkers [3]. The determination of trace water in nonaqueous media represents an important and novel application of organic-phase enzyme electrodes [22]. Such application relies on the strong influence of low levels of water upon the rate of biocatalytic reactions in nonaqueous media. Convenient quantitation of water can thus be accomplished using a tyrosinase electrode in the presence of a fixed substrate (phenol) concentration. Another interesting application involves the biosensing of enzyme inhibitors (e.g. pesticides, thiourea, benzoic acid) in organic solvents [23]. Such operations extend the scope of inhibitor biosensors toward inhibitors with poor water solubility. The new inhibitor and water organic-phase biosensors have been adapted to fast flow injection operation. Operation in microemulsion systems, is currently under investigation in various laboratories.

V CONCLUSIONS

It should be encouraging for analytical chemists to know that enzymes can operate not only in water but also in organic media. While organic-phase biosensors are still in an early developmental stage, there is no doubt that they will play an important analytical role in the future. In addition to enzyme electrodes, we expect to see organic-phase biosensors based on other transducers (e.g. optical or mass devices) or utilizing different biocomponents. The use of natural materials (mushroom and horseradish root plant tissues), instead of isolated enzymes, has already been demonstrated for measuring phenolic and peroxide species in chloroform [10]. Organic-phase immunosensors should be introduced in the near future. New analytical applications are expected based on the isolation or engineering of new enzymes (suitable for organic-phase operation) and the identification of new environments (e.g. supercritical fluids). More fundamental studies on the behavior of enzymes in nonaqueous environments are highly desired to serve as a guideline for future developments of organic-phase biosensors.

ACKNOWLEDGMENT

The generous financial support of the U.S. EPA and NIH is gratefully acknowledged. It is a pleasure to acknowledge the contributions of my numerous coworkers, particularly Y. Lin, N. Naser, A. Reviejo, D. Lopez, L. Angnes and E. Dempsey.

REFERENCES

1. A.M. Klibanov, Chemtech, 16 , 354 (1986).

2. G. Hall, D. Best and A.F. Turner, Anal. Chim. Acta, 213, 113 (1988).
3. S. Saini, G. Hall, M. Downs and A.F. Turner, Anal. Chim. Acta, 249, 1 (1991).
4. F. Schubert, S. Saini and A.P. Turner, Anal. Chim. Acta 245, 81 (1991).
5. J. Wang, B. Freiha, N. Naser, E. Romero, U. Wollenberger, M. Ozsoz and O. Evans, Anal. Chim. Acta, 254, 81 (1991).
6. J. Wang, N. Naser and D. Lopez, Biosensors & Bioelectronics, in press.
7. G. Hall and A.P. Turner, Anal. Letters, 24, 1375 (1991).
8. J. Wang, Y. Lin, A. Eremenko, A. Ghindilis, and I. Kurochkin, Anal. Letters, 26, 197 (1993).
9. A. Miyabayashi, M. Reslow, P. Adlercreutz and B. Mattiasson, Anal. Chim. Acta, 219, 27(1989).
10. J. Wang, N. Naser, K. Kwon and M. Cho, Anal. Chim. Acta, 264, 7(1992).
11. C. Laane, S.Boeren, K. Vos and C. Veeger, Biotech. Bioeng., 30, 81 (1987).
12. P. Aldercreutz and B. Mattiasson, Biocatalysis, 1, 99 (1987).
13. A. Zaks and A.M. Klibanov, Science, 224, 1249 (1984).
14. J. Dordick, Enz. Microb. Technol., 11, 194 (1989).
15. J. Wang, Y. Lin and Q. Chen, Electroanalysis, 5, 23 (1993).
16. J. Wang, A. Reviejo and L. Angnes, Electroanalysis, in press.
17. J. Wang, L. Wu and L. Angnes, Anal. Chem., 63, 2993 1991).
18. J. Wang and Y. Lin, Anal. Chim. Acta, 271, 53(1993).
19. L. Brace, J. Daros and M. LaGuardia, Anal. Chem. 64, 129(1992).
20. J. Wang, A. Reviejo and S. Mannino, Anal. Letters 25, 1399 (1992).
21. J. Wang, Y. Lin and L. Chen, Analyst 118, 277 (1993).
22. J. Wang and A. Reviejo, Anal. Chem. 65, 845(1993).
23. J. Wang, E. Dempsey, A. Eremenko and M. Smyth, Anal. Chim. Acta, in press.

MODIFIED ENZYME ELECTRODES

ANTHONY P. F. TURNER
Cranfield Biotechnology Centre, Cranfield Institute of Technology, Cranfield, Bedford. MK43OAL, UK.

Abstract Our work on redox mediators and their application in amperometric biosensors was reviewed at the last workshop in 1987, which coincided with the commercial launch of the ExacTech™ pen-sized glucose sensor. The microbial activity monitor, enzyme electrode for *in situ* fermentation monitoring and amperometric immunoassay reported in the 1987 paper have also since been commercialised by different companies. The 1st workshop triggered important collaborative relationships which led, for example, to the very successful EC Concerted Action on Chemical Sensors for *in vivo* Monitoring. This contribution will briefly review the above developments. Attention will then be focussed on recent work at Cranfield on modified enzyme electrodes with applications in medicine, the food industry an in environmental monitoring. Three themes will be explored: fabrication, continuous operation and application in harsh or novel environments.

INTRODUCTION

Since the report given to the previous workshop in this series[1], mediator-modified electrodes have been successfully exploited in a range of biosensors incorporating enzymes, antibodies, microorganisms and DNA. The aim has been to produce continuous

G.G. Guilbault and M. Mascini (eds.), Uses of Immobilized Biological Compounds, 263–269.

or semi-continuous digital electronic signals which are proportional to a specific chemical or group of chemicals. Devices have been configured as fixed or portable instruments giving qualitative or quantitative information and several instruments have reached the market. Some of the principal features and benefits of such devices are outlined in Table I.

Table 1 - Features and Benefits of Mediated Amperometric Biosensors

FEATURE	BENEFIT
Targeted specificity	Versatility (from definitive identification of single compound to broad range sensor)
Electronic integration	Elegant and compact instrument design resulting in ease of use
Operate in complex matrices	Reproducibility and simplicity resulting from lack of need for sample preparation
Fast response time	High throughput in automatic analyser. Decreased user time for spot checks
Continuous signal	More accurate understanding of fluctuations & facility to actuate corrective action
Small size	Portable, inexpensive, small sample, less waste, multiple simultaneous assays possible
Mass producible	Inexpensive, disposable, can be widely dispersed

Clinical diagnostics have dominated biosensor development to date and are likely to continue to be the focus for commercial successes over the next five years. Clinical applications are driven by the need for cost effective patient care. Some conditions, for example diabetes, benefit from self-tests usable in the home. Appropriately, the world's most successful biosensor to date is the ferrocene mediated amperometric enzyme electrode for glucose[2] used in conjunction with a pen-shaped instrument (ExacTech™, MediSense, Cambridge, USA & Abingdon, UK). The ExacTech pen-sized glucose sensor was introduced onto the market in 1987. Six years on, the much imitated ferrocene-mediated sensor has become established as the archetypal success with an estimated turnover of $100 million per annum with 15-20 million enzyme electrodes being fabricated by screen printing each month. Diagnostic tests that make sense to perform in the home, however, are few and far between. Pregnancy testing is an obvious example, but the frequency with which these tests are performed makes an instrumented assay uncompetitive with the elegant visual assays now marketed. Other promising analytes that are attracting commercial development of pocket-sized devices include total cholesterol, alcohol and lactate.

SENSOR FABRICATION

A key issue in diagnostics is the need for a reproducible and economic means of mass production. Methods that have been developed for mass production of one-shot devices may find application in reusable or continuous use devices. Printed enzyme electrodes, for example, may be useful in larger machines where the throughput permitted by biosensors can out perform conventional systems or where a competitive advantage can be achieved by broadening the spectrum of analyses offered. Screen printed electrodes have also been used for immunoassays involving the electrochemical detection of the activity of an enzyme label. A product based on this principle was recently launched in Italy (Immunosens, Benevento, Italy). We have built on our experiences with screen printing planar and capillary structures to evolve ink-jet techniques[3] for the fabrication of multilayer or 3-D structures. Marriage of this non-contact process with direct chemical modification offers interesting opportunities for the fabrication of amperometric sensors and diagnostic assays in general. Still wider opportunities can be addressed by harnessing recent advances in optical technology, multichannel integrated optical devices fabricated on single chips offer one way forward in this respect.

IN VIVO MONITORING

A further area of interest is *in vivo* monitoring. The direct insertion of enzyme electrodes subcutaneously is at an advanced stage of experimentation[4,5], but commercial opinion is divided about the commercial risk versus the benefit of such a new product. The 1st NATO workshop triggered important collaborative relationships with the formation of the European Biosensor Group, which later led to the very successful EC Concerted Action on Chemical Sensors for *in vivo* Monitoring (and a number of smaller international research projects principally funded by the EC). Thirty laboratories in 15 European countries participated in this unique collaboration which began in January 1989. The work of the Concerted Action is being continued by a core of 37 laboratories and 54 associated organisations in a new programme funded under BIOMED 1 for a further 2 years from the beginning of 1993. Remaining technical problems are not inconsiderable, but significant progress has been made in recent years on gas, electrolyte and metabolite monitoring.

NON-MEDICAL APPLICATIONS

Much of the technology developed for clinical applications will find other uses in industry, environmental monitoring and defence. Industry is seeking sensors to improve productivity, quality and compliance to legislation. Fermentation monitoring is one area of interest which has broad consequences for the food and biotechnology industries[6]. Simple, devices for food quality or safety testing can be envisaged based on advances in clinical diagnostics. Orion (Boston, MA, USA), for example, have recently launched a bio-selective electrode (***BSE*** ™) for sugar analysis in foods based on a continuously renewable, mediated enzyme electrode. In some specialised cases biosensors can provide information which is not easily obtained using conventional methodology. One example is the meat sensor, developed at Cranfield in collaboration with the Leatherhead Food Research Association, which measures the three dimensional distribution of glucose near the surface of the meat in order to estimate the microbial contamination[7]. An alternative method of mediated amperometric detection of microbial content has also been launched commercially (MIDAS pro, Biosensori, Milano, Italy). In some recent applications, however, such as animal cell culture monitoring monitoring, we have abandoned the use of mediators. In a collaborative project with the GBF in Germany we used FIA in combination with enzyme electrodes based on metalised carbons to achieve the longevity necessary for such applications.

Environmental applications of biosensors are characterised by a desire to protect workers and the public from toxic material, and compliance with legislation. The relevance of biologically-based sensing systems to effects of pollution on both flora and fauna make biosensors ideal candidates for broad range alarms. The exquisite specificity of some enzymes, on the other hand, can distinguish between stereo isomers of the same compound. Increasing concern about chronic exposure to very low levels of hazardous chemicals necessitates the development of ultra-sensitive assays. Our approach to the challenge of broadening the applicability of biosensors took us via mediated organic-phase enzyme electrodes (OPEE's)[8] to operation of enzyme electrodes in the gaseous phase[9]. We are now developing gas-phase biosensors for the detection of phenol vapour, sulphur dioxide and methane.

CONCLUSIONS

As the academic literature on biosensors continues to diversify, a clear trend within industry to critically evaluate and invest in focused areas is emerging. Technical issues concerning the developers include fabrication technologies and a general drive towards miniaturisation. Medicine and food provide the prime focus for new products, but many smaller companies are searching for niche markets in biotechnology, veterinary science, environmental monitoring, industrial safety, defence and research. New biosensor products are emerging quite rapidly and the next decade may well decide the commercial fate of this exciting new technology.

REFERENCES

1. Turner, A. P. F. (1988). Redox mediators and their application in amperometric sensors. *In: Analytical Uses of Immobilized Biological Compounds for Detection, Medical and Industrial Uses* (Eds. G. G. Guilbault and M. Mascini). Reidel, Dordrecht. pp. 131-140.

2. Cardosi, M. F. and Turner, A. P. F. (1990). Recent advances in enzyme-based electrochemical glucose sensors. *In: The Diabetes Annual* **5** (Eds. K. G. M. M. Alberti and L. P. Krall). Elsevier, Amsterdam. pp. 254-272.

3. Newman, J. D., Turner, A. P. F. and Marrazza, G. (1992). Ink-jet printing for the fabrication of amperometric glucose biosensors. *Analytica Chimica Acta* **262**, 13-17.

4. Turner, A. P. F. (1993) *Advances in Biosensors, Supplement 1 - Chemical Sensors for In Vivo Monitoring.* JAI Press, London, 1993.

5. Alcock, S. J. and Turner, A. P. F. (1993). *In Vivo Chemical Sensors: Recent Developments.* Cranfield Press, Cranfield, UK.

6. Brooks, S. L., Higgins, I. J., Newman, J. and Turner, A. P. F. (1991). Biosensors for process control. *Enzyme and: Microbial Technology* **13**, 946-955.

7. Kress-Rogers, E., D'Costa, E. J., Sollars, J. E., Gibbs, P. A. and Turner, A. P. F. (1993). In situ measurement of meat freshness with a biosensor array. *Food Control* **4** (3).

8. Schubert, F., Saini, S., Turner, A. P. F. and Scheller, F. (1992). Organic phase enzyme electrodes for the determination of hydrogen peroxide and phenol. *Sensors and Actuators* **B 7**, 408-411.

9. Saini, S. and Turner, A. P. F. (1992). Media for biocatalytic electrochemical reactions in the gaseous phase. *UK Patent Application* GB 9218376.3

CAN NEW BIOSENSORS BE DEDUCED FROM SENSING IN BIOLOGY ?

H.-L. SCHMIDT
Technische Universität München, Lehrstuhl für Allgemeine Chemie und Biochemie, Vöttinger Straße 40, D-8050 Freising-Weihenstephan

Abstract Natural recognition and transduction systems are checked in regard to their possible immediate or mediate use in biosensors. The majority of the sensors known so far base on enzymes as biological recognition part, and start transduction only after the catalytic formation of a secondary chemical signal. On the other hand in nature transduction is not catalytic but uses a conformational change for direct or indirect - after a chemical amplification cascade - modulation of a membrane permeability. The use of "ion channels" for biosensor construction would in principle be possible, however, suffer from the instability and complexity of the biological system. More promising seem to be the attempts for the development of artificial ions channels. While synthetic ionophors have already successfully been used in sensors, the corresponding application of carrier proteins is just at the beginning, however, the construction of biomimetic systems seems to be more advantageous. Antibodies as most versatile and inexpensive recognition molecules have often been used in biosensors with various transducing principles. A disadvantage is the slow kinetics of the dissociation of antigen-antibody complexes which implies problems with the regeneration of sensors. Therefore in some cases the competition of supramolecular artificial "receptors" may be more promising. The mechanism of odor and taste perception is to-day well understood, but it would not be worthwhile to transfer it to

G.G. Guilbault and M. Mascini (eds.), Uses of Immobilized Biological Compounds, 271–280.

artificial sensors. Nevertheless, the implied adaption of natural receptors to demands of the molecules to be recognized could be a model for computer aided molecular conception of artificial receptors. As the natural selectivity could certainly not be attained, array arrangements could be used for partial compensation. Whenever proteins are integrated in biosensors, one should regard their instability and include into further sensor developments methods for protein stabilization.

I. INTRODUCTION

The organization, development, reproduction and differentiation of organisms, the internal metabolic regulation of cells as well as their communication with other cells and their environment, are controlled by chemical signals, their perception, identification, tranduction and amplification. The biological recognition of molecules inherent in this procedure, is realized through non-covalent reversible interactions between complementary structures, leading to a docking of the molecule to be recognized on a macromolecule. In nature this molecular recognition is in general accompanied by a conformation change of the receptor (protein) leading to an immediate opening of an "ion channel" and a depolarization of a membrane potential, or having for consequence the activation of an enzyme, catalyzing the formation of a "second messenger", which may then initiate a cascade of enzyme modulations with the effect of amplification and spreading of the original signal.[1] Biosensors, though using biological molecular recognition systems, do normally not use these natural transduction principles, but are based on quite different signal production mechanisms. The most common principle is to identify the analyte by a specific enzyme and to produce, by means of its catalytical property, a secondary chemical signal, e.g. H^+ or

H_2O_2. This signal is recognized by a suitable transducer, for example in form of a potential, a current, or the change of an absorption or fluorescence, and then converted to an electrical signal. Thus, on one side biological recognition systems, which do not induce any transduction in nature, have been used in biosensors, while others have not or only tentatively been investigated in this respect. Let us therefore discuss reasons and possible chances.

II. NON-CATALYTIC RECOGNITION AND TRANSDUCTION IN NATURE AND POSSIBLE NON-BIOLOCIAL EQUIVALENTS

A. ION CHANNELS AND RELATED SYSTEMS, PERFECT INTEGRATED SENSING AND TRANSDUCTION SYSTEMS

Among the few directly used biological receptor-transducersystems are ligand-gated ion channels. Membrane fragments with these channels have been isolated by the patch-clamptechnique, and they preserved their natural property of being tuned by ligands and inhibitors.[2] Affinity chromatography and genetic engineering provided the availability of channel proteins in larger amounts.[3] They could, after integration into artificial membranes, function as signal producing systems.[4] However, the construction of these systems is laborious, the ion channels are so far statistically orientated within the membrane, and they are instable. However, they could serve as models for the establishment of more stable semisynthetic and even synthetic systems. Gitler and Yuli integrated synthetic polypeptides, resembling in their structure the bee toxin melittin, into membranes and obtained channels, the permeability of which could be modulated by chemicals.[5] Very recently in the group of J.M. Lehn completely synthetic supramolecular structures have been developed which can be integrated into membranes and behave as selective ion channels.[6] These can still not be modulated, however, they may have the advantage of higher stability and show a way towards new biomimetic transduction systems.

B. IONOPHORS AND CARRIER PROTEINS, RECOGNITION SYSTEMS WITH POTENTIAL TRANSDUCTION

Ionophors are low molecular weight lipophilic ion complexing agents, capable of solvating ions in a lipid membrane. The selectivities of the antibiotics valinomycin for K^+ and nonactin for NH_4^+ ,respectively, have often been used in potentiometric membrane electrodes, and their structure has initated the synthesis of biomimetic chelating agents, mainly crown ethers, for the selective binding of many other ions.
The selective permeability of biological membranes is controlled by carrier proteins involved in the catalyzed active and passive transport. These proteins, on binding ions or molecules, change their conformation, and open their binding centers alternatively to one or to the other side of the membrane. One of the mechanisms realized, mainly for the transport of sugars and amino acids through the intestin and the kidney mucosa, is the so-called co-transport. The binding and transport of one ligand against a concentration gradient is afforded by the simultaneous binding and transport of another ligand in direction of a concentration gradient. In many cases, the second ligand is H^+ or Na^+, and using this fact, very recently several groups constructed biosensors for an organic ligand, combining a carrier protein to a transducer for one of the ions mentioned. Like the systems basing on gated ion channels, the function of these systems with carrier proteins depends on the availability and construction of absolutely dense membranes or vesicles, and principally needs an orientated integration of the proteins within the membrane, which is so far not always satisfactorily realized. In addition, these "sensors" are quite instable. Nevertheless one can assume that more stable artificial carrier molecules will soon be available and serve as base

for biomimetic sensors as already realized on the base of ionophors.

C. ANTIBODIES AND LECTINS, NATURAL RECOGNITION MOLECULES WITHOUT TRANSDUCTION

The most adaptable and versatile biological recognition system Nature has invented is the immuno system. During evolution antibodies have not been conceived for signal production, but for more or less (kinetically) irreversible binding of complementary foreign molecules. The most important feature of the immuno system is that it is capable to adapt fast and easily to any new structure, thus producing recognition molecules of whatever selectivity. Therefore antibodies have very often been used in biosensors as recognition systems, and have been connected to various methods of indication and transduction. With these sensors also enzymatic indicator reactions are used, however, these catalytic devices will not be considered in the present context.

Quite naturally, any system based on the antigen-antibody recognition, many of them using mass measurements as transduction, suffer from poor reversibility, at least under conditions needed for the preservation of the natural protein conformation, and they can hence only be used a few couples of time. A very promising approach, implemented with regeneration possibilities, has recently been proposed by the group of R.D. Schmid.[8] Their system uses an optical fiber, carrying the spacer bound antigen. The saturation by a fluorescent indicator labelled antibody is easily detected, and in presence of the competing free antigen, the fluorescence signal is correspondingly decreased. This system is in principle reversible, but still, due to the slow equilibration kinetics, not continuously usable. A few more partially reversible systems are described in the literature. In this context another type of affinity sensor has to be mentioned, using a natural carbohydrate binding protein, a lectin, which also is based on the competitive displacement of a fluorescence labelled indicator. However, the principle had not found general application.

Why not transfer the principle to synthetic affinants, using an optical or mass indication transduction? The small organic affinant molecules would imply, as compared to proteins, the opportunity of a much denser surface loading, and the possibility of fast and quantitative regeneration. On the other side, the disadvantage of these artificial systems would be that the synthesis of the affinants would often be laborious and expensive, and that the natural selectivity would never be attained.

Nevertheless, in literature more and more examples are reported for attempts to use the experience in supramolecular chemistry for the contruction of sensors. In principle, the transduction methods mentioned above could be used, and in most cases a quite good reversibility of affinant binding, even fast equilibration, can be expected. In our laboratory, it was observed that the reversible binding of phenol or p-nitrophenol by a macro tricyclic ammonium salt can be monitored by means of a pH-electrode.[9] A direct optical response is described for the "recognition" of ethanol by trifluoroacetyl aniline (ETH 6022) implanted into a solvent polymer membrane: The addition of the ligand leads to a hypsochromic shift of the absorption maximum at 305 nm, and the fast equilibrium between the "host" and the "guest" permits the construction of a reversible ethanol sensor.[10]

D. BIOLOCIAL REDOX CHAINS, PERFECT ELECTRON TRANSPORTING SYSTEMS

The respiratory chain in the mitochondria and the electron transport system of the chloroplasts demonstrate perfectly several principles for biosensor contruction: a) A redox potential driven proton transport as an impact for the concept of a poteniometric sensor for redox processes, which has been recently realized in our group[11], b) Molecular switching with orientated small protein molecules, used by H.A.O. Hill for the development of "promotor" molecules, facilitating the direct electrical communication of proteins with electrodes,[12] c) The existence of a loaded conducting membrane functionalizing like a capacitor with charge accumulation and dissipation. The latter principle was the model for the concept and realization of new oxidoreductase electrodes in which these enzymes were entrapped in conducting polymers or polymerized redox centres.[13] We believe that "enzyme wiring" or the entrapment of enzymes and redox catalysts for coenzyme recycling in a conducting nutrix will open perspectives for new redox-enzyme electrodes.

E. BIOLOGICAL CHEMOSENSORS WITH INTELLIGENT TRANSDUCTION ARRAYS

The molecular mechanisms of odor perception has nearly completely been elucidated and understood during the last years.[14] It is now generally accepted that, by the binding of a given molecule to an adequate sensor of an olfactory cell, a conformational change takes place, initiating the production of a secondary messenger, which gates, through a cascade of amplification reactions, the opening of an ion channel, thus starting the depolarization of a membrane. Tasting is in some parts transduced by a similar mechanism, in some other parts by a direct ion channel modulation. The fact that odor perception is capable to recognize and discriminate between a large number of different molecules, is discussed to be either possible through an adaption mechanism of the receptors, comparable to that of the formation of antibodies, or through the processing of signals from several neurons in the sense of an array system.

The first explanation is most fascinating, because it suggests the concept of a versatile sensor developing his recognition structure according to the demand of the molecules to be perceived. The principle is - on a much lower level - imitated by computer aided modelling and development of specific synthetic host molecules. We have tried this kind of optimization with a calixarene molecule originally proposed by Aoyama et.al.[15] for the binding of

glucose. Computer aided modelling permitted to conceive a derivative with probably much better binding properties, which is now being synthesized.[16]
The array principle is well-known as a means to improve selectivity even with low selective sensors of slightly different response characteristics, using a multilinear regression of the signal from several sensors. We have transferred this to flow injection analysis, permitting us to identify and determine the components of mixtures of subtrates for which a completely selective enzyme is not available.[17]
A multisensor system with intelligent signal computing and feedback is the bacterial chemotaxis system, controlling the rotation of the bacterium's flagellum, and hence the movement of the cell as response to concentration gradients of attractants or repellents. This system is to-day completely understood, and it may be the base of some very interesting bioprobes for the general qual ity of a surrounding medium.[18] It would not be reasonable to conceive or construct corresponding biomimetic systems, but better to use the natural bacteria, eventually after suitable genetic modification, and combine them to an adequate transduction system. Several performances have been realized in so-called microbial sensors.

III. LEARNING EFFECT AND CONCLUSIONS

Let us come back to the question put in the title of this contribution: Can new biosensors be deduced from sensing in biology? The answer would be no and yes. "No" would mean that in many cases the immediate use or a direct transfer of biological systems for sensor construction would be very limited and often only possible for the performance of one-shot devices. This will be mainly due the high complexity and instability of the biological systems, many because their function is depending on secondary structures of proteins. A limited enlargement in this context may be possible by

applying more stable proteins from thermophilic bacteria or by improving their stability by special methods.[19]
"Yes" means that many principles demonstrated by nature could be - in addition to the often used catalytic properties of enzymes - transposed to sensor development, for example by developing and using biomimetic channel and carrier molecules, and if necessary compensate the loss of selectivity implied in these non perfect recognition molecules by the application of array arrangements.

REFERENCES

1. H.-L. Schmidt, W. Schuhmann, F.W. Scheller a. F. Schubert, in "Sensors - A Comprehensive Survey" (W. Göpel, J . Hesse, J.N. Zemel, eds.), VCH Weinheim, Weinheim, 1992; Vol.3,pp. 719-817. Specific Features of Biosensors.

2. E.Neher u. B. Sackmann, Spektrum der Wissenschaft 1992(5), 48

3. H. Repke u. C. Liebermann (eds), Membranrezeption und ihre Effektorsysteme, VCH Weinheim, 1987

4. J. Ramsden, F. Hucho a. H. Vogel, in Biosensors; Fundamentals, Technologies and Applications, F. Scheller a. R.D. Schmid (eds), Springer Verlag, Berlin 19.., p. 435

5. C. Gitler a. J. Yuli, note in Model 1992/93, Bericht aus Rehovoth, Weizmann Institute of Science

6. M.J. Pregel, L. Jullien u. J.-M. Lehn, Angew. Chem. 104, 1695 (1992).

7. N. Sugao, M. Sugawara, H. Minami, M. Uto a. Y. Umezawa, Anal.Chem. 1993, in press

8. F.F. Bier, W. Stöcklein, M.Böcher, U. Bilitewski a. R.D. Schmid, Sensors and Actuators B 7, 509 (1992)

9. J. Polster a. F. Schmidtchen, unpublished results

10. U.E. Spichinger, M. Kuratli a. W. Simon, Biosensors and Bioelectronics 7, 715 (1992)

11. T. Vering, D. Seiwald, W. Schuhmann, H.-L. Schmidt, A. Heller, L. Ye a. B. Speiser, manuscript submitted to J.Am.Chem.Soc.

12. J. Higgins a. H.A.O. Hill
Essays in Biochemistry 21 (1985), 119-145

13. L. Ye, M. Hämmerle, W. Schuhmann, H.-L.Schmidt, A.J.J. Olsthoorn, J.A. Duine a. A. Heller, Anal.Chem. 65 (1993) 238-241.

14. H. Hatt, Schering Botschaften aus Forschung and Medizin 7 (1), 73 (1992)

15. Y. Aoyama, Y. Tanaka a. S. Sugahara, J.Am.Chem.Soc. 111 (1989) 5397-5404

16. S. Riehl, R. Medina, W. Schuhmann a. H.-L. Schmidt, unpublished

17. U. Englbrecht a. H.-L. Schmidt, Abstract Dechema Jahrestagung der Biotechnologen, 1993

18. J. W. Lengeler, La Recherche 21, 20 (1990)

19. C.Lehn, J.Danzer, H.-L.Schmidt, M. Mayer u. M. Staudenbauer, Abstract Dechema Jahrestagung der Biotechnologen, 1993

MICROBIAL SENSORS AND MICROBIOSENSORS

ISAO KARUBE AND KENJI YOKOYAMA
Research Center for Advanced Science and Technology,
University of Tokyo

Abstract A novel BOD sensor with a new principle employing a luminous bacterium, *Photobacterium phosphoreum*, was developed. The increased intensity of luminescence resulting from the cellular assimilation of organic compounds in waste water was detected by a photodiode. The BOD response of the bacterial reagent could be obtained within 15 min with 7% error. The temperature condition for optimal BOD response was 18 – 25 °C at pH 7 – 8, indicating that it is possible to measure BOD at room temperature without controlling the temperature of the measuring system. The metabolic characteristics of luminous bacteria were investigated by comparing the BOD values for several pure organic substrates estimated by the BOD sensor with conventional 5–day BOD values. In comparison with the 5–day BOD measurement for some waste water samples, BOD values estimated by the sensor showed a good agreement with the 5–day method. Micromachining techniques were applied to construct a microbiosensor system. The micromachined biosensors have small size, low production cost, and good reproducibility. We made some detection units for flow injection analysis. An electrochemical flow cell was fabricated, and both enzyme immobilized column and electrochemical detector were integrated onto the same chip. A chemiluminescence detector was also fabricated and applied to quantitative analysis of glucose and lactic acid contained in human serum and urine.

I INTRODUCTION

The organic pollution in waste waters and effluents is usually evaluated by determination of BOD. With the conventional test, however, the measurement requires not only many complicated procedures including 5–day incubation but also experienced skill to get reproducible results. A rapid and reliable method for BOD estimation was first developed employing a bioelectrochemical sensor[1]. The sensor consisted of microbial cells immobilized on an oxygen electrode and measured the current decrease resulting from a decrease in dissolved oxygen caused by the oxygen consumption of the microorganism. There have been some reports on BOD sensors with the same principle using other species of microorganisms[2-6]. Luminous bacteria emit light at a constant rate for fairly long time as a result of normal metabolic processes. In this process the reduced substrates produced by catabolic degradation of organic compounds in the cell are shunted to the bioluminescence reaction which is coupled to the electron transport pathway[7]. Thus, when the waste water containing organic substances is supplied, the consumption of organic compounds by luminous bacteria increases the level of light emission. The extent of organic pollution can be thus determined by measuring the increased intensity

G.G. Guilbault and M. Mascini (eds.), Uses of Immobilized Biological Compounds, 281–287.

of emitted light. The BOD measuring procedure in this study employs the bacterial reagent which is a whole cell suspension of luminous bacteria. For measuring light intensities, a simple and economical instrument using a photodiode and amplifier was constructed.

Microfabrication techniques based on integrated circuit technology, such as photolithography and etching, have been applied to the other fields. These techniques which are used to make some small and efficient three-dimensional devices are called micromachining[8-11]. Furthermore, some studies to make miniaturized chemical analysis systems were already reported[12,13]. These analysis systems have many advantages, such as fast response, small amount of sample, low consumption of reagents, as compared with the conventional one. We applied the micromachining technique to make a miniaturized enzyme-based sensor system.

II BOD SENSOR USING LUMINOUS BACTERIA

Microorganisms and measurement of light intensity

Photobacterium phosphoreum IFO 13896 was cultured in a luminous medium on a reciprocal shaker at 20 °C for 16 hours. The resulting culture broth had a cell density of about 5×10^8 cells/mL. Maintenance of the microorganism at 4 °C was on plates of the medium prepared with 1.5% agar.

The bacterial reagent was prepared from the culture broth with appropriate dilution with 0.1 M potassium phosphate buffer (pH 7.0, containing 2.5% NaCl for osmotic protection). An aliquot (0.6 mL) of the bacterial reagent was transferred to a glass pan, and the initial intensity of emitted light from the reagent was measured by a photodiode.

BOD response and calibration

The typical light emitting responses of the luminous bacteria to different levels of BOD (20–160ppm) are shown in Figure 1. The initial light intensity of the bacterial reagent, height A, was obtained, and then a standard BOD solution (20ppm) was injected. The assimilation of organic compounds begins within 2 min and the intensity of emitted light increased from height A to height B (maximum light intensity) within 15 min. Consequently, the difference between height B and A was determined as the value of BOD response for 20 mg–BOD/L solution. The BOD response increased with increasing BOD concentration of standard solution in a linear relationship between the increment of light intensity (BOD response) and the BOD concentration from 0 to 120ppm (Figure 2). Minimum measurable BOD was 5ppm. At 60ppm standard BOD solution, the reproducible response could be obtained within 7% relative error of the mean value. The standard deviation was 3.6ppm.

Influences of pH and temperature on the BOD response

In order to determine the optimum conditions for BOD measurement using the bacterial reagent, the influences of pH and temperature on BOD response were investigated. The maximum response was obtained between pH 7 and 8. The optimum temperature was determined in the temperature range 15 to 30 °C, and

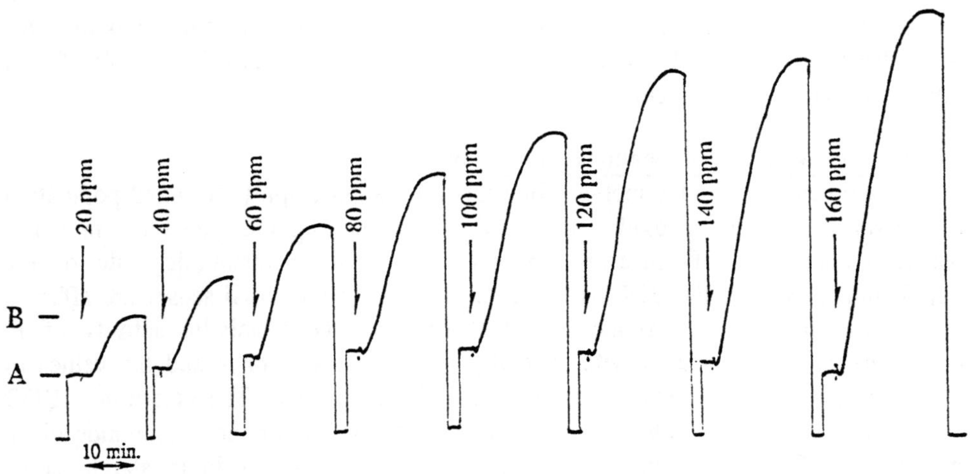

Figure 1 Typical response curve to various BOD concentrations

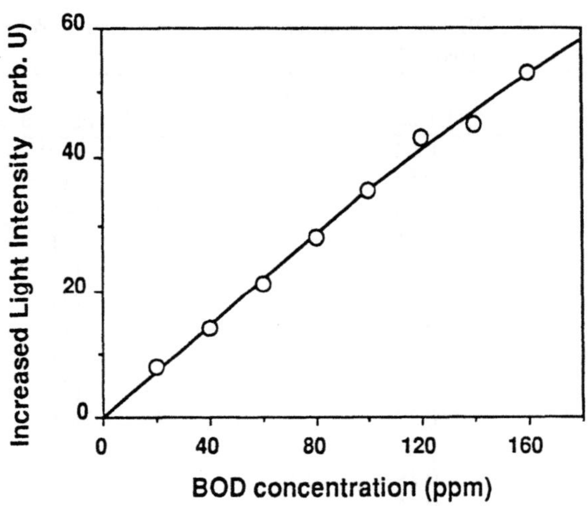

Figure 2 BOD calibration graph based on bacterial luminescence

a constant response of almost equal intensity were obtained between 18 and 25 °C. Accordingly, BOD measurement by luminous bacteria is possible at room temperature without the need for temperature control. This is advantageous over the yeast based (commercial) BOD sensor which must be controlled at 30 °C for optimum BOD measurement.

Comparison with the conventional 5-day BOD

The conventional 5-day method for BOD estimation employs a mixed population of microbes capable of oxidizing the biodegradable organic substrates in waste water. Values of BOD measured reflect the combined metabolic state of the population. However, the BOD values determined by a microbial sensor are affected by the metabolism of a single microbial species. The metabolic activity of *P. phosphoreum* was measured with several pure organic substrates and the values of BOD per unit gram substance were estimated by the sensor. The determined BOD values were compared with reported data of 5-day BOD for the compounds given in Table 1. For the organic substances including lactose, soluble starch, some organic acids and alcohols, the sensor showed lower BOD values compared with the 5-day BODs. The results indicate that the rate of oxidation for such substrates was slower than that of the standard substrates, glucose and glutamic acid. On the other hand, the values for other substrates estimated by the sensor were similar or higher than those of 5-day BOD.

Table 1. Comparison of BOD values estimated by sensor with those determined by 5-day method for various pure compounds.

	BOD (g/g)		
Substrate[a]	BOD sensor	5-day BOD	(Number of data)
Glucose	0.62	0.50–0.78	(12)
Fructose	0.57	0.71	(1)
Sucrose	0.50	0.49–0.76	(6)
Lactose	0.31	0.45–0.72	(5)
Soluble starch	0.02	0.22–0.71	(9)
Glycine	0.50	0.52–0.55	(2)
Aspartic acid	0.48	0.58	(1)
Glutamic acid	0.73	0.64	(1)
Acetic acid	0.27	0.34–0.88	(9)
Lactic acid	0.32	0.63–0.88	(3)
Ethanol	0.25	0.93–1.67	(14)
Propanol	0.28	0.47–1.50	(2)
Glycerol	0.53	0.62–0.83	(7)

a The concentration of tested compounds was 100 mg/L.

Subsequently, the 5-day BODs of some waste water samples were determined by the conventional method for comparison with the BOD values estimated by the sensor. The waste water samples tested were Organization for Economic Cooperation and Development (OECD) synthetic waste water[14], untreated municipal waste water and untreated waste water discharged from a food factory. Each of the waste waters were diluted appropriately (200:1 or 100:1) with dilution water prior to use. As shown in Table 2, relatively good agreements between the two methods were obtained for the test samples. Though the sensor-estimated values were somewhat higher for OECD synthetic waste water and food factory waste water, and lower for municipal waste water than those of conventional BOD, no significant differences were observed. The observed differences between the two methods are likely caused by differences in the organic content of oxidizable and easily assimilable substrates in each of the test waste water samples.

Table 2. Comparison of BOD values measured by sensor with those determined by the conventional 5-day method for some waste water samples.

	BOD(ppm)	
Waste water sample	5-Day BOD	BOD Sensor
Synthetic waste water[a]	13850	15600
Food factory	719	780
Municipal waste water	1400	1328

a OECD synthetic waste water contains peptone 1.5%, meat ext. 1.1%, urea 0.3%, NaCl 0.07%, $CaCl_2$ 0.04%, $MgSO_4$ 0.02%, K_2HPO_4 0.28%.

III MICROMACHINED BIOSENSORS

Electrochemical flow cell

An electrochemical flow cell which has very small inner volume, ca. 20 nL was fabricated. Figure 3 shows a structure of the micro electrochemical flow cell. This flow cell can be used as an electrochemical detector for liquid chromatography or flow injection analysis (FIA). The enzyme-immobilized flow cell can be employed as an electrochemical biosensor. Glucose oxidase was immobilized onto the sample inlet hole of the cell using glutaraldehyde and bovine serum albumin. Glucose was calibrated in the range of 30 mg/dL to 1000 mg/dL when 0.2 μL of the sample was injected.

Integration of enzyme immobilized column and electrochemical flow cell

A long open-tubular column was fabricated on the silicon substrate. A glucose sensor was integrated with both an enzyme immobilized column and an electrochemical flow cell (Figure 4). The column was made by anisotropic silicon etching to be 100 μm wide, 70 μm deep, 1 m long, and total volume of the column is 5 μL. Four gold electrodes were formed on the glass substrate. Both of the two substrates were anodically bonded. Connecting unions to the pump and sample injector, were glued on the inlet and outlet holes with epoxy resin. GOD was immobilized on the inner wall of the column using 3-aminopropyltriethoxysilane and glutaraldehyde. This device was applied in a conventional FIA system.

Integration of enzymatic reactor and chemiluminescence detector

An enzymatic reactor and a chemiluminescence detector were integrated on the same chip (Figure 5). It consists of a silicon and glass substrate. On the silicon

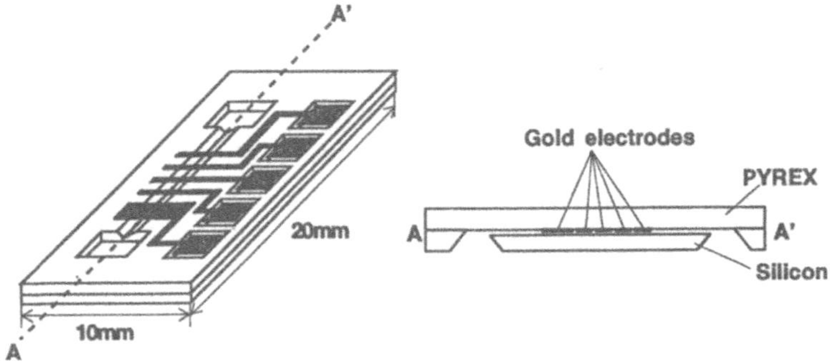

Figure 3 Structure of the electrochemical flow cell.

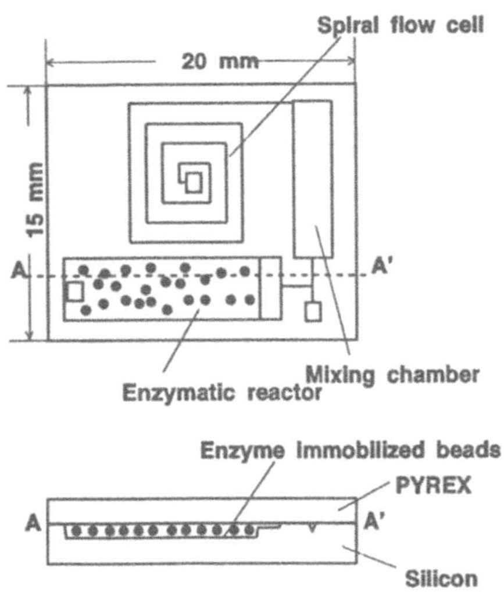

Figure 4 Electrochemical detector integrated with enzyme column.

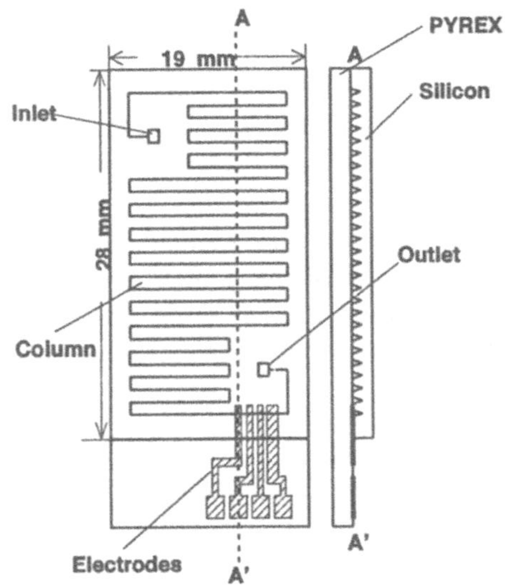

Figure 5 Chemiluminescence detector integrated with enzyme reactor.

substrate, an enzymatic reaction column, a mixing chamber, a spiral flow cell were made by anisotropic etching. The total size of the measuring unit was 15 mm × 20 mm, and the total internal volume of the device was about 15 μL. Enzyme immobilized glass beads were packed into the column, and a photodiode was placed onto the spiral flow cell. Using GOD immobilized glass beads, determination of glucose concentration was carried out in the range of 10 mg/dL to 300 mg/dL. Glucose in human serum and urine was measured by a chemiluminescence detector. The correlation coefficient between this chemiluminescence method and the conventional method was 0.99. Lactic acid contained in human serum was quantitated using the same procedure as the glucose determination. Samples containing L-lactic acid at concentrations from 4 mg/dL to 50 mg/dL could be measured. The correlation coefficient between this chemiluminescence method and the conventional method was 0.98.

REFERENCES

1. I. Karube, T. Matsunaga, S. Mitsuda, S. Suzuki, Biotechnol. Bioeng., 19, 1535, (1977).
2. M. Hikuma, H. Suzuki, T. Yasuda, I. Karube, S. Suzuki, Eur. J. Appl. Microbiol., 8, 289, (1979).
3. N. Kawabata, N. Nakamura, Abstract of the international symposium on New Sensors and Methods for Environmental Characterization,(Kyoto, Japan) (1986).
4. J. Kulys, K. Kadziauskiene, Biotechnol. Bioeng., 22, 221 (1980).
5. K. Riedel, R. Renneberg, M. Kuhn, F. Scheller, Appl. Microbiol. Biotechnol., 28, 316, (1988).
6. S. E. Strand, D. A. Carlson, J. Water Pollut. Control Fed., 56, 464, (1984).
7. J. W., Hastings, C. J. Potrikus, S. C. Gupta, M. Kurfurst, J. C. Makemson, Adv. Microbial Physiol., 26, 235, (1985).
8. J. M. Borky, K. D. Wise, IEEE Trans. Electron Devices, ED26, 1906, (1979).
9. L. S. Fan, Y. C. Tai, R. S. Muller, Sensors and Actuators, 20, 41, (1989).
10. A. Manz, N. Graber, H. M. Widmer, Sensors and Actuators, B1, 244, (1990).
11. L. M. Roylance, J. B. Angell, IEEE Trans. Electron Devices, ED26, 1911, (1979).
12. S. Shoji, S. Nakagawa, M. Esashi, Sensors and Actuators, A21–A23, 189, (1990).
13. S. C. Terry, J. H. Jerman, J. B. Angell, IEEE Trans. Electron Devices, ED26, 1880, (1979).
14. OECD Guideline for Testing of Chemicals, 209, (1981).

STIMULUS-RESPONSE MEASUREMENTS ON PROTEIN CONTAINING MEMBRANES, DEPOSITED ON AN ISFET SURFACE.

P. Bergveld, MESA Research Institute, University of Twente,
Box 217, 7500 AE Enschede, The Netherlands

Abstract

The properties of an amphoteric metal oxide surface as applied for ISFET gatematerials can be described with the occurring acidic and basic surface reactions and the Boltzmann distribution. The same description can be used to explain the chemical behaviour of protein molecules. The main difference in chemical behaviour of the two systems originates from the fact that the surface of a metal oxide is solid whilst a layer of proteins exhibit a diffuse character. Considering these striking similarities and differences, a new view concerning the measurement possibilities arises with respect to the combination of an ISFET and an immobilized layer of proteins deposited on the gate.

Because the surface potential of an ISFET reacts very fast on a pH change of the surrounding solution, including on local changes at a distance of 1-10 nm from the surface, the device is very suitable to measure pH changes inside a membrane with immobilized proteins, which is deposited on the ISFET surface.

Temporarily pH changes in a protein membrane can be induced coulometrically by means of an integrated pH actuator. The reciprocal of the ISFET response to such a pH stimulus delivers the buffer capacity of the protein. Temporarily pH changes can also be induced by a stepwise change in the ion concentration of the liquid in contact with the protein membrane. This so-called ion-step can be regarded as a stimulus at the input of the system, whilst the ISFET reaction to it is the corresponding response. In this way a stimulus-response experiment can be carried out to detect the concentration of an adsorbed protein. In general all affinity based reactions with molecules having titratable surface groups can be monitored using the ISFET stimulus-response system. In addition the concentration of neutral molecules with a charged label can be measured.

G.G. Guilbault and M. Mascini (eds.), Uses of Immobilized Biological Compounds, 289–308.

Experimental results are shown, which support the feasibility of the stimulus-response approach. The reported examples are: the buffer capacity of ribonuclease and lysozyme as function of the solution pH, the response of lysozyme to a KNO_3 stimulus, the response of a charged glucose derivative and the competition with glucose and the response of a protamine loaded membrane and its specific reaction with heparin.

Introduction

Investigating the chemical behaviour of proteins as function of the ionic composition of the solution in order to explain interesting experimental results achieved from ISFETs with a layer of immobilized proteins on the gate, the similarities with the ISFET behaviour was noticed. Therefore in this paper the ISFET behaviour will first be described, but in a different way as usual [1], in order to achieve a description which links with the common description of the acid-base properties of proteins [2, 3]. The latter item will be summarized in the next section and focussed on the dynamic behaviour. Finally the experimental results as mentioned above will be given and analysed.

The buffer capacity of an amphoteric surface.

Inorganic oxides such as SiO_2, Al_2O_3 and Ta_2O_5 have been studied for many decades because of their technological importance, mainly as colloidal particles, but since 1980 also as the gate materials of ISFETs. In colloid chemistry the surface charge of the oxides is the parameter to be determined as function of the solution pH, whilst with ISFETs the oxide surface potential is the measurand. From the colloid chemistry it is known that most of the inorganic oxides have amphoteric surface sites, which means that the same surface hydroxyl group A-OH may dissociate, to yield a proton or take up a proton, according to reactions 1 and 2 respectively.

$$\text{A-OH} \underset{}{\overset{K_a}{\rightleftarrows}} \text{A-O}^- + H_s^+ \qquad (1)$$

$$\text{A-OH} + H_s^+ \underset{}{\overset{K_b}{\rightleftarrows}} \text{A-OH}_2^+ \qquad (2)$$

The protons which stay in the direct vicinity of the surface are denoted in the reactions as H_s^+ and are schematically shown in figure 1 as H_3O^+

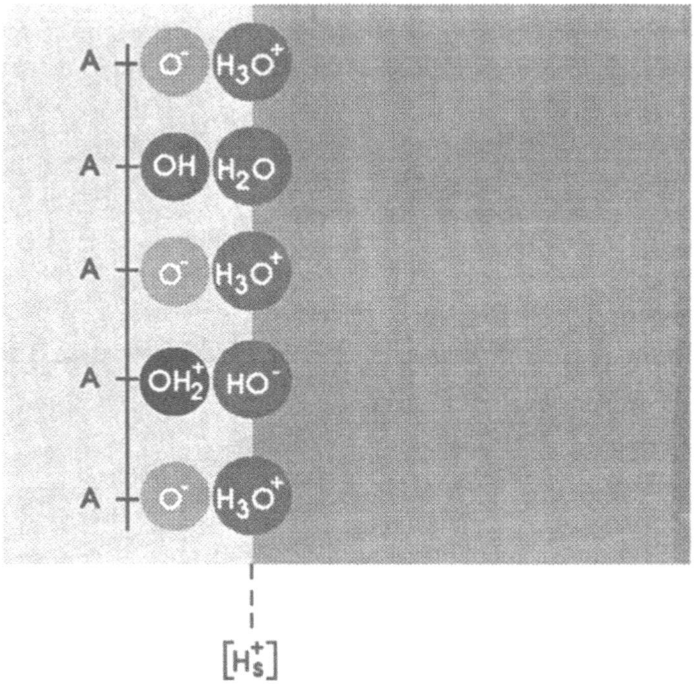

Fig. 1 Schematical representation of an amphoteric surface.

The equilibrium constants of the acidic and basic reactions (1) and (2) are respectively:

$$K_a = \frac{[A\text{-}O^-]\ [H_s^+]}{[A\text{-}OH]} \qquad (3)$$

$$K_b = \frac{[A\text{-}OH_2^+]}{[A\text{-}OH]\ [H_s^+]} \qquad (4)$$

Here [A-OH] is the number of neutral sites per cm^2 and $[A\text{-}O^-]$ and $[A\text{-}OH_2^+]$ the number of negative and positive surface groups per cm^2, randomly divided over the surface of which the values are determined by the respective equilibrium constants and the solution pH.

The value of $[H_s^+]$ follows from equations (3) and (4)

$$[H_s^+]^2 = \frac{K_a}{K_b}\frac{[A\text{-}OH_2{}^+]}{[A\text{-}O]^-} \qquad (5)$$

From equation 5 it can be concluded that $[H^+_s] = Ka/Kb^{1/2}$ for an overall neutral surface ($[A\text{-}OH_2^+] = [A\text{-}O^-]$). In this case $[H^+_s] = [H^+_b]$, the concentration of protons in the solution. The related $pH_b = pH_s$ is called the pH_{pzc} because it determines the point of zero charge at the surface.

The value of pH_{pzc} depends on the type of oxide with its specific values of K_a and K_b: $pH_{pzc} = \frac{1}{2}(pK_a - pK_b)$.

When, starting at $pH_b = pH_{pzc}$, the solution pH is increased, the surface is titrated and according to reaction 1 and 2 a negative surface charge density σ is generated:

$$\sigma = q\,([A\text{-}OH_2^+] - [A\text{-}O^-]) \qquad (6)$$

and consequently

$$[H^+_s]^2 = \frac{K_a}{K_b}\,\frac{\sigma/q + [A\text{-}O^-]}{[A\text{-}O^-]} \qquad (7)$$

The resulting surface potential ψ relative to the bulk potential, which is measured by an ISFET, follows from the Boltzmann equation

$$[H^+_s] = [H^+_b]\exp(-q\,\psi/kT) \qquad (8)$$

or

$$\psi = 2.3\,\frac{kT}{q}\,(pH_s - pH_b) \qquad (9)$$

Assuming that due to the titration pH_s changes with $\Delta pH_s = pH_{pzc} - pH_s$ and introducing a factor

$$\alpha = 1 - \frac{\Delta pH_s}{pH_{pzc} - pH_b} \qquad (10)$$

equation 9 can be described as:

$$\psi = 2.3\,\alpha\,\frac{kT}{q}\,(pH_{pzc} - pH_b) \qquad (11)$$

with $0 < \alpha < 1$.

The relation between the surface charge σ and the surface potential ψ is determined by the Gouy-Chapman relation:

$$\sigma = \psi\left(\frac{C_{st}\,C_{dl}}{C_{st} + C_{dl}}\right) \qquad (12)$$

where C_{st} is the fixed Sternlayer and C_{dl} the diffuse double layer capacitance respectively. In highly concentrated solutions C_{st} is determining the capacitance to be charged, in diluted solutions C_{dl}. In a 1 mM solution C_{dl} determines mainly the capacitance value, being in that case in the order of 1 μF cm^{-2}. This means for equation 7 that, using equation 11 and 12:

$$[H_s^+]^2 = \frac{K_a}{K_b} \cdot \frac{3.6\ \alpha\ 10^{11}\ (pH_{pzc} - pH_b) + [A\text{-}O^-]}{[A\text{-}O^-]} \tag{13}$$

From equation 13 different conclusions can be drawn for the cases that the factor α approaches 0, 1 or is intermediate.

$\alpha \Rightarrow 0$

This is the case that the surface potential ψ as well as the surface charge σ remain zero due to the fact that no surface reactions take place ($[A\text{-}O^-] = [A\text{-}OH_2^+] = 0$). In fact titration of the surface is not possible. $[H^+_s]$ is not determined by the surface but can freely follow $[H^+_b]$. This case is thus not interesting for proton sensors such as ISFETs although one would like to be able to construct such a device as a so-called REFET.

$\alpha \Rightarrow 1$

Surfaces having already large values of $[A\text{-}O^-]$ (and $[A\text{-}OH_2^+]$) at pH_{pzc} compared to $3.6\ 10^{11}\ (pH_{pzc} - pH_b)$ keep the value of $[H^+_s]$ nearly constant during titration of the surface. For metal oxides the upper limit of A-OH sites, delivering the $A\text{-}O^-$ and $A\text{-}OH_2^+$ groups, lies around 10^{15} cm^{-2}. If we assume that 10% is dissociated at pH_{pzc}, $[A\text{-}O^-] = 10^{14}$ cm^{-2}, which means that even with 10 decades pH change ($pH_{pzc} - pH_b = 10$), the value of $[H^+_s]$ is still hardly changed. This phenomenon is in fact the operational mechanism of a buffer, where the pH of the (buffered) solution is also kept relatively constant due to the dependence of the ratio between the acid or base and an added salt concentration, rather than by the acid or base concentration itself.

$0 < \alpha < 1$

In this case the value of $[A\text{-}O^-]$ at pH_{pzc} may not be large enough to neglect the influence of $3.6\ \alpha\ 10^{11}\ (pH_{pzc} - pH_b)$ to $[H^+_s]$. The surface is not able to keep $[H^+_s]$ constant. However far away from pH_{pzc} $[A\text{-}O^-]$ will be increased so much that its value has passed $3.6\ \alpha\ 10^{11}\ (pH_{pzc} - pH_b)$ and consequently, as in the case of $\alpha \Rightarrow 1$, $[H^+_s]$ is relatively constant. Therefore these type of surfaces only buffer far away from pH_{pzc}.

The buffering property of inorganic oxides forms in fact the basis of the ISFET operational mechanism. The Boltzmann equation (8) predicts that the more $[H^+_s]$ = constant, the better ψ is responding in a Nernstian way to the solution pH = - log $[H^+_b]$, with a maximum of 59.3 mV/dec at 25°C. This is also observed in practice. The case α => 1 holds for instance for Ta_2O_5, resulting in an almost maximal pH sensitivity ($\psi \approx$ 58 mV/dec at 25°C) over the full pH range. The case $0 < \alpha < 1$ holds for SiO_2, showing a low sensitivity around pH_{pzc}, increasing to the maximum far away from pH_{pzc}.
Up to now only basic titrations were considered, leading to a negative surface charge. Because of symmetry the same arguments hold for pH values at the other side of pH_{pzc}, leading to a positively charged surface.
When the solution concentration is increased at a constant pH_b by adding a salt, ψ is still determined by pH_b-pH_s and because the value of C_{dl} is increased, in the limiting case to the value of C_{st} (see equation 12), σ will increase proportionally. This additional surface charge density delivered by deprotonation of the still available A-OH sites. With C_{st} = 20 μF cm^{-2}, $[AO^-]$ should be larger than 7.2 α $10^{12}(pH_{pzc} - pH_b)$ which is still possible for most inorganic oxides.
In colloid chemistry one measures indeed an increase of the surface charge as function of added salts, whilst ISFET experiments show hardly any effect on the measured value of ψ. This means that the surface also buffers $[H^+_s]$ against variations in electrolyte concentration.
Only in the case of specific adsorption of small cations in the case of a negatively charged surface and anions in the case of a positively charged surface, the buffer capacity is diminished. In that case the ratio $[A\text{-}OH_2+]/[A\text{-}O^-]$ and thus the pH_{pzc} ($[H^+_s]$) is disturbed. This effect is less for surfaces with a large site density and thus Ta_2O_5 ISFETs are less sensitive to specific adsorption than SiO_2 ISFETs, which effect is also observed in practice.

Note that in the case of membrane covered ISFETs for the development of ion sensors which are selective for one specific ion, the membranes contain the corresponding receptor molecules (ionophores) which serve for maintaining the ions to be measured at a constant level in the membrane. This approach is in fact nothing else than what the amphoteric surface of a pH-sensitive ISFET does in a direct way with respect to hydrogen ions.

Acid-base properties of proteins

The basic components of proteins are α-amino acids. All 20 common natural amino acids have similar structures except for proline and hydroxyproline. The general structure is given in figure 2.

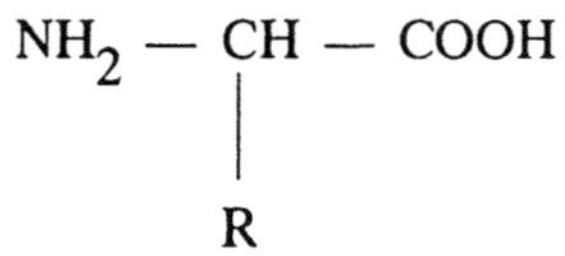

Fig. 2 Notation of an α-amino acid molecule.

All amino acids have an amino group and a carboxyl group. The difference among amino acids is the variation in the side chain R. If the carboxyl group of one amino acid binds with the amino group of another, a water molecule is released and an amide bond is formed between the two amino acids. In this way a dipeptide is formed. Similarly a tripeptide up to a polypeptide can be formed, consisting of an amino and a carboxyl end group with many amino acid residues in between connected in a peptide chain. All amino acid residues contain their own specific side chains R_1 up to R_n as symbolically given in figure 3.

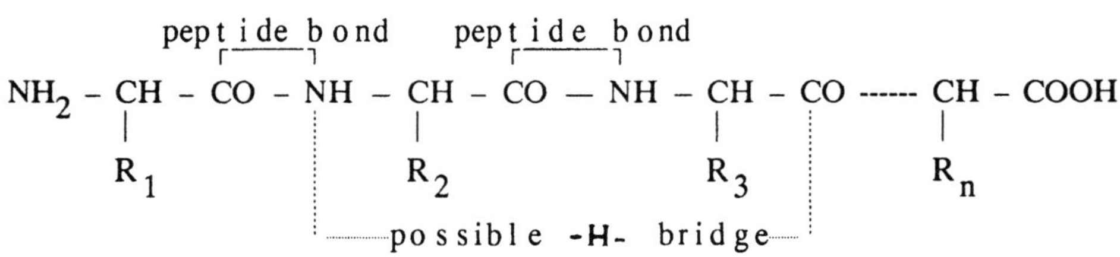

Fig. 3 Notation of a polypeptide with side chains R_1 ..... R_n.

The amino and carboxyl end groups of a polypeptide can bind a proton or release a proton resulting in respectively NH_3^+ and COO^- groups, depending on the solution pH. The side chains R may also contain ionizable groups which can also be positively or negatively charged depending on the solution pH. The result is that the molecule with all the different ionizable groups shows as a whole an amphoteric character, although each particular group is not amphoteric (Surfaces which behave in this way are called Zwitterionic surfaces). Note that in this case different groups are capable to accept a proton and become positive or to release a proton and become negative, while with metal oxides as described in the previous section one and the same group can become positive or negative.

Proteins are macromolecules that are composed of one or more polypeptide chains, linked to each other by secondary bonds or such as disulphide bonds or hydrogen bonds. The final spatial conformation of a protein results therefore mainly from coulombic interaction, covalent bonds and hydrophobic interaction. The sequence of the amino residues of the original polypeptides, each with their specific R chains, determine to a great extent the type of protein. Regardless of the spatial conformation, resulting in a specific biological activity, such as enzymatic activity, or immunological binding activity, the chemical properties of a protein molecule under physiological circumstances are still mainly determined by those of the original polypeptides. From this point of view a conglomerate of protein molecules as for example immobilized on the surface of a sensor can symbolically be represented as shown in figure 4 where for the sake of simplicity only one polypeptide chain is shown.

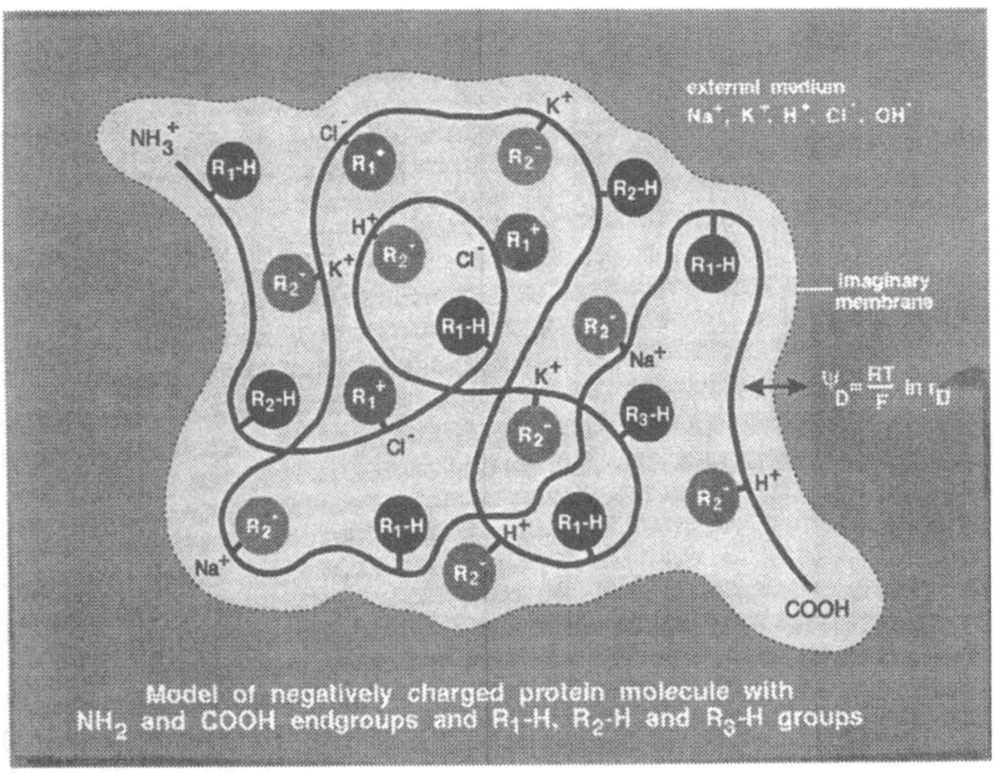

Fig. 4 Model of a conglomorate of protein molecules, with one polypeptide chain drawn.

An imaginary membrane is proposed to surround the conglomerate of molecules, permeable for all ions. The net charge resulting from the titratable groups of the polypeptide chains is a function of the solution pH. The internal concentration of all mobile ions including protons and hydroxyl ions available from the external solution, is determined by the Donnan ratio $r_D = (a_{ie}/a_{ii})^{1/z}$ where a_{ie} and a_{ii} are the ion activities of the mobile ions with valence z, respectively for the external and the internal space. This ion distribution is in equilibrium with the Donnan "membrane" potential ψ_D:

$$\psi_D = \frac{RT}{F} \ln r_D \qquad (14)$$

The Donnan potential is by definition the internal potential with respect to that of the external solution. Note that its value is a function of the external solution concentration.

Comparison between acid-base properties of amphoteric surfaces and proteins.

In the description of the electrochemical behaviour of an amphoteric surface, focussed on the ISFET operational mechanism and the simplified description of the electrochemical behaviour of a conglomerate of protein molecules it is striking that for both systems the external solution pH and ion concentration determine the surface charge density respectively the internal charge density. Due to the amphoteric character of both systems at a certain pH the point of zero charge (pH_{pzc}) is manifesting, for a protein called the isoelectric point (pH_{iep}). Starting at this point the surface of an ISFET can be titrated, resulting in a corresponding surface charge density. However under conditions of large values of $A\text{-}O^-$ or $A\text{-}OH_2^+$ groups, the surface $pH_s = -\log [H^+_s]$ remains constant, as determined by the pK values of the acidic and basic reactions. This phenomenon can be described in terms of the buffer capacity of the surface. Due to the resulting constant value of pH_s, the surface potential depends in a Nernstian way on the external solution pH, hardly influenced by the concentration of the other ions, except for the case of specific adsorption.

The electrochemical behaviour of proteins shows similarities and differences with respect to metal oxides. In the first place, due to the large diversity of ionizable groups, many more reactions of the type (1) and (2) as used in the previous section to describe metal oxides are simultaneously active with as many pK values. So theoretically the system is more difficult to describe but nevertheless the molecules are titratable, show a certain buffer capacity, an internal charge density and a conglomorate of molecules exhibits a "membrane potential", adapting the model represented by figure 4. The model of figure 4 shows the diffuse character of a layer of protein molecules which just implies the difference with respect to the inpermeable metal oxide layers. Due to the diffuse character all type of ions can enter the layer with the result that the "membrane potential" is not only a function of the external pH, but is also influenced by the concentration of all other ions, which can be described with the Donnan theory.

The surface potential ψ of an ISFET is a direct measurable potential on which the existence of ISFETs is in fact based. On the contrary the Donnan potential of a relative dense layer of protein molecules, for instance immobilized in a membrane, is not a measurable potential [4].
Summarizing the comparison between protein and metal oxide layers it has to be concluded that in principle the acid base properties are similar. Both systems contain titratable groups, show a certain buffer capacity and a point of zero charge and develop a surface charge and potential. In the case of metal oxides the surface potential is mainly determined by the external solution pH whilst the surface potential of proteins is a function of the concentration of all types of ion. The surface potential of a metal oxide is measurable by applying the ISFET concept, but the surface potential of proteins is not measurable.

The dynamic behaviour of proteins.

When a layer of protein molecules is in equilibrium with an external solution with a certain $pH \neq pH_{iep}$ and a certain concentration of other ions, the internal charge, pH, ion concentrations and "membrane potential" are fixed by the Donnan effect. For example let an external pH = 7 result in a net negative charge giving tise to a Donnan potential of -3 x 59mV and an internal pH = 4. When at this equilibruim the external ion concentration of for instance KCl is suddenly increased, the Donnan potential will quickly decrease, say to -2 x 59mV. This means that the internal pH has to be adapted to pH = 5. The original pH = 4 must thus increase by releasing protons to the external solution which will therefore in the near vicinity of the protein molecules temporarely decrease to pH = 6. The proton release is supported by the buffer capacity of the neutral groups, especially those having a pK value around pH = 4 up to the moment that the internal pH = 5. The final result will be that temporarily a relative large release of protons occurs, before the new equilibrium is established. These protons will ultimately diffuse into the bulk of the external solution. However, because of the time needed for the diffusion, the proton release causes a temporarily pH decrease (in the example from 7 towards 6) in the direct vicinity of the protein molecules. If these are furthermore immobilized in a membrane which even restrains the diffusion, the temporarily ΔpH is measurable if a pH sensor, having a fast response, is placed in the near vicinity of the protein molecules. In other words: when the proteins are immobilized in a membrane, which is deposited onto an ISFET, the dynamic pH response of the protein molecules to a sudden change in the ion concentration should be measurable. Note that in this case the response of the protein molecules to the concentration stimulus is measurable at a certain distance of the molecules itself, which is very favourable because of the problem of binding proteins to the surface of a sensor which will never succeed within a zero distance. In the next sections results will be shown of stimulus-response experiments with ISFETs and protein molecules in the direct vicinity, focussed on the measurement of the buffer capacity as function of the pH and the proton release as function of an ion concentration stimulus.

Measuring the buffer capacity of proteins.

Considering a buffer system consisting of a weak acid: HA —> $H^+ + A^-$, the buffer capacity is defined as the small amount of base d[B] needed to change the pH by dpH, as given by [5]:

$$\beta = \frac{d[B]}{dpH} = 2.3 \left\{ \frac{K_a C_o [H^+]}{(K_a + [H^+])^2} + [H^+] + \frac{K_w}{[H^+]} \right\} \tag{15}$$

where K_w is the ion product constant for water, K_a is the dissociation constant of the acid HA and C_o is the initial bulk concentration of the acid, i.e. [HA] + [A^-]. The buffer capacity is maximal at pH = pK_a and is the reciprocal of the slope of the titration curve of the acid. So from a measured titration curve the buffer capacity can be calculated as function of the pH. The maximum is found at the deflection point of the titration curve.

Due to the many different titratable groups in the side chains of the polypeptides of a protein molecule, as well as possible specific interactions between the groups, proteins show different intrinsic pKs and thus more maxima of the buffer capacity. Nevertheless the buffer capacity can also in the case of proteins be derived from the titration curve, showing a specific chemical "fingerprint" of the titrated protein. It should however be noted that the registration of a titration curve of a solution containing proteins is a difficult and time consuming procedure and a considerable error is introduced by the differentiation procedure. Therefore we developed an alternative method, using an ISFET provided with a porous gold electrode surrounding and covering the gate area of the ISFET, as schematically shown in figure 5 [6].

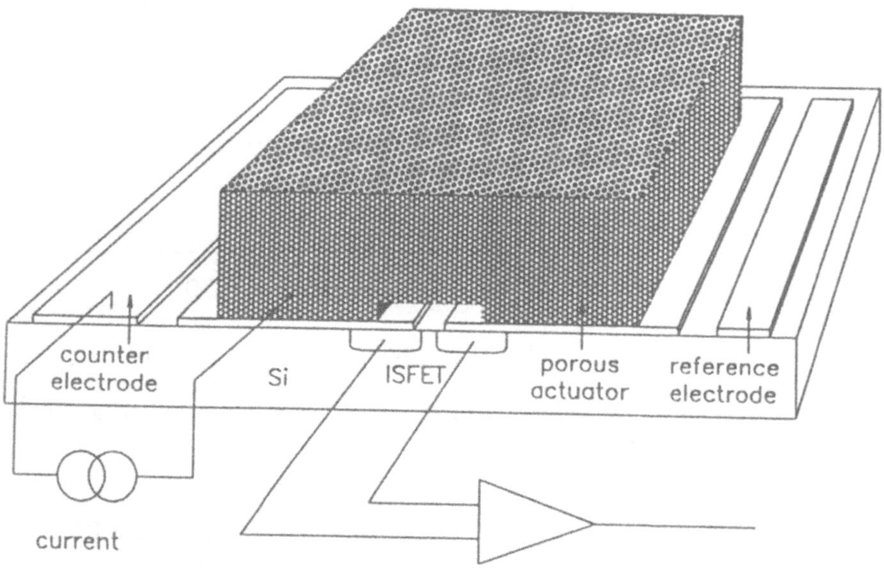

Fig. 5 Schematic representation of the basic components of the coulometric sensor-actuator system based on an ISFET with a gate-covering porous actuator.

The porous electrode is about 100 μm thick and situated at a distance of about 1μm from the gate. The electrode is used as a coulometric proton actuator [ ]. In this case a sinusoidal actuator current of 10 μA is applied with a frequency of 1 Hz, resulting in a sinusoidal pH variation in the pores of the actuator. This procedure is illustrated in figure 6.

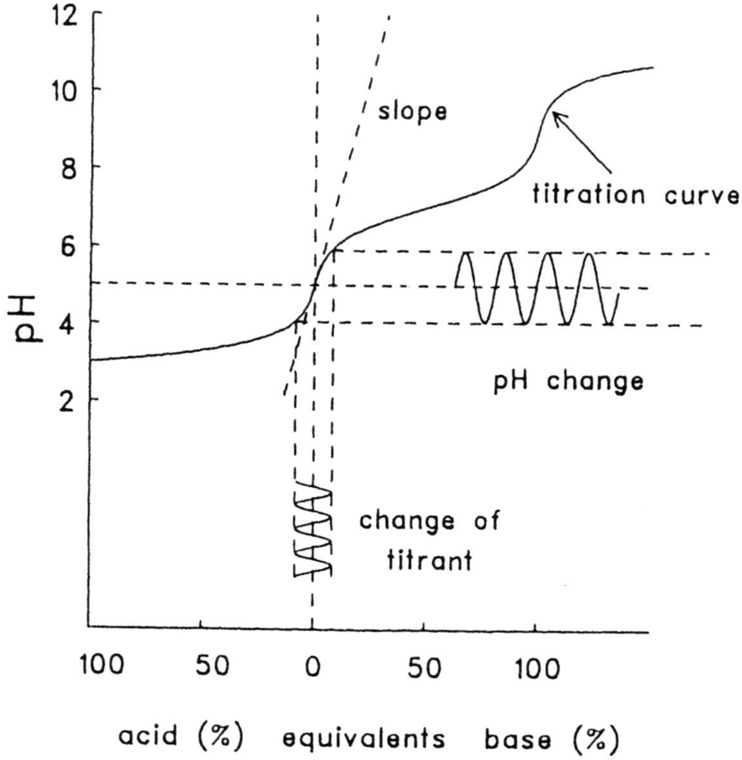

Fig. 6 Illustration of the buffer capacity measurement by means of an applied sinusoidal perturbation of titrant.

The ISFET measures the pH variations and the reciprocal of the RMS value of the ISFET amplifier output voltage is recorded as function of the pH of the solution containing the proteins to be characterized. The proteins may also be adsorbed to the actuator electrode in the pores. As an example the buffer capacity, measured as described above, is recorded from a solution containing 5g/l ribonuclease and lysozyme respectively, corrected for the measurement of the blank solution. The resulting "fingerprints" are given in figure 7.

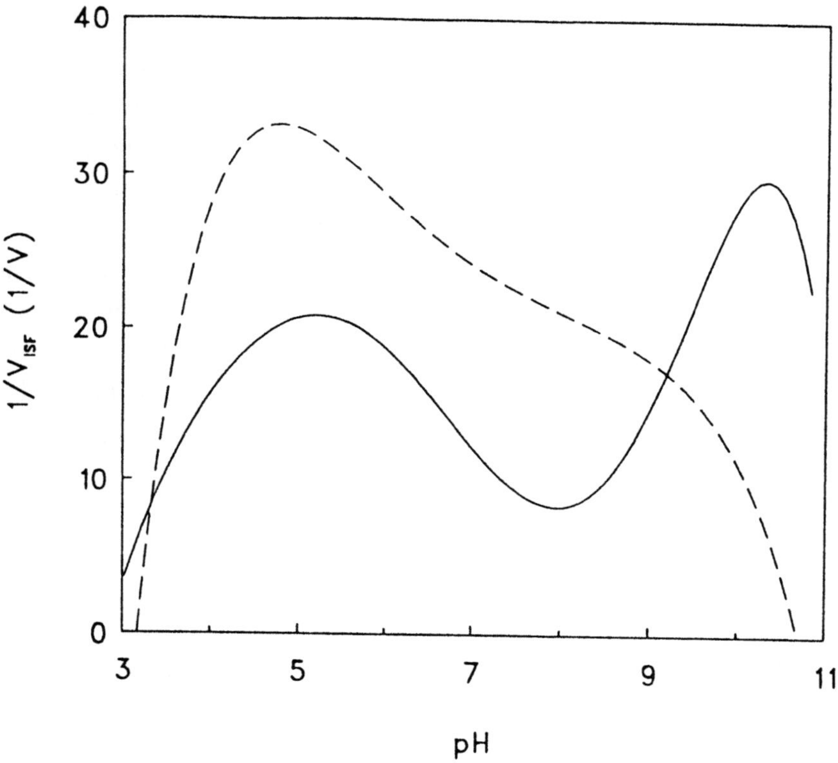

Fig. 7 Buffer capacity measurement of 5g/l ribonuclease (solid line) and lysozyme (dashed line) after correction. [KNO_3 = 0.1 M].

The procedure of measuring the buffer capacity of proteins by means of a local coulometric titration is a clear example of a dynamic measurement in which proteins are combined with an ISFET as proposed to be useful in the previous section.

Measuring the proton release or uptake of proteins upon a concentration stimulus.

For stimulus-response experiments ISFETs have been constructed with a small cavity in the epoxy insulation in which a protein membrane can easily be deposited. The membranes may exist of directly crosslinked proteins or immobilized proteins in a membrane matrix. The membrane covered ISFETs are mounted in a computer-controlled flow-through system which contains a downstream mounted reference electrode. An other option is that the ISFET itself is provided with an Ag/AgCl electrode around its gate, which can act as the reference electrode. The pH of the carrier solution can be varied from pH = 4 to pH = 9 by mixing two buffered solutions which both contain also 10 mM KCl or KNO. This buffered KCl or KNO_3 solution is temporarily replaced by a 50 mM unbuffered KCl or KNO_3 solution within 50 ms, causing a concentration stimulus. After the protein response the ISFET may directly respond to the pH of injected solution if this deviates from the initial solution, however, delayed by the membrane barrier.

As a first test measurements have been carried out using a membrane matrix of polystyrene beads with agarose, incubated with lysozyme, deposited on an ISFET with integrated Ag/AgCl reference electrode. Figure 8 shows a typical result at pH = 4 of the carrier solution, using KNO_3 as the stimulus (from 10 mM to 50 mM). The ISFET response shows a negative value, which corresponds with a proton uptake, caused by the positively charged lysozyme at this pH.

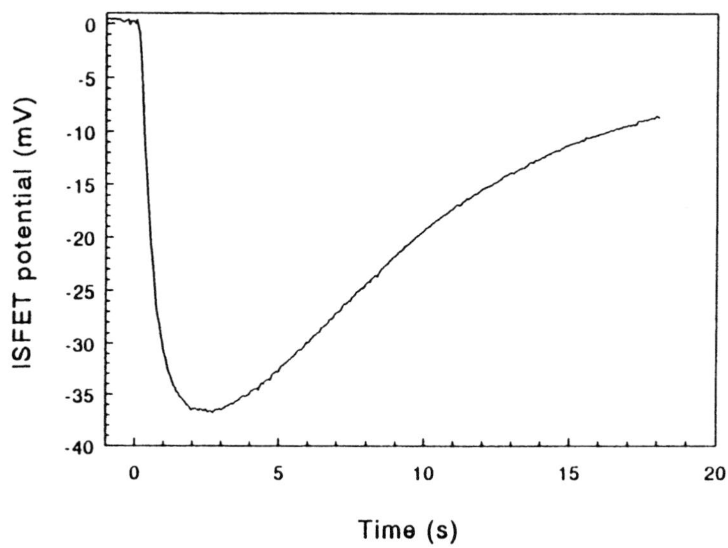

Fig. 8 Response of an ISFET covered with a lysozyme membrane, using a stimulus of 10 mM to 50 mM KNO_3.

The experiment shows unambiquously that around the protein molecules the pH changes temporarely as a response to the concentration stimulus.

Another experiment is related to the development of a glucose sensor, based on a competion reaction between glucose and a charge labeled glucose derivative in a con A containing membrane [7]. For this experiment ISFETs were provided with a sepharose-con A gel, covered with a 70 μm-pore polyester membrane for fixation. Stimulus response experiments with a buffered 1 mM KCl changed to an unbuffered 10 mM KCl solution showed that the con A loaded membranes are positively charged at pH=6 (proton uptake) and negatively at pH=8 (proton release). After 48 hours incubation with 1.5 mg/ml aminophenylglucose-derivatized heparin (hep-APG) the responses are shifted to more positive values, indicating the more negative charge of the membrane. Figure 9, curve 1 shows the result in a pH=7.2 carrier solution. After the proton release the ISFET responds to the pH of the 10 mM KCl solution which is approximately 5.6.

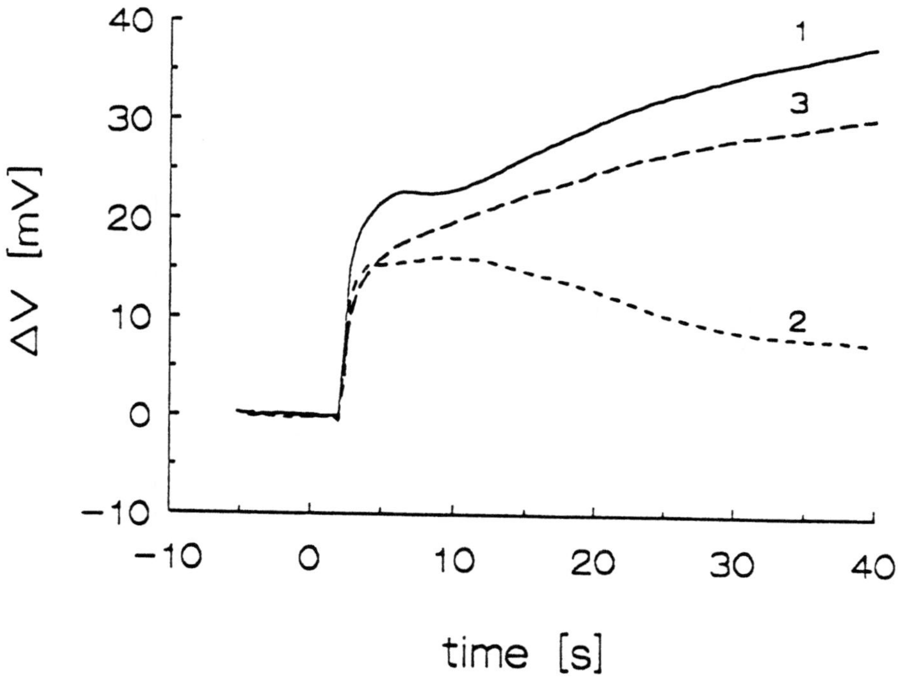

Fig. 9 Response of an ISFET covered with a sepharose-con A membrane towards an ion step from 1 mM to 10 mM KCl at pH = 7.2, subsequently after incubation with 1.5 mg/ml hep-APG (curve 1), incubation with 50 mg/ml glucose (curve 2) and incubation with 1.5 mg/ml hep-APG (curve 3).

Subsequently the devices were placed in a solution containing 50mg/ml glucose for 48 hours, after which the ion-step experiment was repeated. Curve 2 of figure 9 shows the result. Clearly the uncharged glucose has replaced the charge labeled glucose. Finally the devices were again incubated for 48 hours with hep-APG (1.5 mg/ml), resulting in curve 3. Although not all details of the response are completely understood at this moment, the result shows clearly the competition between glucose and hep-APG.

The last example of a stimulus/response experiment is derived from a project aiming to develop an affinity based heparin sensor [8]. In this case ISFETs are provided with a membrane matrix of polystyrene beads with agarose, with covalently bound protamine.

Heparin is at physiological pH values a highly negatively charged molecule, whilst protamine is positively charged. The bottom curve of figure 10 results from a stimulus of 10 mM KCl to 50 mM KCl released to a device containing protamine covalently bound in a membrane consisting of 1.0 μm polystyrene beads. It shows a negative response, corresponding with a proton uptake.

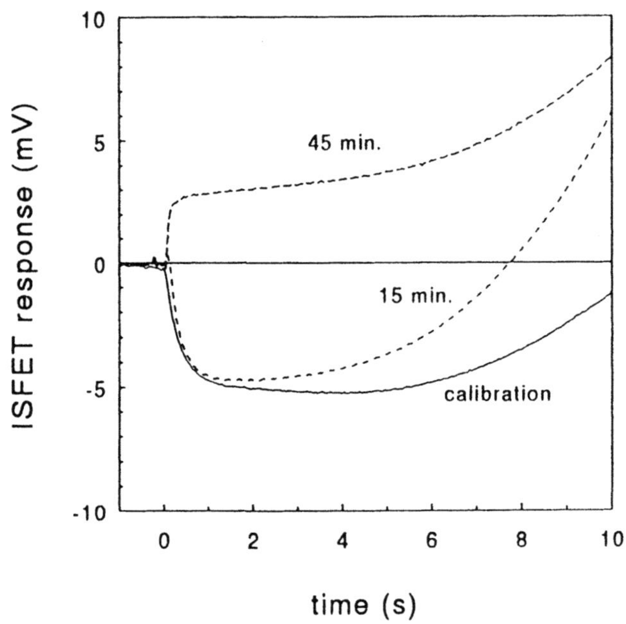

Fig. 10 Response of an ISFET covered with a membrane of 1.0 μm polystyrene beads with covalently bound protamine using a stimulus of 10 mM to 50 mM KCl (calibration curve) and after 15 minutes and 45 minutes incubation with 0.8 Unit/ml heparin.

After subsequent incubation with 0.8 Unit/ml heparine in a PBS solution, the upper curves of figure 10 were recorded after 15 and 45 minutes. The experiment shows that due to the reaction with heparine the net charge of the membrane change from positive to negative, corresponding with a proton release. This change is a function of the heparin concentration. Similar experiments are carried out with incubation of a protamine loaded device in blood plasma with different heparin concentrations. In this case an offset is found due to nonspecific adsorption of plasma components other than heparin.

Conclusions.

The fundamental properties of amphoteric metal oxide surfaces and protein molecules have notably similarities: they have both a certain buffer capacity, exhibit charged groups as function of the surrounding solution pH and a corresponding potential with respect to that solution. Due to these properties ISFETs with a very reactive metal oxide gate material have a Nernstian pH response hardly influenced by the concentration of other ions than protons. On the other hand due to the same properties, protein molecules exhibit a proton release or uptake, induced by a concentration stimulus. Due to the protein buffer capacity, relative small stimuli can cause very large pH changes in the vicinity of the molecules, which can easily be measured by an ISFET which is covered with a membrane with immobilized proteins.

Acknowledgement.

The author wishes to expres his gratitude to the PhD students Jan Eijkel, Joost van Kerkhof, Jans Kruise and Luo Jiang, who carried out the experiments described in the last section.

References

1. P. Bergveld and A. Sibbald.
Analytical and biomedical applications of ion-selective field-effect transistors, Comprehensive Analytical Chemistry, vol. XXIII, 1988, Elsevier Amsterdam.

2. L. Stryer
Biochemistry 2nd ed., 1975, W.H. Freeman and Company, New York.

3. C. Tanford
The interpretation of hydrogen ion titration curves of proteins.
Advances in protein chemistry, 17, 1962, pp. 70-165.

4. R.B.M. Schasfoort, P. Bergveld, R.P.H. Kooyman, J. Greve.
Possibilities and linitations of direct detection of protein charges by means of an immunological field-effect transistor.
Analytica Chimica Acta, 238, 1990, pp. 323-329.

5. D.D. Perrin and B. Dempsey.
Buffers for pH and metal ion control, 1974, Chapman and Hall, London, p. 11.

6. Jiang Luo.
Employment of a porous gold actuator in ISFET-based coulometric sensor-actuator systems. With application to protein characterization. Thesis University of Twente 1993.

7. J. Kruise
Perspectives of glucose sensing based on a charge-modulating competition reaction. Thesis University of Twente 1993.

8. J.C. van Kerkhof, P. Bergveld, R.B.M. Schasfoort.
Development of an ISFET besed heparin sensor using the ion-step measuring method. Submitted to Biosensors and Bioelectronics.

Micro Sensors in Bioelectroanalysis

Géza Nagy, Klára Tóth, Benjamin R. Horrocks# and Allen J. Bard#
Institute for General and Analytical Chemistry,
Technical University of Budapest, Budapest (Hungary)
#Department of Chemistry,
The University of Texas at Austin, Austin, Texas 78712

Abstract The miniaturization of electrochemical sensors, owing to the size and the special properties of micro devices allows and initiates novel, special applications in different fields of experimental sciences. This contribution will show a new trend of the application of micro sensors, namely the possibilities of scanning electrochemical microscopy with amperometric and potentiometric micro tips in studying biocatalytic sites.

INTRODUCTION

The variety of new types of micro or ultra micro electrodes, micro probes, micro sensors is increasing very rapidly today when the research on chemical sensors in general, is very flourishing and progressive. In the sensor research often high technological steps such as photolithography, ion implantation, laser tool machinery, etc. are employed.

Among electroanalytrical micro sensors, micro voltammetric and potentiometric ion-selective electrodes are the most extensively studied. Well established, well described, easy to acquire preparation methods are at disposal. The information about the behaviour and special electrochemical properties of these micro electrodes are close to be incorporated into advanced textbooks.

At different times in the course of our research we employed micro electrodes for studying their properties or for using them in special analytical tasks e.g. in vivo studies. This article is based on our observations obtained in these investigations. Accordingly, in the first part we briefly summarize methods of preparation and specific properties of certain types of micro electrodes and then discuss some of our results obtained studying immobilised enzyme systems with micro electrodes.

PREPARATION AND PROPERTIES OF MICRO ELECTRODES

Considering the single sensing element based electrodes -which are the most common among micro electrochemical sensors- two very different designes are popular. In one the measuring surface is exposed at the end of a very pointy electrode stem. This is the form of the micro pipette type ion selective electrodes or the chemically etched platinum tip holding needle type voltammetric ultra micro electrodes. The measuring surface is not shielded (needle type) or has just a very thin shield around it (micro pipette). The electrodes are ideal for measurements when much care is needed for not influencing the target area adjacent to the measuring tip, like in *in vivo* applications. In other cases the micro sensing element is surrounded by a robust insulating shield, thus the electrode is not very delicate and the measuring disc's surface area is well defined. In electrochemical studies this electrode design is often preferred.

The body of the pointy electrodes is often made of previously well cleaned glass capillaries pulling them down to the required tip size. A skilful glassblower can make this step

G.G. Guilbault and M. Mascini (eds.), Uses of Immobilized Biological Compounds, 309–324.

manually; however, specific electrode pullers are available on the market. They make the pulling easy and reproducible.

In ion-selective micro pipettes the inner wall of the glass tip is coated with a hydrophobic film. The film is formed by reacting the glass surface with an appropriate silylating agent such as trimethyl chlorosilane. The ionophore containing ion-selective cocktail layer is introduced into the tip while the internal filling solution into the barrel. Electrodes can be made with sub micron tip size also in this way. To avoid field potential interferences the reference electrode is also made in the form of a micro pipette, the tip of which is attached to the ion selective electrode measuring tip. In this case double barrel or "theta" (Θ) capillaries are used to start the preparation procedures. Similarly multi-barrel electrode assemblies can be prepared from multi-barrel capillaries by filling the different tips with different ion-selective cocktails.

When a pointy voltammetric electrode is made, the micro graphite fibre or metal wire is glued, cemented or sealed into the tip of the micro pipette. The conductive tip is cut to form a micro disc or etched to give a micro needle. Silver epoxy, mercury or carbon paste contacts are made to a lead wire inside the barrel.

To fabricate a shielded micro disc voltammetric electrode, a conductive fibre (platinum or carbon) is inserted into the closed end of a glass tubing. The end section of the glass tube is carefully heated up by electric coil or micro flame to have the fibre sealed in the centre of the glass along a shorter or longer section. Often, when this electrode preparation step is performed a vacuum pump is connected to the tube, thus the enclosure of air bubbles inside the melted glass can easily be avoided. The next step of electrode preparation is the sanding off the end of the glass tube to expose the electrode disc on the flat end plate of the glass body. The electrode is finely polished afterwards and the electric contact is made to the wire placed inside the tube with mercury, silver epoxy or carbon paste. The preparation procedure is slightly different when a chemical or electrochemical etching is employed. For example, the shielded platinum ultra micro disc electrodes are made of silver-coated platinum wire and after positioning the silver coated fibre inside the glass electrode body a chemical etching step is carried out. The final sealing in step is performed just after the silver removal.

The specific properties of ion-selective micro electrodes are well studied, and described. As a matter of fact the only advantage they possess compared to the conventional ion-selective electrodes is that they can approach small size samples without much environmental disturbance and they can give information with extremely good spatial resolution. Micro ion-selective electrodes can advantageously be used in special analytical tasks. All the other special properties of such electrodes, however are making their use difficult e.g. the high internal resistance which requires specific measuring apparatus and care. Their mechanic delicacy and sensitivity towards electric shock require much care when they are prepared, calibrated, stored or used.

It is far different with voltammetric micro- or ultra micro sensors. They possess very important special characteristics that well can be taken advantage of, even in measurements when the size of the sample does not require small electrode tip. These special characteristics are very well understood and are described in reviews and books. Here we just shortly review some of them.

One important point that because of the relatively increased contribution of the spherical diffusion to the total mass transport, the Faraday current - condenser current ratio is larger. It means that the relatively high intensity of mass transport at the ultra micro electrode surface improves the signal to noise ratio. Furthermore, it can provide additional beneficial effects, namely the intermediate reaction products spend less time in the vicinity of the micro electrode surface. Therefore, the influence of chemical complications in the electrode process (e.g. catalytic regeneration of the reacted species) can have less influence on the shape of the voltammogram. If the electrode is employed under convective diffusion circumstances, e.g. in streaming solutions then the change in the intensity of convection have much smaller influence on the voltammetric current, than it is usual for conventional sized electrodes.

Another remarkable advantage of ultra micro electrodes can be concluded by observing the simple equation describing the time dependence of the condenser charging current (i_{C_t}):

$$i_{C_t} = \frac{\Delta E}{R}\exp[\frac{-t}{RC}] \qquad (1)$$

where ΔE is the potential difference, R is the resistance and C is the capacitance. It is obvious that by decreasing C the faster decay of the capacitive current will occur. This advantage of ultra micro electrodes is well studied and used in fast scan or fast pulse measuring techniques.

Finally the most obvious advantage of ultra micro voltammetric electrodes are mentioned. When an ultra micro electrode is employed, the very small voltammetric current results in a small "iR" voltage drop, accordingly, two electrode measuring apparatus can more often be employed or media with lower conductivity can also be accessible to voltammetric studies.

APPLICATION OF MICRO ELECTRODES

Electrochemical scanning microscopy for local active enzyme site studies

The invention of scanning tunnelling microscopy (STM) in 1981 and the atomic force microscopy (AFM) in 1985 resulted in a new type of surface studying technique, the so called scanning probe microscopy. This technique requires a fine measuring tip, which is very accurately positioned and scanned over the target surface in the very close proximity. A signal is generated by the surface of the probe. The connected computer stores the signal together with the spatial co-ordinates of the probe and composes an image of the surface. By this technique sometimes unbelievably high topographic resolution can be achieved. Images can be created with atomic resolution. A very important advantage of the technique is that no vacuum is necessary for its operation. The probe microscope with all of its belongings in most cases can easily be accommodated on a normal size desk. Electrochemists intensively use STM or AFM to study electrode deposited films or crystals, and electrochemical corrosion processes. Electrochemists also developed their own probe microscopy i.e. the scanning electrochemical microscopy (SECM) which in most of its applications uses shielded micro disc electrode as a measuring tip. It can give some chemical information about the surface studied in addition to the topographical one. However, it could not achieve yet the resolution of STM or AFM. The theory of the technique and different relevant investigations have been carried out with SECM, and described in a series of papers in detail (e.g.[1, 2]).

In our research work the application of SECM for local enzyme activity imaging was investigated. One of the approaches for this can follow the "conventional" working principle of amperometric SECM. It uses a reversible redox-system to create the image. The tip electrode measures the amperometric current. When the electrode is approached to the surface, the amperometric current will decrease or increase. The decrease is caused by the so called negative feed back, which is the result of a steric shielding effect of the surface. The proximity of the surface hinders the mass transport of the electroactive material to the electrode. The current increases if the species reacted on the tip electrode are regenerated in a fast process on the approaching surface. The mass flow of the electroactive species thus increases and we call this as a positive feed-back.

The regenerating reaction can be an enzyme catalysed reaction indicating the presence and activity of an enzyme, a biocatalytic tissue, organelle or micro-organism. This principle has been successfully applied earlier by Bard and co-workers [2].

Another approach can be to detect the concentration of a natural substrate, reaction partner or product with a selective probe, accurately positioned and scanned above the target surface. In our work we investigated this approach. On one hand, amperometric detection with platinum micro disc electrode was used. In these experiments hydrogen peroxide produced in oxido-reductase enzyme catalysed processes was detected by electrochemical oxidation. On the other hand, the applicability of potentiometric detection mode in SECM was studied. In this work pH sensitive potentiometric micro electrodes were used as measuring tips.

It must be mentioned here that while amperometric SECM can rely on the interaction of the system with the chemical species generated at the tip, potentiometric detection requires the existence of concentration profiles between the surface and the bulk of the solution. In other words potentiometric detection is only possible if there are local ion activity profiles in the system under investigation. This is the case if a biocatalyst is on the surface and a chemical process catalysed by the catalyst is proceeding. If the system is at steady state an image of these ion activity profiles can be made by scanning the potentiometric tip across the surface in the usual way.

In life sciences many different designs for small pH electrodes have been successfully employed. In the earlier studies glass membrane micro electrodes gained application, but have since been superseded by the glass micro pipette based sensors. In studies of living systems the location of the measuring tip is determined either by optical microscopy or by the effect of the system on the electrode, e.g. the potential change when the tip penetrates a cell.

An advantage of amperometric SECM is the use of the signal itself to determine the absolute tip to surface distance. This cannot be done with tips such as glass membranes or micro pipettes. The effective use of potentiometric tips in SECM requires an objective distance measurement. In probe microscopy the ideal solution for this would be to have a tip electrode that can be operated in both amperometric and potentiometric modes. After using the amperometric function of the tip to approach the surface and calibrate the distance or do topographical imaging, potentiometric selective chemical imaging can be performed. To achieve this, different approaches can be considered such as multi-barrel sensors, metal/metal oxide sensors, etc. In our preliminary work antimony micro tips were found to be very promising in this respect. Antimony pH sensors are broadly used in different areas as an alternative to glass electrodes, e.g. in fluoride media or in the food industry where the use of glass electrodes is not permitted, to avoid the chance of glass fragment contamination. The properties of antimony electrodes have been reviewed recently [3]. Most of the miniaturised antimony electrode designs involve very thin walled glass capillaries to facilitate cell penetration and minimise invasion [4,5]. For microscopy a more durable polishable design is needed. It is especially important for the exposed electrode surface to have a well defined geometry that allows calculation of the mass transport to the electrode. In this respect the micro disc is the most suitable as the theory for amperometric measurements with this geometry is already available.

EXPERIMENTAL

The basic SECM instrument used in this work has been described in detail previously [1,2]. To allow potentiometric measurements some small modifications were made. A block diagram of the revised microscope is shown in Figure 1. A micro positioning device connected to the PC via a ADC controls the movement of three piezoelectric drivers. The cell is mounted on the horizontal (x-y) stage and the tip is mounted on the vertical (z) piezo. For amperometric measurements the potential of the tip was controlled by an EI-400 bipotentiostat. The output of the

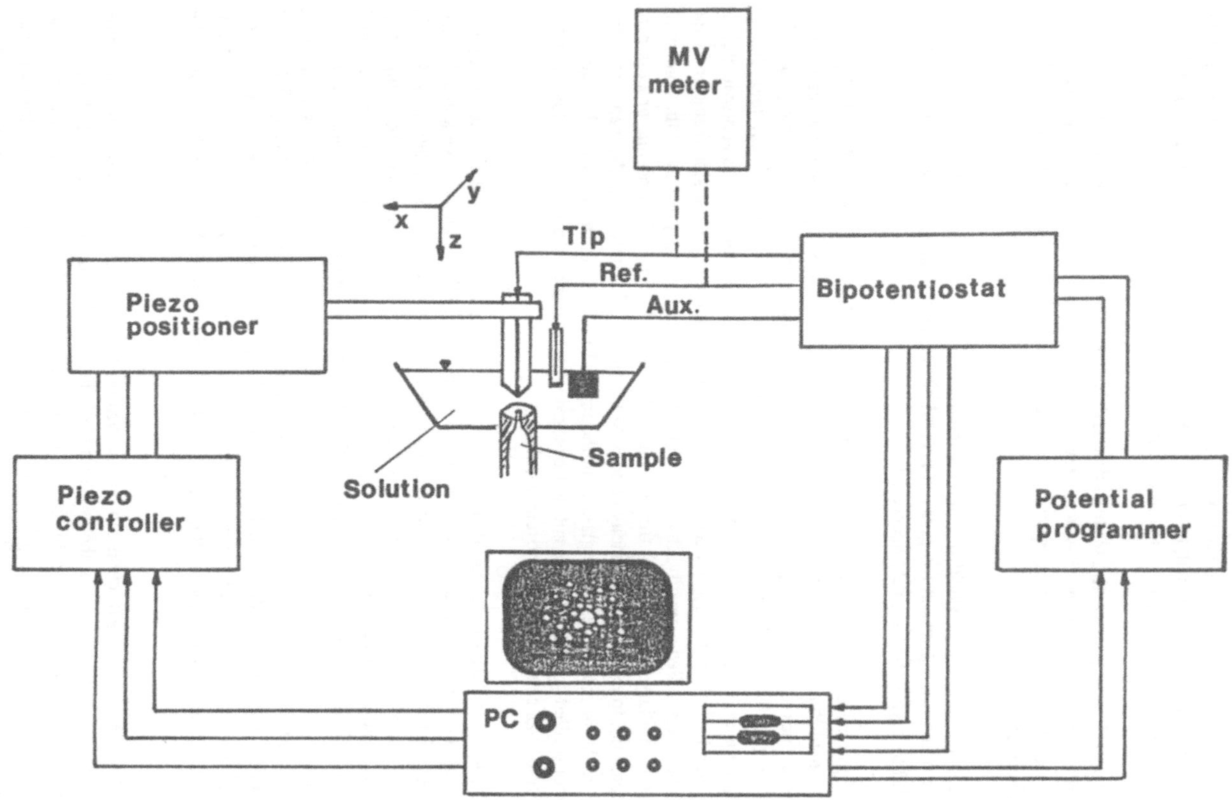

Fig. 1. Block diagram of the scanning electrochemical microscope(SECM)

potentiostat is interfaced to the computer via an ADC board. A home built high impedance voltmeter was used to measure the potential of the tip with respect to a conventional silver/silver chloride reference electrode. The output of the high impedance voltmeter was fed directly into one of the ADC channels to allow acquisition of the data by the computer.

The cell was constructed from Teflon and had a volume of approximately 4 ml. The sample was mounted in the cell base, which screws into the cell body. The auxiliary electrode was a piece of platinum wire and for amperometric measurements a separate silver/silver chloride reference electrode was used.

In the experiments shielded platinum micro disc electrodes with 10μm diameter were used. Their preparation procedure is given in general above. Before use they were finely polished and bevelled to have a 100μm shield diameter.

To prepare the antimony micro electrode antimony shot (3-5 g, 99.999%; Aldrich) was melted in a small Pyrex test tube with a Bunsen flame. Using a 50 ml syringe, molten antimony was drawn into a thick-wall Pyrex capillary (outer diameter, 7 mm, inner diameter, 1 mm) carefully preheated in the flame. The total length of the capillary was typically 25 cm and due to solidification the maximum continuous length of antimony inside the capillary was approximately 5 cm. With standard glassblowing techniques antimony filled capillaries of outside diameter in the range 1-2 mm were pulled manually. In order to avoid building up excess pressure in the column of antimony the capillary was melted at the end and pulled using tweezers. During this process care must be taken not to expose the antimony directly to the flame. After this stage the diameter of the antimony inside, in the centre of the glass is in the range 20 to 100μm. In the second stage 5 cm long antimony filled capillaries were cut and checked under the microscope to ascertain that the antimony fibre was continuous, centric and circular.

These capillaries were placed in the centre of a heating coil. Small weight such as a piece of glass tubing (1-5g) was attached to the end of the antimony capillary. The voltage applied to the coil was set high until the capillary started to extend, then the heat was carefully reduced to obtain a long thin antimony filled capillary. The short stemmed upper section could then be mounted in another glass capillary using heat curing epoxy glue. A well defined disc of antimony was exposed by carefully making a scratch near the end of the capillary and breaking the capillary against a fingertip. Electrical contact was made by mercury introduced through the back of the supporting capillary. The ratio of the total tip diameter of the antimony micro disc was typically 5-10. The relatively thick glass wall of these small capillaries allows careful polishing and the concentric geometry of the tip allows use of previous SECM theory. Antimony fibres down to about 3 μm diameter were made into electrodes in this way. Smaller fibres are not sufficiently robust, cannot be polished.

A more robust construction can be achieved by sealing the very thin fibres in glass tube under vacuum as commonly done in the preparation of shielded platinum micro disc electrodes. For this low melting point glass and careful control of the temperature have to be employed to avoid melting and separation of the antimony fibres to isolated micro beads. This new glass capillary can be pulled again, cut polished and bevelled to give an electrode of around 1μm antimony measuring tip.

Small discs were prepared as model targets for biocatalytic activity imaging. To prepare them glass capillaries were sealed at one end in a micro flame with slow rotation to achieve a gradual tapering of the capillary bore towards the sealed end. The sealed end was then polished back to expose a small circular opening (diameter 25 to 120μm) in the middle of a flat glass surface of diameter roughly 2 mm. These capillaries could then be back-filled with the biocatalyst. In this way a small catalytic micro disc could be prepared.

RESULTS AND DISCUSSION

Amperometric microscopy

To study the applicability of direct amperometric detection of the reaction product for enzyme activity imaging, glucose oxidase solution (3 mg/40 μg) in pH=7 phosphate buffer was introduced into the target tube and it was cross linked there with glutaraldehyde (about 1μl, 12.5% solution). The target was placed in the microscope cell which contained 4 ml of 1 mM glucose solution in pH=7 phosphate buffer (0.1M potassium chloride concentration was employed in the buffer solution to keep the quasi reference electrode (silver chloride coated silver wire) constant. A glass shielded platinum micro disc voltammetric electrode with 10μm diameter was used as probe tip, polarised to 1.0V for amperometric imaging. It was scanned over the target on the horizontal XY plane at about 50μm Z distance. The diameter of the target was 150μm measured with optical microscopy. The amperometric measuring tip in this case detects the hydrogen peroxide formed in the glucose oxidase catalysed oxidation of glucose. The distance calibration in z direction was done in the conventional manner of SECM. The amperometric current of oxygen reduction was recorded approaching the surface. The measured current vs. distance data pairs in the negative feed back section were fitted to the theoretical working curve [6].

Figure 2. shows an image obtained in this experiment. It shows that the direct amperometric method can give information about the location of biocatalytic-active areas on a surface.

Potentiometric microscopy with antimony tip

The electrode potential - pH dependence measured in case of a SECM antimony tip having 6μm measuring disc diameter is shown in Figure 3. The slope of the curve for different electrodes was in the range of 40 to 50 mV/ pH unit. No increase in the pH sensitivity could be achieved by oxidising the antimony by soaking in different oxidising agents as permanganate or dichromate. The electrodes were therefore used for imaging experiments without any pretreatment. However, after use as a voltammetric electrode for tip - distance measurement at negative potentials, the stable pH function was lost. The original pH function could, however be completely restored if the electrode was kept for a few seconds at 0.0 V.

The cyclic voltammetric behaviour of the antimony micro tips was studied in pH 7.0 phosphate buffer. The reduction of oxygen gave a well defined wave. In amperometric mode by setting the electrode potential more negative than -0.7V, a steady-state oxygen reduction current is obtained. This current clearly reflected oxygen concentration changes induced by nitrogen purging or subsequent oxygen bubbling.

Approaching the micro tip to a surface the negative feed back effect could be observed. Figure 4. shows a current - distance curve for a 4μm diameter antimony disc. It can be seen that far from the electrode the current has a constant value. However, on approaching the surface it decreases. This amperometric behaviour is in good agreement with the theory , and it allows the measurement of absolute tip - surface distances. We used the oxygen reduction current for distance measurement in our imaging experiments to avoid addition of external redox couple which could cause further complications.

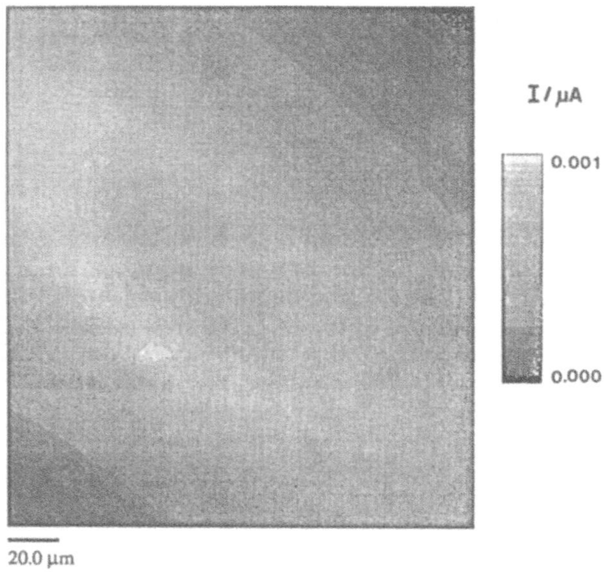

Fig. 2. Amperometric image of a glucose oxidase containing micro disc target. Electron diameter 10 µm, target diameter 150 µm, glucose concentration 1 mM in pH 7.0 phosphate buffer.

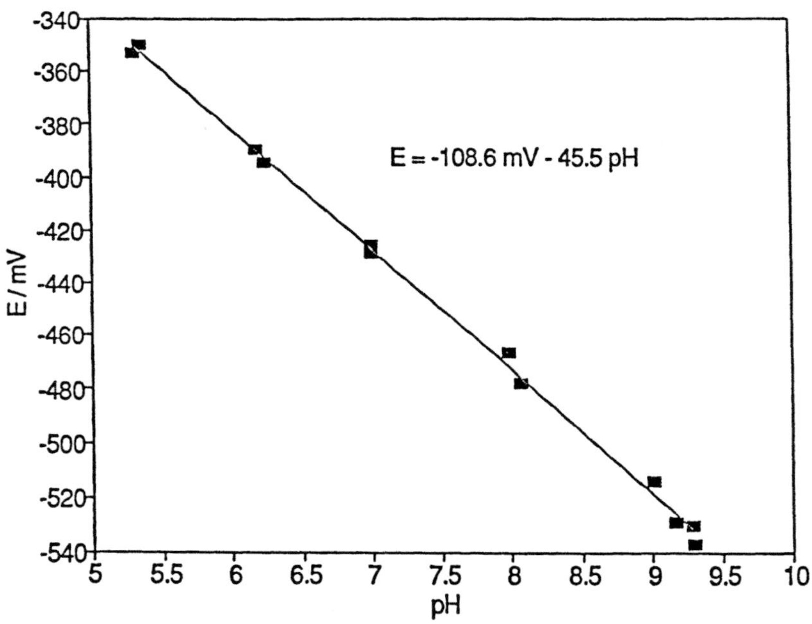

Fig. 3. Electrode potential - pH dependence measured with an antimony micro disk electrode. Diameter 6 µm.

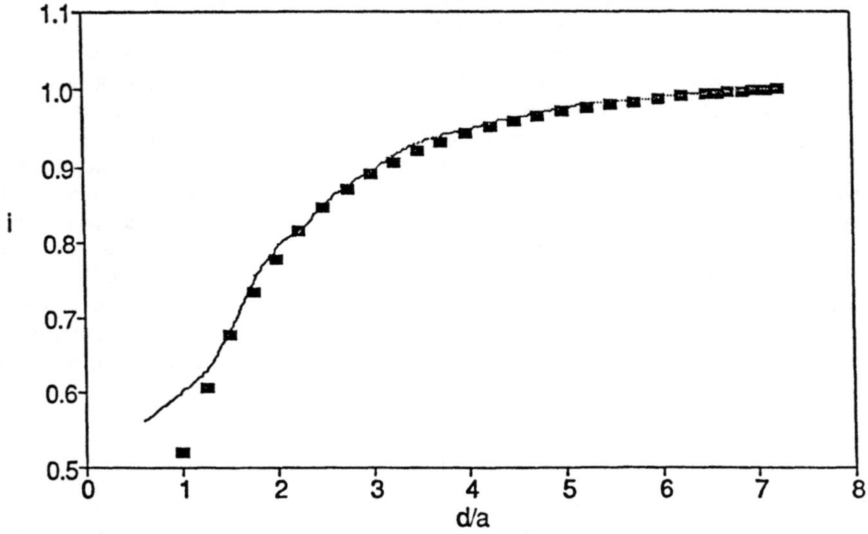

Fig. 4. Amperometric current-distance curve for a 4 μm diameter antimony disk. The electrode indicates the oxygen reduction current at -0.7V potential. The solid line is experimental data and the filled squares are theoretical data.

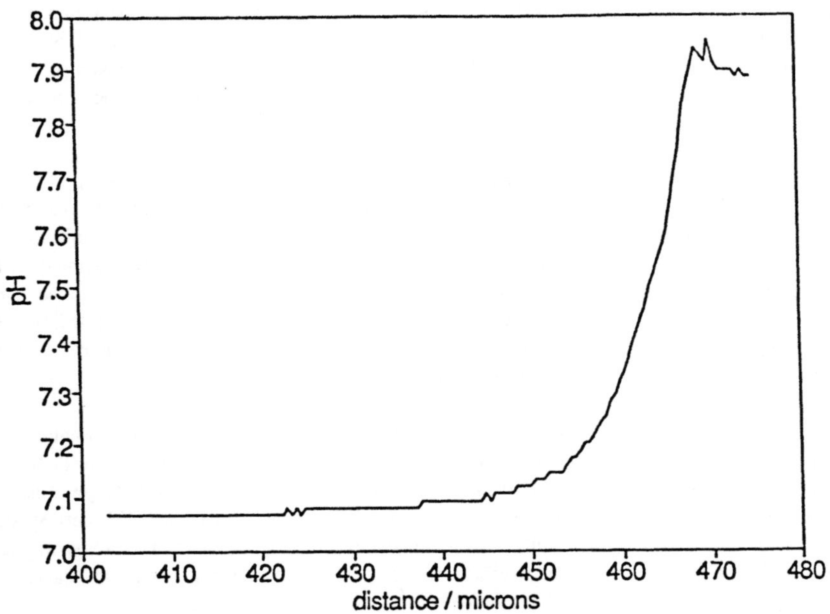

Fig. 5. pH distance profile over the center of a 50 μm diameter immobilized urease target in the perpendicular direction.
The antimony tip diameter was 6 μm and the solution contained 5mM urea, 0.1M KCl and 1 mM pH 7.0 phosphate buffer.

Based on our initial studies the enzyme activity imaging with potentiometric antimony tip was carried out in the following way. First the tip was roughly positioned over the target using the x-y piezo-positioners and was observed through a telescopic lens if necessary. Then air saturated buffer solution was introduced into the cell. The tip was then operated in the amperometric mode at -0.8 V and brought close to the surface using the blocking effect of the surface on oxygen reduction current. By fitting the data to theoretical working curve the absolute distance from the surface can be determined. The tip was then reoxidised by holding its potential at 0.0 V for a few seconds. After the tip had been brought to a known distance from the surface, the tip was switched to the potentiometric mode. Next, the surface was systematically searched using the x and y piezo drivers and the potential of the tip was monitored to observe the pH profile around the target. The approximate centre of the target was located by adjusting the position of the tip in the X,Y plane to obtain the maximum difference between the local and the bulk pH. Then the tip was moved in the Z direction to determine the dependence of pH on the distance from the surface. Finally, images of the pH profile in the X,Y plane at a constant Z height were collected by rastering the tip across the surface in the usual manner.

Imaging of immobilised urease

To investigate the utility of potentiometric SECM in local enzyme activity studies the imaging of local pH changes due to the enzyme-catalysed hydrolysis of urea was attempted. Similarly to the immobilisation of glucose oxidase the urease was entrapped by glutaraldehyde cross linking in the glass capillary target. First the empty capillary was dipped into a 25% aqueous solution of glutaraldehyde, filling the tube with a small amount of glutaraldehyde by capillary action. The tube was wiped on a piece of tissue paper to remove the excess. The cross linked gel was formed by dipping the end of the tube into a solution of urease and bovine serum albumin (BSA). The enzyme solution contained 20 mg BSA and 5 mg urease in 0.2 ml 0.1 M pH=7.0 phosphate buffer. After this procedure, the tube contained a solid gel of cross linked protein. To remove the excess glutaraldehyde, the end of the tube was soaked in 10 mM solution of asparagine in pH 7.0 phosphate buffer. After roughly 30 minutes, the target was ready for use.

The urease target was mounted in the cell, and the SECM investigation was performed as described above. Figure 5. shows a pH - distance profile in the z direction over the centre of a 50μm diameter urease target in 1 mM (pH=7.0) phosphate buffer in the presence of 5 mM urea. No pH profile was observed in the absence of urea or when the tip was far from the enzyme disc. It is well known that the hydrolysis of urea would increase the pH in a low buffer capacity medium of pH=7.0. This can be seen clearly in the figure. The surface of the urease gel is recessed, therefore an absolute distance scale in this case cannot be established. The end section of the pH - distance curve indicates that the tip touches the target surface.

A pH image of a section of a 120 μm diameter urease disc is shown in Figure 6. The whole image could not be captured because the diffusional broadening resulted in a profile that was larger than the maximum scan size of the apparatus. The centre of the enzyme target is near the lower left -hand corner of the image where the potential of the antimony tip is 0.43V corresponding to a pH of roughly 7.5. These results show that potentiometric SECM can be used to determine the distribution of enzyme activity on a surface. In case of different enzymes on the same surface, the latter could be separately mapped in solutions of their respective substrates.

Immobilised yeast colony imaging

An exciting application of SECM with pH sensing tips which can be foreseen is the mapping of pH changes due to the metabolic activity of living systems such as cells, living tissue,

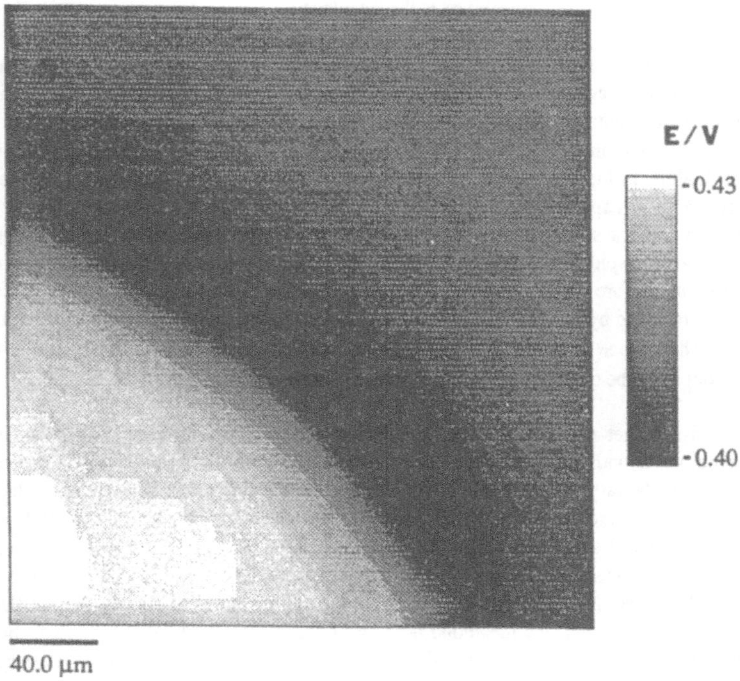

Fig. 6. pH image of a 120 µm diameter urease disk. The tip diameter was 40 um, the tip scan rate was 10 um/sec

Fig. 7. Image of an 80 µm yeast/agarose disk measured with a pH sensitive antimony micro disk electrode.
The tip diameter was 20 µm, the tip scan rate was 10 µm/sec

or associates of micro-organisms. Because of its high metabolic rate, yeast was chosen as the model system to demonstrate the applicability of the technique.

The yeast target was prepared by entrapment of ordinary moist baker's yeast in low gelling temperature agarose. A 3% by weight of agarose in pH=7.0 buffer was heated in a water bath until a clear solution was obtained. About 1 ml of this solution was allowed to cool slowly to 30-40 °C. A small amount of yeast was homogenised in this solution by stirring with a glass rod until a slightly turbid suspension was produced. The turbid suspension was decanted off the sediment. A glass capillary was back-filled with the suspension using a syringe to apply sufficient pressure to force a small droplet out of the end. The target was then allowed to gel at room temperature. To avoid desiccation of the yeast, the target was kept in a sample vial saturated with water vapour. Before use the droplet of the gel protruding from the polished glass surface of the target was removed by carefully wiping with a piece of tissue paper to leave only a small disk of gel exposed. The surface of the gel disc was examined and individual yeast cells (about 6 μm long and 3 μm wide) could be observed and counted with an optical microscope.

The target was mounted in the cell bottom and the experiment carried out in pH=7.0 phosphate buffer with addition of glucose as metabolic supply. A significant decrease of pH was observed near the target surface. No effect was seen in the absence of yeast cells or glucose. This pH decrease was therefore due to the conversion of glucose to acidic metabolites, most likely down to carbon dioxide, by the yeast cells. An image of a section of the pH profile across an 80 μm yeast/agarose disc is shown in Figure 7. The centre of the yeast disc is towards the lower right-hand corner of the image. Near the centre of the target the pH has fallen to about 6.3 from the bulk 7.0 value. This indicates that the metabolic activity of the immobilised yeast target can be mapped. Under optical microscope about 200 individual yeast cells could be counted on the surface of the target disc. Because the pH change observed is fairly large, this technique should be useful in the study of smaller number of cells.

As a **conclusion** of this section we can say that the SECM can be used to image biocatalytic-active spots on surfaces detecting directly the reaction product of heterogeneous catalytic reactions. A more detailed report about the results will be published elsewhere [7].

One dimensional microscopy to study concentration profiles inside the reaction layer of enzyme sensors working in flow injection systems

The biosensor development is receiving considerable attention at the present time. Among the biosensors the biocatalytic electrodes are very well established. The advantage of employing biocatalytic electrodes in flow injection conditions has been recognised quite long ago. Today some kinds of flow injection apparatus are used in most of the research dealing with biocatalytic electrodes.

In our previous work [8,9,10] we studied the influence of different parameters on the flow injection peaks in case of biocatalytic electrode detection. In the investigations model calculations were performed and the calculated functions were compared with experimental results.

It is an interesting question how the thickness of the reaction layer of a biocatalytic electrode influences the flow injection signal, its peak height and peak width. To study the effect experimentally, a special electrode was made in the following way. The flat end of a glass tube with 7 mm outside and about 5 mm inside diameter was covered with a dialysis membrane stretched and

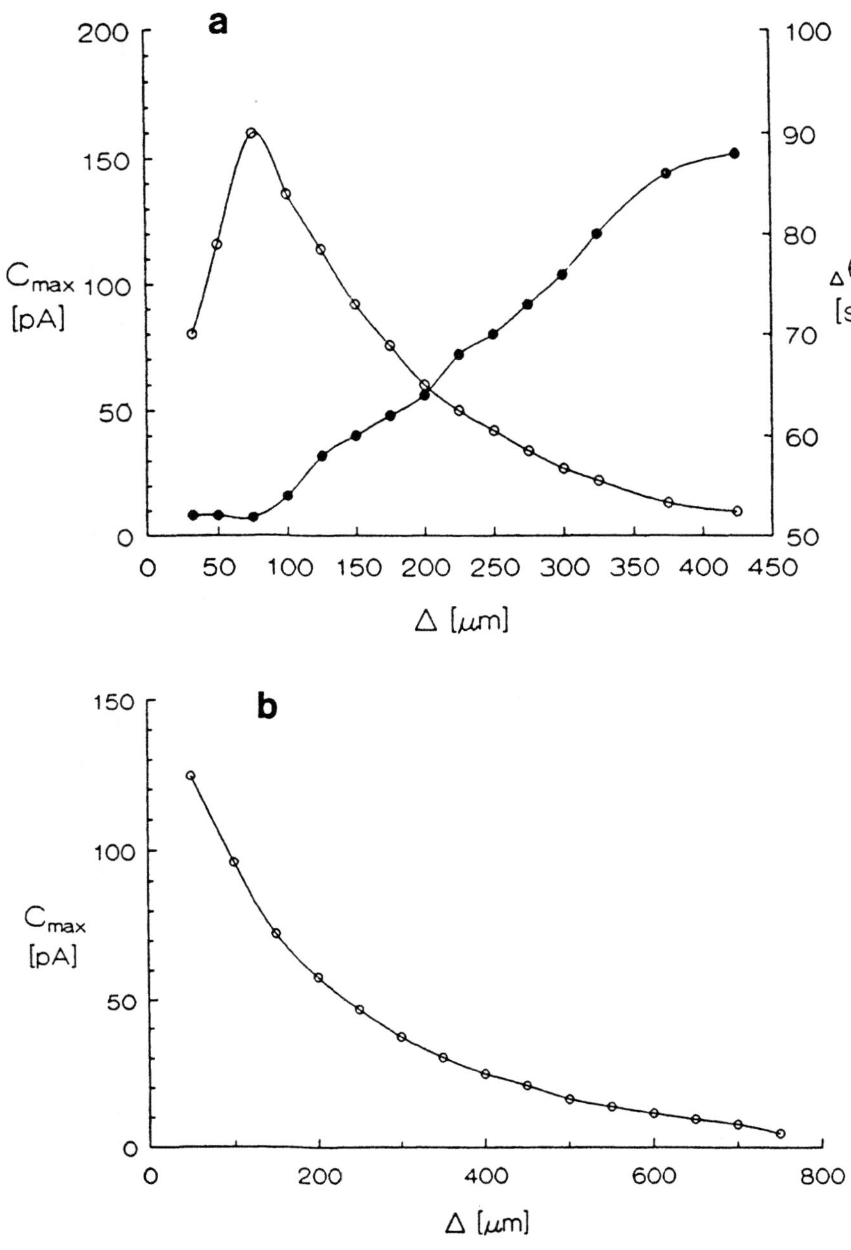

Fig. 8. Influence of the reaction layer thickness(Δ) on the FIA signal for
a) glucose electrode with polyacrylamide gel layer,
b) platinum electrode with a passive membrane
Peak height curve (o); peak width curve (•)

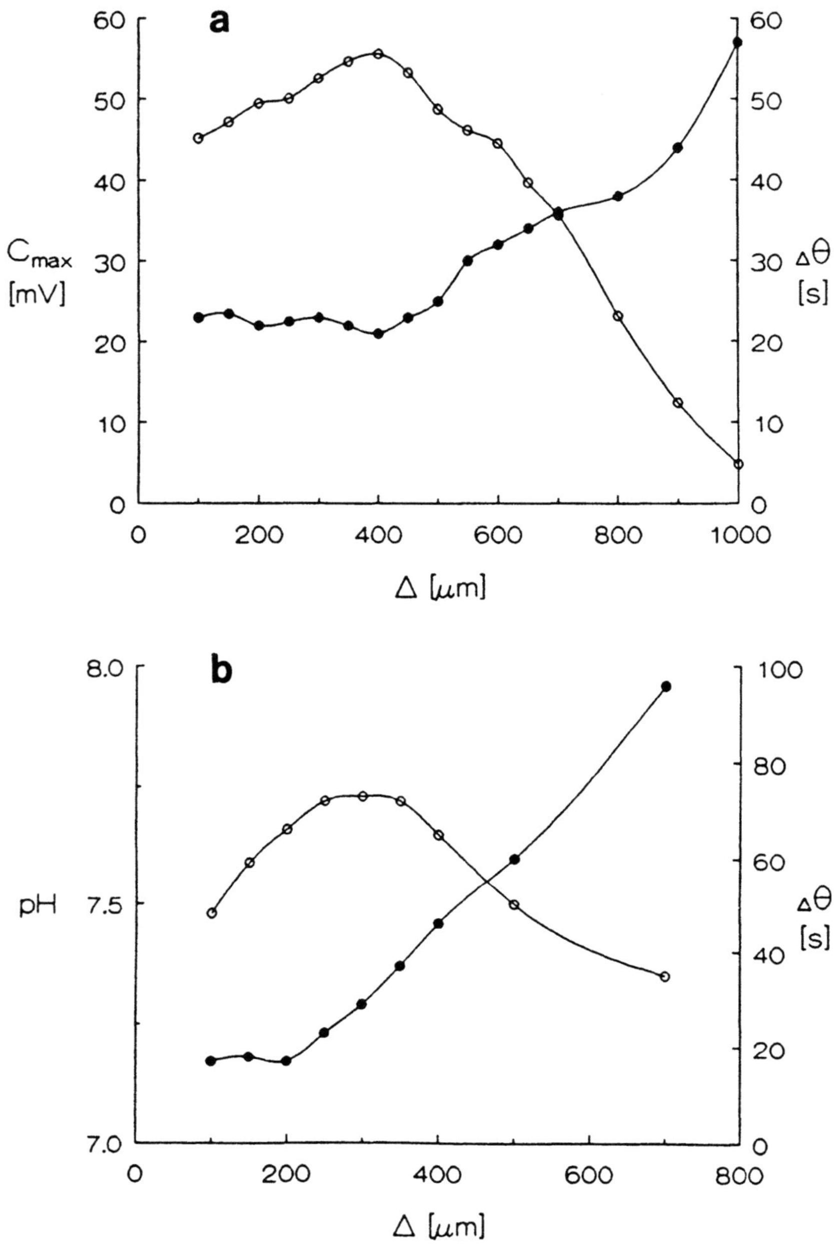

Fig.9. Influence of the reaction layer thickness(Δ) on the FIA signal for a urea electrode with polyacrylamide gel reaction layer and a) ammonium electrode and b) pH base sensor
Peak height curve(o); peak width curve(•)

tied on by thread. A thick layer of enzyme-containing polyacrylamide gel or enzyme containing buffer solution was prepared with light induced polymerisation at the spot or was introduced, respectively. Micro platinum disc electrodes of 80-120 μm diameter with 30μm platinum wire in the centre were prepared. These electrodes were used as base sensor of the biocatalytic electrodes placed inside the glass tube. To detect ammonium ions the platinum micro disc was dip coated with nonactin ionophore-containing plasticized PVC film. It was converted in this way to ion-selective coated micro disc electrode. Before dip coating with the film forming cocktail dissolved in tetrahydrofurane, the platinum micro disc of the tip first was coated with electrolytically deposited silver film, and the silver surface thus obtained was covered electrolytically with a thin layer of silver chloride.The pH monitoring was carried out with with an antimony micro-disc electrode.

The enzyme electrode studied was placed concentrically down in the narrower, lower part of a home made, cylindrical wall-jet measurement cell with its measuring surface placed down centrally at about 0.1 mm from the inlet of the cell. The other electrodes necessary for the measurements were adjusted in the upper internal section of the cell, which was broader. To be able to change the thickness of the reaction layer between the dialysis membrane and the base sensing element the stem of the micro electrode was clamped to a special device which could move the micro electrode inside the outer glass tube parallel to its wall up and down. This special precision micro screw thread motion device was previously a part of a stereotactic tool made for in vivo animal experiments. The resolution of this device was 1μm per revolution scale unit of the screw.

The measurement cell was inserted into a simple flow injection apparatus operating with a peristaltic pump (Labor-MIM Conti Flo 28) of 1.3 ml/min flow rate, a 250μl volume loop injector (Rheodyne Model 7125) and a 100 mm long coiled dispersion section (0.8 mm id.). 30 mg/ml enzyme concentration was employed in the reaction layer.

Glucose oxidase enzyme based reaction layer was used to study the behaviour of the amperometric glucose electrode. In this case the platinum micro electrode was polarised to 0.60V vs. 0.1 M silver/silver chloride reference electrode with PAR Model 173A polarograph to detect amperometrically the hydrogen peroxide produced in the enzyme catalysed oxidation of glucose. pH 6.0 phosphate buffer was streamed and 1 mM glucose solution was injected.

Urease enzyme was employed to investigate the dependence of the FIA signal on the reaction layer thickness of urea electrode. The coated micro disc ammonium electrode was connected to a pH meter (Radelkis, Type OP-208) 10 mM TRIS buffer, pH 7.0 was streamed and 1 mM urea solutions were injected.

The reaction layer thickness dependence of the flow injection signal obtained experimentally for the glucose electrode and for the potentiometric urea electrode is shown in Figures 8. and 9. Passive membrane means polyacrilamide gel containing no enzyme. These experimental results obtained with the "one dimensional scanning electrochemical microscopy" described briefly correlated well with the results of our model calculations.It is obvious that an increase in the passive membrane thickness results a decrease in the FIA signal (Fig. 8 b). More detailed description of the experimental work as well as the calculations are given elsewhere [10].

ACKNOWLEDGEMENT

Spas D. Kolev from the University of Sofia was our co-operating partner in the work shortly described here as "one dimensional microscopy". The help, devotion and efforts of all research scientists mentioned are greatly acknowledged and most appreciated.

The financial support of the Hungarian Science Foundation (OTKA 3115 and OTKA 3116) is greatly acknowledged.

REFERENCES

1. J.Kwak, A.J.Bard, Anal.Chem., 61, 1221, (1989).

2. A. J. Bard, F-R. Fan, D. T. Pierce, P. R. Udvin, D.O. Wipf, F. Zhou, Science, 254, 68, (1991).

3. S. Glab, A. Hulanicki, G. Edwall, F. Ingman, Crit.Rev.Anal.Chem., 21, 29, (1989).

4. H. A. Bicher, S. Ohki, Biochim. Biophys. Acta, 255, 900 (1972).

5. Y. Matsumura, K. Kajino, M. Fujimoto, Membr. Biochem., 3, 99, (1980).

6. M. V. Mirkin, F-R. F. Fan, A. J. Bard, J. Electroanal. Chem. Interfacial Electrochem., 328, 47, (1992).

7. B.R. Horrocks, M. V. Mirkin, D. T. Pierce, A. J. Bard., G. Nagy, K. Tóth, Anal Chem., 65, 1213, (1993).

8. S. D. Kolev, G. Nagy, E. Pungor, Anal. Chim. Acta, 241, 43, (1990).

9. S. D. Kolev, G. Nagy, E. Pungor, Anal. Chim. Acta, 241, 55, (1990).

10. S. D. Kolev, G. Nagy, E. Pungor, Anal. Chim. Acta, 241, 167, (1991).

SURFACE CHARACTERISTICS OF CHEMICAL SENSORS

G. Gabor and G. Frishman
Israel Institute for Biological Research
Ness-Ziona 70450, Israel

ABSTRACT

The role of various surface structure features of optical sensors (optrodes) is related to their efficiencies.
Various O_2 and pH optrode models, in which the support-reagent distance, the polarity and pore size are varied are presented and their contribution to enhance sensitivity and reduce response times is discussed. For a given support - e.g. spherical silica beads with 200μ diameter (Daltosils) optimal surface to pore diameter ratio, allowing sufficient concentrations of the reagent and the analyte respectively, yields the most efficient sensor (re sensitivity, response time etc).
Increasing the support reagent distance results in the enhancement of the sensitivity. For pO2 sensors, the sensitivity expressed as R_I, increases from 39 to 131 (a factor over 3) upon the introduction of 5 more C-C bonds. The sensitivity of the pH sensors, in which the distance is increased by 8 chemical bonds, is increased by a factor of over 4 from 1.74 to 7.4 for HPTS, or from 1.52 to 5.6 for 7-hydroxy-coumarin-3-aceticacid.

INTRODUCTION

The design of extremely small devices (sensors) that permit in situ on line measurements in biological and industrial systems are of major interest[1-10] An efficient optical sensor is based on a fast chemical reagent between the analyte and the reagent immobilized at the sensor tip[3,5,8-10] The analyte is imposed by need, while the reagent is selected, keeping in mind that it should have the following properties:

1) Easily measurable optical signal (absorption or fluorescence) resulting from its interaction with the analyte - meaning high sensitivity.
2) Fast reaction with the analyte - i.e. fast response
3) It should be immobilizible

In this paper we will discuss the region of immobilization, that includes the surface of the support and the reagent phase and its possible contribution to the efficiency in several ways:

a) The reagent-surface distance, ruled by the absence/presence (and length) of a spacer between the reagent and the surface, 1) exposes the reagent to the analyte stream; 2) determines its orientation relative to the surface - by imposing rigidity or permitting free rotation.
b) The polarity of the reactionsite is determined by the vicinal groups, on the surface[11,12] or the spacer. When highly polar silica surfaces - CPG or Daltosil[13] are employed for oxygen sensors, they must be modified to facilitate the approach of the

G.G. Guilbault and M. Mascini (eds.), *Uses of Immobilized Biological Compounds*, 325–334.

none polar O_2 molecule - the analyte - to the bound reagent. A non polar spacer might serve a double purpose.

c) The support's pore size controls the approach of the analyte to the surface bond reagent, whose quantity is ruled by the surface area. This is a trade off.

To demonstrate these features pO_2 and pH optrode models were used.

EXPERIMENTAL

Materials

CPG (Controlled Pore Glass) from Fluka, Switzerland and spherical CPG, Datltosil, with various known pore parameters and sizes from Serva, Germany - were used as supports (or sensor heads). CPG 170 and CPG 550 are 75-125 μm beads with pore diameters of 170 and 550A respectively. The daltosils had 100-200 μm diameters and pore diameters of 200, 600, 1000 and 2500 Å [13].

Pyrenbutyric Acid (PBA) aminopropyltriethoxysilane (APTES), 6-aminocaproicacid (6ACA) and hexamethylenediamine (HMDA) were from Aldrich, USA. Pyrenenonanoicacid (PNA); the coumarin derivatives: umbelliferrone-3-aceticacid (3AU), umbelliferrone-4-aceticacid (4AU) and umbelliferrone-3-carboxylicacid (3CU) as well as hydroxypyrenetrisulfonicacid (HPTS) and its acetate ester were from Molecular Probes, USA.

Measurements

All measurements were carried out by an SLM-500 Spectrofluorometer. CPG or (Daltosil) beads carrying the immobilized reagents were held in front of a fiber bundle in a polyethylene tube. The diameter of the tube was about the same or slightly bigger than the ending of the fiber bundle; it was in fact a sleevelike tube extension, that permitted the injection of the various buffers onto the sensor heads in it. The fiber bundle, purchased from Volpi A.G., Switzerland, contained 100 optical fibers of 200 μm diameter, 50 connected to the light source and 50 to the PMT of the SLM-500.

The fluorescence 1) of the pyrene derivatives was excited at 342 nm, and recorded at 395 nm and 460 nm; 2) of the coumarin derivatives were excited at 370-395 nm and recorded at 445-465 nm; 3) and of HPTS was excited at 470 nm and recorded in the 490-600 nm range.

The quenching efficiency R_I is defined as the ratio of the fluorescence intensities at 0% and 100% oxygen - I/I_{O2}. The relative quantities of monomers and dimers (excimers) is expressed by R_M. R_M is the ratio of the fluorescence intensities in the monomer band-I_M- to that in the excimer band I_{EX} of pyrene monitored at 395 nm and 460 nm respectively $R_M = I_M/I_{EX}$

Immobilization of the reagents

All reagents were immobilized - chemically bond - to the various silica beads "the sensor heads" (supports): silanization of the beads applying APTES according to the Wheetall method[14] was followed by DCC coupling of the various reagents (except HPTS)[15].

To vary the distance between the fluorescent entity and the silica surface pyrenbutyric and pyrenenonanoic acids were used for the pO_2 sensor heads, yielding a distance difference of 5 C-C bonds between

the fluorescent entity and the rigid cross links of the silanizing agent, as well as a less polar microenvironment for the analyte reagent interaction.
The distance between the fluorescent pH indicators: coumarin or HPTS, and the cross link, was extended by the introduction of well known spacers, like 6-aminocaproicacid (6ACA) and hexamethylenediamine (HMDA) between them.

RESULTS AND DISCUSSION

1. Oxygen Optrode
Excitation and emission spectra of PBA and PNA immobilized on CPG550, indicate that 1) the PBA emission originates from the excimer or dimer. (Fig. 1b); 2) Immobilized PBA exists even in the ground state mainly as a dimer that absorbs in the 370-400 nm range[16], and to a smaller extent as a monomer (Fig. 1a), while the pyrene in PNA exists mainly as a monomer, according to the observed fluorescence band in the 390-400 nm range (Fig, 2b), originating in the ground state monomer absorbing at 347 nm[16](Fig.2a). The ratio of the fluorescent intensities R_I and the relative quantities of the monomer and dimer (excimer) R_M, for PBA and PNA, both immobilized on CPG550, are summarized in Table 1. The reversibility of the quenching process as shown in Fig. 3, is excellent.

Table 1. Quenching efficiencies ($R_I=I/I_{O2}$) and relative emissions ($R_M=I_M/I_{EX}$) various sensor heads

Sensor head	λ_M	$R_I=(I/I_{O2})$	$R_M=I_M/I_{EX}$
PBA/CPG	395	39.0	0.089
	460	26.0	
PNA/CPG	395	139.0	6.17
	460	32.5	

Fluorescence exited at 342 nm

The quenching of the monomer is more efficient as expected, due to its longer lifetime in the S_1 state[16,17]. Even for PBA that has a higher excimer peak (R_M=0.089), the quenching as expressed by R_I is somewhat more efficient with the monomer. However the highest R_I values (of 131.0) were obtained from the monomer of PNA. The pyrene residue in this compound is mainly in the monomer form. It should be noted, that the low emission in the 460 nm region for PNA (Fig. 2b), related to the excimer, may result, at least partially, from the overlap of the monomer emission. We suggest that the pyrene residues in PBA bound through four carbons atoms to the silanized silica surface are held parallel to each other, in a way that allows the formation of dimers/ excimers. The pyrene residue in PNA bound through nine carbon atoms, is more distant. Therefore more degrees of freedom of the translational motion of the pyrene residue are allowed, resulting in a

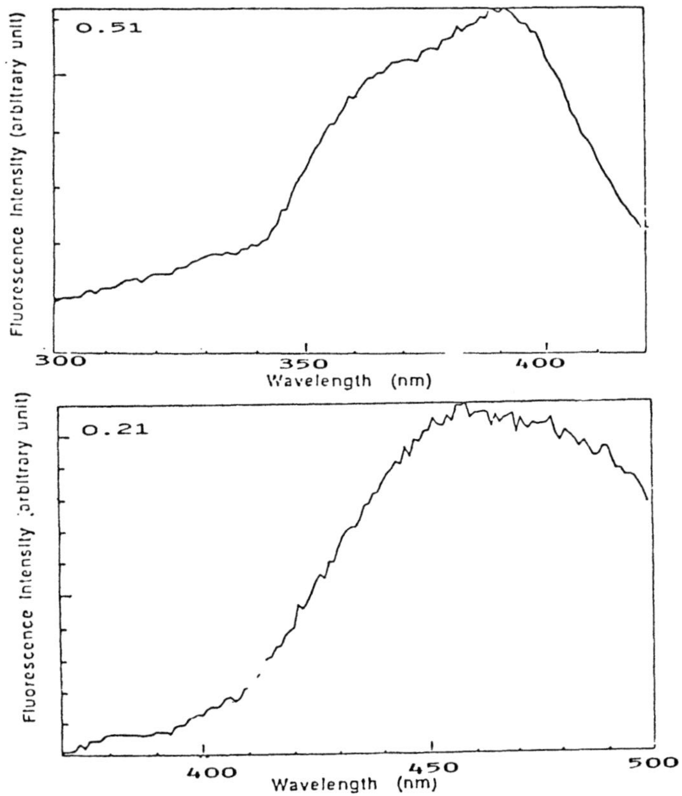

Fig. 1. a and b. Excitation an Emission spectra of CPG 550 bound PBA.
Excitation recorded at 460 nm emission peak.
Fluorescence spectra excited at 342 nm peak.

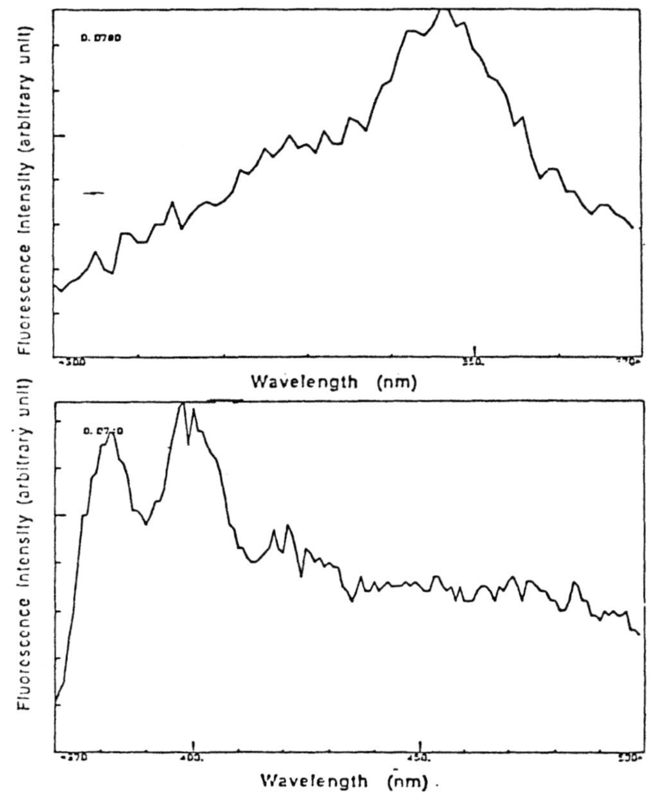

Fig. 2.a and b. Excitation and emission spectra of CPG 550 bound PNA. Excitation at 342 nm, emission at 398 nm.

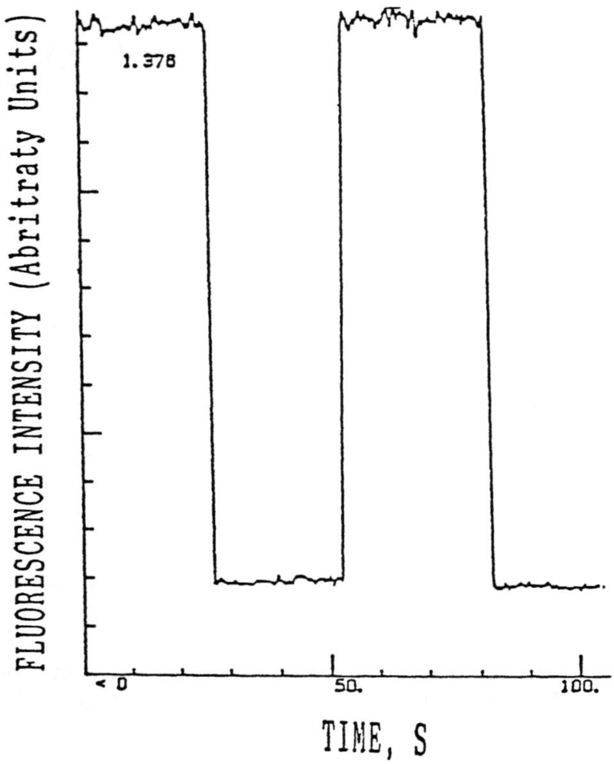

Fig. 3. Typical reversibility of the fluorescence quenching of PNA by oxygen. R_1= 131.0

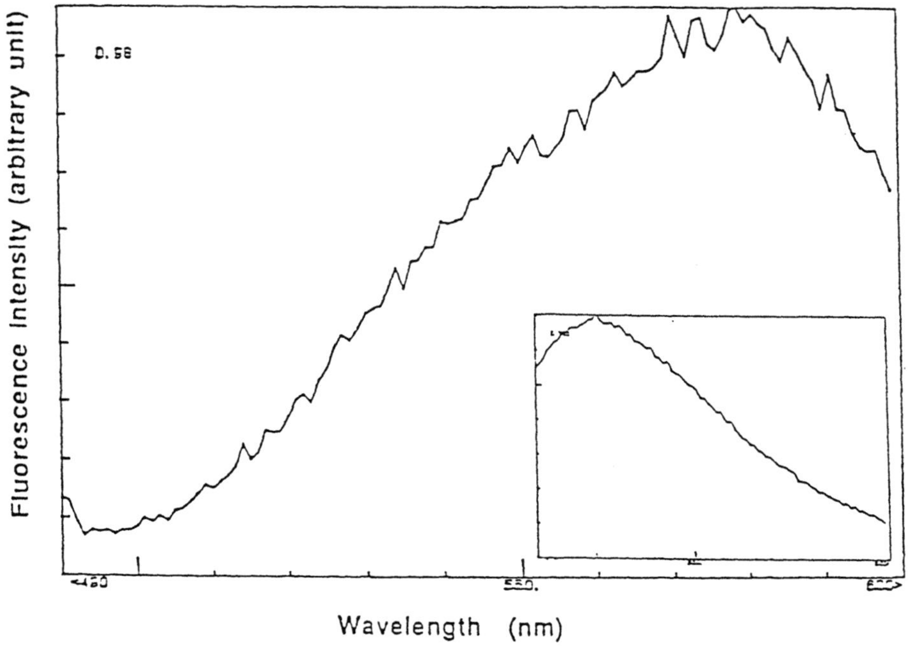

Fig. 4. ACPG-HPTS excimer fluorescence spectrum.Emission at 579 nm max.
Insert=ACPG-HPTS monomer("normal") fluorescence.Maximum 525nm.

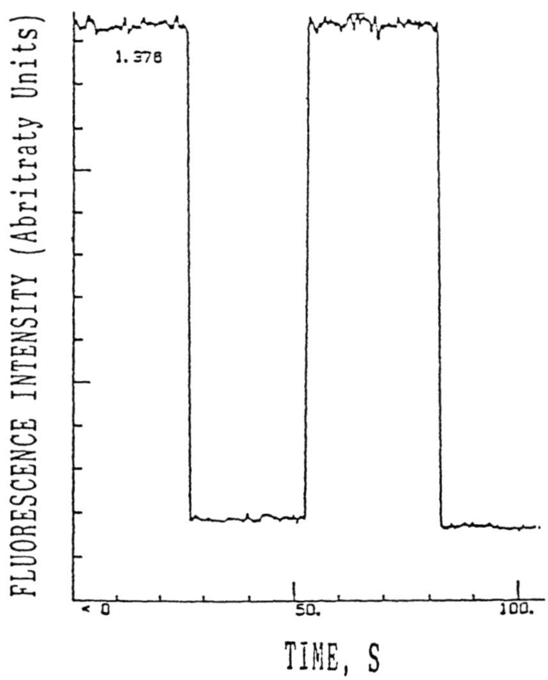

Fig. 5. HPTS-HMDA-CPG. pH response. Excitation at 476nm max,and emission at 525 nm max. R(7.8/6.5)=7.4

random distribution of the fluorescent monomers, and very little, if at all dimer formation in the ground and/or excited states (eq.1).

$$(1)\quad P + P^{*} \rightarrow P_2^{*}$$

The major part of the pyrene residues undergoes fluorescence quenching by oxygen - the analyte - resulting from the collisional bimolecular process having second order kinetics (eq.2) [16].

$$(2)\quad P^{*} + O_2 \rightarrow P + O_2$$

The longer distance supplied by an aliphatic chain increases R_I in three ways: 1) Reduces ground state dimerization and/or excimer (dimer) formation (eq.1) as it does not provide the necessary parallel structure for such a process; 2) Increases the exposure of the fluorescent entity to the quenching analyte flowing through the channels; and 3) Lowers the polarity and thus facilitates the approach of the non polar analyte. This in turn increases the bond reagent - analyte interaction (reaction 2) that also gives rise to less excimer formation (reaction 1), as reactions 1 and 2 are competitive.

2. pH Optrode

The pH sensor heads obtained by binding the common indicators; HPTS or Hydroxycomarin derivatives: 3AU, 4AU etc, directly or through a spacer, 6ACA and HMDA, to various silica surfaces were investigated.
The immobilization of the reagent through a long (7-8 C-C bond length) aliphatic chain facilitates the binding process itself, as the steric hindrance to the approach of the bulky polycyclic reagents to the extended surface becomes negligible. The more distant reagent has more freedom of translational motion, that provides a larger number of encounters of the bimolecular reaction between the reagent and the analyte. The greater the reaction rates the better the extent of the response, as expressed by an increase of R(B/A) the fluorescence intensity ratio in basic to acid solutions, as well as also by the reduced response times. Direct binding of HPTS to ACPG550 (amino CPG) yields a ratio of R (8.0/6.0)=1,74 upon changing the pH from 8.0 to 6.0 and vice versa. Direct binding [18] with physical absorption may result in overloading of this reagent, as shown by the broad fluorescence band peaking at 579 nm (Fig. 4) originating in an excimer. The excimer has, probably, a sandwich like structure - like the well known excimers of pyrene [16]. However, its formation involves probably the hydroxyl groups as well, causing a big shift of the pK_a and a complete stability (insensitivity) to pH in the pH range of 6.0-8.0, and consequently a considerably reduced response.
The binding of this reagent to CCPG550 (Carboxy CPG) through HMDA - an extension of 8 C-C bounds - results in a R(7.8/6.5) of 7.4 monitored in the monomer band - λ_X=476 nm and λ_M=525nm (Fig. 5).
Similar enhancements of the reagent sensitivity were obtained with 3AU by applying HMDA as a spacer between the silica surface and the reagent. The R(8.0/6.5) value increases about 4 times from 1.52 for 3AU-CPG to 5.6 for 3AU-HMDA-CPG, while the response time is reduced from about 60-80s to 2-5s
Another important factor determining the sensitivity of chemical sensors is the pore size - its surface area (S) and diameter (P) - of

the support; CPG or Daltosil. The extent and rate of the bimolecular reaction between analyte and reagent is ruled by their concentrations. These, however, are ruled by opposing factors, arising from the surface structure of the support. The quantity of the reagent increases with the increase of the amount of the binding sites, i.e. the surface area, while the respective concentration of the analyte increases with the diameter of the pore. Consequently the surface to pore diameter ratio must reach an optimum and neither a maximum, meaning many binding sites; nor a minimum that means a high concentration of the analyte in the pores.

Table 2. Sensitivity of sensorheads as a function of the support's pore size.
Sensitivity is expressed by the ratio R(7.7/6.5)

Reagent	Support	Pore size diameter, A	Surface[13] Area m^2/g	R(7.7/6.5)
3AU	DAL 150	200	275	1.15
	500	600	50	2.4
	1200	1000	20	4.4
4AU	DAL 150	200	275	1.4
	1200	1000	20	3.7

The application of large pore size support - 700A - was reported by Hirshfeld previously[19].
As expected the efficiency of the immobilized coumarin derivatives increases with increasing the pore diameter from 200 to 1000 A. (Table 2).
The increase of R(7.7/6.5) with the pore size implies that for these pore sizes (200, 600 and 1000A), the quantity of the bound analyte flowing through the channels is the sensitivity determining factor rather than the quantity of the bond reagent. This implies that even DAL1200 with a surface area of $20m^2/g$ provides sufficient binding sites. Further reduction of the surface area by a factor of 2 (i.e. $10m^2/g$) as in DAL3000 resulted in negligible fluorescence intensity of the immobilized reagent molecules, indicating that the number of the binding sites is not sufficient. We conclude, that if there are enough bond reagent molecules, to provide measurable fluorescence intensities, they also enable the monitoring of the pH fluctuations.

SUMMARY

The role of various surface structure features of optical sensors (optrodes) is related to their efficiencies.
Various O_2 and pH optrode models, in which parameters of the support-reagent region like: distance, the polarity or pore size are varied are presented and their contribution to enhance sensitivity and reduce response times is discussed. For a given support - e.g.

spherical silica beads with 200μ diameter (Daltosils) optimal surface to pore diameter ratio, allowing sufficient concentrations of the reagent and the analyte respectively, yields the most efficient sensor (re sensitivity, response time etc). These conclusions are correct when the reagent is a small molecule. However for optrodes applying immobilized enzymes as reagents, presently investigated by our group, the pore size has a different and more complex influence.

Increasing the support reagent distance results in the enhancement of the sensitivity. For pO2 sensors, the sensitivity expressed as R_I, increases from 39 to 131 (a factor over 3) upon the introduction of 5 more C-C bonds. It should be emphasized that the elongation of the aliphatic spacer has a triple effect: a) it increases the distance of the reagent from the silica surface thus exposing the reagent to the stream of the analyte; b) it reduces the polarity of the microenvironment at the reagent analyte interaction site; c) it disrupts the parallel structure of the pyrene residues, that results in a bigger contribution of the monomeric - oxygen sensitive - structure.
The sensitivity of the pH optrodes, in which the distance is increased by 8 chemical bonds, is increased by a factor of over 4 from 1.74 to 7.4 for HPTS, or from 1.52 to 5.6 for 3AU. Further increase of sensitivity may be achieved by varying the pore size.

It should be pointed out, that although our methods include only optical measurements, our conclusions, as to surface structure properties of sensors may be applied to other (e.g. electrochemical etc) sensors as well.

REFERENCES

1. L.J. Blum and P.R. Coulet, Biosensor Principle and Applications. Marcel Dekker (1991).
2. A.P.F. Turner, I. Karube and G.S. Wilson, Biosensors, Fundamentals and Applications. Oxford University Press (1987).
3. W.R. Seitz, CRC Crit. Rev. Anal. Chem. 19 (1988) 135-173.
4. J. Wang, Chemically and Biologically Tailored Surfaces as Amperometric Sensors. Anal. Proc. 28 (1991) 102.
5. O.S. Wolfbeis, Biomedical Applications of Fiber Optic Chemical Sensors. Intl. J. Optoelectron. 6, (1991) 425-441, and other references quoted therein.
6. O.S. Wolfbeis (ed.) Fiber Optic Chemical Sensors and Biosensors. CRC Press 1991.
7. D. Gibson, Fibre Optics in Laboratory Applications. Intl. Labmate (1992), 51-53.
8. J.I. Peterson and G.G. Vurek. Fiber Optic Sensors for biochemical Applications. Science 224, (1984), 123-127.
9. O.S. Wolfbeis, Fiber Optic Sensors in Biomedical Sciences. Pure and Appl. Chem. 59, (1987), 663-672.
10. W.R. Seitz, Chemical Sensors Based on Fiber Optics. Anal. Chem. 56, (1984), 16A-34A.
11. A.L. Wong, M.L. Hunnicutt and J.M. Harris. Structure of alkylated Silica Surfaces. Anal. Chem. 63, (1991), 1076-1081, and other references quoted therein.

12. S.D. Dowling and W.R. Seitz. Anal. Chem. 57, (1985), 602
13. Serva, Catalogue 1987/8 p. LC3.
14. H.H. Wheetall and A.M. Filbert. Meth. Enzym. 34 (1984) 59
15. J.C. Sheehan et al. J.A.C.S. 87 (1965) 2492
16. R.K. Bauer, P. de Mayo, W.B.R. Ware and K.C. Wu. J. Phys. Chem. 86, (1982), 3781.
17. a) J.B. Birks. J. Phys. Chem. 67, (1963) 2199; 68, (1964) 439
    b) C.A. Parker and C.G. Hatchard. Trans. Far. Soc. 59, (1963), 284.
18. O.S. Wolfbeis et al. Sensors and Actu. 9, (1986), 73
19. T. Hirschfeld, Private Communication.

OPTRODES FOR MEASURING ENZYME ACTIVITY AND INHIBITION

OTTO S. WOLFBEIS
Institute for Organic Chemistry, Analytical Division, Karl-Franzens University, Heinrich St. 28, A-8010 Graz, Austria

Abstract Fiber optics provide a unique means for both invasive and remote detection of chemical and biochemical quantities. Elegant previous work has shown that enzymes are extremely useful "reagents" for (bio)chemical analysis, and we will show here that a combination of enzymatic methods of analysis along with the use of optical fiber sensing schemes can even enlarge the potential of enzyme-based analysis. We will report on how enzyme activities can be measured using optical fibers. In the first approach, a fluorogenic substrate has been immobilized at the tip of the fiber. When brought into contact with an enzyme (such as an esterase), the synthetic substrate undergoes hydrolysis to form a strongly fluorescent fluorophore whose fluorescence intensity is monitored through the optical fiber.

In the second approach, the activity of an enzyme (urease) is determined by a combination of flow injection analysis with optical sensor detection. Enzymatic hydrolysis of urea at pH 7.14 yields ammonium ion which is quantitated using a fully reversible ammonia sensor. Urea is contained in a buffered carrier stream into which samples of various urease activities are injected. The ammonium ions are measured fluorimetrically using an ammonium sensor membrane. The amount of ammonium ion formed is a direct indication of the enzyme activity. The system linearly correlates with urease activity in the range from 1 to 30 U/ml, the dynamic range being between 0.3 to 30 U/ml.

Inhibitors of the enzyme acetylcholinesterase (AChE) can be determined with a fiber optic photometer. The method employs a yellow synthetic enzyme substrate which is hydrolyzed by AChE to give a blue product. In the presence of an inhibitor (such as a pesticide or a nerve agent), the rate of the formation of this blue product is reduced. The resulting signal change can be detected via fiber optics over distances as large as 10 km and may serve as an alarm.

G.G. Guilbault and M. Mascini (eds.), Uses of Immobilized Biological Compounds, 335–344.

I. INTRODUCTION

Enzymatic methods of analysis have found extremely widespread application in almost all areas of analytical chemistry. Mostly, enzyme substrates are determined via their enzymatic degradation and measurement of products resulting from such reactions, e.g. protons, hydrogen peroxide, ammonium ion, reduced NADH, or the consumption of oxygen. A book dated from 1970 still is a most useful source of information[1] .

Aside from the determination of enzyme substrates, it is often important to know the activity of the enzyme, and this has been frequently been performed by making use of natural or synthetic substrates and measuring again the amount of products formed. In constrast to substrate assays, the amount of substrate has to be kept constant in this case. One of the preferred methods of analysis is fluorometry which is both sensitive and selective[2,3] We will show that quite new sensing schemes become possible when use is made of fiber optic chemical sensor technology[4]. This paper will give an account on recent results using fiber optic (remote) sensing schemes based on fluorimetric measurement of enzyme activity or enzyme inhibition, respectively.

II. FIBER OPTIC MEASUREMENT OF ENZYME ACTIVITY USING CHROMOGENIC SUBSTRATES

Fatty acid esters of 1-hydroxyrene-3,6,8-trisulfonate (HPTS) have been shown to be useful substrates for a direct and kinetic assay of esterases[5]. They are commercially available now. The (non-fluorescent) HPTS acetate and butyrate were immobilized on an ion exchange membrane and placed at the tip of an optical fiber[6]. Enzymatic action results in the cleavage of the substrate to give immobilized HPTS which, because of its low pK_a of 7.0, is dissociated by more than 50% at pH 7.2. Both free HPTS and its phenolate fluoresce at 520 nm with quantum yields close to unity. The 460-nm absorption band of the phenolate was chosen for excitation because it is more intense than the 415-nm band of the phenol species. Moreover, fibers have better transmission for blue light than violet light, and the xenon lamp has a higher output at 460 nm than at 415 nm.

Figure 1 shows how fluorescence increases when the tip of a fiber with the HPTS acetate membrane is immersed into solutions containing various concentrations of carboxylic ester hydrolase. The curves are typical for an enzymatic reaction in showing a rapid increase in signal per time in the initial phase, and a slow-down of the signal change as the substrate gets more and more consumed. The initial slope varies linearly with enzyme concentration. Similar results were obtained with the other hydrolases.

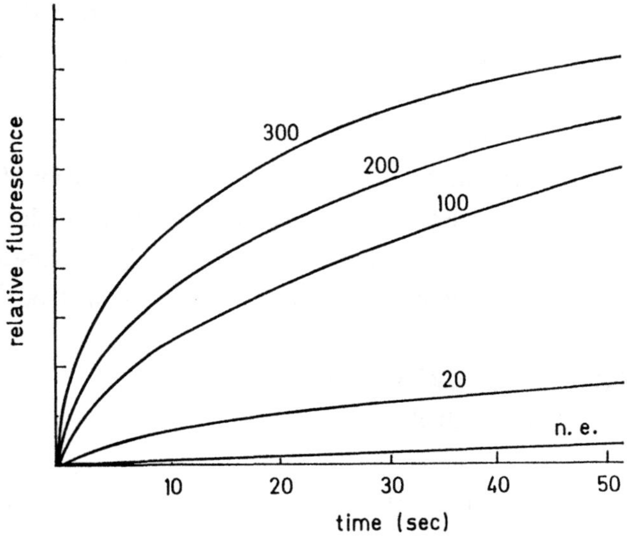

FIGURE 1. Increase in fluorescence intensity with time when immobilized HPTS acetate is immersed into a pH 7.2 solution containing various amounts of carboxyl-esterase. The numbers refer to micrograms of enzyme per milliliter of solution, and n.e. is the increase in signal due to non-enzymatic hydrolysis.

While there is agreement in the relative rates of enzymatic hydrolysis with the data published for cuvette experiments[5], one notes a decrease in the initial reaction velocity by a factor of 3 to 5. The rate of non-enzymatic hydrolysis is very slow at pH's below 7.6, and there is no need for a correction to account for this side reaction at enzyme levels at of above 20 μg/L. The rate of autolysis is distinctly smaller for the HPTS butyrate membrane. Unfortunately, this substrate is hydrolyzed very slowly by all three enzymes tested. Consequently, the HPTS acetate is a much better substrate despite the larger rate of non-enzymic hydrolysis.

The sensitivity of the assay is limited by the rates of the enzymatic and the non-enzymatic reaction. It can be seen from Figure 1 that the limit of detection for carboxylesterase is at about 20 μg/L. Below this level the signal change per time becomes too small (in comparision to autolysis) to be of analytical use, unless a correction is made for the latter process.

This fiber probe works irreversibly, so that its calibration may be more difficult than in the case of reversible senors. Since, however, a single membrane can be used for several short-time measurements (with gradual degradation in performance as the background from hydrolyzed HPTS increases), calibration appears feasible by making an initial measurement using an enzyme solution of defined activity. If separate membranes are to be used, one has to make sure that each sensor of a lot responds in the same way.

A similar approach has been made to measure the activity of amylase[7]. Reference also shall be made to the enzyme-based fiber optic sensor for determination of various substrates[8]. Finally, it has been shown that the concentration of *synthetic substrates* can be determined by this method[9].

III. MEASUREMENT OF UREASE ACTIVITY USING AN AMMONIUM-SENSITIVE OPTRODE

The basis for this measurement scheme[10] is the enzymatic hydrolysis of urea at pH 7.14 which yields ammonium ion according to the following equation:

$$H_2N\text{-}CO\text{-}NH_2 + 2\,H_2O + H^+ \Longrightarrow 2\,NH_4^+ + HCO_3^- \qquad \text{(equ. 1)}$$

The ammonium ion formed is quantitated using a pvc-based ammonium sensor membrane containing a neutral ammonium carrier (nonactin) and di-octadecyl rhodamine (DODR) acting as the proton carrier and possessing a pH-dependent emission.

The flow-injection manifold shown in Figure 2 was employed for serial measurements[11-14]. It is composed of an autosampler, a peristaltic pump, an injection valve with a 200-μL sample loop, and a flow cell containing the ammonium-sensitive membrane placed at the end of an optical fiber. The components of the manifold were connected with silicone rubber tubings. Optical measurements were performed with an Oriel 3090 fiber optic photometer equipped with a xenon lamp pulsed at 9 Hz. A 515-nm interference filter was used to isolate the excitation light, and fluorescence collected at 570 nm.

The sensing scheme is straightforward. Urea is contained in the carrier stream into which samples of varying urease activity are injected. The concentration of the ammonium ions, formed as a result of enzymatic hydrolysis of urea according to equ. (1), is measured with the ammonium fluorosensor, and is a direct indication of enzyme activity. In order to establish a calibration graph, 2-μL volumes of urease solutions of various activities (0.3, 1.0, 3.0, 10.0 and 30.0 U/mL) were injected into the carrier stream (a Tris-HCI buffer, 50 mM, pH

7.14), and mixed with a 10 mM urea solution. Hydrolysis occurred on the way to the flow-through cell and the ammonium ion formed changed the fluorescence intensity of the ammonium sensor. Figure 3 shows the relative signal change following injection of increasing amounts of urease activity.

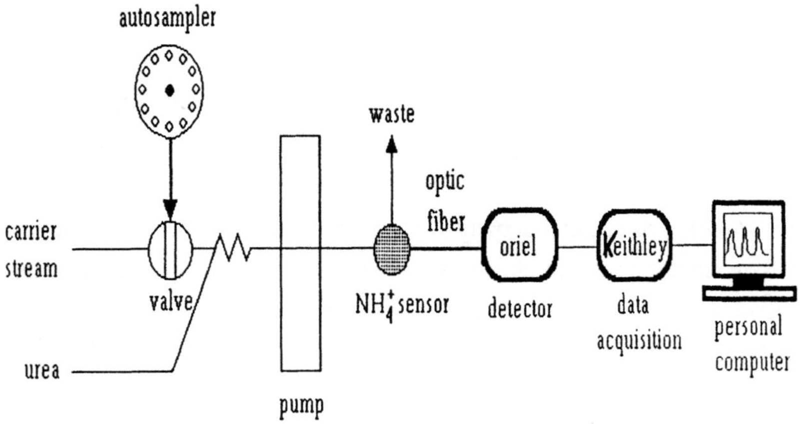

FIGURE 2. Schematic diagram of the FIA manifold. The sample is injected into the pH 7.14 buffer carrier steam in the injection valve, mixed with the urea solution, and passes the ammonium optrode detector, whose fluorescence is monitored by the fiber-optic fluorimeter. The whole flow-injection system is controlled by a PC.

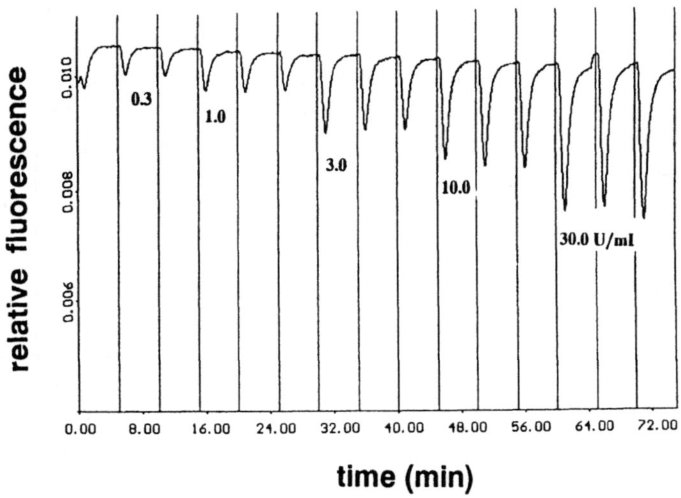

FIGURE 3. Determination of enzyme activity by FIA using an ammonium optrode as the transducer. Urease solutions (with specific activities of 0.3, 1.0, 3.0,10.0 and 30.0 units per mL) were injected (in triplicate) into the carrier stream (a 10 mM urea solution of pH 7.14), and the formation of ammonium ion was measured with an ammonium sensor membrane.

Ureases exhibit absolute specificity[15] Of a large number of compounds tested by various workers, including substituted ureas, only urea was metaboluzed. The enyme is competively inhibited by thiourea, ammonia and large concentrations of Na^+ and K^+. Inorganic phosphate is an activator. As these species can be present in samples were enzyme activities are to be determined, they are potential interferents.

IV. REMOTE MEASUREMENT OF PESTICIDES AND NERVE AGENTS VIA INHIBITION OF ACETYL-CHOLINESTERASE

The *inhibition* of enzyme activity[16] has been exploited in a fiber-optic method for the determination of inhibitors of acetylcholinesterase (AChE). The sensing scheme is based on the use of a synthetic chromogenic enzyme substrate, 2-(2-acetoxy-5-methyl-phenylazo)-N-methyl-1,3-thiazolium methosulfate (AMPT), described first by Hooidonk *et al.*[12].

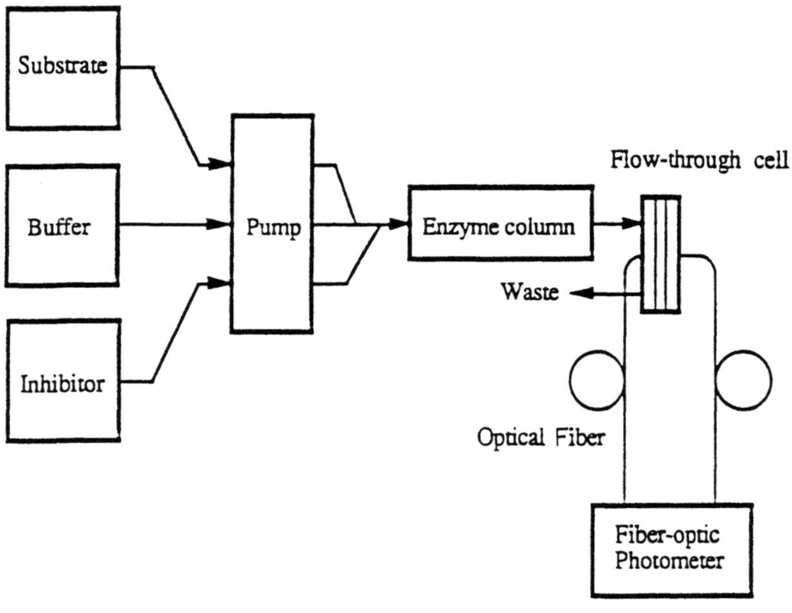

FIGURE 4. Flow system for monitoring inhibitors of acetylcholinesterase. Sample solution containing an inhibitor is mixed with the buffer and the substrate solution and transported through the enzyme column where the substrate is converted into a blue product whose absorption in measured in the flow-through cell via optical fibers. The rate of formation of the blue dye is reduced in the presence of an inhibitor.

The experimental set-up (a flow system with a detector cell) is schematically shown in Figure 4. Enzyme substrate solution was mixed with buffer and inhibitor solution via a peristaltic pump and pumped through an enzyme reactor. There, the yellow substrate (AMPT) is hydrolyzed by the enzyme to form the blue hydroxy form (HMPT). The absorbance was monitored at 580 nm. A home-made fiber-optic photometer was used for the remote fiber measurements. Optical fibers were connected to the yellow LED source (580 nm) to guide the light to the flow-through cell. Another fiber carried transmitted light back to the detector photodiode. AChE was immobilized on controlled-pore glass (modified by aminopropyl groups) by both the glutaraldehyde and the carbodiimide method.

In the absence of any inhibitor, the substrate is completely hydrolyzed to form blue HMPT. In the presence of an enzyme inhibitor, the amount of blue dye being formed is reduced. Figure 5 shows the inhibition of AChE immobilized on aminopropyl CPG by the reversible inhibitor BW 284 C51. A steady state was observed after about 5 to 15 min. The regeneration times varied from 8 to 25 min (depending on the concentration of the inhibitor). The detection limit was approximately 10 nM of BW 284 C51.

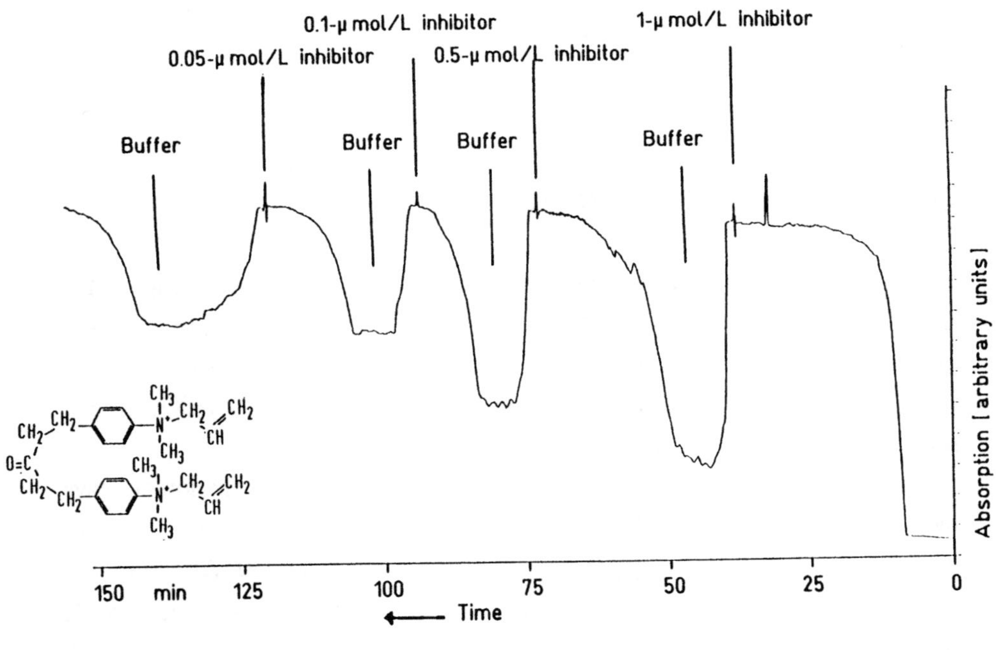

FIGURE 5. Time dependence of the inhibition of AChE, immobilized on aminopropyl CPG, by BW 284 C51 (pH 8.0; flow rate 1.2 mL/min).

Numerous inhibitors of the enzyme AChE are known[16] We choose one representative example out of each of the three most important inhibitor groups, namely (a), an organophosphorus compound (paraoxon), (b), a carbamate (pyridostigmine bromide), and (c), a quaternized ammonium ion (BW 284 C51). The detection limits found for paraoxon (200 nM; equal to 55 μg/L) and BW 284 C51 (10 nM, 5.7 μg/L) are promising, while that for pyridostigmin (1 μM, 126 μg/L) is unsatisfactory.

The method described here does not differentiate between single inhibitors. It rather measures a sum parameter. The system not only detects hazardous pesticides[19] but also other inhibitors such as heavy metal ions. It is therefore ideally suited as a means for describing the quality of water in terms of a sum parameter for total inhibitor activity. Unlike in HPLC and in immunoassay, it does not require each single known (and potentially unknown) inhibitor to be identified and quantitated. In fact, the occurrence of unknown (or new) inhibitors, for which in HPLC no standards are known (or available) and in immunoassay no antibodies are available, is one of the major limitations of HPLC methods and immunoassays. The fact that the present system responds differently to each inhibitor is also the reason why no general calibration curve for pesticides can be given.

V. DETERMINATION OF HEAVY METALS VIA THEIR INHIBITORY EFFECT ON UREASE

Urease is sensitive to traces of heavy metals. Silver(I), mercury(II), copper(II), cadmium(II), cobalt(II), nickel(II), manganese(II) and lead(II) have all been quantified by using data from the inhibition of the urea/urease reaction by the metal. The urease inhibition is relatively selective for mercury(II)[20] with a selectivity of 100 (or greater) over the other.

Yerian *et al.*[12] report on two kinds of approaches for determination of heavy metals: In the first, immobilized urease was exposed to various levels of mercury(II) contamination in the carrier stream (without EDTA), and the response to a 10 mM urea standard was monitored. At low levels of mercury(II) (0.2 - 2.0 μM), injections of 10 mM urea showed an exponential decrease in peak height. The other mode of sample monitoring would be alternate injections of urea standard and mercury samples. This, however, was not investigated further by the authors. Conceivably, both methods are suitable for continuous assay of mercury in environmental samples.

VI. CONCLUSION

We think to have demonstrated that a combination of methods of enzymatic analysis with optrode technology can result in very powerful analytical techniques for measuring enzyme activity and inhibition. Its potential has not been fully exploited yet. Because both the enzyme and the optrode detector can be very specific, selective methods of analysis become possible. If coupled to fluorescence assay, both sensitivity and selectivity may be increased. Sensors can be used as such, but may as well be used in conjunction with flow injection analysis (FIA) which can provide the benign environment that is preferred when making use of of enzymatic methods of analysis with their sensitivity to pH, ionic strength, temperature, all of which affect enzyme activity. The FI system, in addition, provides the possibility of frequent sensor calibration. Therefore, the FIA/optrode hyphenation indeed looks like an almost ideal combination.

ACKNOWLEDGEMENT

This work was supported by the Austrian Science Foundation (FWF) within projects S5701-TEC and S5702-TEC which is gratefully acknowledged.

REFERENCES

1. G. G. Guilbault, *Enzymatic Methods of Analysis,* Pergamon, 1970.
2. G. G. Guilbault, *Practical Fluorescence: Theory, Methods, and Techniques,* Marcel Dekker, 1973
3. O. S. Wolfbeis (ed.), *Fluorescence Spectroscopy: New Methods and Applications,* Springer Verlag, Heidelberg, 1992.
4. O. S. Wolfbeis (ed.), *Fiber Optic Chemical Sensors and Biosensors,* CRC Press, Boca Raton, Florida, 1991.
5. O. S. Wolfbeis and E. Koller, Fluorimetric assay of hydrolases at longwave excitation and emission wavelength with new substrates possessing unique water solubility, *Anal. Biochem.* **129**, 365 (1983).
6. O. S. Wolfbeis, Fiberoptic Probe for Kinetic Determination of Enzyme Activities, *Anal. Chem.* **58** , 2875 (1986).

7. Z. Zhang, W. R. Seitz and D. O'Connell, Amylase activity assay via immobilized fluorogenic amylose, *Anal. Chim. Acta* **236**, 251 (1990).

8. M. A. Arnold, First enzyme-based fiber optic sensor for determination of substrate concentration, *Anal. Chem.* **58**, 1137 (1986).

9. G. Freeman and L. Bachas, *Biosensors & Bioelectronics* **7**, 49 (1992).

10. H. Li and O. S. Wolfbeis, Determination of urease activity by flow injection analysis using an ammonium-selective optrode as the detector, *Anal. Chim. Acta* **276,** 115 (1993).

11. T. D. Yerian, G. D. Christian and J. Ruzicka, Flow injection analysis as a diagnostic technique for development and testing of chemical sensors, *Anal. Chim Acta* **204**, 7 (1988).

12. T. D. Yerian, G. D. Christian and J. Ruzicka. Enzymatic determination of urea in water by opto-sensing flow injection analysis, *Analyst* **111**, 865 (1988).

13. O. S. Wolfbeis, Optical sensors in flow injection analysis, *J. Mol. Struct.* **292**, 133 (1993).

14. M. Valcarcel and M. D. Luque de Castro, Fluorescence Detection in Flow Injection Analysis. In: *Fluorescence Spectroscopy: New Methods and Applications*, O. S. Wolfbeis (ed.), Springer, Berlin, 1992.

15. Th. Barman, *Enzyme Handbook*, vol. II, Springer, New York, 1969, p. 648.

16. H. Zollner, *Handbook of Enzyme Inhibitors*, VCH, Weinheim, 1989, pp. 3 and 55

17. C. v. Hooidonk, C. de Borst, F. A. Mitzka and C. C. Groos, Chromogenic substrates for cholinesterases, *Anal. Biochem.* **48**, 33 (1972).

18. W. Trettnak, F. Reininger, E. Zinterl and O. S. Wolfbeis, Fiber optic remote detection of pesticides and related inhibitors of the enzyme acetylcholine esterase, *Sensors & Actuators* **B11,** 87 (1993).

19. H. U. Bergmeyer (ed.), *Methods of Enzymatic Analysis*, Vol. XII: Drugs and Pesticides, VCH, Weinheim, 1986, pp. 404-438.

20. L. Ögren and G. Johansson, Determination of traces of mercury(II) by inhibition of an enzyme reactor electrode loaded with immobilized urease, *Anal. Chim. Acta* **96**, 1 (1978).

LUMINESCENCE DETECTION OF BIOLOGICAL REACTIONS

P. R. COULET, L. J. BLUM and S. M. GAUTIER
Laboratoire de Génie Enzymatique, EP 19 CNRS-Université Lyon 1
43 Bd du 11 Novembre 1918, 69622 Villeurbanne Cedex France

Abstract A highly sensitive and specific fiberoptic biosensor based on bioluminescence or chemiluminescence reactions has been developed in our group. The sensor is made of a fiber optic bundle associated with a polyamide enzyme membrane connected to a luminometer, enabling the emitted light to be processed. Contrary to other optical devices, this one does not require a light source or monochromators, making the system simple and attractive for remote analysis. It can be associated with flow injection analysis and allows unpretreated samples to be assayed. ATP and NAD(P)H could be easily determined at the picomole level using the bioluminescent enzyme systems from the firefly and from marine bacteria, respectively. Enzyme activities, especially dehydrogenases, could also be monitored with the same device. A reagentless design, involving a loosely embedded co-reactant continuously released in the bound enzyme microenvironment, enabled 30-35 assays to be performed with the same loading.

I INTRODUCTION

Since the discovery by the end of the nineteenth century in Lyon by the French physiologist Raphaël Dubois that it was possible to produce light by mixing two compounds extracted from a light emitting beetle, bioluminescence reactions were found to occur in numerous species especially from marine origin. In the last decades the related enzyme reactions involving luciferases were extensively studied, especially in the group of Mac Elroy[1], enabling an ultrasensitive test for ATP to be widely available in biochemistry laboratories.

An increasing interest appeared for optical transduction in recent years but little attention was devoted to bio- or chemiluminescence[2,5]. Our main goal was to develop a fiberoptic biosensor involving light emitting reactions with different types of luciferases. Two systems were selected : the first one was the monoenzyme system from the firefly *Photinus pyralis* specific for ATP and the second one was the bienzyme system found in marine bacteria specific for NAD(P)H.

Luciferase (EC 1.13.12.7) from the firefly *Photinus pyralis* catalyzes the reaction :

G.G. Guilbault and M. Mascini (eds.), Uses of Immobilized Biological Compounds, 345–350.

$$ATP + LH_2 + O_2 \xrightarrow{\text{firefly luciferase / } Mg^{2+}} AMP + PPi + Oxyluciferin + CO_2 + h\nu$$

The presence of luciferin LH_2 and Mg^{2+}, which serve as co-reactants, and of molecular oxygen enables the enzymic reaction to take place and the light emission to occur. The intensity of emitted light ($\lambda max = 560$ nm) is proportional to the ATP concentration in a wide linear dynamic range.

For NAD(P)H monitoring, the enzymatic systems from *Vibrio harveyi* and *Vibrio fischeri* were used. In this case, two consecutive reactions take place involving the specific enzymes NAD(P)H : FMN oxidoreductase (EC 1.6.8.1) and luciferase (EC 1.14.14.3). The light emission intensity at $\lambda max = 490$ nm is directly related to the NAD(P)H concentration.

$$NAD(P)H + FMN \xrightarrow{\text{oxidoreductase}} NAD(P)^+ + FMNH_2$$

NAD(P)H, which is the target analyte, is oxidized, whereas the coenzyme FMN is reduced into $FMNH_2$ which is a substrate for luciferase.

$$FMNH_2 + R\text{-}CHO + O_2 \xrightarrow{\text{bacterial luciferase}} FMN + R\text{-}COOH + H_2O + h\nu$$

This second enzyme requires both molecular oxygen and a long chain aldehyde, generally decanal.

II DESIGN AND PERFORMANCE OF THE LUCIFERASE-BIOSENSOR

The principle is shown in Fig. 1. The bioactive membrane with the selected enzyme system for the light emitting enzyme reaction (LER) is associated with the waveguide, here a fiberoptic bundle connected to the photomultipier of a luminometer used for signal processing.[6] The bioactive membrane was prepared through direct enzyme covalent immobilization, using preactivated polyamide membranes from Pall Co., USA. In brief, simply wetting the membrane with a few microliters of concentrated enzyme solution enables any bioactive membrane to be prepared in a few minutes.[7]

FIGURE 1 Principle of the Bioluminescent sensor

It is possible to prepare multienzyme systems through the co-immobilization of auxiliary enzymes, enabling extension of the potentialities of the biosensor to a variety of analytes. The sensing end of the fiber optic bundle can be either connected to a light-tight thermostatted vessel in batch analysis or to a black PVC flow-cell for use with FIA[8]. In the batch procedure, 10-50 microliters sample volumes are injected through a septum and a multichannel peristaltic pump delivers the different reagents when using FIA. Using the batch procedure, ATP could be measured over a wide linear dynamic range from 2.8 x 10^{-10} M to 1.6 x 10^{-6} M ATP.
For NADH, the linear dynamic range was between 1 x 10^{-9} M and 3 x 10^{-6} M. For NADH, the linear dynamic range was between 1 x 10^{-9} M and 3 x 10^{-6} M. Using the same membrane with the two bioluminescent systems coimmobilized, one or the other bioluminescent system could

function using the appropriate reagents. Measurements in the FIA mode could be performed on a wide linear calibration range with detection limits as low as 0.25 pmol ATP and 5 pmol NADH, respectively.
With dehydrogenases coimmobilized with the bacterial bioluminescent system alone, several metabolites could be determined with the batch system[9]: ethanol from 4 x 10^{-7} M to 7 x 10^{-5} M, D-sorbitol from 2 x 10^{-8} M to 2 x 10^{-5} M , and oxaloacetate from 1 x 10^{-9} M to 3 x 10^{-6} M

Determination of dehydrogenase activities

The possibility of monitoring enzyme activities with the bioluminescent sensor[10] was investigated with lactate dehydrogenase (EC 1. 1. 1. 27) due to its importance in both clinical analysis and biotechnological processes. The principle of measurement is given in Fig. 2.

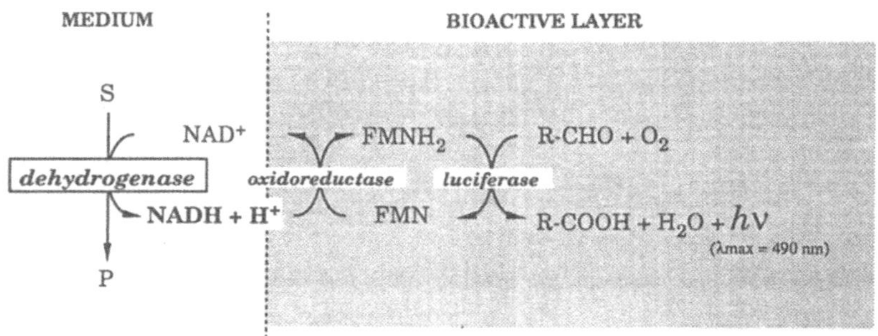

FIGURE 2 Determination of dehydrogenase activity with the bioluminescent sensor equipped with the bacterial bienzyme system.

The dehydrogenase activity is revealed by adding into the medium the enzyme substrate and the coenzyme. The oxidation of L-lactate (substrate S) into pyruvate (product P) occurs with the concomitant reduction of NAD^+ into NADH. Conditions had to be optimized and the effect of NAD^+ on the oxidoreductase-luciferase system carefully investigated. The FIA approach was chosen, which essentially allows the pH to be adjusted for both the LDH reaction and the bioluminescence detection. Using serum samples, the detection of LDH could be achieved in the range 5 -250 IU . l^{-1} with a cycle time depending on the activity and comprised between 45 s and 4 min.

Reagentless biosensor

The main drawback when using complex enzyme systems is the compulsory addition of coreactants. In principle, a type of reagentless biosensor could be designed by embedding co-reactants in the bioactive tip for subsequent continuous release in the bound enzyme microenvironment. In this way we used a second membrane with loosely embedded co-reactant stacked in the sensing tip . The bacterial bioluminescent system was chosen which requires FMN and the long chain aldehyde R-CHO as coreactants. Only FMN was embedded, whereas the aldehyde was added in the reaction medium as shown in Fig. 3.

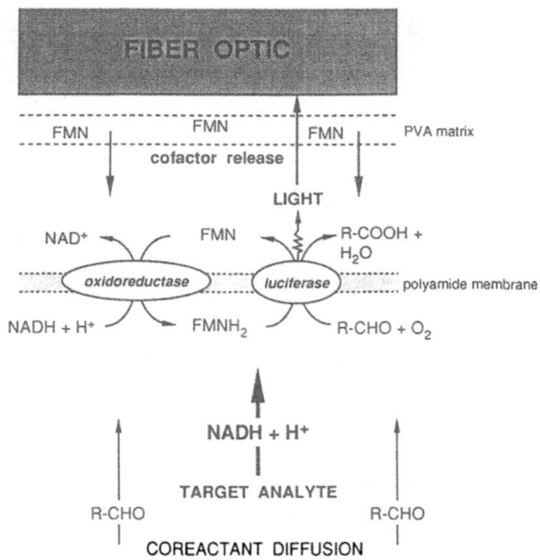

FIGURE 3 Sensing tip of the Bioluminescent sensor with the stacked enzyme membrane and the FMN - containing membrane.

Polymerization of polyvinyl alcohol (PVA) was performed in the presence of FMN in several ways. First, glutaraldehyde as cross-linking reagent was used. Then photosensitive functionalized PVA was also found suitable for entrapment and subsequent release of FMN, after exposure to UV radiation. Good results were obtained in both cases with 30-35 reliable assays easily performed, but the most promising matrix was the photo cross-linked PVA[11].

III PROSPECTS

Refering to the recent development of biosensors, optical transduction appears to be really promising. Furthermore, bioluminescence is particularly attractive due to its very high sensitivity. ATP and NAD(P)H, involved in a number of metabolic reactions, can be monitored with specific luciferases, and the required instrumentation is simple. This led us to design a fiber optic biosensor based on the bioluminescent enzyme systems already mentioned. Extension to chemiluminescent systems, especially those based on the luminol - hydrogen peroxide reaction was possible. Many areas are concerned with the development of such biosensors in biotechnology, biomedical engineering, environment control and defense.

REFERENCES

1. W. D. McElroy, Proc. Natl. Acad. Sci. USA, 33, 341, (1947).
2. G. G. Guilbault (ed.), Practical Fluorescence, Marcel Dekker, New York (1991).
3. D. L. Wise and L. B. Wingard Jr. (eds.), Biosensors with fiberoptics, Humana Press (1991).
4. O. S. Wolfbeis (ed)., Fiberoptic chemical sensors and biosensors, CRC Press, Boca Raton, (1991).
5. L. J. Blum and P. R. Coulet (eds)., Biosensor Principles and Applications, Marcel Dekker, New York, (1991).
6. L. J. Blum, S. M. Gautier and P. R. Coulet, Anal. Lett., 21, 717 (1988).
7. C. H. Assolant-Vinet and P. R. Coulet, Anal. Lett., 19, 875 (1986).
8. L. J. Blum, S. M. Gautier and P. R. Coulet, Anal. Chim. Acta, 226, 331 (1989).
9. S. M. Gautier, L. J. Blum and P. R. Coulet, J. Biolum. Chemilum., 5 57(1990).
10. S. M. Gautier, L. J. Blum and P. R. Coulet, Anal. Chim. Acta, 266, 331 (1992).
11. S. M. Gautier, L. J. Blum and P. R. Coulet, Anal. Chim. Acta, 243, 149 (1991).

PROGRESS IN FIBER-OPTIC BASED BIOSENSORS AT THE UNITED STATES NAVAL RESEARCH LABORATORY

J. MATTHEW MAURO[a,b], JOEL P. GOLDEN[a], GEORGE P. ANDERSON[a], ROBERT A. OGERT[a], DAYA WIJESURIYA[a,b], LISA C. SHRIVER-LAKE[a], AND FRANCES S. LIGLER[a]

[a]Center for Biomolecular Science and Engineering, Naval Research Laboratory, 4555 Overlook Avenue SW, Washington, DC 20375; [b]GEO-CENTERS, INC., 10903 Indian Head Highway, Fort Washington, MD 20744

abstract A fiber-optic based biosensor has been developed which integrates a novel array of biological, optical, and electrical components. Distally tapered, chemically activated glass fibers are coated with antibodies specific for desired analytes. A sandwich immunoassay is performed by exposing a fiber to a solution of analyte containing a second, analyte-specific and fluorescently labeled antibody. Fluorescent light emitted from antibody/analyte complexes bound within the evanescent region of the laser illuminated tapered fiber is optically filtered and electronically quantitated. Assays for botulism and ricin toxins, as well as specific detection of fluorescently-stained *Bacillus anthracis* cells, are described.

I INTRODUCTION

Fiber-optic based technology is an important element of the ongoing biosensor development program at NRL. The current generation of biosensors utilizes immobilized antibodies and other receptor molecules to promote specific binding of fluorescent analytes to laser illuminated optical fibers. Several relevant aspects of these sensors will be discussed, including selected examples of antibody dependent assays presently in use.

II GENERAL APPROACH

Under conditions required for "total" internal reflection, the electric field of light propagating within a waveguide does not abruptly drop to zero at the waveguide boundary, but decays exponentially as a function of distance from the waveguide surface [1]. Excitation of and consequent light emission by fluorophores located within this approximately 100 nm deep *evanescent region* form the basis for antibody-based assays of solution analytes. In a typical *sandwich immunoassay* [FIGURE 1], the fiber optic probe is coated with analyte- specific antibody, and the fiber is immersed in a solution of the analyte containing a second analyte-specific and fluorescently labeled antibody. The analyte/fluorescently-tagged-antibody complex is recognized and bound

G.G. Guilbault and M. Mascini (eds.), Uses of Immobilized Biological Compounds, 351–357.

by the fiber immobilized antibody. In this way, the fluorescent complex is concentrated within the evanescent region of the illuminated fiber; emitted light that is captured by the fiber is propagated to the detection/quantitation components of the system [2].

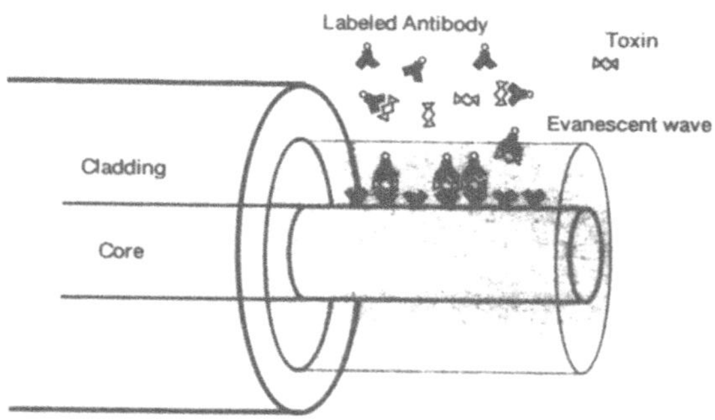

FIGURE 1 Localization of fluorescent antibody/analyte complex within the evanescent region of the illuminated fiber in a sandwich immunoassay

III SENSOR COMPONENTS

The type and arrangement of optical components have been studied and refined at NRL in order to optimize signal coupling efficiency and to minimize the effects of stray light and noise [FIGURE 2]. The system includes a 50 mW Argon-ion laser, an off-axis parabolic mirror, a biconvex fused silica focusing lens, beam chopper phased by means of a lock-in amplifier to a photodiode detector, and the fiber sensing element.

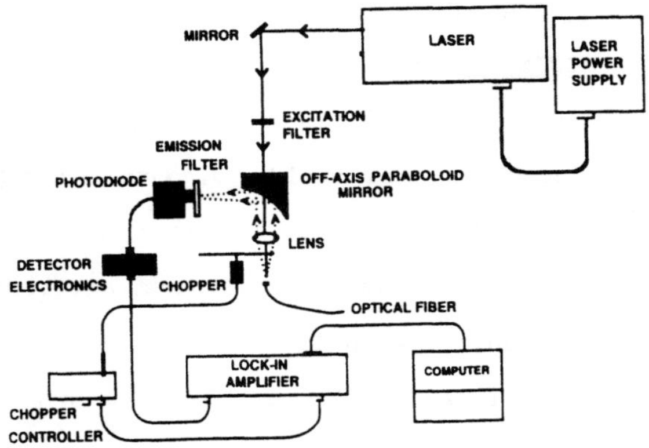

FIGURE 2 Schematic of the fiber-optic biosensor components

Sensing elements are prepared in a three-step procedure beginning with commercially obtained long plastic-clad fused silica fibers (radius 100 microns: 1) Plastic cladding from several centimeters of the distal end of each fiber is removed, and computer controlled HF dip-etching is conducted to form an exposed fiber core having a continuously tapered radius. This tapering process minimizes the V-number mismatch that occurs as light enters the clad portion of the fiber from the solution-exposed sensing region. Tapering results in an up to 50-fold enhancement in detector signal strength compared to signals obtained from untapered fiber cores [3]. 2) A layer of reactive thiol groups is formed on the surface of the clean, tapered fibers by treatment with mercaptopropyltrimethoxysilane in dry toluene; immediate immersion of the thiolsylated fibers into a buffered solution of heterobifunctional crosslinking agent such as N-g-maleimidobutyryloxy succinimide ester (GMBS) completes the activation of the surface for subsequent protein attachment. 3) Purified mono- or polyclonal antibodies are allowed to react with the GMBS activated fiber surface. In general, glass surfaces bearing reactive N-hydroxysuccinimide esters have been the most useful substrates for immobilizing active antibodies [4].

IV SELECTED EXAMPLES OF ASSAYS

BOTULISM AND RICIN TOXINS

Sandwich immunoassays have been developed for detection and quantitation of the neurotoxin produced by *C. botulinium* (botulism toxin) [5] and for agglutinin I and II from *R. communis* (ricin)[6].

In preparation for immobilization on fibers, antibodies derived from serum of horses hyperimmunized with *C. botulinium* were affinity purified using Sepharose 4B immobilized botulinium toxin A. In a typical two-step assay, an anti-toxin A derivatized fiber is incubated with a solution of 1 to 200 ng/ml botulinium toxin A for 10 minutes, followed by addition of tetramethylrhodamine labeled horse anti-toxin A IgG at 5 ug/ml. FIGURE 3 shows the detector response vs. time for various concentrations of the toxin.

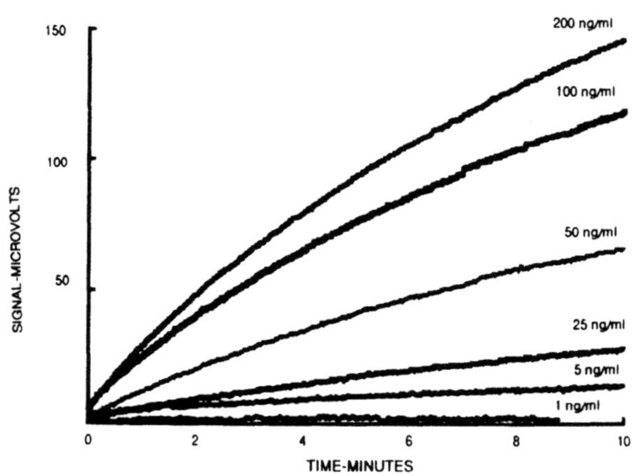

FIGURE 3 Botulinium toxin assay; time vs. detector response at various toxin concentrations

The current lower limit of sensitivity of the fiber optic assay for botulism toxin is on the order of 1 ng/ml; ongoing efforts are aimed at improving system optical components and probe tapering technology to allow this technique to approach the sensitivity of the mouse acute toxicity test (0.005 ng/ml). The key advantage of the fiber optic assay is the short analysis time required for obtaining analytical results (2-3 minutes) compared with the 4 day time interval necessary for obtaining results using the *in vivo* mouse assay.

A similar sandwich immunoassay for ricin having a sensitivity limit of about 1 ng/ml has also been developed [6]. FIGURE 4 illustrates the results of assays carried out with river water and urine spiked with known amounts of ricin. It is clear that substances present in both types of samples can have strong effects on observed signal strength. The source(s) of these effects are presently unknown.

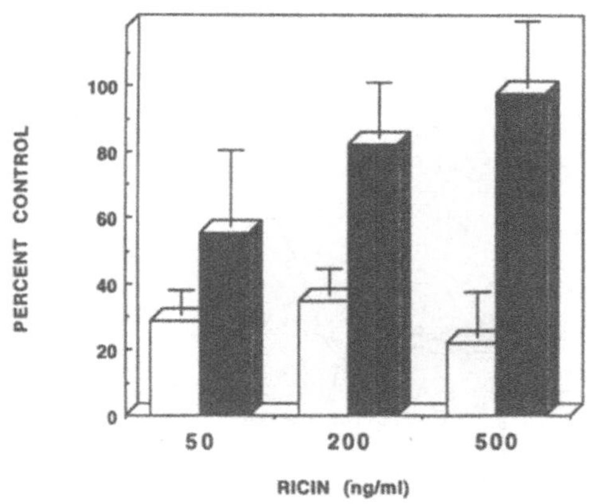

FIGURE 4 Fiber optic immunoassay of ricin in urine and river water. 50, 200, and 500 ng/ml ricin was assayed in urine (white bars) and river water (black bars) using a two-step assay. Values are the percent of control in ricin-spiked PBS. Errors are S.D. values for 3 independent assays. Signal was observed 2 min post incubation with TRITC-goat and anti-ricin IgG.

DETECTION OF STAINED *B. ANTHRACIS* CELLS

Rapid and specific detection of bacterial cells in environmental samples is important in monitoring pollution sources, evaluating bioremediation efforts, and, conceivably, in biological warfare scenarios. Initial efforts at NRL have been focused on fiber optic immunoassay of *Bacillus anthracis* cells. In these assays, bacterial suspensions are initially treated with a fluorescent dye capable of incorporating into the bacterial surface, resulting in highly fluorescent cells. After this non-specific labeling step, the fluorescently-labeled cell suspensions are exposed to fibers coated with monoclonal antibodies specific for the *B. anthracis* outer coat elements; the specificity of the assay depends upon the ability of the antibodies to recognize and bind to the surface of the stained bacilli. Bound cells stained with Nile Red give an immediate signal upon exposure to the antibody derivatized fiber, as demonstrated in FIGURE 5.

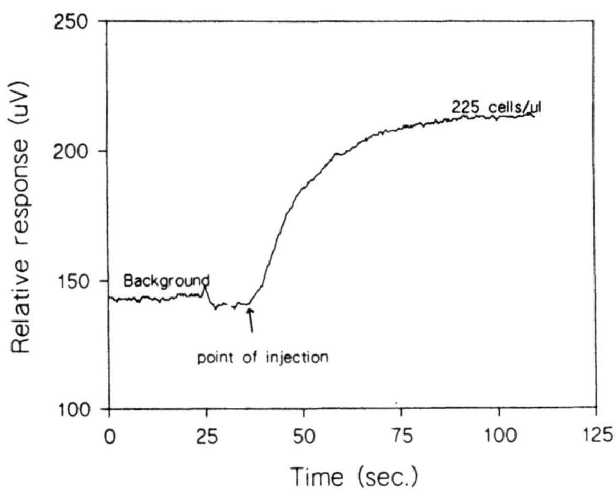

FIGURE 5 Detector response versus time in a fiber optic immunoassay of *B. anthracis* cells stained with Nile Red.

At present, a lower limit of approximately 3,000 cells/ml is detectable with this system. Experiments examining the effects of interfering substances and the presence of competing bacteria are currently in progress.

V CONCLUSIONS

Advantages of the fiber optic immunoassay format that has been developed at NRL include simplicity, rapidity, good detection limits and specificity. Efforts aimed at improving sensitivity and expanding our repertoire of assays for chemicals, toxins, and bacteria continue.

VI REFERENCES

1. M.N. Kronick and W.A. Little, *J. Immunol. Meth., 8*, 235, (1975).
2. J.P. Golden, L.C. Shriver-Lake, G.P. Anderson, R.B. Thompson, and F.S. Ligler, *Optical Engineering, 31(7)*, 1458 (1992).
3. G.P. Anderson, J.P. Golden, and F.S. Ligler, *Biosensors and Bioelectronics*, (in press, 1993).
4. F.S. Ligler, J. Georger, S.K. Bhatia, J. Calvert, L.C. Shriver-Lake, and R. Bredehorst, *U.S. Patent No. 5,077,210* (1991).
5. R.A. Ogert, J.E. Brown, B.R. Singh, L.C. Shriver-Lake, and F.S. Ligler, *Analytical Biochemistry, 205*, 306, (1992).
6. R.A. Ogert, L.C. Shriver-Lake, and F.S. Ligler, *SPIE Advances in Fluorescence Sensing Technology, 1885*, 11, (1993).

IMMOBILIZED BIOLOGICAL COMPOUNDS IN DEFENSE APPLICATIONS

SHEILA J. WOOD
Chemical and Biological Defense Agency, Edgewood Research Development and Engineering Center, Aberdeen Proving Ground, MD. 21010 USA

Abstract Capture and characterization of biological materials within a fluid stream using four different immobilization strategies will be discussed. The moieties and strategies are as follows: 1) DNA-DNA hybridization using avidin-biotin linked probes, 2) a 170 kDa protein using carboxyl-amine coupling, 3) a 15 residue acidic peptide by thiol linkage, and 4) an acrylamide gel layer for tethering molecules with free amino groups. All molecules were immobilized so as to enable their reactive groups to freely interact with biological counterparts. Measurements were made in real time as reactions occurred within a fluid stream.

INTRODUCTION

Various strategies are implemented within detection devices to encourage capture and recognition of biological materials of interest. Classic immobilization methods, such as silanization and gluteraldehyde treatment, bind antigens and antibodies to flat surfaces. Such methods often leave the molecules incompetent to react with their biological counterparts or render their reactive portions inaccessible. The methods and immobilization strategies presented here encourage competent and complete interactions of molecules as a result of their stationary position, yet unhampered ability to interact. Two monitoring systems for analysis of these interactions will be discussed.

The Biospecific Interaction Analysis system, BIAcore, (Pharmacia Biosensor, Piscataway, N.J.,USA) uses surface plasmon resonance to detect changes in a sensor surface as bio-molecular

G.G. Guilbault and M. Mascini (eds.), Uses of Immobilized Biological Compounds, 359–368.

reactions occur. Changes in light reflection are a direct result of changes in mass occurring at the sensor surface, shown in Figure 1. The reaction surface can be made biospecific by immobilizing the ligand of choice onto a derivitized (COOH) dextran layer. Controlled sequential injections of reactants make it possible to measure each binding event as it occurs.[1]

A reaction surface is prepared in BIAcore by covalent coupling of a ligand to a derivatized dextran matrix located on the sensor chip surface. This reaction surface is layered onto to a 50nm deep gold layer which rests on glass. Monochromatic light, directed at the sensor surface, is reflected internally due to differences in the refractive indices of water and glass. Concomitantly, evanescent wave energy from within the metal surface is generated and results in a portion of minimally reflected light. The change in the angle of minimally reflected light is a direct result of mass changes on the metal surface. Changes are monitored electronically and recorded graphically as reactions occur. A response, expressed as resonance units (RU), of 1000 corresponds to approximately 1 ng per mm^2 monoclonal antibody (ml wt 100,000 to 120,000). Other RU values corresponding to weight vary, depending on the protein used and its complexity. [1]

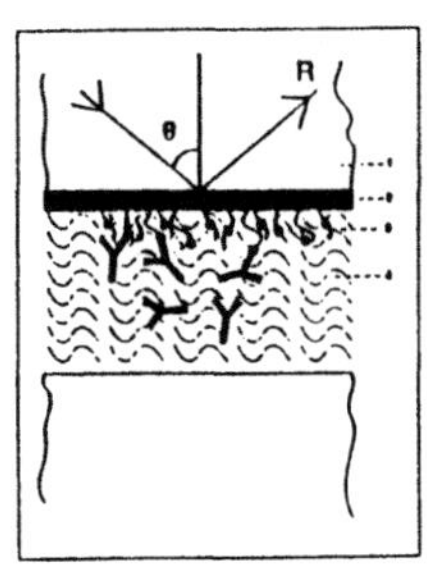

Figure 1. Change in angle of minimal reflectance as a result of mass change.

Another monitoring system designed and assembled by Dr. Daniel Hammer, (Cornell University, Ithaca, N.Y., USA) analyzes molecular interactions within a flow chamber positioned beneath a phase contrast, inverted microscope. Video images of rolling cells are analyzed using image software. The immobilization surface consists of a derivitized acrylamide gel onto which molecules are tethered by a free amino terminal group.[2]

IMMOBILIZATION AND ANALYSIS

Oligonucleotide Analysis Using BIAcore

Hybridization reactions were carried out in BIAcore using oligonucleotides provided by Dr. Kim Rogers, (Environmental Protection Agency, EMSL, Las Vegas, NV, USA). Avidin served as the primary layer on the dextran surface. Biological materials consisted of a 20 mer biotinylated probe, BUNI; a 40 mer probe, BETA, containing base pairs complementary to 20 base pairs in BUNI; and a 40 mer unrelated oligonucleotide, FETA.

Avidin (0.44U) was coupled to the matrix using conventional carboxyl-amine coupling and gave an RU response of 22,083. The biotinylated 20 mer probe (105 ng) gave a response of 5703 RU and, shown in Figure 2, the 40 mer complementary oligo (105 ng) gave a response of 3333 RU. BETA (105 ng) was inoculated over an avidin surface as a control for nonspecific reactions. No non-specific interactions occurred. Shown in Figure 3, the 40 mer unrelated oligo was inoculated over the avidin/BUNI surface and no response was observed. A second challenge with BETA resulted in a response of 3553 RU. BUNI and BETA, the complementary oligos, were incubated at the T_{hyb} of 51°C for 30 min prior to injection. When 2.8 ng and 105 ng were injected, responses were commensurate with concentration.

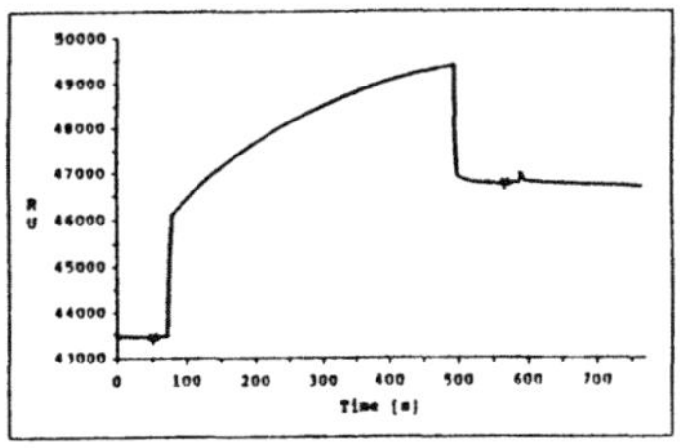

Absorbance Response	Relative Response	Identity
43438	0	Baseline/Avidin/Buni
46771	3333	BETA

Figure 2. Hybridization of a 40 mer oligo to a 20 mer biotinylated oligo on an avidin base.

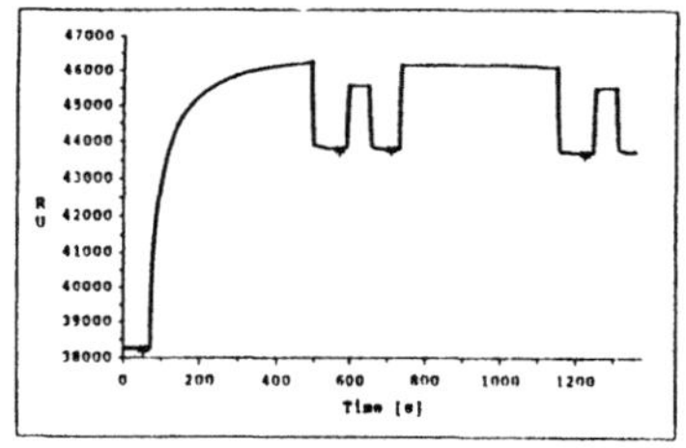

Absorbance Response	Relative Response	Identity
38254	0	Baseline/Avidin
43813	5559	BUNI
43789	-24	Buffer wash
43707	-82	FETA

Figure 3. Challenge with a 40 mer unrelated oligo.

Using the conservative estimate that 40-45% of the molecules introduced to the sensor bound to the surface, estimates were obtained for the amount of DNA remaining on the surface after reactions were complete. Following the conversion of mg/ml DNA to µmol/L DNA, second order rate kinetics were used to analyze the results. It was estimated that the hybridization reaction, 40 mer to 20 mer, was 1 part in 1.664 parts complete in 480 seconds. The reactions were carried out at room temperature under low stringency conditions. Although T_m and T_{hyb} values were calculated for these oligos based on AT and GC content, the calculations may be closer to assumptions than to realistic evaluations. Under conditions of low stringency for new methods, experimental determinations of T_m and T_{hyb} may be more accurate than theoretical determinations.[3]

Peptide Analysis Using BIAcore

Microorganisms carry cell surface components responsible for the activation of the alternate complement pathway, beginning with C3, and generate cidal C5b-9 membrane attack complexes. During an immune response, membrane attack complexes can damage host cell surfaces. One mechanism that mammalian cells use for protection is the complement binding region known as CD59 which protects the cells from lysis by binding the late components of complement.

Dr. William Petri, Jr,(University of Virginia, Charlottesville, Va USA) has isolated an adherence protein from the surface of Entamoeba histolytica which has been implicated in disease production. It has been shown to play a role in abrogation of lysis by the immune system in in vitro assay systems. It also binds to the late components of complement, C8 and C9 in reduced form. Anti-CD59 antibodies directed against the protective portion of mammalian cell membranes, cross-react with a portion of this lectin.[4] In addition, Dr. Steve Tomlinson, (New York University, N.Y. USA) has shown that CD59 chooses a 15 residue portion of C9 to bind to in its abrogation of mammalian cell lysis. (personal communication). We asked the following question. Does the amebic lectin bind this segment of C9, possibly protecting the amoeba from elimination by the immune system?

The C9 segment was tethered to the dextran matrix in BIAcore via a terminal cysteine residue by thiol coupling. N-hydroxysuccinimide/ethylenediamine-carbodiimide (NHS/EDC) treatment of COOH groups on the matrix resulted in activation of the sensor surface

(COO-). Cystamine was injected to derivitize the surface with disulfide groups. Dithiothreitol was inoculated to reduce the disulphides to thiols. When the ligand in question, containing an SH group, was introduced, it became bound to the matrix via a disulphide linkage. PDEA [2-(2-pyridinyldithio)ethaneamine hydrochloride] was then added to remove free ligand and to cap free sulfhydral groups.

The C9 peptide, tethered to the matrix, was challenged with lectin in buffers at pH 6.0, 7.4, and 8.3. No reactivity was observed. However, three of seven monoclonal antibodies (mAb) with unique binding sites for the lectin bound to the C9 peptide, as shown in Figure 4. When these mAb's were combined with lectin and then reacted with the C9 peptide, binding diminished. This suggests that an epitope on the surface of the amoeba is capable of producing mAb's which cross-react with C9. The organism uses molecular mimicry as a means to avoid elimination by the immune system. It is projected that regions similar to CD59 on mammalian cells and C9 in the immune system allow the organism to illicit the production of antibodies which prolong its survival in the body.

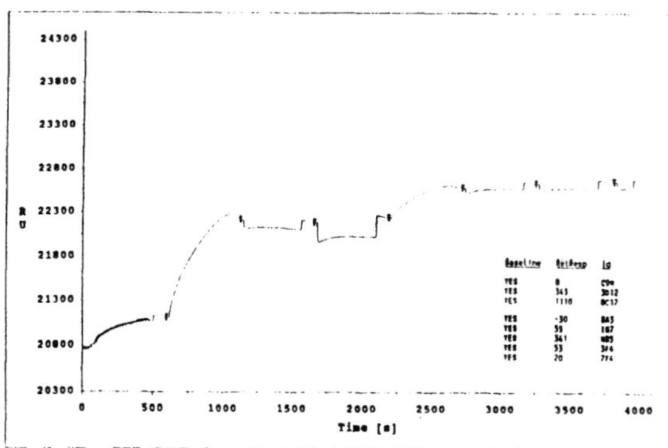

Figure 4. Three of seven mAb's to the galactose-inhibitable adherence lectin of <u>Entamoeba histolytica</u> bound to the C9complement peptide.

Protein Analysis in BIAcore

The 170 kDa adherence lectin of Entamoeba histolytica was immobilized to the matrix via carboxyl-amine coupling. EDC/NHS were used in the coupling reaction and ethanolamine was used to cap reactive groups remaining on the matrix. In vitro assays by Dr. Petri have shown that two (3F4, 8A3) of seven mAb's enhance adherence of amoeba to CHO cells and human colonic mucins.[5] The adherence is inhibitable by galactose. These in vitro assays are performed using at least 3 components at once. Therefore, the empirical results are difficult to control for nonspecific interactions. In order to appropriately control these experiments, it should be shown, unequivocally, that no interaction of galactose with lectin and/or mAb's alone precedes the adherence enhancement and/or binding inhibition step with mammalian cells.

The immobilized lectin was challenged with mAb's in a solution of buffered 50 mM galactose as a control for mAb integrity. Also, the lectin surface was bathed in 50 mM galactose and challenged with mAb as a control for lectin integrity. No effect on binding in either circumstance was observed. These findings add credence to the hypothesis that a conformational change is initiated by mAb 3F4 and mAb 8A3 which enhances binding of amoeba to mammalian cells. Enhanced binding, presumably a result of the structural change, is inhibitable by galactose. Neither lectin alone or mAb alone are affected by galactose preceding the binding process.

Ligand/Receptor Binding Using A Derivatized Polyacrylamide Surface

Dr. Daniel Hammer, (Cornell University, Ithaca, N.Y., USA), uses custom developed gels for tethering molecules to the surface of an acrylamide layer which is positioned within a monitoring microscope. The video monitor and computerized analysis are used to observe cellular activity within the flow chamber. Acrylamide is polymerized using a bifunctional linker, n-succinimidyl acrylamidohexanoic acid, which renders it active for the aminolysis coupling of 2,4 dinitrophenol-e-lysine (DNP-lysine). The gel, then derivitized with COO- groups, can accept any free terminal amine, shown in Figure 5.

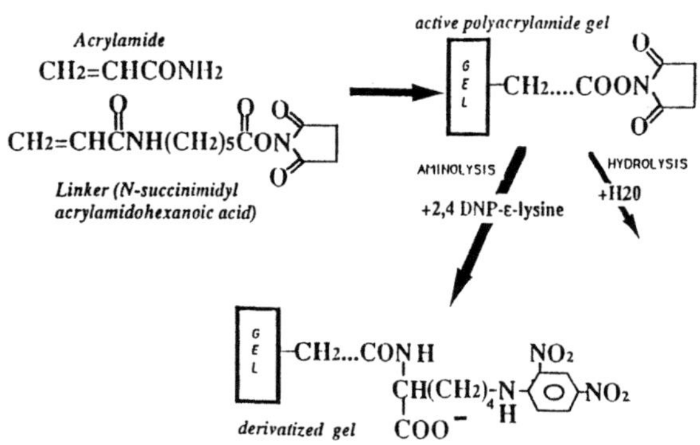

Figure 5. Chemistry involved in the synthesis of polyacrylamide ligand-coated gels.

Rat basophilic leukemia cells which harbor Fc receptors on their surfaces were incubated with antibody to DNP. Rolling velocity of the cells was both theoretically projected and measured directly. Close agreement of both determinations showed that at a fluid velocity of 1.4 cm/min, and a shear rate of 57 (1/sec), approximately seven bonds were necessary to affect capture of the cell. The velocity of cells with antibody bound to their surfaces was slowed to 1/2 the velocity of cells without antibody under these conditions of flow. Adjustable and measurable parameters, paralleled with a mathematical projection model, allowed predictions of capture of biological materials within a fluid stream. This model system will be expanded for detection of pathogens.[2]

SUMMARY

The four immobilization procedures outlined in this paper allowed us to see molecular interactions as they occurred in real-time. DNA-DNA hybridization was made possible by the use of an initial avidin layer to capture the biotinylated probe which then hybridized with a complementary oligonucleotide. The hybridization was specific and non-specific interaction with the matrix was not noted. Reactions were evident in 3.3 minutes and were 1 part in 1.644 parts complete in 8 minutes at room temperature within a fluid stream.

Thiol coupling of a portion of C9, the ninth component of complement, showed interaction of the peptide with antibodies directed against a 170 kDa adherence lectin of <u>E.histolytica</u>. This finding suggests molecular mimicry as a strategy used by the amoeba to evade elimination by the immune system. Indirect evidence exists that similar mechanisms may be used by bacteria to evade the immune system.[6]

Carboxyl-amine coupling of the 170 kDa protein from E.histolytica enabled us to perform controls which lend credence to the hypothesis that a conformational change is initiated by the binding of monoclonal antibodies 3F4 and 8A3 which enhance galactose-inhibitable binding to mammalian cells.

Derivatized acrylamide gels to which ligands were tethered were used in a flow analysis system to assess the number and density of ligands necessary to affect binding within a fluid stream. Actual measurements and theoretical projections were in close agreement. Studies will be extended to assess pathogen capture.

ACKNOWLEDGEMENT

This work was supported by the Department of Defense, Chemical and Biological Defense Agency, Edgewood Research Development and Engineering Center, MD. USA.

REFERENCES

1. Pharmacia Biosensor AB. Application notes 102,201,202. (1990).
2. D. Hammer, s. Apte. Biophys. J., 63(1),35-57,(1992).
3. S. Wood. Microchemical J., 47, 000-000, (1993).
4. L. Braga, et al. J. Clin. Invest., 90, 1131-1137, (1992).
5. W. Petri, et al. J. Immunol., 144(12), 4803-4809, (1990).
6. C.Parsot, E.Taxman, J.Mekalanos. Proc. Natl. Acad. Sci, 88(5), 1641-1645, (1991).

BIOSENSORS IN THE DETECTION OF BIOLOGICAL AGENTS

MARTIN McDONNELL

Chemical and Biological Defence Establishment,
Porton Down, Salisbury, Wiltshire, SP4 0JQ, UK.

Abstract Biosensors based on optical detection such as the Surface Plasmon Resonance (SPR) and Resonant Mirror (RM) systems show great promise in the specific detection of bacteria, viruses and toxins. These techniques monitor the changes in refractive index when an analyte binds to a recognition element immobilised on the sensor surface. The binding of the agent is measured directly resulting in rapid response times. The assays need only a single reagent coupled to the optical surface. In contrast, most other transduction systems require the further addition of a second antibody labelled with an enzyme, fluorophore, or some other tag to make the binding event detectable. These SPR and RM devices are compact with simple fluidics, have no moving parts and are easy to ruggedise. Multi-analyte monitoring should be possible with multiple detection spots on the optical surface, each with specificity for a different analyte or as controls.

I INTRODUCTION

Agents of biological origin which may be used by an aggressor in an attack cover a wide spectrum from pathogenic micro-organisms, which may be used to cause disease in man, animals and plants, to the poisons produced by micro-organisms, the proteinaceous toxins. The Geneva Protocol (1925) exists to prohibit the use of these agents. There is, however, a need to develop rapid and easy techniques for the detection of agents of biological origin. This paper describes some technologies which may be suitable for application in an automated biological agent detector.

The Light Addressable Potentiometric Sensor (LAPS) is an example of a technique[1] that is used for biological agent detection. The LAPS device is an effective way to measure small changes in pH and, when combined with an immunoassay using urease labelled antibody, has the capability for the specific detection of bacteria, viruses and toxins.

The LAPS assay is fairly complex with 2 antibodies - one labelled with urease enzyme and the other labelled with biotin. The analyte solution is mixed with the reagent solution containing the labelled antibodies and streptavidin - a protein which can bind up to four biotin groups. After one minutes incubation the resulting complex is filtered through a nitrocellulose membrane to which is bound a biotin labelled protein. After washing, the surface with the immobilised complex is exposed to a urea solution. The urease enzyme

G.G. Guilbault and M. Mascini (eds.), Uses of Immobilized Biological Compounds, 369–375.

converts the urea to ammonia resulting in a change in pH which is detected by the LAPS system (figure 1).

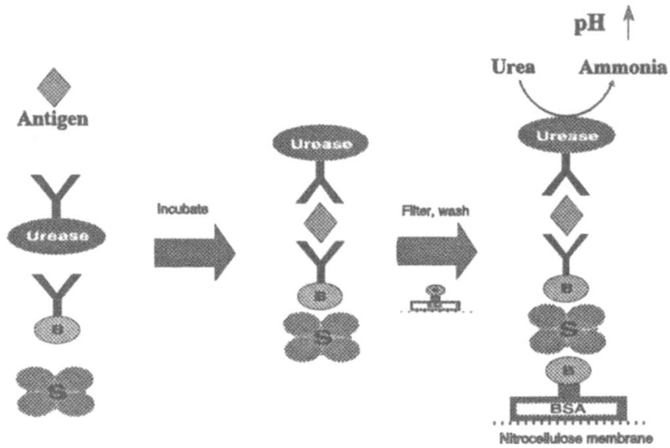

Figure 1 LAPS assay scheme.

The complexity of the immunoassay leads to many problems. Apart from the logistics problem of supplying the assay components, the manual system is very labour intensive and the technique does not lend itself to easy automation due to the complexity of the resulting fluidics system for a multi-analyte detector. There is a requirement for an automated, continuous multianalyte specific biological agent detector. Some of the desirable characteristics of the detection system are :

i) Sensitive, selective and rapid
ii) Multianalyte detection capability
iii) Capable of continuous monitoring
iv) Able to be automated
v) Little or no wet chemistry
vi) No sample pre-processing
vii) Simple to operate, rugged and compact
viii) Low power requirements

Optical biosensor systems including surface plasmon resonance (SPR) and Resonant Mirror technologies have the potential to meet these characteristics.

II Surface Plasmon Resonance

If a beam of light strikes the interface between two transparent media of different refractive index (eg glass and water), light coming from the side of higher refractive index undergoes total internal reflection for incident angles above the critical angle. At the point of reflection, an electromagnetic field component of the light called the evanescent wave penetrates a short distance into the medium of lower refractive index. The depth of this penetration is of the order of a wavelength of the incident light.

If light is p-polarised and the prism surface is coated with a thin metal layer (eg silver or gold ca. 50 nm thick), then at a precise incident angle this evanescent wave can stimulate oscillations of the surface electrons in the metal. Energy is transferred from the light beam to these surface electrons, resulting in a decrease in intensity of the reflected beam (Figure 2). The angle at which this resonance occurs is highly sensitive to the refractive index of the medium directly above the metal surface within the range of the evanescent field.

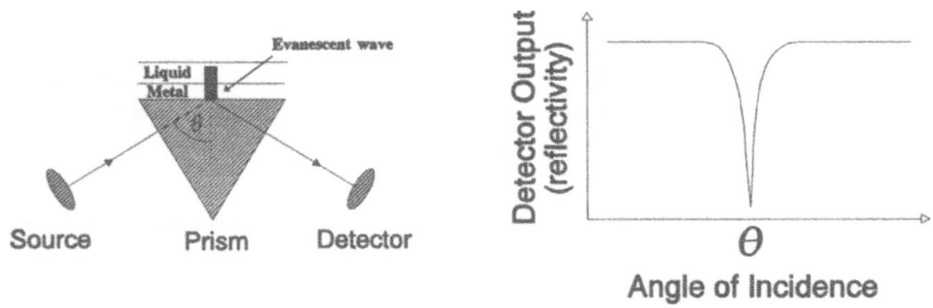

Figure 2 Surface plasmon resonance. a) Light undergoing total internal reflection at the interface between prism and external medium. b) Reflectivity plot showing how the intensity of the reflected light depends on the angle of incidence, θ.

To make a detector using this phenomenon, antibodies are coated onto the surface of the metal film. If an analyte solution containing antigens specific to these antibodies is passed over the surface then binding will occur. This binding event results in a change in the refractive index at the surface and will be detected as a shift in the resonance angle (Figure 3a). A typical SPR response for an injection of an antigen containing solution onto a surface with immobilised antibody is shown in figure 3b. In this case a commercial SPR device was used - the BIAcore from Pharmacia Biosensor AB, Uppsala, Sweden. The analyte solution contains a toxin at 40 μg/ml while an antibody specific to that toxin is immobilised on the sensor surface. 1000 response units on this device corresponds to a shift of 0.1°. After the injection of toxin the position of resonance has shifted by approximately 6500 RU (0.65°). SPR directly monitors the antigen-antibody binding event and so requires a very simple assay procedure (figure 4). Alternative capture agents could be immobilised on the sensor surface, for example gene probes or biological receptors.

The SPR device would work as a continuous monitoring detector intended to give rapid warning of relatively high concentrations of biological agents. If high affinity antibodies are used there should be accumulation of bound agent on the surface with time giving the potential for the detection of very low levels of agent over extended time periods. Use of

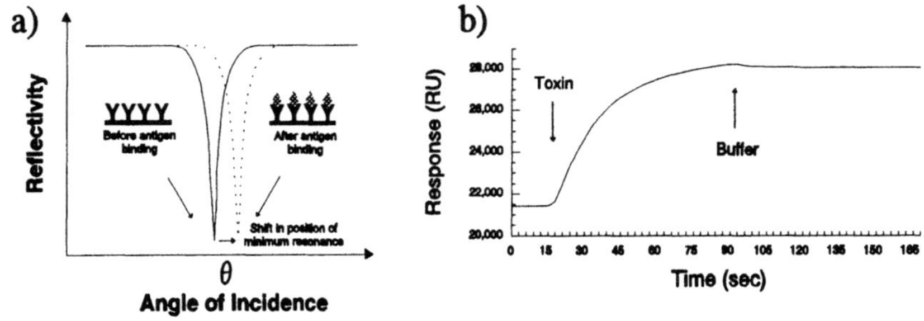

Figure 3 SPR biosensors a) Shift in position of reflectivity minium on binding antigen b) typical SPR response curve. Analyte solution contains 40 μg/ml of a toxin, the sensor surface is coated with an antibody specific to that toxin.

these high affinity antibodies may require the development of procedures to regenerate the surface after a positive challenge.

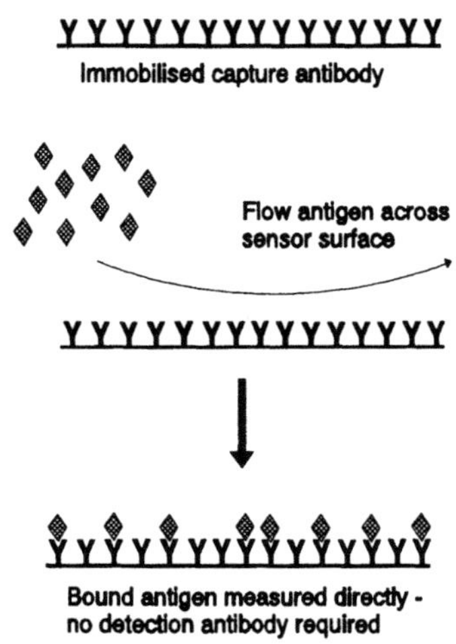

Figure 4 SPR assay scheme.

A practical SPR device for use as a biological agent detector would probably have an optics system similar to that shown in figure 5. This comprises an LED source with a 'fan-beam' arrangement producing angles of incidence over approximately a 6° range, the intensity of the reflected light being monitored with an array detector. The device is compact, has low power requirements and has no moving parts.

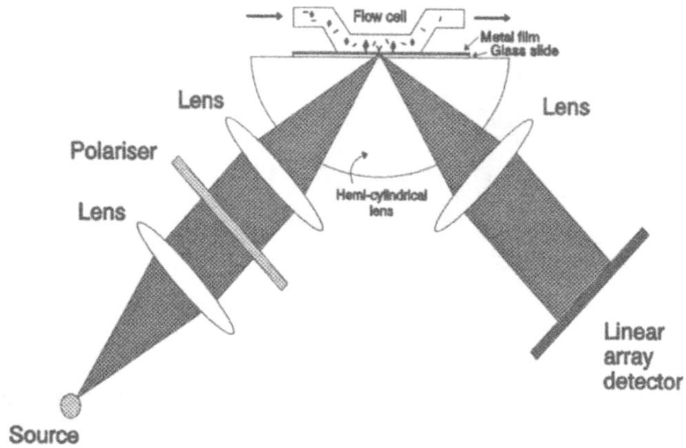

Figure 5 Possible configuration of a practical SPR device.

III Resonant Mirror

The resonant mirror biosensor is intended to combine the sensitivity of waveguide sensors with the simplicity of the SPR technology[2]. As with SPR, the Resonant Mirror technique involves total internal reflection of a beam of light at the sensing surface using a glass prism. The thin metal layer, however, is replaced by a dielectric matrix consisting a 1 μm thick layer of silica and a 100 nm thick layer of high refractive index (eg. titania). The evanescent wave produced at the point of total internal reflection penetrates into the silica layer and couples into the high index resonant layer (figure 6). Coupling of the light only occurs for incident angles where phase matching of the incident beam with the resonant modes of the high index layer is achieved. At the point of resonance the light propagates along the sensing interface before coupling back into the prism. As with SPR the angle of this resonance is highly sensitive to small changes in the refractive index of the layer immediately adjacent to the sensing surface.

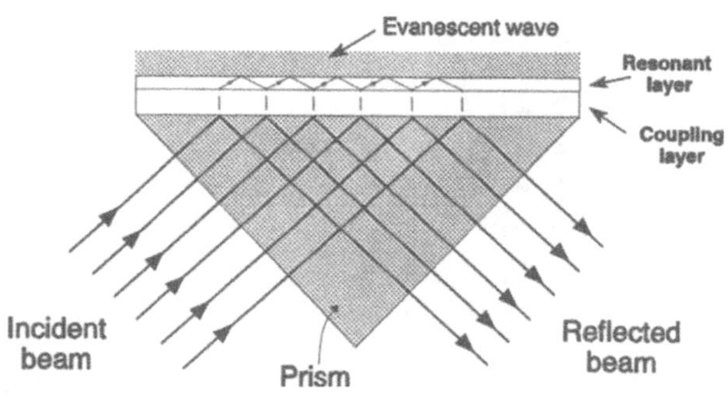

Figure 6 Structure of Resonant Mirror sensor.

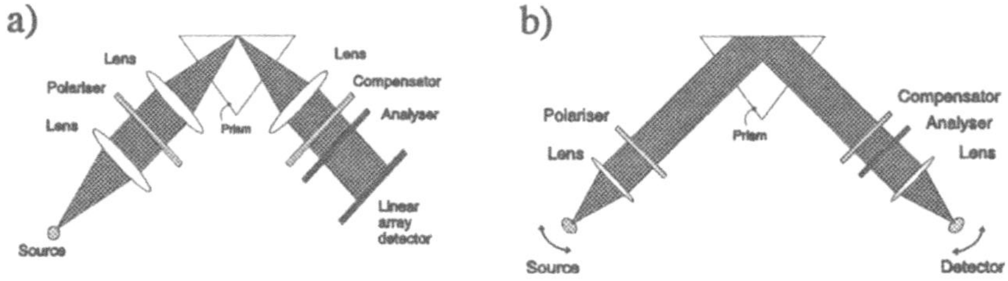

Figure 7 Configurations for a Resonant Mirror device a) focused beam, b) rotating beam.

In the resonant mirror technique the point of resonance is detected by a change in phase of the reflected light. Resonance occurs for both TE and TM polarized light, unlike SPR where only the TM polarized mode undergoes resonance. Binding events at the sensing surface are monitored by observing the shift in resonance angle for either the TM or TE mode, the position of the phase change being measured interferometrically. The possible instrument configurations are shown in figure 7. A resonant mirror response using the commercially available IAsys device (Fisons Applied Sensor Technology, Cambridge, UK) is shown in figure 8. The device allows simultaneous monitoring of two detection channels. On one channel an antibody specific to a particular toxin has been immobilised, while on the other a human IgG antibody has been immobilised as a control for non-specific binding, bulk refractive index changes and possible temperature drifts. Simultaneous injection of a 30 μg/ml solution of the toxin over both surfaces shows a large specific response with a small bulk refractive index change detected on the control channel.

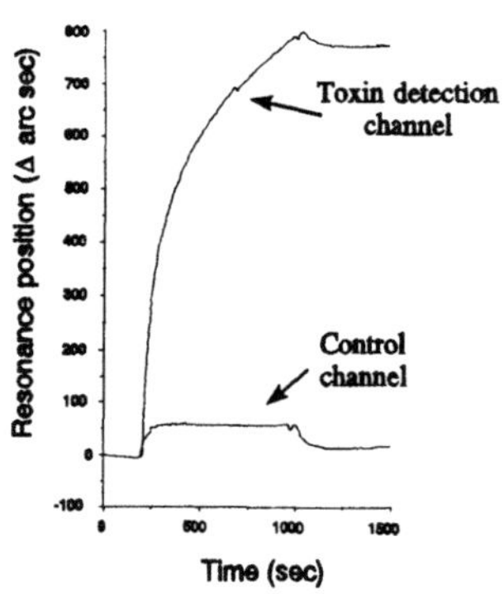

Figure 8 Resonant Mirror response to an injection of 30 μg/ml toxin solution. Detection channel has immobilised antibody specific to that toxin, control channel has immobilised human IgG as a non-specific antibody.

IV CONCLUSION

Biosensors based on optical detection such as the Surface Plasmon Resonance (SPR) and Resonant Mirror (RM) systems show good potential for use as biological agent detectors. They should be capable of continuous monitoring with a rapid, selective and sensitive response. Their direct assay procedures result in fairly low logistics requirements and the uncomplex fluidics system should make automation simple. Multi-analyte detection should be possible using a wedge beam source together with a two-dimensional array of photodiodes (figure 9). This would allow monitoring of multiple detection spots on the detection slide, each with specificity for a different analyte or for monitoring interferents, non-specific binding, and the operating state of the instrument.

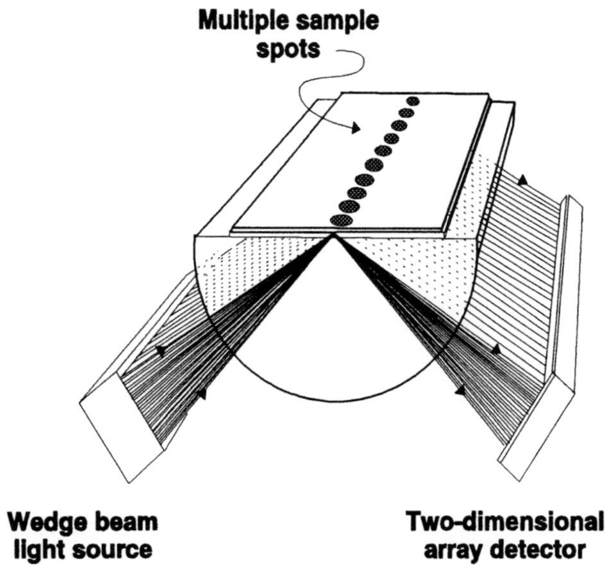

Figure 9 Possible arrangement for multi-analyte detection.

REFERENCES

1. D.G. Hafeman, J.W. Parce, and H.M. McConnell, Science, 240, 1182, (1988).

2. R. Cush, J.M. Cronin, W.J. Stewart, C.H. Maule, J. Malloy, and N.J. Goddard, Biosensors and Bioelectronics, in press.

BIOSENSORS FOR TOXINS AND BACTERIA

G.G. GUILBAULT, D.E.T. SIBLEY, R.M. CARTER & G.J. LUBRANO

UNIVERSAL SENSORS, INC.
5258 VETERANS BLVD., SUITE D
METAIRIE, LA 70006

ABSTRACT. The rapid and sensitive detection of bacteria (*Escherichia coli* and *Listeria*), pesticides (herbicides [Atrazine]) and marine toxins (brevetoxin, saxitoxin) are important in the food industry. Bacterial organisms cause severe food poisoning, thyphoid fever and other gastrointestinal infections. Pesticides which contact food in the production process may cause serious illness. Toxins like brevetoxin, saxitoxin or domoic acid contaminate seafood, causing illness or death if ingested.

In this chapter, piezoelectric (PZ) crystal and electrochemical (EC) immunobiosensors will be described for the assay of bacteria, pesticides and toxins. Polyclonal and monclonal antibodies to E. coli, Atrazine and other pesticides were produced, immobilized and used as layers on a transducer. The formation of the immune complex is measured either via a change in mass (PZ) or an amperometric signal (EC), due to an electroactive product of the enzymatic reaction. Antibodies to brevetoxin and saxitoxin were conjugated to glucose oxidase. Toxins were immobilized to a solid support membrane and immunoelectrodes were prepared for the assay of these toxins at ppb concentrations. The probes are specific, fast and reusable.

1. Introduction

Rapid, sensitive detection of food and water contaminants is a crucial element in the protection of consumers from potentially dangerous bacteria, pesticides and toxins. Immunosensors have been developed using an immunochemical sandwich assay principle and amperometric or piezoelectric detection of Atrazine, marine toxins and several pathogens such as *Escherichia coli* and *Listeria* Type 4. Microbial and bacterial contamination of food products, such as, *E.*

G.G. Guilbault and M. Mascini (eds.), Uses of Immobilized Biological Compounds, 377–385.

coli and *Listeria monocytogenes* (listeriosis) are potentially hazardous and lethal.[1] Pesticides and toxins cause numerous forms of food poisoning and gastrointestinal irritations.

The Enterobacteriaceae family includes *E. coli* and *Salmonella typhimurium*. Most enterobacteria are saprobic organisms of the human intestinal tract, but can be found living outside the human body in soil, water, plants and food. Conventional methods of microbiological cell count determination are lengthy (48 - 72 hrs) and expensive.[2] The availability of antibodies against enterobacterial common antigen (ECA), a glycophospholipid of the outer bacterial cell wall membrane common to all enterobacteria, enables the development of a group-specific immunosensor for detection of an entire family of bacteria.[1]

Pesticides are an integral part of the commercial food industry for acheiving high agricultural yields. Usually, the classification of pesticides is according to the agent against which they are specific, comprising herbicides, fungicides, insecticides and pesticides. It is evident that pesticides enter ground and drinking water in small amounts.[3] However, trace amounts accumulate in tissues, which in high concentrations can cause diseases in animals. Until recently, the concentration level of pesticides in drinking water, 0.1 - 0.5 ppb, has only been detectable by sensitive HPLC, GC or GC-MS, which all require tedious pre-test manipulation of the sample.

Toxin-producing dinoflagellates, *Ptychodiscus brevis* (PbTx) and *Gonyaulax sp.* (STX), of marine origin pose a threat to not only marinelife, but also the consumers of fish and shellfish which feed on the dinoflagellates. Development of dependable methods for marine toxin detection is critical with the increasing reports of shellfish poisoning. At present there is no known antidote for such food poisonings. Brevetoxin, a neurotoxin, causes massive fishkills along the Florida coasts, commonly referred to as red tide. The threat to human health is not only caused by eating contaminated seafood, but also by contact with contaminated seaspray which acts as an eye and nose irritant producing coughing and sneezing.[4,5,6] Saxitoxin, a paralytic seafood toxin, is considered the most frequently encountered of the marine toxins. Poisoning from saxitoxin has been reported along the east and west coasts of the United States. These toxins have been known to produce deadly paralysis if ingested.[7]

The need for rapid determination of toxins warrants the development of a quick, user-friendly immunobiosensor. As a precursary reference for development of electrochemical immunobiosensor devices, Universal Sensors, Inc. has developed sensitive ELISA for brevetoxin and saxitoxin.

2. ELISA Technology

One of the most common methods for immunodiagnostics is the enzyme-linked immunosorbent assay (ELISA). ELISA plates have traditionally been a one-time, disposable unit. Concern over stripping the immobilized antibody or antigen from the microtiter well and subsequent loss of antigenic capacity has prompted this pattern. Universal Sensors, Inc. has developed a reusable, microtiter plate, incorporating proprietary adhesives, to conserve the time and money expended with a single-use ELISA plate. Immobilized ELISA plates are stable for one month if stored at 4 °C in buffer prior to use. Plates can be chaotroped and reused at least ten times without significant loss of response. Storage of the pre-immobilized plates in aa aqueous buffer at 4 °C for 3-7 days between use gave little or no loss in antigenic specificity from use to use.[8]

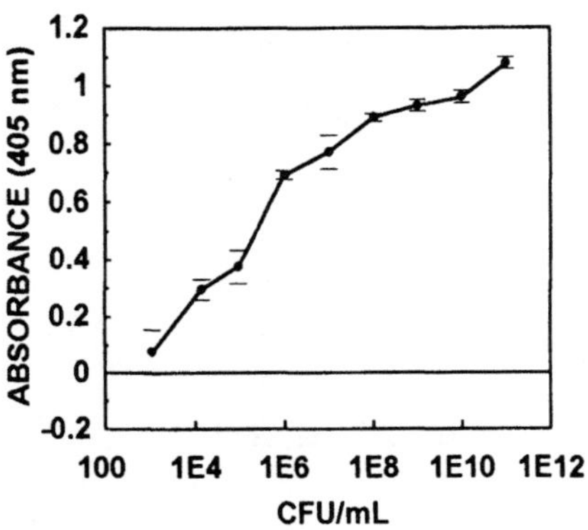

Figure I Detection of *Escherichia coli* by reusable ELISA format (n=6)

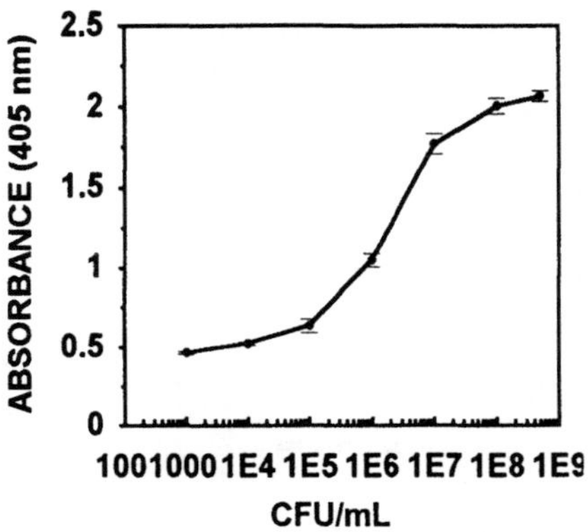

Figure II Detection of *Listeria O* by reusable ELISA format

Immunological assays have been developed for a variety of antigenic targets. These ELISAs have been developed as a prerequisite reference for concurrent development of electrochemical immunosensors. A rapid and reliable screening for bacteria in food and water supplies is very important. Bacterial ELISAs for *Escherichia coli* and *Listeria* have reached detection levels of $1x10^3$ and $1x10^2$ cfu/mL, respectively. One of the most common causative agents of gastrointestinal infections is *Escherichia coli*. A sandwich assay for *E. coli* has been developed with detection limits of 10^3 - 10^4 cfu/mL. Since one *E. coli* cell weighs 3 pg,[2] the assay can detect 300 pg *E. coli* in a 100 μL microtiter well. Small Business Innovation Research (SBIR) funding from the U.S. Army enabled the development of a reusable, sensitive assay as demonstrated in Figure I,[9] and has supported the development of sensitive assays for a variety of bacteria, such as *Listeria, Brucella, Bordetella, and Francisella*. Figure II illustrates the detection of *Listeria Type* 4 with the reusable plate.

Specific, sensitive ELISA for the detection and quantitation of brevetoxin and saxitoxin have been developed using polyclonal goat and burro antibodies, respectively.[7] The standard ELISA method was used as a

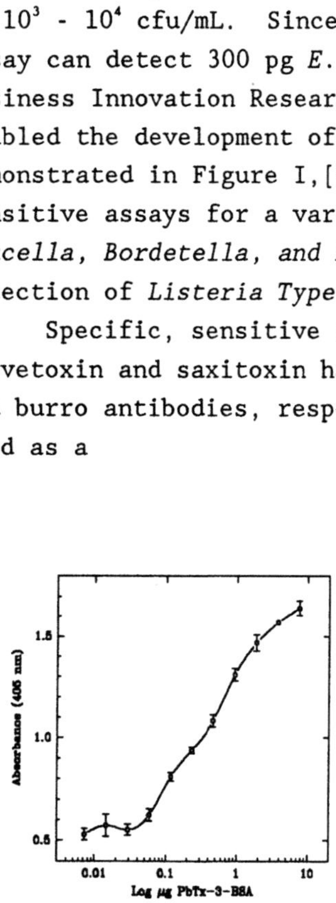

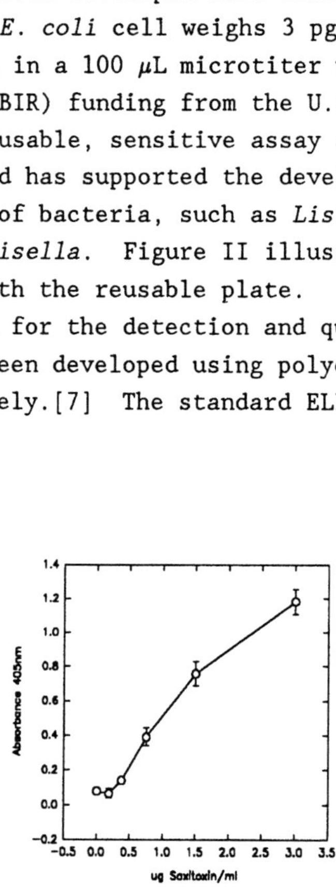

Figure III. Brevetoxin ELISA on Nunc Microwell Plate.

Figure IV. Saxitoxin ELISA on Nunc microwell plate.

reference for concurrent development and testing of electrochemical immunosensors. Detection limits for both assays were approximately 1.5 - 5 ppm. Figures III and IV show the ELISA detection of brevetoxin and saxitoxin, respectively.

3. Piezoelectric Detection

A few American companies have invested heavily in the development of piezoelectric detection devices. Universal Sensors, Inc. has developed and introduced an inexpensive ($4500) PZ Immunobiosensor Detector, PZ 106, which offers 10 MHz crystals coated with any antibody or receptor system. Research is currently being conducted on reusable, crystals immobilized with antibodies for a variety of analytes. A wide range of antigenic materials ranging from low molecular weight compounds to large macromolecules can be determined using this methodology, with sensitivities as low as nanogram amounts, PZ crystals are coated with antigen or a protein mixture containing a given amount of free antigen and antibody. Unlike previous immunoasssay methods, piezoimmunosensors are a non-enzyme label assay of some enterobacteria or toxins. Time and effort are reduced by this direct detection method. The decrease in frequency is related to the antigen concentration.[10,11] The frequency-to-mass relationship is described by the Sauerbrey equation:

$$\Delta F = -2.3 \times 10^{-6}\ F^2\ \Delta M/A$$

where ΔF is the fundamental frequency change, F is the resonant frequency, ΔM is the change in the mass deposited on the crystal and A is the area of the quartz crystal which is coated.

3.1 Atrazine

Based on various polyclonal antibodies specific for atrazine and a competitive assay format with peroxidase-labelled haptens, Guilbault, Hock and Schmid have developed a reusable atrazine assay with detection limits 0.03 - 100 ppb.[3] Protein A was precoated onto the gold electrode of a 10 MHz piezoelectric quartz crystal.

Sheep polyclonal antibodies for atrazine were layered onto the protein A-coated crystal. The sensor is reusable for about 8 - 9 assays.

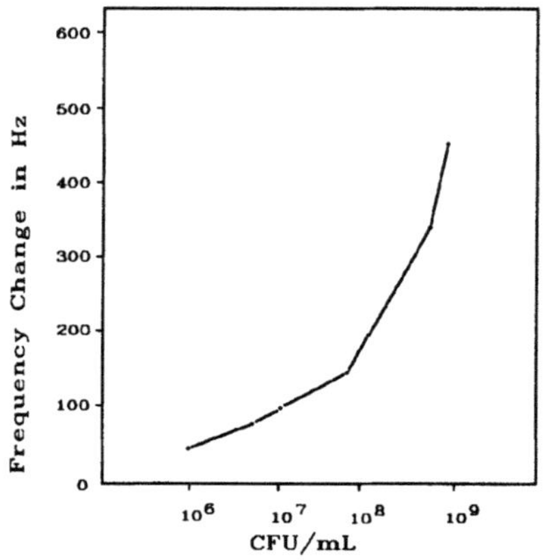

Figure V. Response of a 5 mg/mL α-ECA antibody-coated crystal (F_1) to *E. coli* K12 (CFU/mL). Figure reprinted with permission of the authors.[12]

3.2 Bacteria

A piezoelectric crystal immunosensor has been developed for enterobacterial detection. A response is observed for 10^6 - 10^9 cells/mL of various members of the Enterobacteriaceae family, such as *Escherichia coli* K12, *Citrobacter freundii*, *Enterobacter aerogenes*, *Proteus mirabilis* and *E. coli* (wild strains). Anti-enterobacterial (α-ECA) antibody is applied by protein A immobilization onto the gold electrode of a 10 MHz quartz crystal.[12] Anti-ECA antibody-coated crystals and reference crystals coated with anti-atrazine antibody (which showed no cross-reactivity to enterobacteria) were dipped into a solution of bacteria and incubated. The crystals were washed with PBS and distilled water and dried in a P_2O_5 dessicator connected to the measurement chamber of a PZ 106. The cell concentration was calculated by reading the F_1 [ΔF_M (measuring crystal frequency) - ΔF_R (reference crystal frequency)] and comparing this to a linear calibration plot of F_1 vs cells/mL. Figure V shows the response of a 5 mg/mL anti-ECA crystal to *E. coli*.

4. Electrochemical Biosensors

Immunoelectrochemical biosensors for brevetoxin and saxitoxin have been developed.[13] The reusable biosensors are constructed on a solid support membrane mounted on a conventional electrode. Conjugation of brevetoxin (PbTx-3) or saxitoxin (STX) antigen to BSA was performed by succinylation and then covalently-linked to an

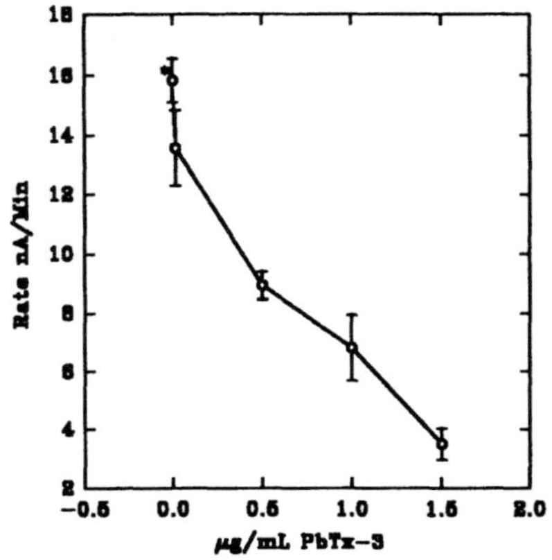

Figure VI. Brevetoxin immunosensor response. *No competition zero point

Immobilon® (Millipore®) membrane through the lysine moiety of the BSA. Antibody to PbTx-3 or STX was conjugated to glucose oxidase (G.Od.) by the glutaraldehyde method, and purified by Protein G chromatography. The toxin-immobilized membrane was incubated with enzyme-conjugate. An antibody-antigen reaction occurs and the biosensor reacts to β-D-glucose by amperometric detection of H_2O_2. Electrodes immobilized with PbTx-3 or STX and incubated with antibody-enzyme without competition from free PBTX3 or STX produced the maximum signal. Competitive reactions between free toxins and immobilized toxins in the presence of enzyme-conjugated antibody will decrease the generated signal in an inversely proportional manner to the concentration of free brevetoxin or saxitoxin. The biosensors were slightly more sensitive that the standard ELISA, with a shelf life of approximately six months without deterioration. Reusability of the biosensor makes it a rapid, inexpensive alternative to conventional ELISA methods.[13] Figure VI depicts a typical brevetoxin electrode response to free brevetoxin.

5. Summary

Development of reusable, immunochemical, detection devices, such as piezoelectric crystals and electrochemical biosensors, may greatly aid the environmental and food industries constant struggle to protect the public from contaminants. Pesticides, bacteria, pollutants and toxins can now be detected quickly and easily. It is no longer necessary to collect extensive samples and transport them back to a laboratory where lengthy, tedious procedures for detection of contaminants must be performed. With the advancement of antibody-based piezoelectric and amperometric detection for bacteria, pesticides and toxins, testing can be performed in the field with little or no pretreatment of the samples. The PZ106 and the ABD developed by Universal Sensors are portable, battery-operated instruments which can be used to detect a wide range of bacteria, toxins and other analytes. Sensitivity and selectivity are now available in a real-time matrix for a variety of environmental and food samples.

6. References

1. Peters, H., Jurs, M., Jann, B.K., Timmis, K. and Bitter-Suermann, D. (1985) Am. J. Microbiol. 50, 459-466.

2. Muramatsu, H., Watarabe, Y., Hikuma, M., Ataka, T., Dubo, I., Tamiya, E. and Karube, I. (1989) 'Piezoelectric Crystal Biosensor System for Detection of *Escherichia coli*', Anal. Lett. 22(9), 2155-2166.

3. Guilbault, G.G., Hock, B. and Schmid, R. (1992) 'A Piezoelectric Immunosensor for Atrazine in Drinking Water', Biosensors & Bioelectronics 7, 411-419.

4. Baden, D.G., Mende, T.J., Poli, M.A. and Block, R.E. (1984) in Seafood Toxins (E.P. Ragelis, ED.) ACS Symposium Series 262, Washington, D.C., 359-365.

5. Pierce, R.H. (1986) Toxicon 24, 955-965.

6. Ahmed, F.E. (1991) 'Naturally Ocurring Seafood Toxins' in J. Toxicol - Toxin Reviews 10(3), 263-287.

7. Carter, R.M., Sibley, D.E.T., Poli, M.A., Lubrano, G.J. and Guilbault, G.G. (1993) 'ELISA and Immunoelectrochemical Detection of the Marine Toxins: Brevetoxin and Saxitoxin', IBEX Abstract, San Francisco, CA.

8. Sibley, D.E.T., Ramsay, G., Lubrano, G.J. and Guilbault, G.G.

(1993) 'Stability and Reusability of Enzyme-Linked Immunosorbent Assay (ELISA) Plates', Anal. Lett. 26(8), (in press).

9. Sibley, D.E.T., Ramsay, G. and Lubrano, G.J. (1991) 'Natural Marine Adhesive-Based Electrochemical Immunosensors for In-Field Pathogen Detection', U.S. Army SBIR Phase I Final Report.

10. Oliveira, J.R. and Silver, S.F. (1989) U.S. patent #424096.

11. Rice, T.K. (1980) U.S. patent # 4236893.

12. Plomer, M., Guilbault, G.G. and Hock, B. (1992) 'Development of a Piezoelectric Immunosensor for the Detection of Enterobacteria', Enzy. Microb. Technol. 14, 230-235.

13. Carter, R.M., Poli, M.A., Pesavento, M., Sibley, D.E.T., Lubrano, G.J. and Guilbault, G.G. (1993) 'Immunoelectrochemical Biosensors for Detection of Marine Toxins Brevetoxin and Saxitoxin', Immunomethods, (in press).

REUSABLE FIBER OPTIC BIOSENSORS
FOR DETECTION OF DRUGS AND TOXICANTS

Mohyee E. Eldefrawi[1], Amira T. Eldefrawi[1], Nabil A. Anis[2] and James J. Valdes[2]

[1]Department of Pharmacology and Experimental Therapeutics, School of Medicine, University of Maryland, Baltimore, Maryland 21201 and [2]Biotechnology Division U.S. Army Research and Development and Engineering Center, Edgewood, Maryland 21010.

Abstract Use of fluorescein derivatives of specific analytes, that are bound reversibly by biological sensing elements (e.g. receptors or antibodies), proved to be a successful strategy for fiber optic biosensors. Two biosensors, one using nicotinic acetylcholine receptor (nAChR) and the other using antibodies, detected anaytes in real time and were easy to regenerate for multiple measurements.

I. INTRODUCTION

The pharmaceutical industry relies heavily on radioactive-ligand-receptor binding assays in screening protocols as part of their drug discovery programs. These binding assays are expensive, use hazardous radioisotopes, generate hazardous waste and require highly trained personnel and expensive instruments to perform. Toxicologists and analytical chemists who used radioimmunoassays, faced similar problems and found them poor substitutes for instrumental analysis (e.g. by HPLC or GC/MS). Enzyme-linked immunosorbent assays (ELISA) were an improvement over radioimmunoassays, but still required trained personnel and were slow.

The application of biosensor technology to these problems held the promise of developing faster, less hazardous, and less costly, yet precise analytical methods. In our laboratory, we have been investigating the use of a fiber-optic evanescent wave technology

G.G. Guilbault and M. Mascini (eds.), Uses of Immobilized Biological Compounds, 387–395.

to develop microsensors that are suitable for detection of drugs and toxicants. A major feature that is emphasized in these biosensors, is that they are reusable and are suitable for flow injection so as to be applicable to monitoring purposes. The following presentation will focus on description of two fiber optic biosensors: one using a nAChR protein and the other antibodies as sensing elements.

II. A RECEPTOR-BASED BIOSENSOR FOR THE DETECTION OF RECEPTOR SPECIFIC LIGANDS

The nAChR protein is a well characterized heteromeric protein that regulates the response of neuronal and skeletal muscle responses to neuronal activation. The nAChR is a 250 KDa protein, made of 5 subunits ($\alpha_2\beta\gamma\delta$), and is found in high concentrations in electric organs of electric fish. The nAChR binds the neurotransmitter acetylcholine (ACh), which results in receptor activation[1]. The nAChR also binds numerous drugs and toxins with nM affinities, including peripheral neuromuscular relaxants (e.g. α-tubocurarine) and venoms of the banded krait snake toxin α-bungarotoxin (α-BGT), the cobra, α-*Naja* toxin (NT) and the cone snail conotoxin (CNTX).

The nAChR protein, purified from the electric organ of *Torpedo nobiliana* by affinity chromatography, was immobilized by adsorption on quartz fiber (1 x 60 mm) and used to detect receptor ligands[2]. α-BGT, an 8 KDa peptide toxin isolated from the venom of *Bungarus multicinctus*, was labeled with fluorescein isothiocyanate and the fluorescinated α-BGT (FITC-α-BGT) was used as an optical signal generator. A portable fluorometer, designed and built by ORD Inc. (N. Salem, NH), was used to monitor receptor-ligand interactions. The results were impressive in terms of the ability of the nAChR-coated fiber to bind and detect α-BGT, and other cholinergic ligands without interference from nonspecific binding.[3]

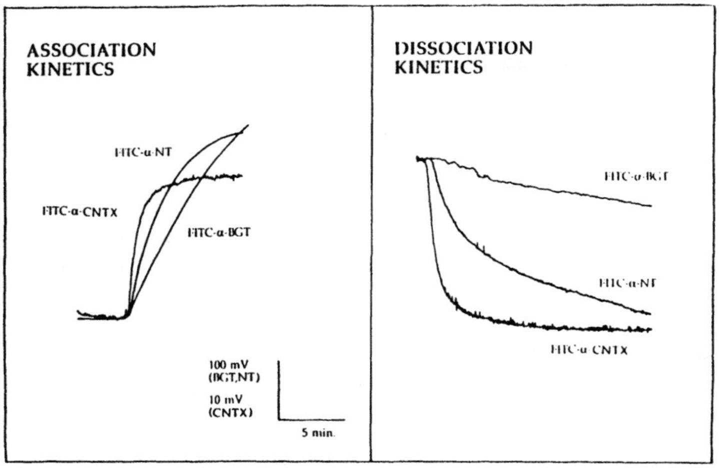

FIGURE 1. Comparative association and dissociation time courses for binding of FITC-labeled toxins to nAChR-coated fibers. FITC-toxin concentrations were 5 nM and 1 μM for FITC-α-BGT. FITC-α-NT and FITC-α-CNTX, respectively in phosphate buffered saline (PBS) + 0.1 mg ml^{-1} bovine serum albumin (BSA). After a 30 min treatment with the labeled toxins, dissociation time courses were initiated by perfusing the flow cells with PBS + BSA in the absence of labeled toxin.

The fiber, however, was useful for only a single measurement, because α-BGT bound to the nAChR almost irreversibly (i.e. very high affinity).[4] Thus, to meet the demand of the assigned task (i.e. to construct a biosensor for toxic peptides that is suitable for fast and sensitive detection), a more reversible optical signal generator was sought. Fluorescein-tagged α-*Naja* toxin (FITC-α-NT) proved to be better because its interaction with the nAChR is more reversible.[4] It was possible for the nAChR-coated optic fiber, equilibrated with FITC-α-NT, to detect nanomolar concentrations of αBGT in seconds. FITC-α-CNTX was even a better probe to use, because of its reversible binding to the nAChR[5]. When compared, the three fluorescein-tagged toxins, α-BGT, α-NT and α-CNTX displayed decidedly different association and dissociation kinetics (Fig. 1) and the affinity constant (K_A) values, calculated from the rate constants, were similar to those obtained from radioactive-ligand binding assays and cellular responses (Table 1)[3].

TABLE 1
Kinetic parameters for binding of fluorescently labeled toxins to the nAChR-coated fibers

Toxin	$k_{+1}(M^{-1}\ min^{-1})$	$k_{-1}(min^{-1})$	$K_d{}^a(M)$
FITC-α-BGT	8.4×10^6	7.9×10^{-3}	9.4×10^{-10}
FITC-α-NT	6.0×10^6	4.8×10^{-2}	7.3×10^{-9}
FITC-α-CNTX	1.4×10^6	8.0×10^{-1}	5.7×10^{-7}

[a]K_d values were determined using the relationship $K_d = k_{-1}/k_{+1}$. The K_d value reported in bioassays of α-CNTX is 2 X 10^{-7}M (Hashimoto et al., 1985), for the binding of α-BGT to rat muscle homogenates is 10^{-10}M (Colquhoun & Rang, 1976) and for binding of α-NT to *Torpedo* membranes is 4 X 10^{-9}M (Johnson & Taylor, 1982).

The FITC-αCNTX was the most useful probe to use with nAChR-fibers to construct a biosensor for detection of peripheral neuromuscular relaxants or other nAChR-specific drugs. Exposure of the nAChR-coated fiber to FITC-α-CNTX for 20 minutes, was sufficient to reach a steady state receptor-toxin complex. Subsequent exposure of the fiber to receptor-specific drugs, such as tubocurarine, resulted in dose-dependent displacement of the FITC-α-CNTX. Eventually, the fiber was washed, and FITC-α-CNTX-nAChR complex was regenerated (Fig 2)[3].

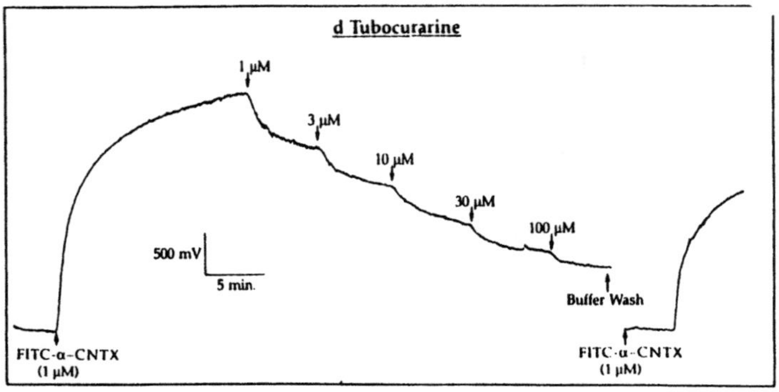

FIGURE 2. Inhibition of reversible binding of FITC-α-CNTX to nAChR immobilized on a quartz fiber by addition of d-tubocurarine. The nAChR-coated fiber was preincubated with PBS + 0.1 mg ml^{-1} BSA prior to addition of various concentrations of d-tubocurarine. After equilibrium was established upon addition of 100 µM d-tubocurarine, the receptor-coated fiber could be reused after a 10 min was with PBS.

III. A REGENERABLE IMMUNOSENSOR FOR DETECTION OF THE HERBICIDE PURSUIT

Antibodies are more specific in their interactions with chemicals than receptors and immunoassays are used routinely at present for the detection and quantitation of drugs and toxicants[6-8]. Furthermore, polyclonal and monoclonal antibodies have been used to construct a variety of biosensors[9,10]. Immunosensors can be generated by dissociating analytes from the immobilized antibody[11]. However, this usually requires low pH and high ionic strength buffers which usually lead to quick deterioration of the immunosensor.

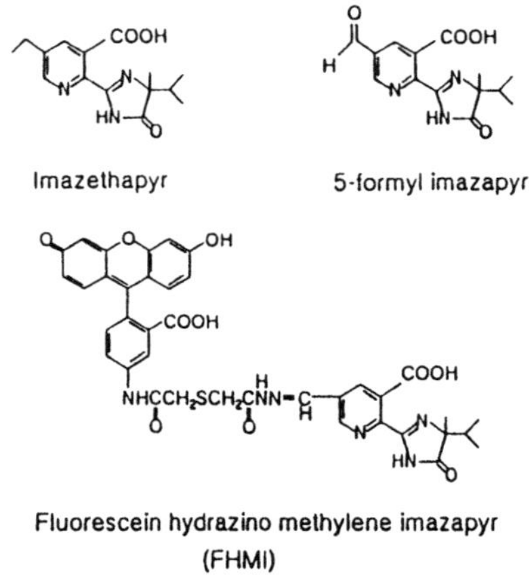

FIGURE 3. Structure of imazethapyr, 5-formyl imazapyr, the analog compound for fluorescein label preparation, and fluorescein hydrazino methylene imazapyr (FHMI).

Imazethapyr is a selective imidazolinone herbicide used successfully to control weeds in fields of soybeans and other legumes (Fig. 3). We constructed an immunosensor using a partially purified IgG fraction from a sheep immune serum containing polyclonal antibodies against the herbicide imazethapyr.[12] The purified antibodies were immobilized on quartz fibers and a fluorescent-tagged imazethapyr derivative (designated FHMI - see Fig. 3) was used at the optimal concentration of 25 μg/ml. When unlabeled imazethapyr was added to the buffered saline containing 25 nM FHMI, the initial rate of binding of the latter was decreased as expected from a competitive reversible binding assay (Fig. 4). More importantly, when the antibody-coated fibers were equilibrated with 25 nM FHMI (which took 10 min), the fluorescence was reduced slowly by washing the fiber with buffer saline alone.[12]

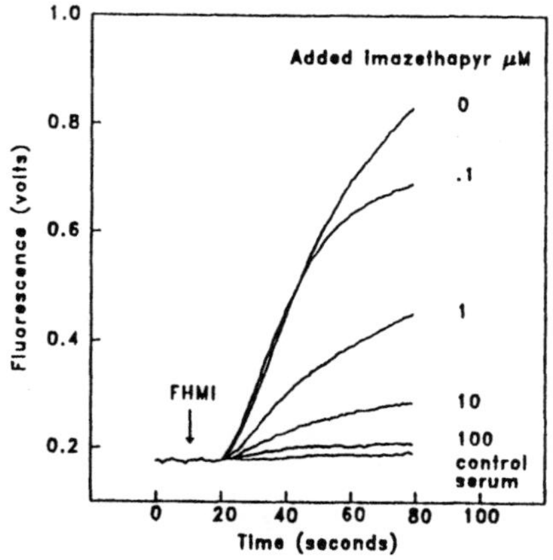

FIGURE 4. Inhibition of FHMI association by imazethapyr. Varying concentrations of imazethapyr were added to the perfusion solution (25 nM FHMI in casein/PBS). Each perfusion solution reacted with a new antibody coated fiber. All fibers were prepared on the same day.

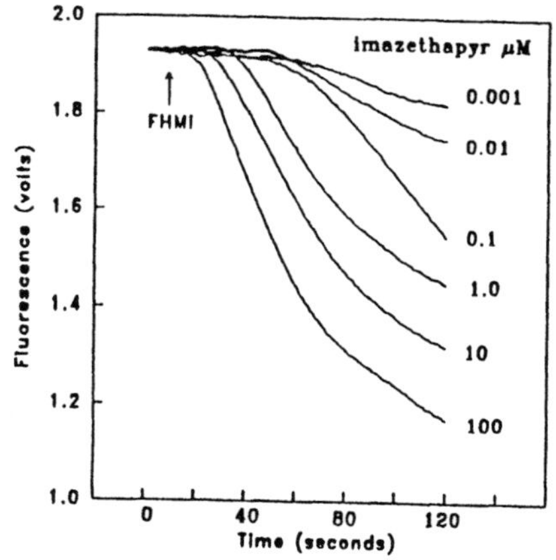

FIGURE 5. Displacement of bound FHMI by imazethapyr. A 25 nM FHMI in casein/PBS perfusion solution was used to establish a steady-state level of fluorescence. The perfusate (FHMI, casein/PBS) was then changed to contain different concentrations of imazethapyr (0.001-100 μM). The decrease in fluorescence was measured in seconds.

However, the rate of reduction of fluorescence increased in a concentration dependent manner if unlabeled imazethapyr was added in increasing doses to the perfusion buffer containing 25 nM FHMI (Fig. 5). A single fiber was used to generate all the data presented in Fig. 5 using imazethapyr concentrations ranging from 1nM to 10 μM. The fiber was easily regenerated simply by perfusion with buffered saline containing only the fluorescent probe (Fig. 6).

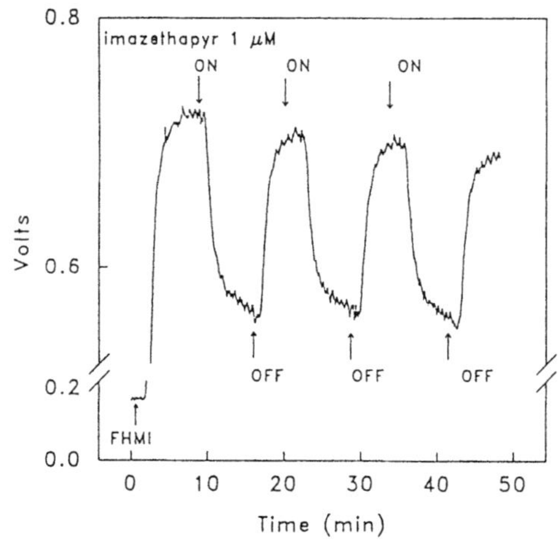

FIGURE 6. Reusability of the fiber sensor. A 25 nM FHMI in casein/PBS solution was perfused to reach a steady state of binding (about 5 min). At the point indicated as "ON", 1 μM imazethapyr was introduced in the perfusate (FHMI, casein/PBS). As a considerable amount of FHMI was displaced, the perfusion solution was switched back to 25 nM FHMI casein/PBS (indicated as "OFF"). The ON and OFF process was repeated.

IV. PROSPECTS

Reusable fluorometric biosensors can be extremely useful for monitoring purposes. The basic requirement is a sensing element that has good affinity for a fluorescent derivative of the analyte and a better affinity for the analyte. Our studies demonstrated that a receptor protein can be used effectively as a sensing element with the appropriate fluorescent probe, and may be effective tools in screening programs. However, immunosensors hold the brighter future because of their extremely high specificity. Techniques used to obtain sufficient amounts of pure receptor proteins are limited, even though most receptors have been isolated and cloned. On the other hand, the techniques to clone the genes for antibodies or their variable fragments, such as the single chain variable fragment, are available and would allow production of large quantities of the sensing antibodies as well as selection of antibodies with the appropriate affinities for the analyte and the fluorescent probe.

ACKNOWLEDGEMENTS

This research was partially supported by funds from U.S. Army contract DAAA-89-15-0007, NRC fellowship to Dr. Nabil Anis.

REFERENCES

1. F. Hucho, Eur. J. Biochem. 158, 211, (1986).

2. K.R. Rogers, J.J. Valdes, and M.E. Eldefrawi, Anal. Biochem. 182, 353, (1989).

3. K.R. Rogers, M.E. Eldefrawi, D.E. Menking, R.G. Thompson, and J.J. Valdes, Biosensors and Bioelectronics 6, 507, (1991).

4. E.X. Albuquerque, A.T. Eldefrawi and M.E. Eldefrawi, in "Snake Venoms" ed. C.Y. Lee Handbook of Experimental Pharmacology, Springer Verlag, Berlin, 52, 377 (1979).

5. D. Cooper and E. Reich, J. Biol. Chem. 247, 3008, (1972).

6. G.G. Guilbault and R.D. Schind, Biotechnology and Applied Biochem. 14, 133 (1991).

7. B.M. Kaufman, M. Clower, Jr. Assoc. Off. Anal. Chem. 74, 239, (1991).

8. D.E. Dixon-Holland, in "Analysis of Antibiotic/Drug Residues in Food Products of Animal Origin, V.K. Agawal ed. Plenum Press, NY, pp. 57 (1992).

9. R.O. Harrison, A.A. Brimfield, and J.O. Nelson, J. Agr. and Food Chem. 37, 958, (1989).

10. R. Karlsson, A. Michaelson, and L. Mattson, J. Immunological Methods 145: 229 (1991).

11. F.V. Bright, T.A. Bettes, and K.S. Litwiller, Anal. Chem. 62, 1065, (1990).

12. N.A. Anis, M.E. Eldefrawi and Wong, R.B., J. Agr. and Food Chem., 41, 843, (1993).

THE BIOSENSOR TECHNOLOGY PROGRAM AT THE NATIONAL INSTITUTE OF STANDARDS AND TECHNOLOGY (NIST)

HOWARD H. WEETALL
Biotechnology Division
Biosensor Technology Group
National Institute of Standards and Technology
Gaithersburg, Maryland 20899
USA

Abstract The biosensor technology program at NIST has been in existence for less then two years. During this time we have initiated projects using both optical and electrochemical methods for analyte detection. Projects presently underway include: DNA intercalation for carcinogen detection using evanescent wave technology, electrochemical detection of glucose using a mediator containing electrode, detection of bacteria by natural fluorescence, characterization of lipid membranes for two dimensional protein movement, α-toxin pore characterization, a light-activated enzyme switch, and a series of studies on bacteriorhodopsin films, sol-gel glasses, and membranes for several future applications.

INTRODUCTION

Recently NIST has initiated a major effort in the area of Biotechnology. As part of this effort we have organized a biosensor technology program. This program is directed toward examining generic biosensor technologies applicable to a wide variety of industries and to transfer of this technology to industry as it becomes available.

Although the Biosensor Technology Group is part of a larger and diversified Biotechnology Division, the programs within the group either directly involve biosensors or have application to biosensor technology.

G.G. Guilbault and M. Mascini (eds.), Uses of Immobilized Biological Compounds, 397–414.

DNA intercalation for detection of Carcinogens

This project utilizes the ability of most of the major carcinogens to intercalate into DNA. The assay is carried out using an architecture similar to that commonly used for immunoassay. The carcinogen of interest is permitted to compete with and displace a fluorescent intercalating dye. The quantity of carcinogen can be determined, using a standard curve, from the quantity of the fluorescent dye remaining intercalated after the system reaches equilibrium. Figure 1 shows in cartoon form, the process of intercalation upon which this assay is based.

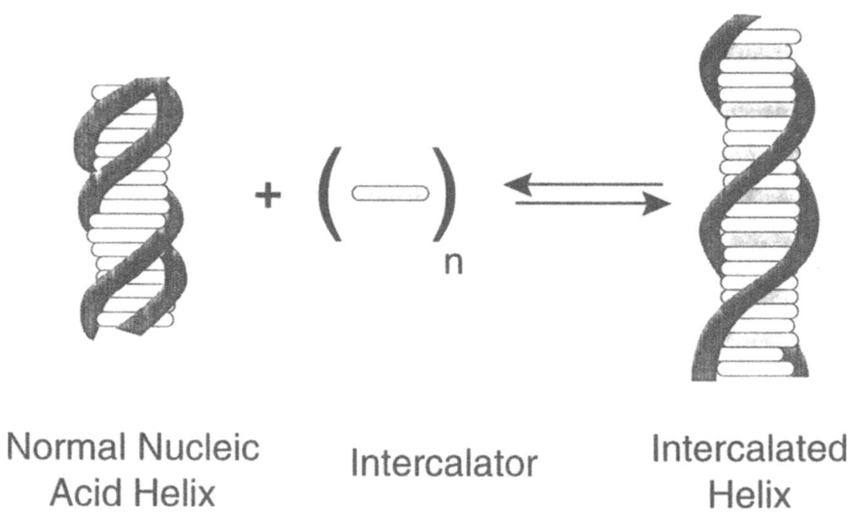

FIGURE 1. Cartoon representation of DNA intercalation.

Initial studies were carried out using fluorescence polarization. The intercalating dye when bound into the DNA shows strong polarization. Equilibrium in this system requires only 1-3 minutes. Figure 2 shows a example of the results obtained using this system in a competitive mode and displacing the dye (acridine orange) with a carcinogen. Using this technique we have detected carcinogens in the parts per trillion region. In addition to detection we have been able to determine the apparent binding constants for several of the carcinogens detected. We are presently attempting to relate the binding constants to degree of carcinogenicity.

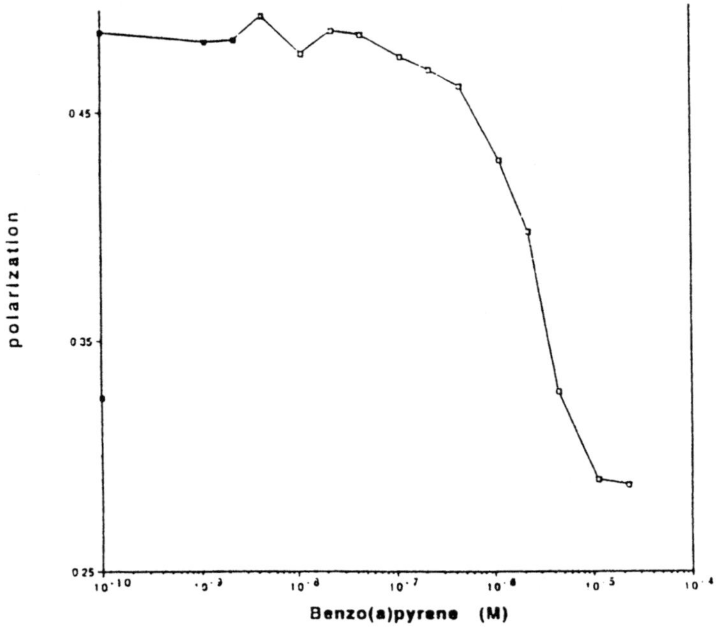

FIGURE 2. Competitive binding curve for benzo(a)pyrene using acridine orange. Data collected by fluorescence polarization.

Our immediate plans call for converting this system to an evanescent wave approach. We believe that such a "real time" system will have even greater sensitivity, greater speed, and it should be portable. We have such a system on loan from the Naval Research Laboratory. This system uses optical fibers. We are presently examining methods for the attachment of double stranded DNA to the fibers.

Electrochemical Detection of Glucose using a Mediated Electrode

This system uses flow injection with a graphite paste electrode containing glucose oxidase and 7,7,8,8-tetracyanoquinodimethane (TCNQ) mixed with mineral oil to decrease water solubility. The electrode is prepared by filling a 1 mm X 3 mm hole in a small plastic block having an electrical connection between the paste and the external circuit. The system uses a Ag/AgCl reference electrode and the metal electrode holder as the counter electrode. Figure 3 shows the effect of flow rate at constant glucose concentration using an 8 X 10^{-6}M solution of glucose.

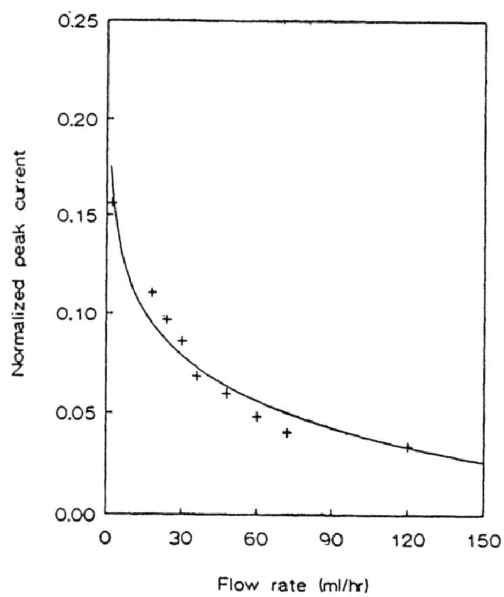

FIGURE 3. Effect of flow-rate at constant glucose concentration.

Figure 4 represents the results observed when the flow rate is held constant and the injection volume is varied. In this figure the 8×10^{-6}M glucose was added to a 250 μL loop either in a 25 μL or a 250 μL aliquot.

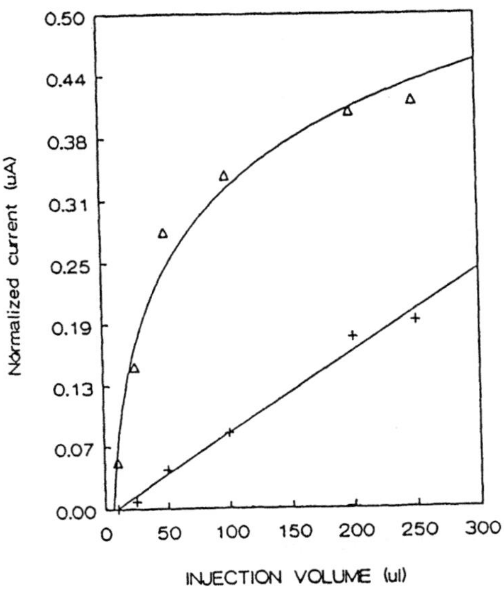

FIGURE 4. Effect of glucose content at constant flow-rate. Data was collected using two glucose concentrations diluted into a 250 microliter loop. The data presented using triangles was five times the concentration of the data presented using the plus sign. The data at the lower concentration of glucose shows linearity.

Figure 5 represents a typical standard glucose curve. Using standards prepard in 6% bovine serum albumin.

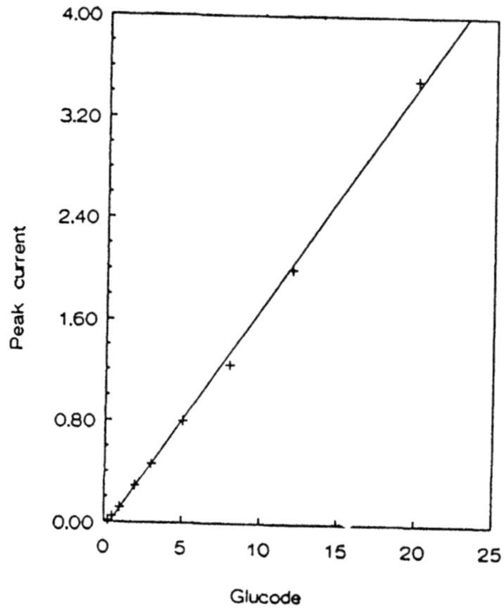

FIGURE 5. Standard glucose curve prepared in 6% bovine serum albumin solutions.

Figure 6 shows the comparison between a commercial glucose kit and the electrode for whole human serum. The R^2=0.998 for the samples tested.

This electrode system was also tested with ascorbate, a known interferant. At concentrations ten times higher than found in the blood, no added current was observed. The reason for the lack of interference is that the addition of the TCNQ permits the use of a very low over voltage which is not high enough to oxidize the ascorbate

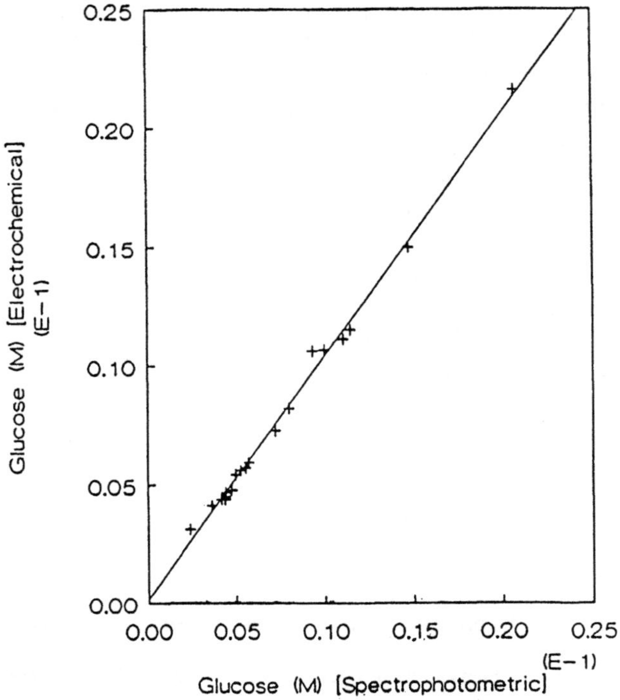

FIGURE 6. Corrolation between glucose electrode and a commercial kit.

Detection of Bacteria by their Natural Fluorescence

The National Aeronautics and Space Administration was interested in detecting bacteria in a closed water system. In our initial study we have simply taken water spiked with bacteria and passed the fluid through a 0.2 micron silver filter. Bacteria deposited on a filter were excited with filtered light. The emission was detected using a microscope with a photon counter attached. The number of photons detected were related to the number of organisms in the field of view. Figure 7 lists the typical excitation and emission wavelengths for compounds one might expect to find in a living bacteria.

EXCITATION AND EMISSION WAVELENGTH MAXIMA
TYPICAL OF BACTERIAL FLUOROPHORES

COMPOUND	EXCITATION MAX	EMISSION MAX
Tryptophan	290 nm	330-340 nm
Pterin	353 nm	440 nm
Pterin-6-carboxylic acid	356 nm	438 nm
Biopterin	358 nm	442 nm
Neopterin	356 nm	443 nm
Folic acid	361 nm	442 nm
Lumazine	331 nm	458 nm
NADH	346 nm	457 nm
FMN	467 nm	520 nm
FAD	467 nm	520 nm

FIGURE 7. Excitation and emmision wavelength maxima typical of bacterial fluorophores. This data was collected from several published sources.

Figure 8 presents date taken from a single field using an excitation wavelength of 300 - 400 nm and an emission of everything over 500 nm. The limit of detection was approximately 300 organisms.

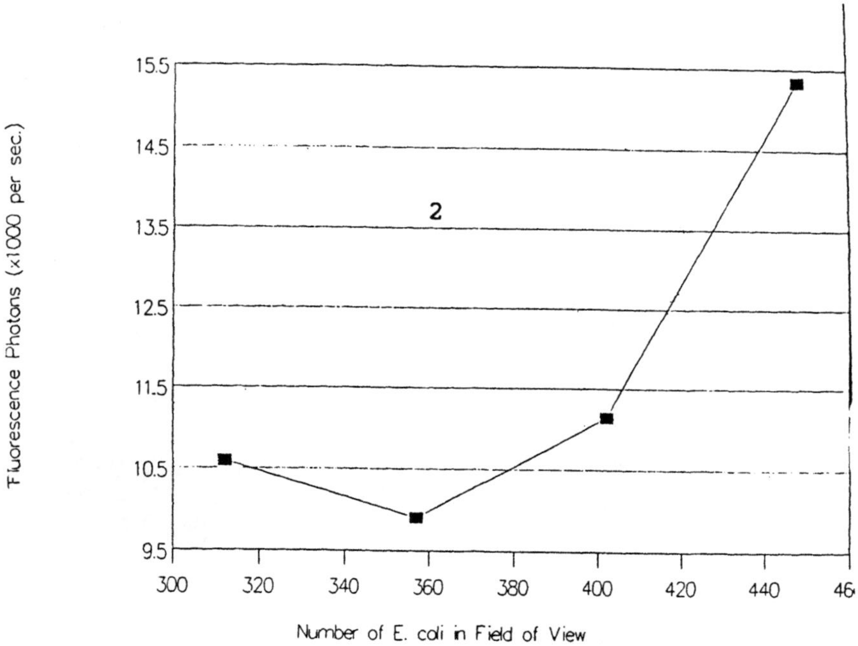

FIGURE 8. Number of E. coli dectected in one field of view.

Lipid bilayer for the study of planar motion of proteins

Planar models of cell membranes allow the study of membrane structure and membrane - protein interaction. This information is of interest for future biosensor components. We have developed methods for preparing self-assembling planar bilayers which have characteristics similar to cell membranes but are simple to prepare. Alkylthiols can be laid down on gold surfaces forming a single lipid like layer. Addition of liposomes to this monolayer under proper conditions causes the liposome to open and form a bilayer on the gold surface. Figure 9 presents a diagram of the mechanism for formation of the planar bilayer.

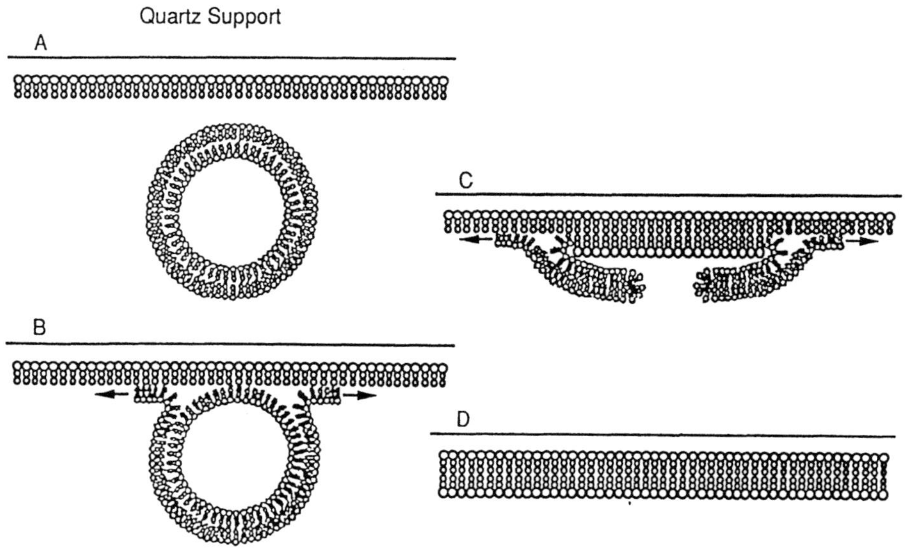

Adapted from Kalb, et al., Biochi. Biophys. Acta 1103 (1992) 307-316

FIGURE 9. Depiction of the mechanism of bilayer formation between the alkylthiol surface and a liposome.

Conformation of the bilayer formation was obtained by capacitance measurements (Figure 10) and cyclic voltametery. In this case the protein melettin was added to the bilayer. This protein binds to phospholipid membranes and inserts into the hydrophobic portion of the bilayer. Aggregation of several melittin molecules results in the formation of a pore through the membrane. The voltamogram shows increased current flow upon addition of the protein.

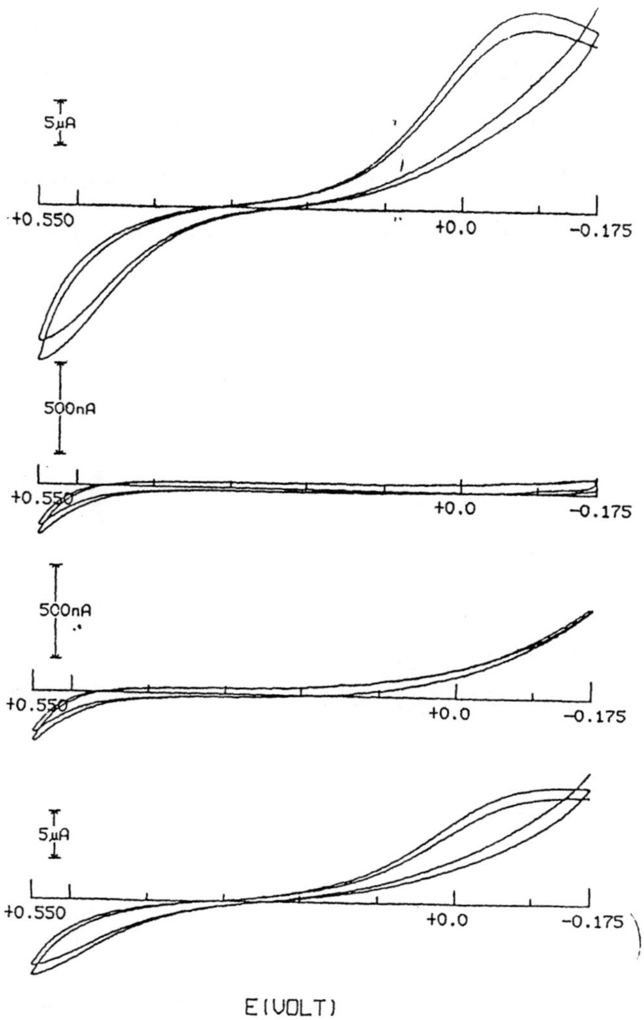

FIGURE 10. Voltamogram showing the effect of the protein addition. From top to bottom: decanethiol monolayer; decanethiol/Phospholipid bilayer; bilayer plus melettin; bilayer plus 10X meletin over previous.

α-Toxin Studies in Lipid membranes

The object of this project is to explore the possibility of using ion channels as practicable nanoscale devices. The methods used to characterize these channels involve the self-assembly of *Staphloccocus aureus* α-toxin into pores across a lipid membrane used for the study. The device used is shown in figure 11.

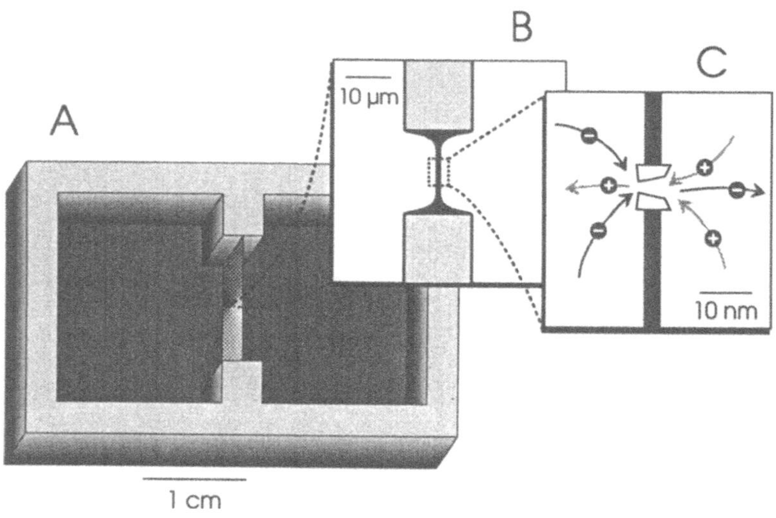

FIGURE 11. Schematic of device used for study of α-toxin.

The pores can be monitored by current flow. In addition these pores can be opened and closed by increasing the potential across the membrane (Figure 12). The eventual goal of the project is to develop methods of inserting unique compounds into these pores which will give them ion specificity which can be monitored electronically producing a nano-sized sensor.

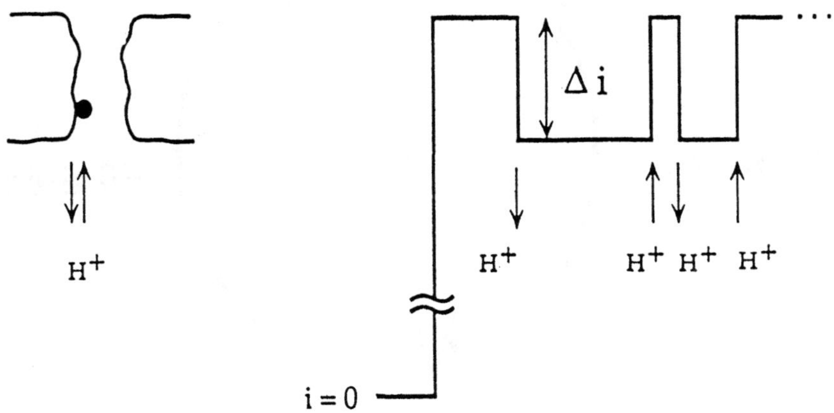

$$G_{open1} + H^+ \underset{k_{off}}{\overset{k_{on}}{\rightleftharpoons}} G_{open2}$$

FIGURE 12. Electrical effect of the opening and closing of a pore.

Bacteriorhodopsin (BR) Research

The object of this research is to better understand the mechanism by which this light sensitive protein pumps protons across a membrane and to use this information in several applications including optoelectronics storage and processing, holography, and gating. The simplified mechanism for the BR photocycle is presented in figure 13.

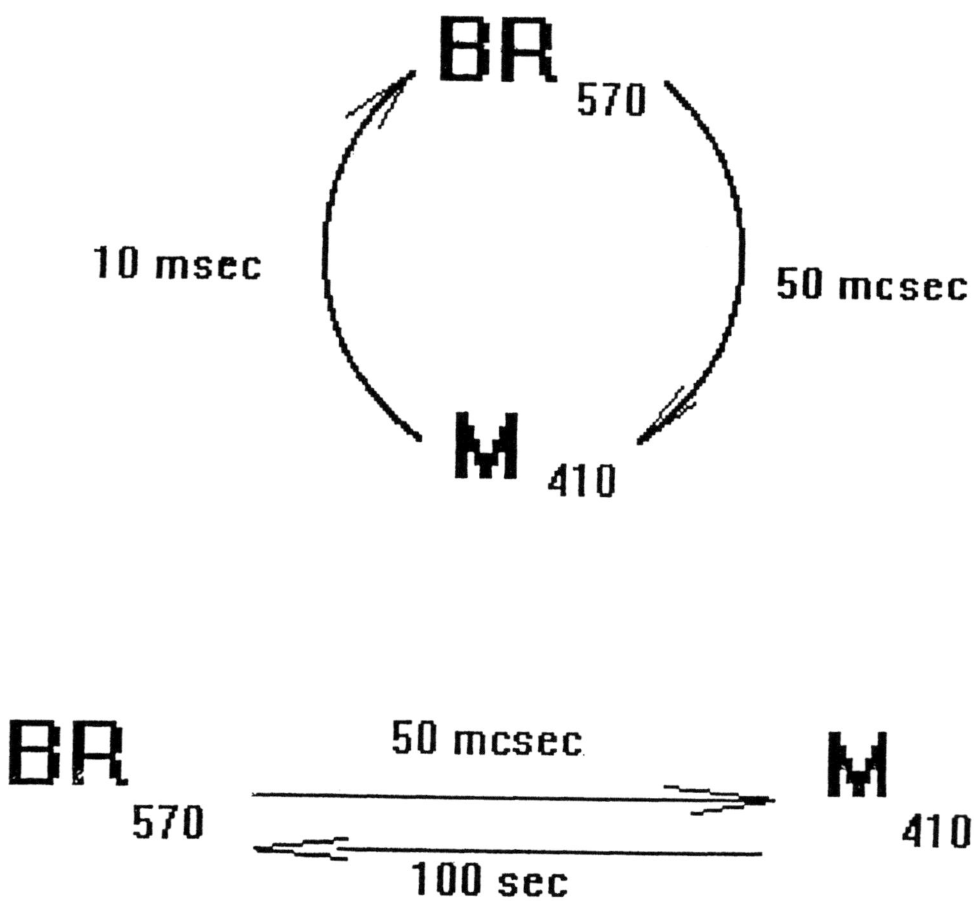

FIGURE 13. Schematic representation of the BR photocycle.

The spectral changes which occur are most easily observed in monitoring the rate at which the BR moves from the M state back to the ground state. When the BR is prepared in a gelatin or polymeric film it shows identical properties. However the rate at which it returns to ground state from the M state can be slowed by the addition of a variety of chemical additives. A typical example of these effects are presented in figure 14. These films are stable and can be maintained for several years.

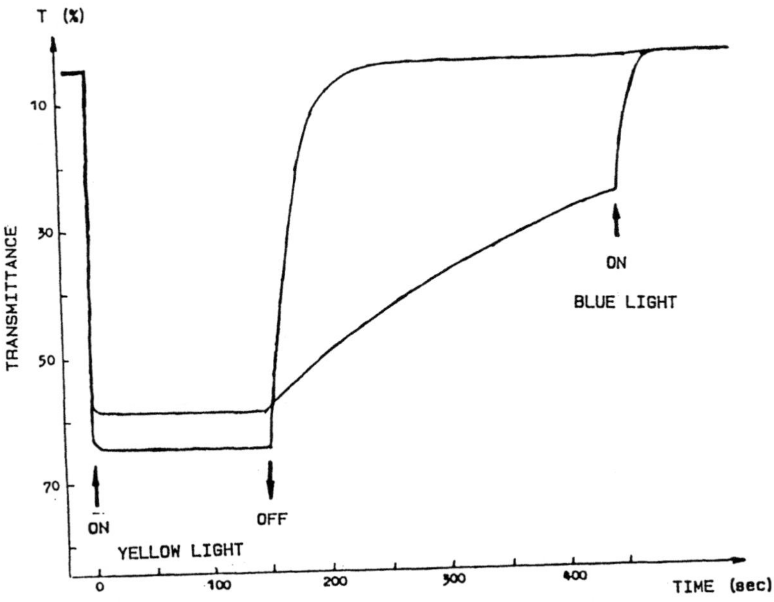

FIGURE 14. Comparison of BR entrapped in Gelatin film in the presence and absence of amine additives.

We have also entrapped the BR in sol-gel glasses. Unlike the films which require chemical additives to slow the conversion from M state to ground state, these sol-gels do not. Data indicates that the presence of the BR in the more confined and charged material somehow decreases the rate at which it moves from M state to ground state (Figures 15 and 16).

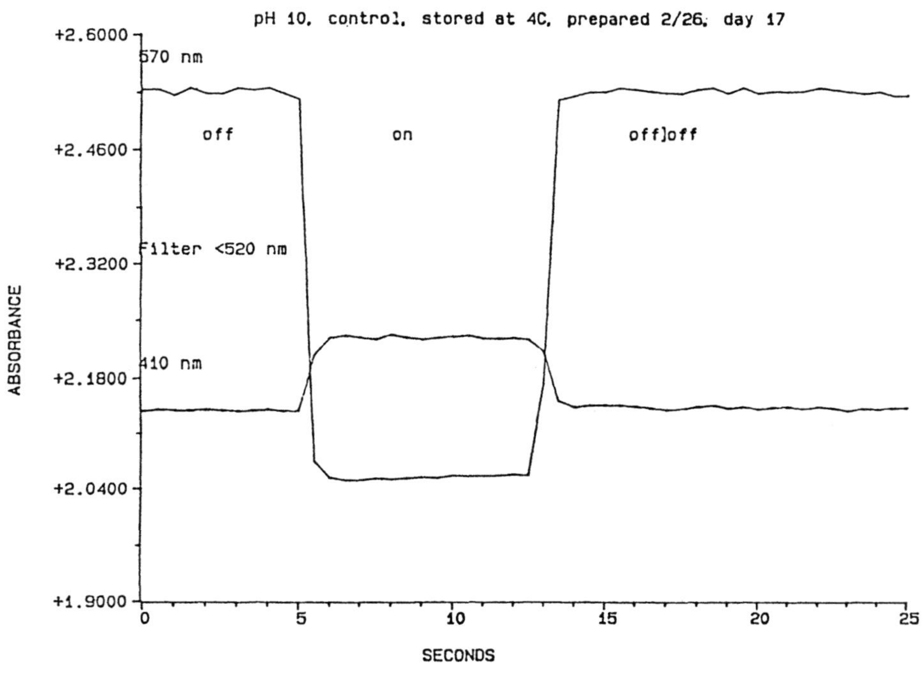

FIGURE 15. Control, pH 10 borate buffer, stored at 4°C, day 17

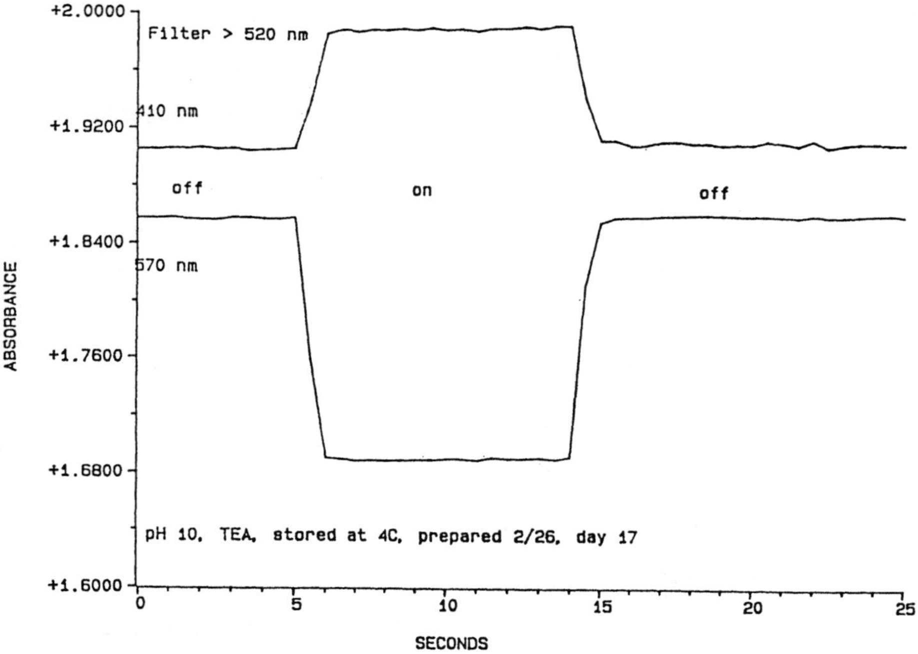

FIGURE 16. Experimental sample, pH 10 containing triethanolamine, stored at 4 ^{o}C, day 17.

In addition to our basic studies on the characteristics of the BR immobilized in glasses and films, we are attempting to develop a light activated enzyme switch. The switch utilizes BR attached to an enzyme. When the BR is light activated, it pumps protons into the environment of the enzyme causing a change in the pH and a change in the enzyme kinetics. Our initial studies indicate that BR when attached to β-galactosidase can change the reaction rate and shift the bulk pH profile of the enzyme.

SUMMARY

The projects presented in this report were the work of several persons who deserve the credit. The persons and the projects are as follows:

* DNA Intercalation - J. Horvath
* Glucose electrode - P. Pandey
* Bacterial detection - S. Glazier
* Lipid bilayer on gold - A. Plant
* α-Toxin membrane studies - J. Kasianowicz
* BR films - N. Vsesolodov and T. Dyukova
* BR sol-gel glasses - H. Weetall
* BR enzyme switch - M. McCurley

Each of these projects will yield information which we believe will be useful for the development of new biosensors and biosensor technology.

FIA IMMUNOASSAYS WITH ELECTROCHEMICAL AND OPTICAL DETECTION

KARL CAMMANN, AXEL WARSINKE, MONIKA WORTBERG,
CORNELIA MIDDENDORF, UWE KARST AND REINHARD RENNEBERG

Institut für Chemo- und Biosensorik und Lehrstuhl für Analytische Chemie der Westfälischen Wilhelms-Universität,
Wilhelm-Klemm-Str.8, D-4400 Münster, Germany

Summary: Different approaches have been made for developing novel immuno FIA (FIIA) systems for herbicide detection : antigens have been labeled with ferrocene and enzymes, anti-herbicide antibodies were labeled with fluorophores (Europium chelates).

First attempts have been made to develop FIA based immunoassays and immunosensors for the herbicides 2,4-dichlorophenoxyacetic acid (2,4-D) and 2-chloro-4-ethylamino-6-isopropylamino- 1,3,5-triazine (atrazine).

Different approaches have been used for generating electrochemically and optically detectable signals:

1. Ferrocene - labeled herbicides for immunoassays

Different ferrocen-derivatives of herbicides have been synthesized and successfully used in competitive immunoassays. The main idea was to detect ferrocen-labeled herbicides using differential pulse voltammetry and to reach lower detection limits by signal averaging
(if the voltammetric signals are measured repeatedly and averaged afterwards).

The ferrocene-labeled herbicides were tested both in competition and replacement assays. The replacement assays uses the displacement of already bound labeled antigens (with low affinity towards the antibody) by the high-affinity analyte antigen .

The ferrocene-labeled herbicides could be used successfully for immunoassays. However, the detection limit of these assays was several orders of magnitude above the limits for drinking water (0.1 µg/l). In addition, the poor stability of the derivatives in the oxidation state +III (ferricinium cation) made them uneffective for practical immunoassays.

2. Enzyme-labeled herbicides

In contrast, with glucose oxidase (GOD)- labeled herbicides could be used for an extremely sensitive immunoassay for triazines (Fig.1) The detection limit for terbutryn was 3 ng/l , which is 10 times below the described limit for triazine assays.

G.G. Guilbault and M. Mascini (eds.), Uses of Immobilized Biological Compounds, 415–419.

These findings are used now for constructing an immuno-FIA with electrochemical detection of hydrogen peroxide, the reaction product of the herbicide label GOD.

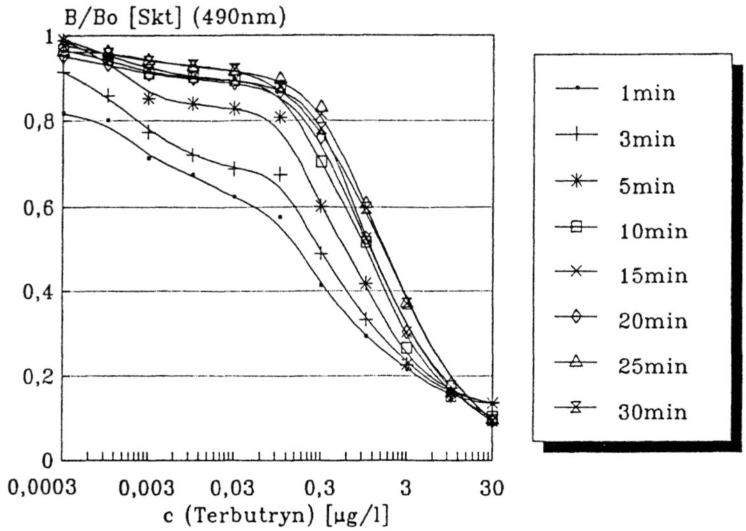

Fig.1 Calibration curves for a sequential saturation immunoassay for the determination of terbutryn using different incubation times for the GOD-labeled herbicide

Membrane immunosensors (Fig.2) based on enzyme-labeled herbicides have been developed for detection of herbicides like 2,4 D..

The conventional competition ELISA (labeled herbicide and unlabeled compete for binding site of the antibody) was combined with FIA and used for detection of 2,4 D.

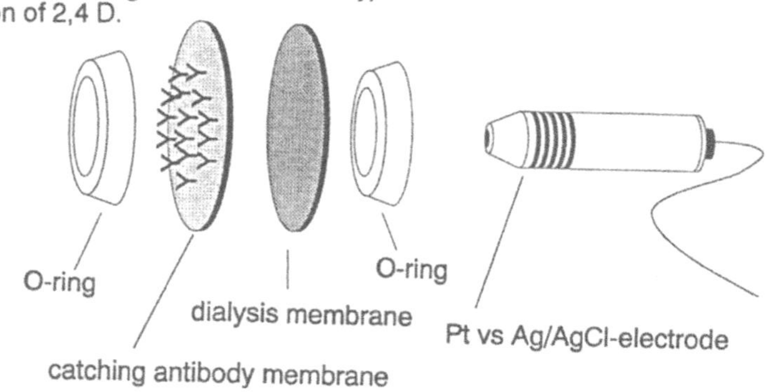

Fig.2

For labeling alkaline phosphatase (aP) was used with p-aminophenylphosphate as substrate. The product p-aminophenol could be followed easily at +5o mV vs.S.C.E. with a carbon electrode.

Fig.3 shows the principle of sequential saturation used in the approach.

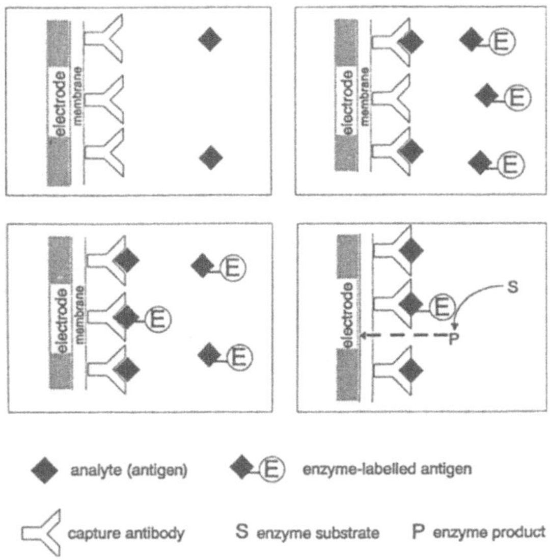

Fig.3 Principle of sequential saturation

The FIA setup is shown in Fig.4 and the sequence of FIA directed steps is shown in Fig.5.

After 10 min incubation with 2,4 D and 4 min with 2,4 D- aP conjugate the equilibrium was established. Together with washing steps a total measurement time of 22 min was achieved.
This can be shortened furthermore by using non-equilibrium conditions. As the membrane was used only one times the reproducibility was not high (23 %).

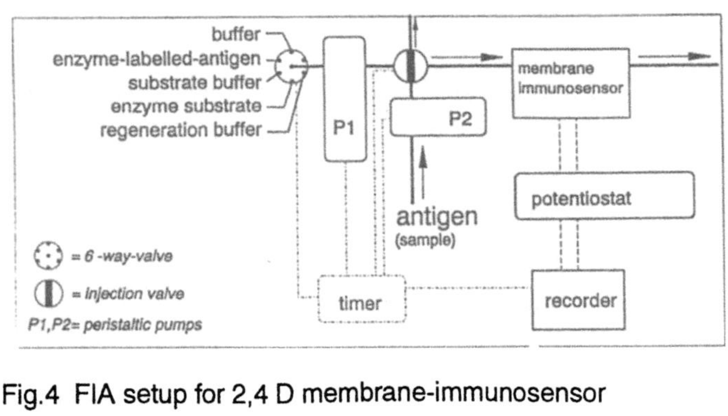

Fig.4 FIA setup for 2,4 D membrane-immunosensor

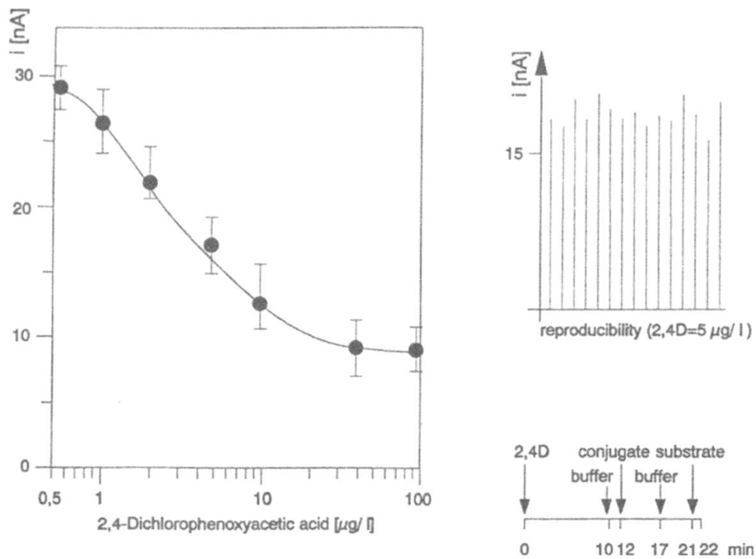

Fig.5 Membrane-immunosensor for determination of 2,4 D

3. Fluorophore-labeled anti-herbicide antibodies

Another approach being tested now is the replacement principle leading to an alarm sensor (Fig.6) .

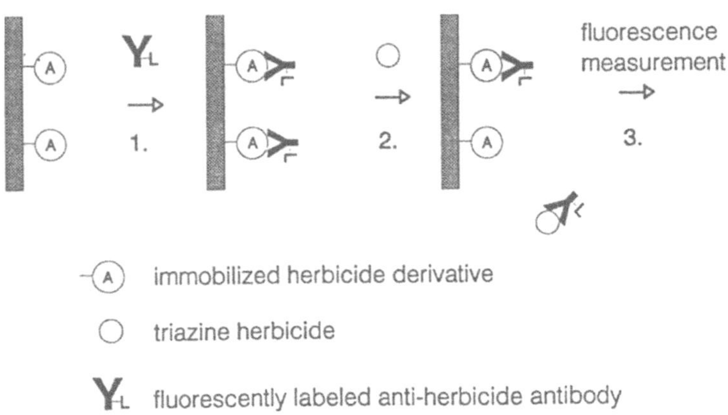

Fig.6 Principle of the replacement assay

An atrazine derivative is bound to the matrix of the carrier and is saturated with fluorophore-labeled antibodies. As fluorophores Europium (III) chelates have been used.

When the analyte is added the matrix-bound atrazine is displaced and the labeled antibody can be washed off followed by fluorescence measurements.

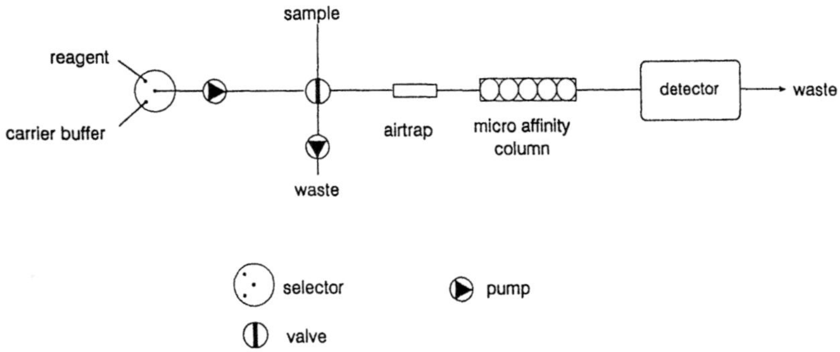

measuring cycle: 1. load column with fluorescently labeled antiibody (reagent)

2. wash out excess of reagent

3. inject herbicide sample

4. measure fluorescence of replaced antibodies

Fig.7 FIA setup and parameters for the fluorophore-labeled approach

The novel approaches can be performed within minutes. It has a high reproducibility , a stability of more than 2 weeks and for more than 25 determinations . The lowest detection limit was 1 µg atrazine /l. The general aim is to set up a continuously working alarm system where a steady flow of samples is checked and only after reaching a threshold of a special pesticide the alarm is given.

Parameters of the FIIA

- column volume 50 - 125 µl
- carrier: polymer beads (250 µm in diameter)
- triazine derivative covalently immobilized
- flow rate 0,25 ml/min
- assay buffer PBS
- concentration of labeled antibody: 10 µg/ml
- cycle:15 min loading, washing,

20 min replacement, measuring etc.

IMMUNOASSAY DEVICES FOR EXTRA-LABORATORY MEASUREMENTS OF TOXIC CHEMICALS BASED ON CAPILLARY MIGRATION AND LIPOSOME AMPLIFICATION

RICHARD A. DURST, SUI TI A. SIEBERT, MATTHEW A. ROBERTS AND STUART G. REEVES

Analytical Chemistry Laboratories
Department of Food Science and Technology
Cornell University, Geneva, New York 14456 USA

Abstract Simple, single-use biosensing devices are being developed based on the immunologic recognition of specific environmental and food commodity contaminants in combination with signal amplification strategies using marker-loaded liposomes to provide high sensitivity. Solution flow in the devices is controlled by capillary action and detection can be based on marker color or electroactivity. Several areas of investigation are being addressed: 1) immobilization of antibodies onto the solid phase of the capillary bed; 2) liposome surface density of analyte molecules conjugated to phospholipids; 3) capillarity-driven timing sequence for optimum flow and interaction of components; 4) concentration/measurement zone based on an avidin-capture mechanism; and 5) evaluation of several modes of detection/quantitation. Preliminary laboratory testing of several prototypes of this device based on optical detection has demonstrated the feasibility of this approach.

I. INTRODUCTION

Details of an automated, flow-injection immunoassay system based on liposome amplification of the signal resulting from the competitive binding reaction between the theophylline and theophylline-tagged liposomes have already been published (1-4). While this computer-controlled system has been used very successfully in the laboratory, the basic principles have the potential for extra-laboratory (*in-situ*) sensing applications. By appropriate simplification of this system, the important features of high sensitivity, specificity, non-isotopic labels and analytical speed can be retained in a portable system for field monitoring. The features sacrificed in developing the remote-sensing device are automation and reusability. In many ways,

G.G. Guilbault and M. Mascini (eds.), Uses of Immobilized Biological Compounds, 421–432.

however, the single-use character of the immunomigration devices minimizes the possibility of errors due to fouling and cross-contamination which is more likely to occur in a extra-laboratory environment.

While many of the basic principles and procedures on which these field devices will be constructed have already been demonstrated in practice, several important areas of development have to be optimized before this type of sensor will be feasible: 1) immobilization of antibodies and avidin onto the solid phase of the capillary bed; 2) liposome surface density of analyte molecules conjugated to phospholipids; 3) capillarity-driven (rather than computer-controlled flow-injection) timing sequence for optimum flow and interaction of components; 4) concentration/measurement zone based on an avidin-capture mechanism; and 5) evaluation of several modes of detection/quantitation of the markers.

In contrast to the automated flow-injection system, in these devices solution flow is controlled by capillary action, similar to clinical immunochromatographic methods (5), and detection is either electrochemical or optical. The use of electroactive markers or visual dyes significantly simplifies the detection and quantitation process compared to fluorescence, although at some reduction in sensitivity.

Although several assay formats are being studied, the current design of the disposable sensor, illustrated in Figure 1, consists of a strip of plastic-backed, porous nitrocellulose containing some, or all, of the following zones: wick, sample/liposome introduction, immobilized antibody, and concentration/measurement. An analysis consists of the following steps:

1) Test (sample or control) solution and liposome/buffer solution are placed in the carrier solution tube. (Alternatively, the sample solution may be spotted onto the zone between the wick and immobilized antibody zone.);
2) The wick is inserted into the carrier solution tube;
3) Wicking buffer transports the sample and analyte-labeled liposomes into the antibody zone where competitive binding occurs;

4) Unbound liposomes (proportional to the analyte concentration) are carried into the concentration/measurement zone where they are affinity concentrated, or ruptured to release the entrapped electrochemical marker; and
5) Liposomes or released marker molecules are quantified optically by measuring the color intensity, or electrochemically (by potentiometry or amperometry).

In addition, each test strip may contain "sub-strips" on which one or two control(s) can be run simultaneously with the sample to verify response and to increase analytical accuracy. These arrangements are illustrated in Figures 2 and 3 for test strips designed for optical detection.

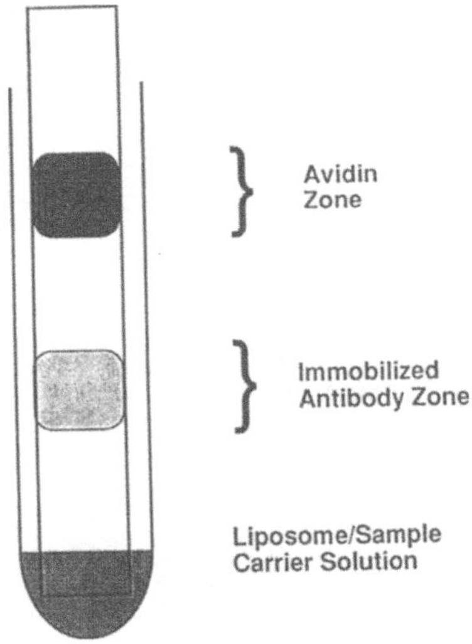

Fig. 1.
Diagram of the "single channel" test strip in which sample and liposome carrier solutions are added to the test tube prior to test strip insertion and development. In this design, avidin is immobilized in the measurement zone to capture all of the liposomes which were not bound in the antibody competition zone.

The feasibility of the liposome-immunomigration approach has been demonstrated using polychlorinated biphenyls (PCBs) and the herbicide, Alachlor, as model compounds in evaluating and optimizing the prototype biosensing device.

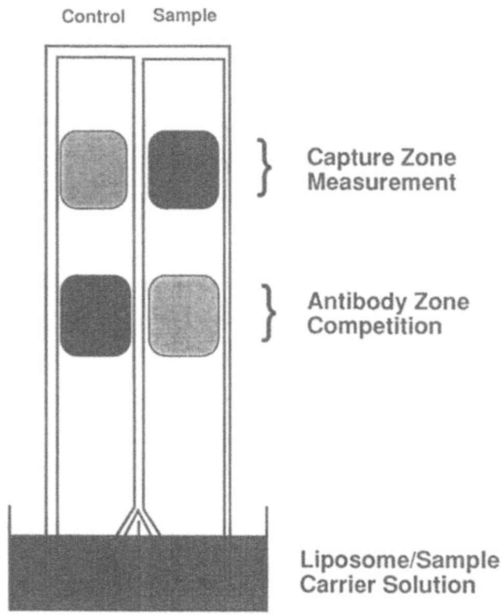

Fig. 2.
Diagram of the "double-channel" test strip in which the sample and tolerance-level control are added directly to the separate chambers containing the carrier (liposome) solution. Measurement takes place in the avidin capture zone of the test strip.

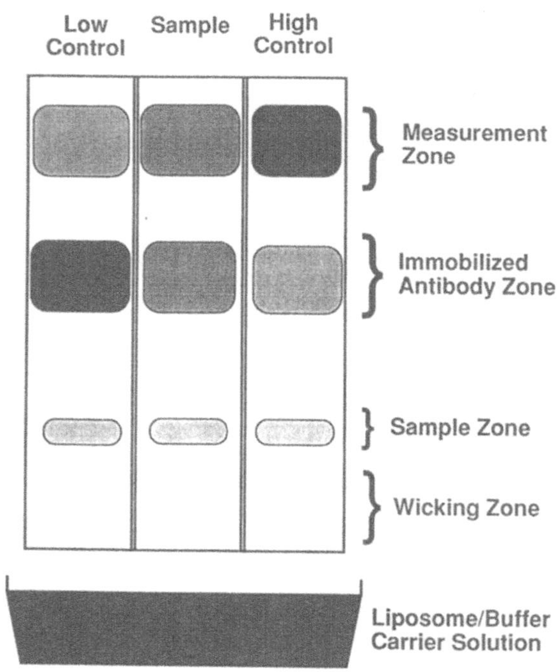

Fig. 3.
Diagram of the "triple-channel" test strip design for the simultaneous measurement of the sample plus two controls to provide linear interpolation and verification of response. Sample and high- and low-level controls are spotted onto the sample zones prior to insertion of the strip into the carrier solution. After migration of the solution to the end of the strip, color intensity is quantified in the measurement zone indicated in the diagram.

II. EXPERIMENTAL

Production of Conjugates and Antibodies

For the purpose of these studies, two analytes (Alachlor and polychlorinated biphenyls) are representative of food and environmental contaminants and were chosen to demonstrate the feasibility of the immunoassay device. The haptens Alachlor and 2,4,5 trichlorophenoxyacetic acid were conjugated to BSA and to DPPE, respectively, by modification of published methods, as described elsewhere (6). The 2,4,5 trichlorophenoxyacetic acid was also conjugated to BSA using similar methods. Using these immunogenic conjugates, antibodies were raised in New Zealand white rabbits by standard procedures at the Cornell College of Veterinary Medicine which then provided the antisera. The antibodies produced were purified by the caprylic acid/ammonium sulfate precipitation method (7).

Immobilization of Proteins

For these assays, a protein-binding membrane with a plastic backing to provide rigidity was required, and nitrocellulose membranes supported in this manner were found to be the most suitable. Several approaches were tried for immobilization of the proteins onto the nitrocellulose solid phase. The most effective immobilization for both the antibody and egg white avidin was achieved by the use of an air brush to spray the respective solutions onto the desired areas of the strip (which had been prewetted in 10% methanol in TBS and air dried). The strip was then air dried, immersed in the blocking agent (2% polyvinylpyrrolidone containing 0.002% Tween-20) for one hour, and allowed to dry under vacuum before use. This approach has been successfully demonstrated and is being further evaluated. The air brush application yields a very uniform and precisely defined antibody and avidin zones on the strips. Applied in this manner, it was found that the adsorbed or entrapped antibodies and avidin are physically constrained to the application area and covalent or other binding methods are not necessary.

Conjugation of Analyte to Phospholipid

For the competitive assay format, it is necessary to provide antigenic sites (epitopes) on the surface of the liposomes. To do this, a conjugate must be formed between the analyte molecule and a lipid (DPPE) which is then incorporated into the liposome bilayer. There are a variety of possible conjugation approaches that can be used, for example, bifunctional covalent coupling agents and strong non-covalent binding, such as biotin-avidin coupling. In a modification of reported procedures (8,9) for conjugation of the Alachlor to DPPE, a thiolating reagent, N-succinimidyl-S-acetylthioacetate (SATA), was chosen as the coupling agent.

Confirmation of the successful conjugation reaction was achieved by a combined thin-layer chromatographic and enzyme immunostaining method (10). The procedure involves the duplicate chromatographic analysis of the reaction mixture on an appropriate thin layer chromatographic plate and solvent system. One of the TLC plates is dried, blocked, washed and placed overnight in a solution containing the antibody to Alachlor. The plate is washed and placed in a solution containing a goat anti-rabbit alkaline phosphatase conjugate for two hours. The plate is again washed and developed with the substrate for alkaline phosphatase (nitroblue tetrazolium in aqueous DMF with magnesium chloride and 5-bromo-4-chloro-3-indolyl phosphate in DMF). When color development is complete, the plate is washed in distilled water and dried. A purple spot indicates the presence of Alachlor. The other TLC plate is examined under a UV lamp (254 nm), then exposed to iodine vapor and finally sprayed with molybdenum blue reagent (0.65% molybdenum oxide in 4.2 M sulfuric acid) which is specific for phospholipids. The Alachlor-DPPE spot appears purple with the alkaline phosphatase substrate stain and blue with the molybdenum blue spray reagent. Thus, by this method, the location on the two TLC plates corresponding to these two colors identifies the desired analyte-lipid conjugate.

The approach used for the conjugation of the PCBs to DPPE was to couple the hapten, 2,4,5-trichlorophenoxyacetic acid, to DPPE via preactivation of the hapten to an N-hydroxysuccinimide ester in the presence of N,N´-dicyclohexylcarbodiimide (11). Confirmation of the successful conjugation reaction was performed by the TLC/immunostaining method and will also be confirmed by the use of radiolabeled 2,4,5-trichlorophenoxyacetic acid.

Preparation of Dye-Encapsulated Liposomes

Liposomes were formed by a reverse-phase evaporation method (12). This was preferred to the alcohol injection method (13) as it produced a high yield of concentrated liposomes in suspension. The liposomes were prepared from a mixture of dipalmitoyl phosphatidyl choline (DPPC), cholesterol, dipalmitoyl phosphatidyl glycerol (DPPG), and dipalmitoyl phosphatidyl ethanolamine-Alachlor (DPPE-Ala) conjugate or dipalmitoyl phosphatidyl ethanolamine-2,4,5 trichlorophenoxyacetic acid conjugate (DPPE-PCB) in a molar ratio of 5:5:0.5:0.01 (14). Forty-three micromoles of this mixture were dissolved in 4.2 mL of a solvent mixture containing chloroform : isopropyl ether : methanol (6:6:1 by volume). This solution was warmed to 45 °C and 0.7 mL of the dye solution was added with swirling. This mixture was sonicated for 5 min under a low flow of nitrogen. The organic phase was removed under vacuum on a rotary evaporator at 40 °C until all frothing had stopped. An additional 1.3 mL aliquant of the dye solution was added, and the liposomes were then sequentially extruded through polycarbonate filters of decreasing pore sizes of 1.0 μm and 0.4 μm. This procedure resulted in a preparation of average diameter 0.6 μm (85% of the preparation being between 0.51 and 0.76 μm) that migrated more evenly on the solid support than did the unextruded preparation. Finally, to remove any unencapsulated dye, the liposomes were gel filtered on a 1x14 cm Sephadex G-50 column and dialyzed overnight against TBS at 4 °C. When stored at 4 °C, there was no measurable leakage of dye over a period of 9 months.

Sulforhodamine B (14) was chosen as the dye for encapsulation because of its high molar absorptivity which gives high sensitivity for visual and/or photometric quantitation. To prepare the dye for encapsulation, 20 mM tris(hydroxymethyl)aminomethane (Tris) was used to buffer the dye solution. The pH was adjusted to ca. 7.0 with NaOH to effect dissolution. The final solution contained 200 mM dye in 20 mM Tris at a pH of approx. 7.0 and an osmolarity approximately equal to Tris buffered saline (TBS). TBS is the buffer routinely used in all aqueous operations of the experiments.

Furthermore, because sulforhodamine B is highly fluorescent and this fluorescence undergoes self-quenching when encapsulated, the integrity of the liposomes can be determined by measuring fluorescence intensity before and after lysis. The total lysis of the liposomes was effected by addition of a solution (final concentration = 30 mM) of n-octyl-ß-D-glucopyranoside at room temperature. For these fluorescence experiments, the dye was excited at a wavelength of 543 nm and fluorescence measured at the emission wavelength of 596 nm.

To test the various optical detection modes, other non-fluorescent dyes, such as Isosulfan Blue and Fast Green CFC, were also encapsulated in liposomes of the same composition and prepared by the same reverse-phase evaporation method.

Test Strips

For adequate capillary action, it was found that plastic-backed nitrocellulose strips (4 x 75 mm) having a nominal pore size of 3 µm (Schleicher & Schuell, Keene, NH) provided a good flow rate for the sample/liposome solution. Following antibody and avidin immobilization, the polyvinylpyrrolidone/Tween-20 solution was used as the blocking agent. Although there was some nonspecific binding of the liposomes to the test strip, this did not cause significant problems in quantitation, but may compromise the ultimate detection limits of the devices.

Detection/Quantitation

Detection/quantitation can be effected in either of two zones. If the measurement is performed in the antibody-binding zone, the concentration of bound liposomes will be inversely proportional to the analyte concentration, thus producing a color intensity that is greater for the lower concentrations of analyte. This is the response observed in most competitive-assay approaches. By using a measurement zone in which the unbound liposomes are concentrated by avidin capture before measurement, a much more rational response is obtained in which the color intensity (or as discussed below, the electroactivity) is directly proportional to the analyte concentration.

The two principal approaches for the detection of the competitive binding reactions, between the analyte molecules and the tagged liposomes for the immobilized antibody binding sites, are electrochemical and optical. In the proposed device using electrochemical detection, the antibodies will be lysed by surfactant and the electroactive marker (ferrocyanide) will be accumulated at the surface of an ion-exchange polymer-modified electrode prior to amperometric quantitation (15). Since this approach is based on a system successfully used as a homogeneous immunoassay, it is expected that few difficulties will be encountered in adapting this analysis strategy to the heterogeneous, competitive-binding assay. In fact, the use of surfactant lysis of the liposomes, in contrast to the original method which used complement lysis, should significantly simplify and improve the reliability of the method. However, because the optical detection approach should be technically simpler to implement, especially if visual estimation of the color intensity is used, this direction is the one which has received our initial attention.

For optical detection, two approaches are being investigated simultaneously: visual and reflectometric quantitation. The most significant differences between the two modes of detection are that visual detection does not need instrumentation for quantitation of the dye intensity, but does require higher concentrations of liposomes for estimation of the color levels. Conversely, for instrumental (reflectometer/photometer) detection, it is expected that, as a benefit of the additional cost and complexity, it should be possible to achieve greater accuracy, precision, and sensitivity.

III. RESULTS and DISCUSSION

Analysis Format

Using one of the optical-detection assay formats as an example, the analysis protocol consists of the following procedure. Two drops (100 μL) of the sample solution, one drop (50 μL) of a 3x concentration of TBS and one drop (50 μL) of the liposome solution are dispensed into a 10 x 75 mm glass test tube. The test strip is inserted into the tube, and migration of the sample/liposome solution is allowed to proceed to the end of the strip (approximately 8 to 10 min). The intensity of the dye color in the immobilized antibody zone or in the capture zone is measured with a computer scanner which converts the intensity at designated regions on the strip into gray-scale readings for quantitation. While this instrumental measurement will not be performed in the field, it provides a more accurate measurement than visual quantitation and is thus very useful for development studies. In addition, some initial success has been achieved in the photometric quantitation performed using portable single-channel reflectometers with digital readout. However, these reflectometers were commercial units designed specifically for blood glucose measurements and not optimized for the spectral characteristics of our particular dyes.

Since there are numerous permutations of how this type of assay can be performed, these initial measurements are designed to prove principle and demonstrate feasibility. Future research will investigate some of the variations discussed above. For example, improved accuracy should be achieved when this device is used in the dual-strip mode in which a control is run simultaneously with the sample and the readout of a dual-channel reflectometer provides a difference reading between the concentrations of the sample and tolerance-level control.

A further extension of this approach will entail the use of a triple-strip in which high and low controls will permit the more accurate interpolation of the sample concentration. A specially designed three-channel photometer with an appropriate interpolation microprocessor will be required for this application.

Finally, preliminary experiments using the egg white avidin collection zone approach have been very promising. In this mode of operation, the color intensity (or electroactivity) of the marker is directly proportional to the analyte concentration and provides higher response sensitivity for the analyte.

In conclusion, preliminary laboratory testing of an early prototype of this device based on optical detection has demonstrated the feasibility of this approach. Using Alachlor standards, it was found that the device response was linear over the concentration range from 0.02 to 1.0 μg/mL (ppm) with an estimated detection limit of 0.01 μg/mL (ppm). Current research is directed toward evaluating various assay formats and optimizing the characteristics of the prototype. Future plans include the packaging of the strip in a "user-friendly" device in which only the wick, measurement zone, and strip end are visible. In this way, the protected strip will be conveniently manipulated and will be used in conjunction with a simple readout photometer or electrometer.

ACKNOWLEDGMENTS

The authors acknowledge the partial support of this research received from: the Cooperative State Research Service, U.S. Department of Agriculture (CSRS/USDA), under Subcontract No. USDA-TPSU-CU-7119-757 through The Pennsylvania State University; the CSRS/USDA Water Quality Special Grants Program, project no. 93-01360; and the National Institute of Environmental Health Sciences, National Institutes of Health, DHHS, under the Superfund Research and Education grant no. 1 P42 ES05950-01. One of us, M.A.R., also acknowledges NIEHS/NIH for pre-doctoral training support under the Superfund Research and Education grant. Any opinions, findings, conclusions, or recommendations expressed in this publication are those of the authors and do not necessarily reflect the views of the U.S. Dept. of Agriculture, the U.S. Dept. of Health and Human Services, The Pennsylvania State University or Cornell University.

REFERENCES

1. Locascio-Brown, L., Plant, A. L., Horvath, V. and Durst, R. A. (1990) Liposome Flow Injection Immunoassay: Implications for Sensitivity, Dynamic Range, and Antibody Regeneration, Anal. Chem. **62**, 2587-2593.
2. Durst, R. A., Locascio-Brown, L. and Plant, A. L. (1990) Automated Liposome-Based Flow Injection Immunoassay System, Flow Injection Analysis Based on Enzymes or Antibodies, R.D. Schmid, ed., GBF Monograph Series, **14**, VCH Publishers, Weinheim, FRG, 181-190.
3. Yap, W. T., Locascio-Brown, L., Plant, A. L., Choquette, S. J., Horvath, V., and Durst, R. A. (1991) Liposome Flow Injection Immunoassay: Model Calculations of Competitive Immunoreactions Involving Univalent and Multivalent Ligands, Anal. Chem. **63**, 2007-2011.
4. Locascio-Brown, L., Plant, A. L., Chesler, R., Kroll, M., Rudell, M. and Durst, R. A. (1993) Liposome-Based Flow-Injection Immunoassay for Determining Theophylline in Serum, Clin. Chem., **39**, 386-391.

5. Zuk, R. F., Ginsberg, V. K., Houts, T., Rabbie, J., Merrick, H., Ullman, E. F., Fischer, M. M., Sizto, C. C., Stiso, S. N. and Litman, D. J. (1985) Enzyme Immunochromatography - A Quantitative Immunoassay Requiring No Instrumentation, Clin. Chem. **31,**1144-1150.
6. Reeves S. G., Roberts, M. A., Siebert, S. T. A. and Durst, R. A. (1993) Investigation of a Novel Microtiter Plate Support Material and Scanner Quantitation of Proteins, Phospholipids and Immunoassays, Anal. Lett. ***26*** (7).
7. McKinney, M. M. and Parkinson, A. (1987) A Simple, Non-chromatographic Procedure to Purify Immunoglobulins from Serum and Ascites Fluid. J. Immuno. Methods. **96**, 271-278.
8. Duncan, R. J. S., Weston, P. D. and Wrigglesworth, R. A. (1983) New Reagent which May Be Used to Introduce Sulfhydryl Groups into Proteins, and Its Use in the Preparation of Conjugates for Immunoassay, Anal. Biochem. **132**, 68-73.
9. Feng, P. C., Wratten, S. J., Horton, S. R., Sharp, C. R. and Logusch, E. W. (1990) Development of an Enzyme-Linked Immunosorbent Assay for Alachlor and Its Application to the Analysis of Environmental Water Samples, J. Agric. Food Chem. **38**, 59-163.
10. Mattsby-Baltzer, I. and Alving, C. R. (1984) Lipid A Fractions Analyzed by a Technique Involving Thin-Layer Chromatography and Enzyme-Linked Immunosorbent Assay, Euro. J. Biochem. **138**, 333-337.
11. Schmidt, D. J., Clarkson, C. E., Swanson, T. A., Egger, M. L., Carlson, R. E., Van Emon, J. M. and Karu, A. E. (1990) Monoclonal Antibodies for Immunoassay of Avermectins, J. Agric. Food Chem. **38**, 1763-1770.
12. Szoka, S., Olsen, F., Heath, T., Vail, W., Mayhew, E. and Papahadjopoulos, D. (1980) Preparation of Unilamellar Liposomes of Intermediate Size (0.1-0.2 μm) by a Combination of Reverse Phase Evaporation and Extrusion Through Polycarbonate Membranes, Biochim. Biophys. Acta **601**, 559-571.
13. Batzri, S. and Korn, E. D. (1973) Single Bilayer Liposomes Prepared Without Sonication, Biochim. Biophys. Acta **298**, 015-1019.
14. O'Connell, J. P., Campbell, R. L., Fleming, B. M., Mercolino, T. J., Johnson, M. D. and McLaurin, D. A. (1985) A Highly Sensitive Immunoassay System Involving Antibody Coated Tubes and Liposome Entrapped Dye, Clin. Chem. **31**, 1424-1426.
15. Durst, R. A. and Kannuck, R. M. (1989) Chemically Modified Electrode for Liposome-Mediated Homogeneous Immunoassay, Ion-Selective Electrodes, **5**, E. Pungor, ed., Pergamon Press, NY, 65-79.

DIRECT IMMUNOASSAYS: LIMITATIONS ON THE PHYSICAL PRINCIPLES AND ON THE TECHNOLOGY STATE OF THE ART

M. ALVAREZ-ICAZA, A. GEBBERT, R.D. SCHMID
Gesellschaft für Biotechnologische Forschung (GBF)
Department of Enzyme Technology, Mascheroder Weg 1
3300 Braunschweig, Germany

Abstract. In order to directly detect an immunological reaction not only the connection to a physical variable is important but also the state of development of the technology to measure that variable and to produce the necessary devices. The measurement of changes in dielectric layer of a surface (capacitance sensors) due to immunological reactions exhibit analogous technological advantages to those shown by piezoelectric crystals. Sensors based on capacitance changes fabricated by producing a dielectric layer on a tantalum electrode are limited by background noise. Further investigation of the origin of this noise relates it to the tantalum purity and dissolved oxygen variations in the buffer solution.

INTRODUCTION

For the development of biosensors there are basically two possibilities: to use a biochemical component that catalyzes a reaction (enzymes, for example) or to use one that specifically binds to an analyte (antibodies or receptors). Through turnover, catalytic components can use many molecules of the analyte to produce a large signal. Extending the integration time the sensitivity of the system can be increased practically without limit. The signal produced by affinity components is, in contrast, limited by the number of binding places present in the measuring system. The maximal possible signal is equal to the number of binding places.

Measuring methods based on affinity components can be modified, using catalytic components to amplify the analytical signals. This can be done, for example, using

G.G. Guilbault and M. Mascini (eds.), Uses of Immobilized Biological Compounds, 433–441.

enzyme conjugated antibodies. The antibody provides the specific binding and the enzyme the catalytic amplification. Unfortunately, these methods (indirect) require an intensive use of reactives, and although they can be easily implemented using flow injection systems (FIA) they are not the solution for every analytical situation. Specifically, for one step disposable immunosensors, miniature solid state sensors or on-line monitoring of a variable, direct detection of the immuno reaction is advantageous.

For the direct detection of immuno reactions, a physical variable has to be used. The final signal depends on the relationship between the physical signal and the changes produced by the immuno reaction, but also on the state of the art of the technology to measure that particular variable.

DIRECT DETECTION OF AN IMMUNOLOGICAL REACTION

If a surface where antibodies have been immobilized is immersed in a solution where antigen is present, and the right conditions of temperature and pH are given then, even after the removing this surface from that solution, physical changes should be detected. After incubation on the surface with antibodies there are changes in mass or superficial density, probably in viscosity due to a different interaction of the antigens in the interface, in the surface free energy due to the spontaneous binding of the antigens, in the thickness of the protein layer, in the refractive index of a layer just above the antibodies, or making the observation at low frequencies, in the dielectric constant of the protein layer over the surface.

From the instrumentation point of view, to investigate which variables are convenient to measure it is interesting to analyze how the fundamental units are defined in the SI of measurements. This convention defines seven fundamental units: Meter, Kilogram, Second, Ampere, Kelvin, Mol, and Candela. All of these except the Kilogram, the Kelvin and the Mol require in their definition (directly or indirectly) measurements of time or frequency. Surely, a consequence that time can be measured with high precision. The definition of second, for example, is given [1] with a precision of 1 part in 10^{10} and today is possible to measure it with precision better than one part in 10^{14}. With today available technology, no other physical variable can be measured better than time. Therefore, if the immunological reactions can be converted to changes in time or frequency, a clear advantage (from the instrumentation point of view) is gained. An example of this is the advantage has been demonstrated with the use of piezo-crystals to detect immunological reactions.

Piezoelectric crystals can also be connected in a differential mode which provides an ideal configuration to detect very small variations. Unfortunately, the variation in oscillation frequency depends strongly on viscosity, and from the Sauerbrey equation

$$\Delta F = - \frac{k\, F_o^2\, dm}{A}$$

where k is a constant that depends on the crystal, F_o is the initial oscillation frequency, A is the crystal surface area and dm is the change in mass. When crystals are operated in aqueous media, F_o is strongly reduced (even to start the oscillations may be difficult). The sensitivity of the device is greatly reduced.

To be used in aqueous media, new devices and physical variables have to be chosen. Measurements of change of in the refraction index, by optical methods, are clearly a good option and its feasibility has been demonstrated experimentally and even as a commercial instrument. However, because for mass production of instruments the electric and electronic industry is more developed than the optical industry, if the detection of immunological reactions can be directly related to an electric principle and electric devices, it may represent (at the present time) an advantage.

The measurement of changes in dielectric constant can be performed through a typical electric measurement, capacitance. In addition, capacitance changes can be easily converted to frequencies. There are very good commercial instrumentation to measure them, for example, with a precision of 1 part in 10^6. It is possible to measure small variations using differential configuration. The technology to manufacture capacitors is well established for a number of years and finally, to determine the capacitance no DC currents are required which eliminates the need of reference electrodes and the measurement can be performed in a spectral region with lower level of electrical noise.

In particular, electrolytic capacitors appear as ideal devices for sensitive determination of dielectric constants of layers in aqueous media. The structure of these capacitors is a thin dielectric layer placed between a metal and a liquid electrodes. The latter formed by an electrolyte (aqueous media). In addition, because the dielectric layer is very thin, the behavior of the structure can be modeled using a parallel plates capacitor independently of the macroscopic shape of the metal electrode. Hence, the capacitance depends only on the characteristics of the material between the two electrodes.

TECHNICAL DEVELOPMENT OF A CAPACITIVE IMMUNO-SENSOR

In order to develop an immunosensor based on the structure of a electrolytic capacitor, the surface of the metal electrode has to be protected against corrosion and has to be modified in order to have active groups to fix the antibodies. But, on the other hand, the sensitivity of the sensor decreases with the thickness of a layer on the metal electrode. Therefore, a very thin (comparable to the size of an antibody) and stable layer has to be produced on the surface of the metal electrode.

Thin dielectric layers, on the surface of metal electrodes, can be produced by electrochemical oxidation. Specifically, a layer of tantalum oxide Ta_2O_5 is produced on the surface of tantalum when the metal is anodically oxidized in acid solutions (10% phosphoric acid, for example). This technology not only leaves a hydrated surface where antibodies can be covalently attached (using a aminosilane compound followed by carbodiimide)[2] but also has the advantages that the thickness of the dielectric layer of tantalum oxide is determined by the electrical potential [3] used during the oxidation and the produced film is transparent and uniform. Hence, its thickness can be determined by interferometric methods.

With this method an immunosensor sensor was constructed and was found that it was possible to observe the immuno reaction of relatively large (MW 150 000 or larger) molecules (Figure 1). [2].

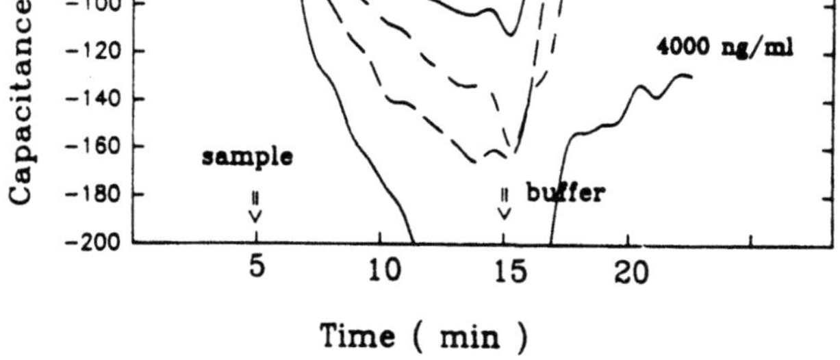

Figure 1.- Response of a capacitance immunosensor to different concentrations of IgG-ß galactosidase. (Reproduced from Gebbert et al. *Anal. Chem.* 64 997(1992))

In this measurements we found that background noise reduced considerable the signal to noise ratio. The whole potential of the instrument (1 part in 10^6 for a Hewlett Packard HP4284A Precision LCR-meter) was not being used. For this reason, we have investigated the origin of this background noise.

The background noise was also observed in capacitance sensors without antibody. Hence, it should have a purely physical origin. In a first attempt, the mechanical fragility of the tantalum oxide layer was suspected as responsible for the noise. However, this possibility was ruled out after testing sensors manufactured on a thin tantalum layer sputtered on a rigid ceramic substrate ; instead of the original thin [4] tantalum foil. The movement of liquid was also ruled out because the background noise was not dependant on the speed of the flow.

The next element to be examined was the tantalum oxide layer itself. We found in the literature that from normal commercial tantalum the tantalum oxide layer [5] produced by anodic oxidation has a complex structure (Figure 2).

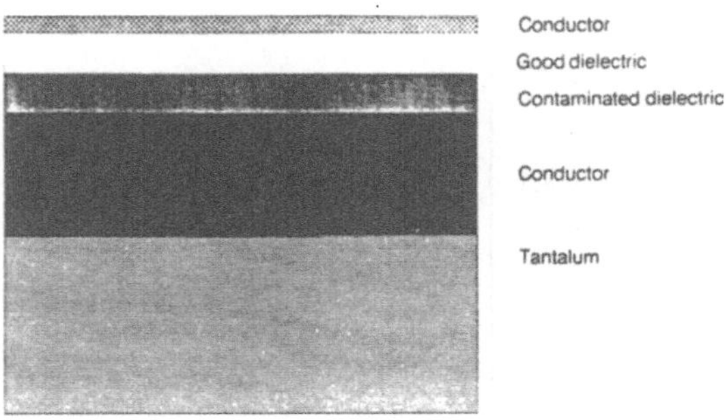

Figure 2.- Structure of the profile of an anodically grown layer of tantalum oxide on a tantalum electrode

The most external part of the layer has a conductive nature, but immediately bellow this interfacial section, a region with good dielectric properties due to the formation of Ta_2O_5 is found. Immediately bellow this layer another one due to impurities in the original metal is found. The latter is a layer not permeable to ions and oxygen which precludes the formation of Ta_2O_5. Other forms of tantalum oxide (Ta_2O_X where $X<5$) which are conductive are produced. The same authors demonstrated that the use of high purity tantalum avoids the formation of the oxygen impermeable layer and with it the formation of the deeper conductive layer. We have tried the formation of the dielectric layer with high purity tantalum (Trace elements below 25ppm except for oxygen with 51ppm from NRC Inc. Newton, MA, USA). A reduction of the background noise was observed but it was not eliminated.

The dielectric layer has to be permeable to oxygen to allow the formation of a good dielectric layer. Hence, the presence of oxygen during the measurements may have influence in the capacitance. We tested the capacitance between buffer (borate pH 9) saturated with oxygen and oxygen free (Figure 3).

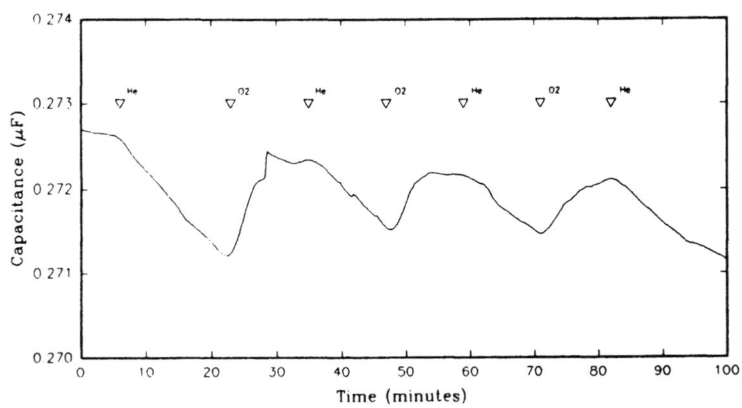

Figure 3.- Variation of the measured capacitance observed in a flow-through capacitance cell when oxygen saturated and oxygen free buffer (borate pH 9) was passed through it

The capacitance was sensitive to oxygen and may also be a contributor to the background noise. A dielectric surface, formed with pure tantalum and tested in oxygen free buffer showed variations (due to background noise) of ± 1 in the fifth significative figure. This level of noise allows a better use of the available instrumentation. Improvement of the immobilization methods and ways of reducing the dependance of the sensor to the amount of oxygen in the buffer are currently under investigation.

ACKNOWLEDGEMENT

We thank Dr. Klaus Andersson from HC-STARCK GmbH & Co. KG. Goslar Germany, for providing us with high purity tantalum.

REFERENCES

1. H.Boer, in *Die SI-Basisenheite Definition Entwicklung Realiserung* Physikalisch Technische Bundesanstalt. Braunschweig (1990) p. 19

2. A. Gebbert, M. Alvarez-Icaza, W. Stöcklein and R.D. Schmid. *Anal. Chem.*, **64** 997(1992)

3. Vermilyea, D. A. *Acta Metall*, **1**,282(1953)

4. A. Gebbert, M. Alvarez-Icaza, H. Peters, V. Jäger, U. Bilitewski and R.D. Schmid. Submited to *J. of Biotechnology*.

5. F. Climent, R. Capellanes, E. Garcia and J. Gil. *An. Quim.* **84B**, 231(1988)

COMPARISON OF DIFFERENT SOLID PHASES FOR AN ELISA BASED DETERMINATION OF AFLATOXIN B

MARIA PESAVENTO

Dipartimento di Chimica Generale, V. Taramelli 12, 27100 PAVIA, (Italy)

and

BOB CARTER

Universal Sensors, 5258 Veterans Blvd., Suite D., Metairie, LA 70006

Abstract- An ELISA has been set up for the determination of Aflatoxin B1, based on a competition reaction, in which the antigen, covalently bound to a solid phase, competes with the antigen in the sample for binding a limited amount of antibody in solution. The solid phase is a commercially available hydrophilic membrane chemically activated, which can be re-used at least 30 times without losing its ability of linking the specific antibody, after dissociation of the Ag-Ab complex with a 0.1 M glycine solution at pH=2.3. The equilibrium of the Ag-Ab complex formation was studied both in the solid and in the solution phase, and the results obtained in the case of the covalently coated solid phase compared with those obtained with a conventional polystyrene surface on which the antigen is physically sorbed.

I. INTRODUCTION

The use of immunoreagents linked to a solid phase makes easy the separation of the complex antigen-antibody from the free reagent. For instance, the ELISA method (Enzyme linked immunosorbent assay), characterized by the fact that one of the immunoreagents is linked to a solid phase and the marker is an enzyme, has been used for a large number of assays since its first description by Engvall and Perlmann [1]. Moreover, some more recently developed devices, for instance immunosensors, require the use of an immunoreagent immobilized to a solid support. Thus, the characterization of the equilibria involved in the formation and dissociation of the antigen-antibody complex between two distinct phases is of great importance for improvement of well established methods and future developments. This point has been here investigated by an ELISA procedure applied to the determination of a particular antigen, making use of two different solid phases.

Usually ELISAs are carried out in polystyrene microwell plates, onto which the immunoreagent is first fixed by physical adsorption techniques. The wells are, however, not reusable, since during the assay procedure or during the dissociation they are in contact with solutions of

G.G. Guilbault and M. Mascini (eds.), Uses of Immobilized Biological Compounds, 443–452.

different composition (pH, ionic and proteic composition,...), which causes a partial desorption of the physically adsorbed immunoreagent. On the contrary some solid phases have been recently introduced, and are commercially available from a number of companies, to which the immunoreagent can be linked covalently, so giving coated solid phases which are at least in principle reusable. These solid phases are potentially very useful for routine applications, even in laboratories which are not particularly specialized in immobilization techniques. Moreover they are well suited for use in immunosensors. As example, two commercially available membranes can be cited here, the Immobilon AV membrane, by Millipore, which has been considered in the present investigation, and is a hydrophilic polyvinylidene difluoride solid phase, and the membrane Immunodyne, consisting of chemically preactivated nylon 66, commercialized by Pall. The nature of the reactive group is not known, however big or small molecules can be linked to the membrane through primary aminic groups. Many other solid phases are now commercially available and can be used for covalently linking immunoreagents, for instance beads for affinity chromatography or microplates preactivated for covalenly binding proteins. Some of these commercially available activated surfaces have been recently examined[2] for the formation and dissociation of Ag-Ab complex in the solid phase. It has been shown that the surfaces to which the immunoreagent (a protein in that case) is fixed by a covalent bond give the best retention of proteins, and also the best reassociation.

The present investigation is devoted to the comparison of an usual ELISA for an apten on microplate (Nunc-Immunoplates Maxisorp) with one performed by covalently binding the immunoreagent to one of the preactivated membranes which are now commercially available, Immobilon AV.

II CHARACTERISTICS OF THE CONSIDERED ANALYTE

The particular assay here examined is developped for the determination of aflatoxins, some of the most cancerogenous substances so far known, at least for some species of animals. Thus their content must be severely controlled in food and animal feed. Aflatoxins are products of the metabolism of some fungi of the Aspergillus type, and they are often but not always present in agricultural commodities contaminated by these fungi. In the present investigation Aflatoxin B_1, probably the most dangerous among the different aflatoxins, has been considered. It is a low moleculare weight substance (PM=300) with very low solubility in water. The problem is that the contamination is quite erratic, and a correct sampling procedure requires the analysis of many sub-samples. Thus an analytical procedure rapid, inexpensive, but accurate and with a low detection limit is required. As the

ELISAs present these characteristics, many different protocols based on this method have been proposed for aflatoxins determination.[3] Some of them, which are commercially available, are spot tests, which only indicate the presence of the toxin above or below a given concentration, while others give quantitative results, and are carried out on conventional microtiter plates.

Due to the low solubility in water of aflatoxins, they must be extracted from solid samples with organic solvents, and the extracted solution should be analyzed preferably after a limited number of further purification steps. The effect of the presence of an organic solvent in the samples which must be finally assayed by the ELISA procedure on microplate and on membrane must be tested.

III IMMOBILIZATION PROCEDURE

In the ELISA test here considered the immunoreagent bound to the solid phase is the antigen. Aflatoxin B_1 is a small molecule, which can't be physically adsorbed on polystirene surfaces, and does not carry any group that can be covalently bound to Immobilon AV. Instead its conjugate with BSA was used for the immobilization on both surfaces.

The immobilization and blocking procedures were those suggested by the seller, both in the case of microplates (Nunc-Immuno Plates MaxiSorp) and of Immobilon AV (Millipore).

The membrane can be used simply by cutting it into small squares (5x5mm) or circles, and putting them in polyethylene disposable tubes for following treatments. They can also be fixed at the top of a plastic stick by a O-ring, and then used by immersion in plastic tubes. The coating in both cases is performed simply by contacting the microwell or the membrane with a small volume (0.010-0.100 mL) of a solution containing BSA-Aflatoxin B_1. The solution was DPBS (Dulbecco phosphate buffered saline, pH=7.4). The incubation time was overnight for the microtitration plates and 20' for Immobilon AV, at room temperature.

The residual active sites were blocked with 1% BSA in DPBS for 60' at room temperature. Then the antibody can be linked to the solid phase mainly by specific interaction with the antigen, as demonstrated by the low blank values obtained (Fig.1). The Immobilon AV membrane can be blocked not only with proteins, as microplates, but also with small molecules containg primary amines. This could allow a modulation of the surface properties, for instance its charge, which can be useful in some applications. However it has been found that in the case of Aflatoxin B_1 test, the best blocking agents were those carrying a negative charge in the considered solution (DPBS; pH= 7.4). Thus finally the blocking with albumin was adopted, the same used for microplates.

IV DETERMINATION OF THE ANALYTE BY COMPETITION ELISA PROCEDURE.

A competition ELISA was used for the quantitative determination of Aflatoxin B_1. The assay consisted in the competition of Aflatoxin B_1 in solution with that bound to the solid phase for a limited amount of antibody anti-Aflatoxin B_1. A secondary antibody (anti rabbit-IgG from goat) labeled with alkaline phosphatase was used for the determination of anti-Aflatoxin B_1 antibody linked to the solid phase after the incubation time.

Table 1

Protocol for the determination of Aflatoxin B_1 by competition ELISA on microplate activated with BSA-Aflatoxin B_1, 5-0.5 ng in 0.100 mL DPBS (Dulbecco phosphate buffer saline, pH= 7.4), overnight incubation at room temperature.

Each determination is made in triplicate.

I incubation with antibody anti-Aflatoxin B_1 (15'-180' incubation at 37 C):

0.05mL of DB containing stock Ab diluted 1:5000-1:20000

+

0.05mL solvent (water/acetonitrile=80/20): direct ELISA

or

0.05mL solvent (water/acetonitrile=80/20) containing AflatoxinB_1:competition ELISA

—

washing three times with 0.100 mL DB

—

II incubation with anti-IgG(rabbit) labeled with alkaline phosphatase (60' incubation at 37 C):

0.1mL of DBT containing stock anti-IgG diluted 1:250-1:1000

—

washing three times with 0.100 mL DB

—

incubation with substrate solution (10 mg PNP per mL of carbonate buffer at pH=9.8), (15' incubation at 37 C): 0.1mL of substrate

Reaction is stopped by adding the substrate solution to 1.8 mL of 1 M NaOH

Absorbance readings at 405 nm, l=1 cm.

DB: DPBS+ 1% BSA .

The protocol for the microplate assay is reported in Table 1. That for Immobilon AV is only slightly different.

The enzymatic activity of the solid phase inversely depends on the concentration of Aflatoxin B_1 in solution. It is measured by contacting the solid phase with a substrate solution for alkaline phosphatase, and measuring its absorbance after a proper incubation time.

As an example in Fig.1 some dose/response curves are reported, obtained by assaying standard solutions containing incresing amounts of Aflatoxin B_1. They show that actually the Aflatoxin B_1 in the solid phase, both microwell and Immobilon AV, is able to compete with Aflatoxin B_1in solution.

The curves reported in Fig 1 can be used as calibration curves for the determination of Aflatoxin B_1 content of samples, in the dynamic range from around 0.1 to 100 ppb of Aflatoxin B_1, which can be useful for practical applications. The dynamic range greatly depends on the assay conditions, for instance the amount of antigen bound to the solid phase, as shown in Fig. 1, but also the dilution of the antibody, and the incubation time.

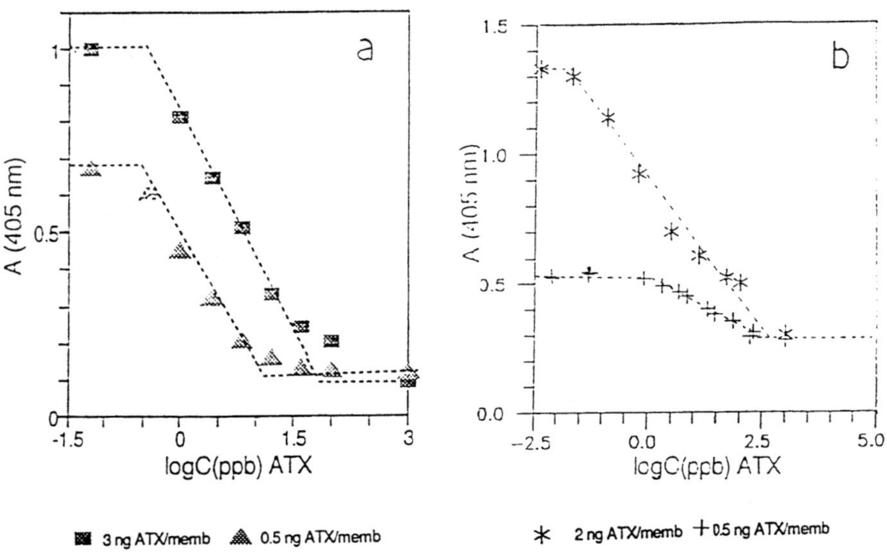

Fig. 1

Competition ELISA in microplate (a) and in Immobilon AV (b).

Ab dilution: 1:5000; I incubation: t=180'.

The incubation with the specific antibody was performed in the presence of acetonitrile, which is often the organic solvent of choice for the extraction of aflatoxins from solid samples. An important difference between the two solid phases was detected when the presence of a non aqueous solvent was considered, which is a point of importance when dealing with all the analytes with low solubility in water. With microplates the presence of amounts of acetonitrile higher than 20% in the incubation with the antibody reduces the formation of the complex Ag-Ab, while with Immobilon AV the final absorbance slightly increased.

Considering the good performances of the Immobilon AV membranes, the assay with the covalently bound antigen was applied to the determination of Aflatoxin B_1 in real samples of spices. Apparently there was not any matrix effect when using Immobilon AV membranes, even when analyzing directly the raw extract. However this point is not of general validity, and should be tested for any particular protocol and sample.

V REUTILIZATION OF THE COVALENTLY ACTIVATED IMMOBILON MEMBRANE

The activated membranes can be used many times, after treatment with a dissociation solution, which is able to dissociate the complex Ag-Ab, while leaving the conjugate BSA-Aflatoxin B_1 linked to the solid phase. This possibility has been previously investigated by others [2] in the case of the immobilization of an immunoglobulin, obtaining good results with a number of different dissociating solutions, both in term of dissociation of the complex and of reassociation. 0.1 M glycine at pH=2.3 was successfully used here. 30' soaking in this solution at room temperature completely dissociates the Ag-Ab complex, while maintaing the coating of the membrane by BSA-Aflatoxin B_1 for subsequent determinations. It has been observed that these membranes can be used for many determinations, at least 30 times in a period of a month.

VI CHARACTERISATION OF THE FORMATION OF Ag-Ab COMPLEX IN THE SOLID PHASE

The antibody combines to Ag in the solid phase according to the equilibrium:

$$Ab + n\,T' <==> AbT_n' \qquad (1)$$

with a formation coefficient:

$$K' = \frac{[AbT_n']}{[Ab]\,[T']^n} \qquad (2)$$

where Aflatoxin B_1 is indicated by T, the prime ' indicates the species in the solid phase, and n can be 1 or 2, if Ab is a monomeric antibody, as IgG are. The concentrations in solution are expressed in mol per L (molarity, M), while those in solid phase in arbitrary units, for instance mmol per g of dry solid. The grams of dry solid will be indicated with w.

Relationship 2 is similar to that proposed by Scatchard[4] for the combination of small molecules with proteins, but the unit which combines with the antigen is here assumed to be the whole IgG molecule, not the single combination site. This is convenient because the quantity finally determined in the ELISA procedure here considered is the absorbance of the substrate solution (A), which is directly proportional to the molar amount of Ab linked to the solid phase (AbT_n').

a) Direct ELISA experiments

The values of n and K' can be determined by experiments in which the coated solid phase is assayed directly with Ab in solution not containing antigen (direct ELISA method). In this case the following mass equilibria hold :

$$[AbT_n']\, w = c^* A \qquad 3$$

$$[Ab] = c_{Ab} - c^* A/V \qquad 4$$

$$[T']\, w = c_T'\, w - n\, c^* A \qquad 5$$

where A is the absorbance of the substrate solution at 405 nm for the test here described, and c^* is the proportionality coefficient which relates the final absorbance to the total amount of Ab in the solid phase. c_T' is the concentration of the antigen in the solid phase, and c_{Ab} is the concentration of Ab referred to the solution volume V. It is given by: $c_{Ab} = c_s/d$, where c_s is the concentration of Ab in the stock solution, and d is the dilution factor. The direct ELISA experiments are carried out by keeping constant c_{Ab}, while the total amount of toxin used for immobilization ($c_T'w$) is varied. At high amounts of T in solid phase, all the antibody present in solution is quantitatively linked to it, so that the enzymatic activity of the surface depends on the total amount of Ab, and is independent of the amount of toxin. As a consequence, a plateau value of absorbance, A_T, in the curves A vs $c_T'w$ is reached which is related to total amount of Ab ($T_{Ab} = c_{Ab}\, V$) by the following relationship:

$$c_{Ab}\, V = c^* A_T \qquad 6$$

From the plateau value the quantity $c_s/c^* = A_T\, d/\, V$ can be evaluated. It is a measure of the titer of the antiserum in the particular protocol adopted. However a value of c_s/c^* independent of the Ab dilution is obtained only at Ab dilutions higher than around 5000. Probably a monolayer of Ab is formed on the solid surface at high concentration of Ab, which sterically prevent any further combination. The values obtained for the titer of the antibody stock solution (c_s/c^*) are 1.6 10^5 and

4.0 10^4 respectively for the microplate and the Immobilon AV procedure. This difference is ascribable to the sligtly different protocols adopted (the dilution of the labeled antibody is 1:250 in the plates, and 1:1000 with Immobilon AV; the incubation volumes are higher with Immobilin AV). Small and not controlled changes in temperature, time, and other casual errors, cause variations on c^* from an experiment to another, both in the case of microplates and Immobilon AV.

The fraction of Ab bound to the solid phase is given by

$$B_{Ab}/T_{Ab} = \frac{[AbT_n'] \, w}{[AbT_n'] \, w + [Ab] \, V} = \frac{1}{1 + \dfrac{V}{K' \, [T']^n \, w}} \qquad 7$$

where B_{Ab} is the amount (mmol) of Ab bound to the solid phase. As

$$[T'] \, w = c_T' \, w - n \, c^* A \qquad 8$$

eqn 7 can be written as

$$(A/(A_T-A))^{1/n} / c_T' \, w = (K'/w^{n-1}V)^{1/n} - (K'/w^{n-1}V)^{1/n} \, n \, c^* A / c_T' \, w \qquad 9$$

The term to the left can be evaluated from

$$T_{Ab}/B_{Ab} - 1 = c_S V / d \, c^* A - 1 \qquad 10$$

where the ratio c_S/c^* is known from the previous direct ELISA experiments.

In all the experiments here considered $(A/(A_T-A))^{1/n} / c_T'w$ was independent of $A/c_T' \, w$, which means that $n \, c^* A / c_T' \, w$ was much lower than 1, i.e. that the total amount of Aflatoxin B_1 in solid phase is much higher than the Ag-Ab complex, so that $[T'] = c_T'$. In this case the data can be also examined by a logaritmic relationship obtained from eqn. 7

$$\log (T_{Ab}/B_{Ab} - 1) = - \log K'w^{1-n}/V - n \log c_T'w \qquad 11$$

from which n and $\log K'w^{1-n}$ can be obtained simultaneously.

Actually straight lines were obtained in the Aflatoxin B_1 test when the antigen amounts in solid phase were higher than $10^{-8.5}$ -$10^{-8.8}$ mmoles. An important difference between the two solids is the stoichiometry of the complexes: while in the plate $n=1$, in Immobilon $n=2$. The values obtained for $K'w^{1-n}$ are reported in Table 2.

Useful informations can also be obtained by testing the same amount of toxin in solid phase with different dilutions of Ab. In this case eqn 7 can be written as following

$$A = c_S/(d \, c^* \, (1/V - 1/K' \, (w \, c_T' - n \, c^* A)^n)) \qquad 12$$

By plotting $\log A$ vs $\log 1/d$, a straight line with slope +1 should be obtained if w

c_T' is much higher than $n\,c^*\,A$, i.e. $[T'] = c_T'$. In the case of the Aflatoxin B_1 test straight lines with slope +1 were obtained for dilutions higher than 5000. Moreover two different amounts of toxin linked to the solid phase have been tested (3 and 0.5 ng of Afaltoxin B_1), obtaining points on the same straight line. This indicates that $1/V$ is much higher than $1/K'(w\,c_T)^n$. Thus c_s/c^* can be evaluated from the intercept of the straight line. The values obtained are respectively 2.6 10^5 and 5.8 10^4 for microplate and Immobilon AV.

b) Competition ELISA experiments

In the presence of antigen in solution, the distribution equilibrium of Ab between the solid phase and the solution can be described by a relationship similar to eqn 7, but taking into account also the formation of Ag-Ab complex in aqueous solution:

$$B_{Ab}/T_{Ab} = \frac{[AbT_n']\,w}{[AbT_n']\,w + [AbT_m]\,V + [Ab]\,V} \qquad 13$$

The equilibrium for the formation of the Ag-Ab complex in the solid phase is that reported above, while that in solution is

$$Ab + m\,T \;<==>\; AbT_m \qquad 14$$

with a formation coefficient: $K = \dfrac{[AbT_m]}{[Ab]\,[T]^m}$

Thus eqn 13 can be written as following

$$B_{Ab}/T_{Ab} = \frac{1}{1 + \dfrac{K\,[T]^m\,V}{K'\;[T']^n w} + \dfrac{V w^{n-1}}{K'\;[T']^n\,w^n}} \qquad 15$$

This relationship is very similar to the usual competition equation. From it the following relationship, linear in $[T]^m$, is obtainable:

$$T_{Ab}/B_{Ab} - 1 = \frac{K\,[T]^m\,V}{K'\;[T']^n w} + \frac{V w^{n-1}}{K'\;[T']^n w^n} \qquad 16$$

The term to the left can be evaluated as described above by eqn.9.

Eqn 16 can be used for the evaluation of both K' and K if the concentration of the free toxin [T] is proportional or equal to c_T, as it is probably true when the Ab concentration is much lower than that of the antigen. By plotting $T_{Ab}/B_{Ab} - 1$ against c_T straight lines were actually obtained in the case of microplate and Immobilon tests for Aflatoxin B_1.

This experiment allows to evaluate at the same time the complex formation coefficient of the Ag-Ab complex in solution, and that in the solid phase if the stoichiometric coefficient *n* is known. The results are reported in Table 2, and are in agreement with the complex formation coefficients in the solid phase evaluateby the direct ELISA experiments, considering the statistic uncertainty. The value obtained for *K* is very similar for the two solid phases, and in the case of microplate is also very similar to *K'*, the complex formation coefficient in the solid phase. This could indicate that the hydrophobic nature of the surface does not influence the formation of the Ag-Ab complex.

The values of *K* and *K'* obtained in different experiments are remarkably constants, while c_s/C^* varies, more in the case of the microplate assay than of Immobilon AV.

Table 2

Formation coefficients of Ag-Ab complex in solution and in solid phase for Aflatoxin B_1 test.

		direct ELISA	competition ELISA	
solid	n,m	$K' w^{1-n}$	$K' w^{1-n}$	K
microplate	1,1	$5\ 10^7$	$7.04\ 10^7$ (dil 1:5000)	$8.93\ 10^7$
	1,1	$5\ 10^7$	$8.47\ 10^7$ (dil 1:20000)	$1.58\ 10^8$
Immobilon AV	2,1	$4.8\ 10^{16}$	$2.51\ 10^{16}$	$2.67\ 10^7$

ACKNOWLEDGMENT

Work supported by the CNR (National Research Council of Italy)

REFERENCES

1. E. Engvall, P. Perlmann, Immunochemistry, 8, 871 (1971).
2. G.C. Blanchard, C. G. Taylor, B.R. Busey, M.L. Williamson, J. of Immun. Methods, 139, 263, (1990).
3. J.J. Pestka, J. Assoc. Off. Anal. Chem., 71, 1075 (1988).
4. G. Scatchard, Ann. N. Y. Acad. Sci., 51.699 (1949).

Polystyrene and Poly(styrene/acrolein) Latexes with Immobilized Proteins as a Basis of the Diagnostic Agglutination Test

Teresa Basinska, Stanislaw Slomkowski
Center of Molecular and Macromolecular Studies, Polish Academy of Sciences, Sienkiewicza 112, 90-363 Lodz, Poland

Properties of the poly(styrene/acrolein) latexes (PSA), with controlled surface concentration of aldehyde groups, as carriers of proteins (human serum albumin (HSA) and gamma globulins (γG)) are described. It has been found that the maximal surface concentration of attached proteins decreases with increased fraction of polyacrolein in the surface layer. This suggests that initially proteins are attached flatly to the latex particles. At the later stages the already adsorbed protein macromolecules change their orientation, become packed protruding from the surface, and make possible accommodation of new protein macromolecules coming from the solution. Such rearrangement is suppressed in the case of proteins attached covalently. Determination of the surface concentrations of proteins, adsorbed and covalently immobilized onto the PSA latexes, indicate that for the latexes with higher content of polyacrolein the surface concentration of adsorbed protein is lower. From the adsorption and immobilization isotherms for HSA it follows that the physical adsorption is preferred. Apparently, the hydrophobic patches, suitable for adsorption, are covered first and only thereafter the covalent immobilization begins to play an important role. For γG the different picture has been observed. Namely, the areas suitable for the covalent immobilization are covered first. Presumably, denaturation of the "soft" HSA macromolecules, at the latex-solvent interface, facilitates the hydrophobic protein-latex interactions. In the case of γG macromolecules, which are less prone to denaturation, the covalent immobilization competes successfully with adsorption. The described latexes were used in the model test for detection of anti-HSA. Some were used for manufacturing of the diagnostic test for determination of fibrinogen degradation products in the blood serum and in urine.

INTRODUCTION

Polymeric latexes found many applications in medical and veterinary diagnostics, mainly in the agglutination tests [1]. In the simplest form the agglutination test consists on mixing the droplet of the suspension of latex particles with attached antibodies with the droplet of liquid to be analyzed. In the presence of appropriate antibodies the interactions of antigens with antibodies attached to the latex particles result in latex aggregation. Aggregation is usually detected with the naked eye. In spite of the inferior sensitivity to the contemporary ELISA tests the simple agglutination tests are still being widely used. This is due to the unusual simplicity of equipment required for the analysis, often reduced to the test-tubes and/or glass or plastic slides provided together with the kit containing latexes with appropriate proteins and control solutions, and to the short time of analysis (ca 5 min). The disadvantage of conventional latex tests is related to the difficulty in the automatization and the necessity to store latex suspensions in refrigerator. Nevertheless, often when the simple, fast, and inexpensive diagnostic test is required the latex test is chosen. It is worth to note that the recent progress in application of latexes resulted in the so called "dry latex test"

G.G. Guilbault and M. Mascini (eds.), Uses of Immobilized Biological Compounds, 453–462.

combining in one device all reagents in the dry form. In these systems the only liquid is the analyzed liquid and the device can be designed to provide the answer in the simple form of "+" or "-" sign. Such tests, based on the polypyrrole particles placed onto the nitrocellulose blotting paper, were designed for detection of the surface antigen of the hepatitis B virus and anti-HIV antibody [2].

Till now the polystyrene latex has been used most often in the diagnostic tests. To the pure polystyrene particles proteins could be attached only by simple adsorption and, in effect, at certain conditions leak during prolonged storage. Thus, in many instances, the covalent immobilization of protein macromolecules onto the latex particles is preferred to the physical adsorption. Usually $-NH_2$ groups from lysine or arginine moieties of protein macromolecules are involved into reactions resulting in the covalent immobilization of proteins. Latexes with various active groups (e.g. amino, hydroxide, carboxyl, and aldehyde), suitable for the covalent immobilization of protein macromolecules have been synthesized [3,4]. The hydroxyl and carbonyl groups require activation prior to the protein immobilization (e.g. the first with cyanogen bromide and the latter with carbodiimide). In the case of protein immobilization onto latexes with amino groups the dialdehyde linkers have to be used. The advantage of latexes with aldehyde groups stems from the possibility to use these particles for protein immobilization without any activation (formation of Schiff base). Reaction of aldehyde and amino groups leading to Schiff base is reversible. However, when the protein molecule is attached via several imino linkages the immobilization becomes, for practical purposes, permanent. Always, when it is required, the imino groups of the Schiff base could be reduced to amino groups in reaction with cyanoborohydride.

Latexes with aldehyde groups have been synthesized by polymerization of acrolein [5,6,7], copolymerization of acrolein and hydroxyethyl methacrylate (HEMA) [8], and polymerization of acrolein [3,9] or p-formylstyrene [10] on seeds of the polystyrene latex. However, there are data related to the immobilization of horseradish peroxidase onto the latexes from poly-HEMA and polyacrolein indicating that proteins, upon immobilization to particles with high fraction of polyacrolein, lose their activity. Of course, grafting of polyacrolein onto the polystyrene latex results in particles with polyacrolein shell. For the incoming proteins these particles are not different from the pure polyacrolein latex.

Recently, we described the mutual polymerization of styrene and acrolein giving the possibility to obtain the poly(styrene/acrolein) latexes with controlled fraction of the polyacrolein in the surface layer [11]. In this paper we are going to describe the results of our recent studies on the attachment of chosen proteins to these particles. Human serum albumin (HSA), one of the common proteins in the blood serum, and gamma globulins (γG), important protein of the immune systems were chosen for our studies. Latexes were used also for immobilization of the rabbit polyclonal antibodies against the degradation products of human fibrinogen (FgDP), element of the agglutination test for detection of FgDP.

Experimental part

Polystyrene (PS), poly(styrene/acrolein) (PSA), and polyacrolein (PAC) latexes were synthesized and characterized as it was described in our earlier paper [11]. Concentration of the anionic groups at the surface of latex particles was determined conductometrically. Fraction of polyacrolein in the surface layer (PAC) was determined by using the XPS method and the surface concentration of aldehyde groups, accessible for reaction with amine, was measured by using the characteristic analytical reaction of aldehyde groups with 2,4-dinitrophenylhydrazine.

Human serum albumin (HSA) (Sigma, Cohn fraction V), human gamma globulins (γG) (Cohn fractions II and III), and goat anti-HSA serum (Sigma) were used as received. FPLC traces (superose-12 column, eluent phosphate buffer pH = 7, flow rate 0.5 ml/min) indicated that HSA contained protein macromolecules in the form of monomer, dimer, and oligomers in proportion 77.2%, 16.1%, and 6.7% respectively.

Human fibrinogen (Fg), plasmin induced degradation products of human fibrinogen (FgDP), and polyclonal rabbit antibodies against FgDP (anti-FgDP) were obtained from Prof. C.Cierniewski (Medical Academy, Lodz, Poland).

HSA and γG were attached to the latexes by incubation of the required amounts of protein and latex in PBS (usually 10 ml) at room temperature during 24 h. In some experiments anti-FgDP was attached to the latex in 0.1 **M** glycine buffer (pH = 8.2). The amount of protein attached to the latex has been determined by using the modified Lowry method [12,13].

RESULTS

Properties of the Poly(styrene/acrolein) Latexes

Fraction of polyacrolein in the surface layer (PAC), surface concentrations of the anionic ($-SO_4^-$ groups from potassium persulfate used for initiation of the polymerization in the synthesis of latexes), surface concentration of the accessible aldehyde groups ([-CHO]) (accessible for the 2,4-dinitrophenylhydrazine label), values of number average diameters D_n, and polydispersity factor (D_v/D_n) are collected in Table 1.

Table 1. Properties of the poly(styrene/acrolein) latexes.

Latex	f(PAC)	$\bar{D}_n$ *) μm	$\bar{D}_v/\bar{D}_n$	$[-SO_4^-]$ x10^6 mol/m^2	[-CHO] x10^6 mol/m^2
PS	0	0.52	1.005	3.39	0
PSA1	0.50	0.51	1.010	2.52	2.20
PSA2	0.63	0.49	1.002	1.38	2.28
PSA3	0.78	0.36	1.002	2.57	1.76
PSA4	0.84	0.38	1.007	2.14	2.51
PAC	1.00	0.31	1.003	3.24	1.96

*) $\bar{D}_n$, $\bar{D}_v$ number and volume averaged diameters determined from SEM micrographs

The detailed description of the synthetic procedures and methods used for the latex characterization are given elsewhere [11]. Here we would like to mention only that the polydispersity index (D_v/D_n) close to 1 indicates that latexes are monodisperse. There is correlation between the fraction of hydrophilic polyacrolein in the surface layer (f(PAC)) and the diameter of the latex particles. The higher is value of f(PAC) the smaller is the diameter.

It might be surprising that the surface concentration of the aldehyde groups accessible for 2,4-dinitrophenylhydrazine label is independent from f(PAC) (for f(PAC) > 0.5). However, one have to remember that not all of the aldehyde groups on the surface could be involved in reaction, due to the steric hindrance from the already immobilized 2,4-dinitrophenylhydrazine molecules.

Attachment of HSA and γG onto the PS, PSA, and PAC latexes

The dependence of the surface concentration of attached protein (Γ) on the concentration of protein in solution has been studied in a series of experiments with the constant concentration of the latexes (3 mg/ml) and with varied concentration of protein ([HSA] and/or [γG]).

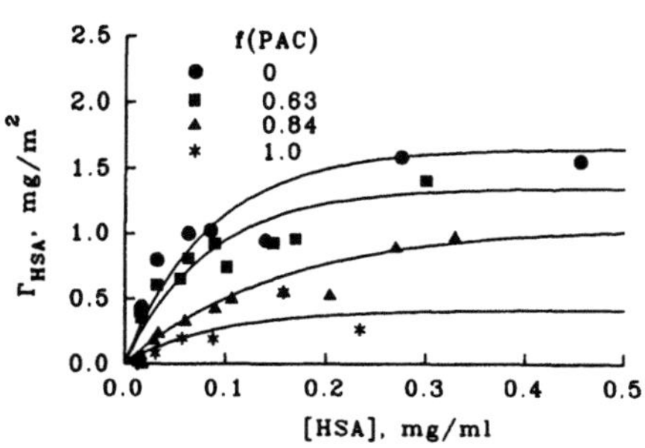

Fig.1 Dependence of Γ(HSA) on [HSA], [Latex] = $3 \cdot 10^{-3}$ g/ml

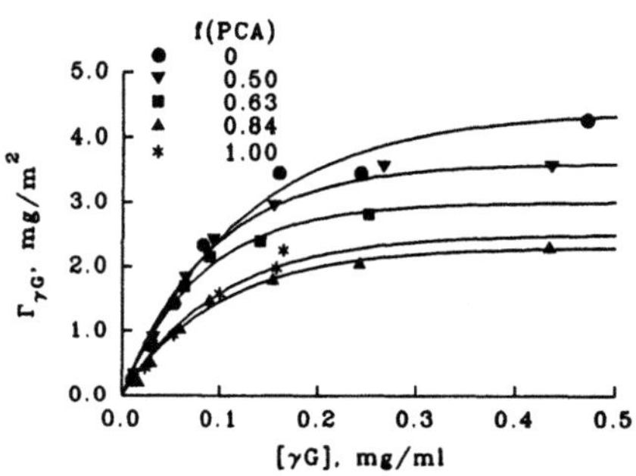

Fig.2 Dependence of Γ(γG) on [γG], [Latex] = $3 \cdot 10^{-3}$ g/ml

Plots, collected in Fig. 1 and Fig. 2, clearly indicate that for the same concentrations of protein in solution the surface concentration of attached protein is lower for latexes with higher fraction of polyacrolein in the surface layer. Presumably, initially the protein macromolecules became attached flatly, maximizing their contact with the polymer surface. The new protein macromolecules could be accommodated only when the already attached ones change their orientation making space for them. Such reorientation, rather easy for proteins adsorbed due to the physical interactions, becomes impossible for macromolecules immobilized covalently. Thus, for latexes with higher fraction of polyacrolein, providing aldehyde groups for the covalent immobilization, the lower surface concentration of attached proteins could be expected.

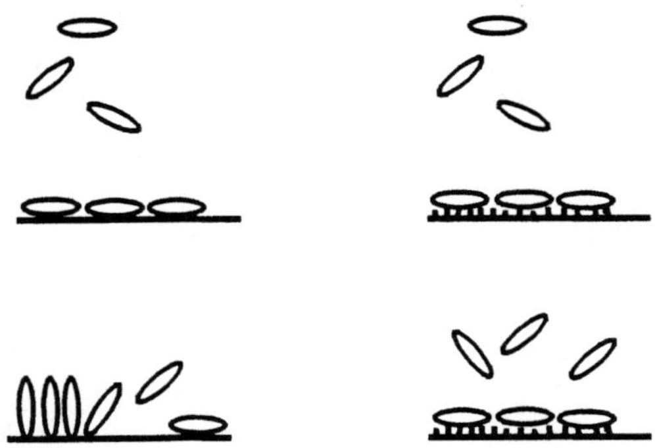

Adsorption Covalent immobilization

Fig.3 Schematic illustration of protein attachment by adsorption and covalent immobilization.

Figures 1 and 2 illustrate the total attachment of HSA and γG to the PS, PSA, and PAC latexes. However, for any particular kind of particles it is essential to know the proportion between the adsorbed and covalently immobilized proteins. It is known that the surface active agents are capable to remove adsorbed proteins, leaving on the surface the covalently immobilized molecules. We found that within 2 h the 3% solution of sodium dodecylsulfate (SDS) is capable to remove whole HSA from the PS latex. In the case of γG and PS latex the solution of 3% SDS removed the main part of attached protein. However, in all experiments the small part of γG remained undetached, regardless from the time of treatment with SDS (up to 24 h) and sonication. The surface concentration of γG attached permanently to PS latex never exceeded $2 \cdot 10^{-4}$ g/m^2 ($1.25 \cdot 10^{-9}$ mol/m^2). It is possible that the polystyrene chain ends equipped with $-SO_4^-$ groups function as haptens.

In the case of HSA attached to the PS, PSA, and PAC latexes we assumed that only the adsorbed protein could be removed by SDS. Thus, determination of the total surface concentration of attached protein (Γ_t) and of the surface concentration of protein which could be removed with SDS (Γ_a determined by measuring the amount of protein desorbed from the given amount of latex) made possible calculation of the surface concentration of HSA immobilized covalently ($\Gamma_c = \Gamma_t - \Gamma_a$). Similar reasoning could be used also for γG. However, in this case Γ_c will represent protein macromolecules immobilized due to their reaction with aldehyde groups of polyacrolein and strongly attached to some other specific sites. Fortunately, the latter constituted not more than 15% of γG immobilized to any of the PSA latexes.

The typical plots of Γ_t, Γ_c, and Γ_a vs the protein concentration in solution, registered for HSA and γG attached to the PSA2 latex, are illustrated in Fig. 4 and Fig. 5. It is worth to note that in the case of HSA the surface concentration of adsorbed protein (Γ_a) approaches plateau for lower [HSA] than the surface concentration of protein immobilized covalently (Γ_c). In the case of γG the reverse relation is true. Apparently, the "soft" HSA macromolecules with the hydrophobic domains responsible for transport of lypophilic compounds [14,15] can change easily their conformation on the hydrophobic polymer -

solvent interface and in effect are preferentially adsorbed onto the hydrophobic patches of polystyrene. Gamma globulins, less prone to the conformational changes than HSA [16], are less suitable for adsorption and more effectively immobilized by covalent imino linkages.

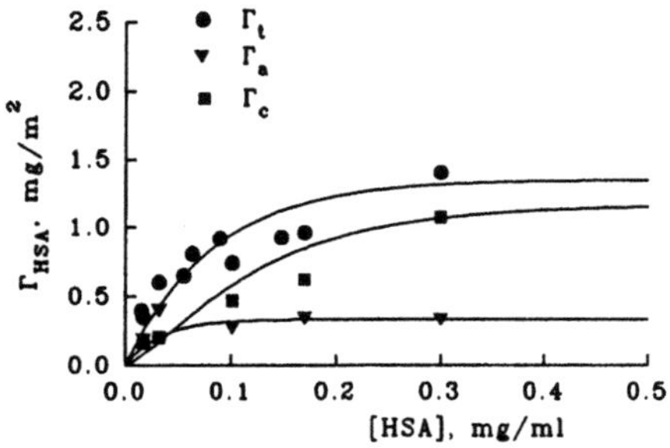

Fig. 4 Isotherms for the overall attached (Γ_t), adsorbed (Γ_a), and covalently immobilized (Γ_c) HSA. [PSA2] = 3 mg/ml

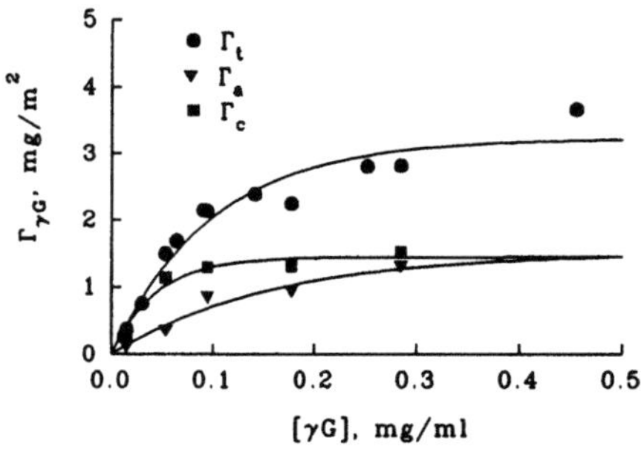

Fig. 5 Isotherms for the overall attached (Γ_t), adsorbed (Γ_a), and covalently immobilized (Γ_c) γG. [PSA2] = 3 mg/ml

We expected that the maximal surface concentrations of covalently immobilized and adsorbed proteins will be related to the fraction of polyacrolein in the surface layer.

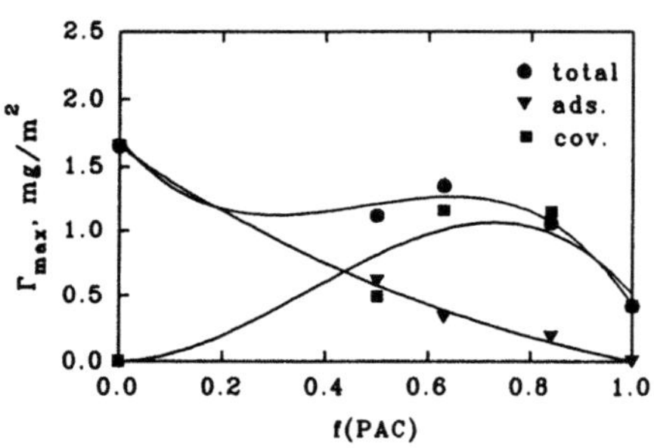

Fig. 6 Attachment of HSA. Dependence of Γ_t(max), Γ_a(max), and Γ_c(max) on the fraction of polyacrolein in the surface layer.

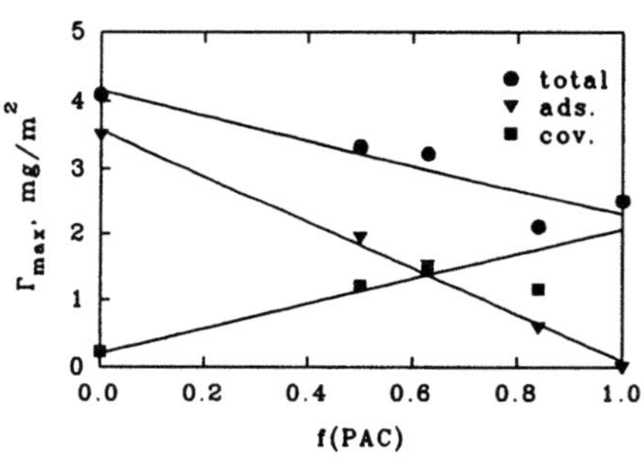

Fig. 7 Attachment of γG. Dependence of Γ_t(max), Γ_a(max), and Γ_c(max) on the fraction of polyacrolein in the surface layer.

Plots collected in Figures 6 and 7 clearly indicate that the maximal surface concentration of adsorbed proteins steadily decreases with increasing f(PAC). The maximal surface concentration of covalently immobilized γG increases proportionally with f(PAC). It is worth to note that in the case of HSA the relation between the maximal surface concentration of covalently immobilized protein and the fraction of polyacrolein in the surface layer is more complex. For HSA Γ_c(max) initially increases with increasing f(PAC) then, after passing through maximum, begins to decrease. This observation conforms to the model assuming higher degree of the conformational changes of HSA at the polymer-solvent interface than in the case of γG. When f(PAC) is still low its increase facilitates covalent immobilization of higher number of HSA macromolecules. However, when the fraction of polyacrolein is high the further increase of f(PAC) leads to more tight immobilization of denaturated HSA macromolecules, reducing their ability to change conformation making free space for new protein macromolecules. This conforms to the picture given in Fig. 3 illustrating reduction of the protein attachment by covalent immobilization in comparison with much more lose hydrophobic adsorption.

Model agglutination test based on the poly(styrene/acrolein) latexes

The potential applicability of the poly(styrene/acrolein) latexes for the slide agglutination test has been verified by using suspensions of particles with attached HSA for detection of anti-HSA in the serum from goat immunized with HSA. In these experiments we used the PS, PSA1, and PSA2 latexes. Latexes were incubated with HSA at room temperature for 24 h. Concentrations of PS latex and of HSA were equal to 15 mg/ml and 0.9 mg/ml. In the case of the PSA1 and PSA2 latexes concentration of latex was 7 mg/ml and concentration of HSA was equal to 8.75 mg/ml. After incubation, particles with attached HSA were purified from unbound protein by centrifugation, separation from supernatant, and washing with fresh portions of PBS buffer. This procedure was repeated five times. The last supernatant was free from protein. The purified latexes with attached HSA were resuspended in PBS. Their concentrations were determined gravimetrically (the known volume of the latex suspension was dried to the constant weight). The observation under optical microscope indicated that the suspensions of latexes ready to be used for detection of anti-HSA did not contain any aggregated particles. Before using in further experiments the latex-HSA suspensions were stored at +4 °C.

Sample (A1151 - Sigma) of the goat anti-HSA serum was dissolved in 2 ml PBS and from this stock solution several solutions with lower concentrations of anti-HSA have been prepared.

In each agglutination experiment 20 μl of latex and 20 μl of anti-HSA were mixed on the test slide. In the control experiment 20 μl of latex was mixed with 20 μl of PBS alone. After 4 min. the mixtures were checked for aggregation of the latex. Latexes with added PBS (control) gave stable suspensions without any sign of aggregation. In the case of latexes with attached HSA mixed with solutions of anti-HSA aggregation was related to the anti-HSA concentration. Results are given in Table 2.

Data collected in Table 2 indicate that PSA latexes, with aldehyde groups suitable for covalent immobilization of proteins via the Schiff base linkages, could be successfully used in the agglutination diagnostic tests for the detection of anti-HSA. It is worth to note that the sensitivity of the test based on PSA2 (fraction of polyacrolein in the surface layer f(PAC) = 0.63) is even higher then for tests involving the polystyrene latex. In test with PSA2 the detection limit corresponds to the 4000-fold dilution of anti-HSA serum whereas for PS the highest dilution at which anti-HSA could still be detected was 2000-fold.

Table 2 Activity of latexes PS-HSA and PSA-HSA towards anti-HSA

Dilution of Sigma A-1151 anti-HSA serum	PS latex 12.8 mg/ml	PSA1 latex 15.9 mg/ml	PSA2 latex 8.9 mg/ml
1:2	++	+	+
1:4	++	+	++
1:8	+++	++	+++
1:16	+++	+++	+++
1:32	+++	+++	+++
1:64	+++	+++	+++
1:128	+++	+++	+++
1:256	+++	+++	+++
1:256	+++	+++	+++
1:512	+++	++	++
1:1024	+++	+	++
1:2048	+	+	++
1:4096	—	—	+
1:8192	—	—	—

"+++" Denotes the very strong aggregation, "++" well developed aggregation, "+" weak aggregation, and "—" absence of aggregation.

Latex agglutination test for detection of the fibrinogen degradation products (FgDP)

Thge latexes used in model test were used also for preparation of the diagnostic tests for detection of FgDP in blood serum and in urine. Fragments D and E of FgDP and rabbit polyclonal anti-FgD and anti-FgE were provided by Prof. C.Cierniewski from the Medical Academy of Lodz. Fibrinogen degradation products were obtained by plasmin induced degradation of fibrinogen derived from the patient sera. Fragments D and E were isolated from the mixture of FgDP by chromatography (column lys-sepharose). Rabbits were immunized by 30 subcutaneous injections, each containing 50 μg of FgD and FgE 1:1 mixture. Polyclonal antibodies anti-FgD and anti-FgE were isolated from the rabbit blood sera by affinity chromatography, on agarose with immobilized FgD and FgE. These antibodies were immobilized on latexes by incubating 1 volume unit of latex (concentration 80 mg/ml) in glycine buffer (pH = 8.2) with 1 volume of anti-FgD and anti-FgE (2:1) with protein concentration equal to 0.5 mg/ml. Incubation was carried out during 2 h at 37 °C.

After removal of the unbound protein the latex was transferred to the BGA buffer (glycine buffer pH = 8.2 with 0.1% BSA). These latexes were used for detection of FgDP in solutions containing various concentrations of this antigen in the PBS buffer and for detection of FgDP in the positive and negative control sera. Tests were carried out in the similar manner as it has been described above for the model. Experiments with Boehringer latex diagnostic kit were performed for comparison. Results ate collected in Table 3.

Table 3. Activity of latexes with antibodies anti-FgD and anti-FgE

Antigen	Latex from Boehringer diagnostic kit	PS latex
control serum (—)	—	—
control serum (+)	+++	+++
FgDP - 400 μg/ml	+++	+++
200 μg/ml	+++	+++
100 μg/ml	+++	+++
50 μg/ml	+++	+++
25 μg/ml	+++	+++
12.5 μg/ml	++	++
6.25 μg/ml	++	++
3.12 μg/ml	+	+
1.56 μg/ml	—	+
0.78 μg/ml	—	—

The lowest concentration of FgDP which could be detected (1.56 μg/ml) corresponds to 30 ng of antigen in 20 μl of the analyzed sample.

CONCLUSIONS

Experiments described in this paper indicate that the poly(styrene/acrolein) latexes with various fractions of polyacrolein in the surface layer can be used for the attachment of protein with the controlled proportion of protein macromolecules adsorbed by physical interactions and covalently immobilized due to the reaction of NH_2- groups from protein with aldehyde groups from polyacrolein. These latexes with the attached appropriate proteins can be used for preparation of the latex agglutination tests.

The work was supported by the KBN grant No 2 2687 92 03.

Literature

1. Polysciences Inc., Data Sheet No. 238, **1984** with over 70 references to original papers on application of latexes in medical diagnostics.

2. P.J.Tarcha, D.Misun, D.Finley, M.Wong, J.J.Donovan in *Polymer Latexes; Preparation, Characterization, and Applications*, Eds: E.S.Daniels, E.D.Sudol, M.S.El-Aasser, ACS Symp.Ser. **492**, 347 (1992).

3. A.Rembaum, S.P.S.Yen, R.S.Molday, *J.Macromol.Sci.,Chem.* **A13**, 603, **1979**.

4. A.Rembaum, S.P.S.Yen, E.Cheong, S.Wallace, R.S.Molday, I.L.Gordon, W.J.Dreyer, *Macromolecules*, **9**, 328, **1976**.

5. L.Marcus, M.Offarim, S.Margel, *Biomat., Med.Dev., Art.Org.*, **10**, 157, **1982**.

6. S.Margel, U.Beitler, M.Offarim, *J.Cell.Sci.* **56**, 157 **1982**.

7. D.Horowitz, S.Margel, T.Shimoni, *Biomaterials*, **6**, 9, **1985**.

8. M.Chang, M.,Colvin, A.Rembaum, *J.Polym.Sci., Polym.Let.Ed.* **24**, 603

9. I.A.Gritskova, S.A.Gusev, E.Grzywa, E.N.Teleshov, S.M.Shumanskii, I.Grzywa-Niksinska, *Polimery*, **36**, 229, **1991**.

10. B.Charleux, P.Fanget, C.Pichot, *Makromol.Chem.* **193**, 205, **1992**.

11. T.Basinska, S.Slomkowski, M.Delamar, *J.Bioactive Biocompatible Polym.* accepted.

12. T.Basinska, S.Slomkowski, *J.Biomater.Sci., Polymer Edn.* **3**, 115, **1991**.

13. S.Slomkowski, T.Basinska, in *Polymer Latexes; Preparation, Characterization, and Applications*, Eds: E.S.Daniels, E.D.Sudol, M.S.El-Aasser, ACS Symp.Ser. *492*, 328, **1992**.

14. W.Norde, F.MacRitchie, G.Nowicka, J.Lyklema, *J.Colloid Interface Sci.* **112**, 447, **1986**.

15. X.M.He, D.C.Carter, *Nature*, **358**, 209, **1992**.

16. D.Kowalczyk, S.Slomkowski, F.A.Wang, *J.Bioactive Biocompatible Polym.* submitted.

AN INTEGRATED OPTICAL IMMUNOSENSOR

P.J.R. LAYBOURN[1], Y. ZHOU[1*], R.M. DE LA RUE[1], W. CUSHLEY[2], C. MCSHARRY[3] AND J.V. MAGILL[1]

[1]Department of Electronics and Electrical Engineering,
University of Glasgow, Glasgow G12 8QQ, Scotland, UK

[2]Department of Biochemistry,
University of Glasgow, Glasgow G12 8QQ, Scotland, UK

[3]Department of Bacteriology and Immunology,
Western Infirmary, Dumbarton Road Glasgow, G11 6NT, Scotland, UK

* present address :
School of Electrical and Electronic Engineering,
Nanyang Technological University, Nanyang Avenue, Singapore 2263

Abstract

The paper describes a new optical sandwich immunosensor using patterned ion-exchanged waveguides. The one step device only requires the addition of a sample solution into the sensor cell and no washing step is needed. In addition the sensor is capable of detecting several different analytes in a single test. Fabrication is based on an ordinary glass microscope slide and is therefore cheap to produce and could be disposable. Results are in agreement with conventional ELISA measurements.

I INTRODUCTION

Immunological binding between an antibody and an antigen can be detected and measured by a number of established techniques. However, there is still a demand for a simple, rapid and inexpensive diagnostic test and new methods continue to be investigated. Optical evanescent-wave immunosensors offer advantages over other techniques, and several arrangements have been described [1-5].

G.G. Guilbault and M. Mascini (eds.), Uses of Immobilized Biological Compounds, 463–470.

An evanescent wave device uses the limited interaction distance near to the optical interface supporting a guided light wave to discriminate in favour of only those molecules which are bound close to the interface. Other unbound molecules in the bulk of the sample solution are not detected and hence need not be washed away. In a solid-phase fluorescence immunoassay, for example, only the fluorescently-labelled antibodies which are bound to the solid phase will be excited to emit fluorescent light.

Optical evanescent waves are associated with light guided along optical fibres or integrated optical waveguides, and are also present during total internal reflection, within a prism, for example. Integrated optical waveguides formed on a planar substrate allow multi-analyte sensing and internal references[1]. The ion-exchange technique of making thin-film waveguides on glass slides is low-cost and very suitable for making a disposable immunosensor cell [6].

We present here a development of an earlier design [3] and discuss possibilites of improving the optical system to fulfil the requirements for simplicity of use and low cost, with the aim of producing a cheap and sensitive disposable sensor cell.

II PRINCIPLE OF OPERATION

Figure 1 shows a schematic diagram of the patterned waveguide immunosensor. The glass sensor substrate contains a buried waveguide which is located immediately below a series of wells etched into the surface of the glass. A particular antigen or control protein is immobilised in each well. A capillary cell is constructed above the wells with a fluorescently labelled antibody attached reversibly to the upper surface of the cell. When the test sample is introduced at the edge of the cell, capillary action draws the liquid in to fill the cell. The fluorescently labelled antibody becomes detached from the upper surface and binds to the appropriate antibody in the test sample. Finally the antibody pair will bind in any of the etched wells containing the specific antigen. As a result the antibody contained in the test

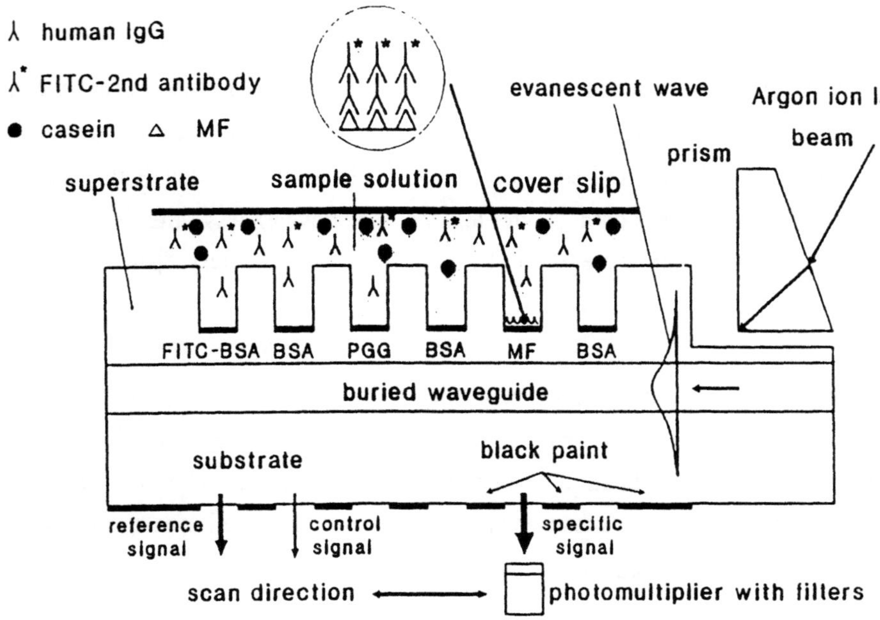

Fig. 1 Diagram of the Wash-Free Sandwich Immunosensor
BSA: bovine serum albumin, PGG: pigeon gamma globulin
MF: micropolyspora faeni, FITC: fluorescein isothiocyanate

sample is sandwiched between an immobilised antigen and the fluorescently labelled antibody which have mutually high specific binding properties. If the fluorescently labelled antibody is specific to the Fc (fragment crystallisable) part of the test antibody then the sensor can be used for multi-analyte detection.

When light of the appropriate wavelength is coupled into the buried waveguide, the evanescent field associated with the guided mode penetrates into the wells thereby exciting the fluorescence of the immobilised antigen-antibody sandwich. The intensity of this fluorescent emission is easily detected using a photomultiplier and is directly related to the concentration of the test antibody in each well.

III EXPERIMENTAL DETAILS

A two-step thermal ion exchange process is used to produce buried planar waveguides in standard glass microscope slides [1-3]. The slides are immersed in, firstly, potassium nitrate, and secondly, sodium nitrate molten salt baths each at 490°C, for 30 minutes per bath. After cleaning to remove excess salts the substrate contains a buried planar waveguide at a depth of approximately 8μm below the substrate surface[6].

Wells are etched into the surface using a standard photolithographic process. The mask used contains six rectangles corresponding to the well dimensions. Following photoresist application and development, wells are etched to a depth of approximately 6 μm into the substrate surface. In addition a small area at one end of the slide is etched to facilitate in-coupling of light from a prism.

Antigens and control proteins are immobilised into the wells covalently using aminosilane and glutaraldehyde linking molecules. In this experiment the species immobilised in the six wells were; fluorescein isothiocyanate labelled bovine serum albumin (FITC-BSA), BSA, pigeon gamma-globulin (PGG), BSA, micropolyspora faeni (MF) and BSA. FITC-BSA is used to monitor the excitation light intensity and BSA as a control protein to allow for any non-specific binding and background noise. The serum samples used were taken from patients exposed to either PGG or MF and thus were expected to contain high concentrations of immunoglobulin G (IgG) specific to one of these antigens. The fluorescently labelled antibody used was Fc specific FITC labelled swine anti-human IgG (FITC-IgG).

Figure 2 shows in diagrammatic form the illumination and light detection arrangements. The light source used was a small 488nm argon ion laser, chopped at 250Hz. A series of slits and long pass filters was used to reduce stray light and remove the excitation wavelength. The detection system consisted of a photomultiplier and lock-in amplifier.

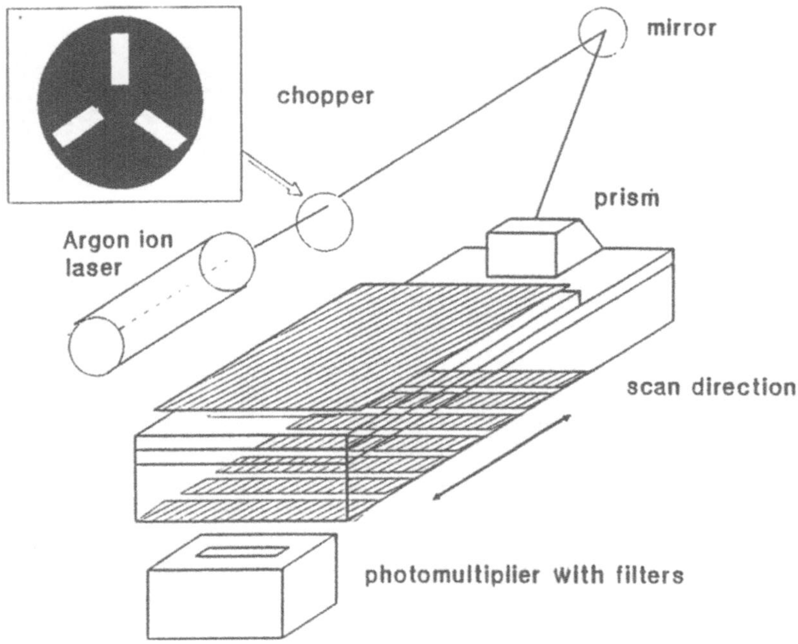

Fig.2 Experimental arrangement for the measurement of fluorescence in the wells

IV RESULTS AND DISCUSSION

Figure 3 shows the relative evanescent fluorescence measured from the etched wells when different dilutions of serum samples were added. The relative fluorescence is defined as the signal from a control or antigen well divided by the signal from the FITC-BSA reference well.The horizontal axis lists the immobilised antigens and the control proteins. The data corresponds to average data for a number of experiments.

It can be seen clearly that the signal from the well containing an antigen is strongest when a serum sample containing the specific antibody is added. The signal rises with increasing dilution, until a peak is reached, after which the signal falls as

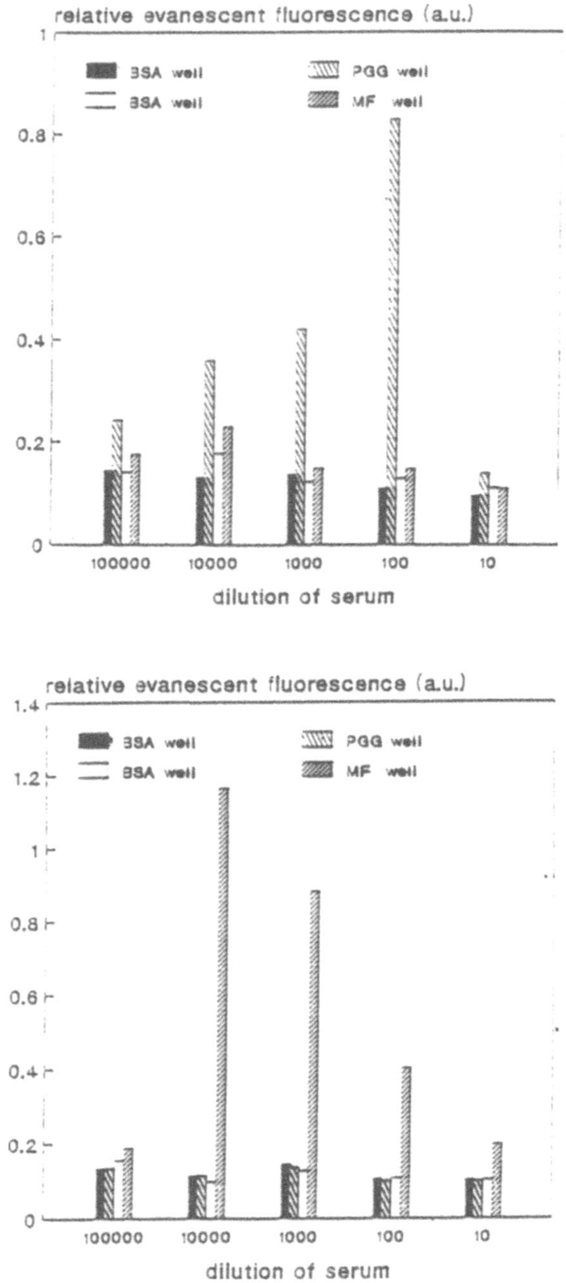

Fig. 3. Measured fluorescence with added human serum:
Avian disease (above); Farmer's lung (below)

the specific antibody concentration is further reduced. The non-proportional relation between fluorescent signal level and serum concentration can be explained as follows. At low serum concentrations there is sufficient labelled antibody to label all the serum and the number of antigen binding sites occupied by labelled molecules reflects the serum concentration. The site occupancy will eventually saturate as serum concentration is increased, and at even higher concentrations the fraction of serum being labelled will fall as the labelled antibody supply is limited, resulting in reduced fluorescence. The results for the two particular antibodies have been mirrored in the results from ELISA measurements, shown in figure 4.

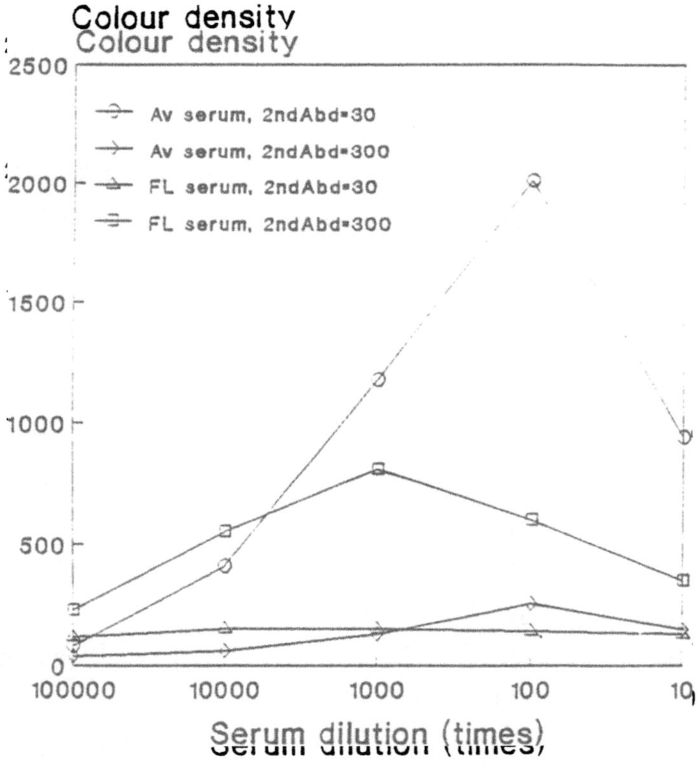

Fig. 4. ELISA results using the same serums

The sensor cell and measurement system are still at an experimental stage. The fluorescent light collection system is being improved, and for ease of use by relatively unskilled operators an improved light input coupling system is required.

The prism coupler could be replaced by the provision of a diffraction grating coupler formed by lithography and etching at the start of the buried waveguide.

V CONCLUSION

We have successfully demonstrated that one-step sandwich immunoassays can be performed using our patterned waveguide immunosensor. By incorporating a fluorescently labelled Fc-specific antibody, the sensor can be used for multi-analyte diagnostic testing of human blood serum.

ACKNOWLEDGEMENTS

The work was supported by SERC under grant no. GR/E 18704 and by Unilever plc. Dr Zhou was supported under the Sino-British Friendship Scheme.

REFERENCES

1. Y Zhou, J V Magill, W Cushley, R M De La Rue, and P J R Laybourn, Sensors and Actuators B, 11, 245-250 (1993)
2. Y Zhou, J V Magill, P J R Laybourn, W Cushley, and R M De La Rue, J Molecular Electronics, 7, 135-149 (1991)
3. Y Zhou, P J R Laybourn, J V Magill, and R M De La Rue, Biosensors and Bioelectronics, 6, 595-607 (1991)
4. M T Flanagan, A N Sloper, and R H Ashworth, Anal. Chim. Acta., 213, 353-355 (1988)
5. B I Bluestein, I M Walczak, and S Y Chen, Trends in Biotechnology, 8, 161-168 (1990)
6. Y Zhou, P J R Laybourn, J V Magill, and R M De La Rue, IEE Proceedings-J, 139, 223-227 (1992)

Immobilized Biological Compounds in Food Analysis

Saverio Mannino

Department of Food Science and Technology

University of Milan, Via Celoria 2, 20133 Milano (Italy)

Abstract: After a short review on the use of biosensors to address analytical problems in food chemistry, results obtained for the determination of polyphenols and peroxide value in vegetable oils by using organic-phase enzyme electrodes are presented.

I. Introduction

There are various types of analytical methods applied to food analysis that can provide information concerning physical and chemical characteristics of food, including data about major and minor constituents. However, these methods are often complicated, time consuming and require analyte separation prior to detection.

The food industry has a general need for methods suitable for process and quality control that are simple, rapid and having characteristics such as accuracy, repeatability and sensitivity.

Biosensors, devices which convert a selective biochemical response into a quantifiable electrical signal, can offer many unique opportunities for rapid monitoring of food production, packing, storage and transportation. Furthermore, because of their inherent specificity, a selective determination can be carried out in a multicomponent matrix thereby alleviating the need for any difficult and time-consuming separation. Such characteristics greatly facilitate the application of

G.G. Guilbault and M. Mascini (eds.), Uses of Immobilized Biological Compounds, 471–476.

As it can be seen, the enzyme electrode responsed rapidly to these concentration changes and the rapid return to the baseline indicates facile transport of the substrate and washout of the reaction product.

Oil samples were diluted 1:1 with the chloroform carrier solution prior to injection in the flow system.

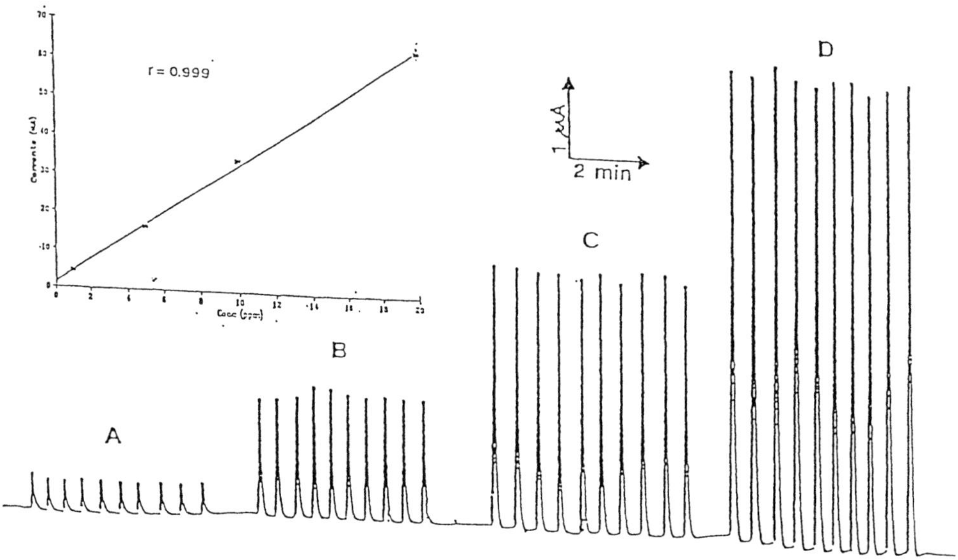

Figure 1.: Flow injection peaks for chloroform solutions containing increasing level of tyrosol A, 1; B, 5; C, 10; D, 20 ppm. Flow rate, 0.6 ml/min.; operatin potential -0.20 V vs Ag/AgCl. Also shown the resulting calibration plot.

Results obtained from the analysis of oils of different origins, cultivars and extraction technologies are reported in Table 1, together with those obtained using the Folin-Ciolcalteau method.

As expected, results obtained with the two methods are not in agreement, since the Folin-Ciolcateau method use a different calibration standard (caffeic acid) and, as reported above, tends to overestimate polyphenol levels. Hovewer, the results obtained using the tyrosinase electrode, capable of measuring more effectively mono-phenols and biosensors to on-line monitoring, providing real-time information on processed food.

This report points out the advantages of using biosensors in food quality control and presents some results obtained in our laboratory using organic-phase enzyme electrodes for the analysis of vegetable oils.

II. Polyphenols in olive oils

Polyphenols are compounds present in olive oils that have received considerable attention in recent years, because of their importance to taste and stability of the oil to autooxidation [1-4]. The total polyphenol content is usually determined by a conventional colorimetric method using the Folin-Ciolcalteau reagent that is not specific and often overestimate its level.

Considering the ability of enzymes to operate in organic solvents and the suitability of organic-phase tyrosinase electrode for a rapid monitoring of phenols in olive oils [5], we applied this enzyme electrode to a large number of extravirgin olive oils of different origins, cultivars and extraction technologies in order to verify if there was a correlation between biosensor response and stability of olive oils to autoxidation.

Experimental conditions and electrode preparation were as reported by Wang et al. [5], except that a flow-through electrode and an HPLC pump were used as sensor and solvent deliver system. Amperometric detection was performed at room temperature by applying a potential of -0.20 V vs Ag/AgCl. Calibration plots were obtained using standard solutions of tyrosol, the main olive oil phenol.

Flow injection peaks for chloroform solutions containing increasing levels of tyrosol, along with the resulting calibration plot are reported in Figure 1.

Table 1. Characteristics of the analyzed samples and their poliphenol content obtained with the official and the proposed method.

Samples*	Variety	Ripeness degree	Folin-Ciolcateau	Biosensor
1	Coroneiki (G)	unripe	89.1	11.7
2	Coroneiki (G)	normal ripeness	109.2	13.4
3	Coroneiki (G)	over-ripe	168.7	26.1
4	Coroneiki (G)	normal ripeness	51.6	13.9
5	Tzunnati (G)	normal ripeness	97.2	23.9
6	Coratina (I)	unripe	150.1	9.5
7	Coratina (I)	normal ripeness	140.7	6.7
8	Coratina (I)	over-ripe	165.1	13.5
9	Coratina (I)	normal ripeness	186.6	23.5
10	Cima di Bitonto (I)	normal ripeness	66.5	10.4
11	Pigual (S)	unripe	166.2	46.4
12	Pigual (S)	normal ripeness	90.4	32.4
13	Pigual (S)	over-ripe	83.2	28.7
14	Arbequina (S)	unripe	25.8	11.5
15	Arbequina (S)	normal ripeness	22.3	8.6
16	Arbequina (S)	over ripe	16.5	5.2

G= Greece; I = Italy; S = Spain

* All samples were obtained by centrifugation except for samples 4 and 9 that were obtained by percolation and pressing, respectively.

ortho-substituded phenols, are very well correlated with the real antioxidant power of the oil since these compounds were shown to be the major responsable of oil stability. In fact, a good correlation exists between the results obtained with the proposed biosensor and the amount of ortophenols (tyrosol and oxityrosol) determined in the same analyzed samples by a method based on HPLC with amperometric detection [6].

III. Peroxide value in edible oils

Another interesting application of biosensors in organic media is the determination of the so-called peroxide value in vegetable oils and fats. Lipid oxidation, defined as the oxidative deterioration of polyunsaturated fatty acids, is a process that can give rise to several degenerative conditions and can cause the qualitative deterioration of fatty foods producing off-flavor and off-odor effects. Since lipid hydroperoxides are the primary products of lipid oxidation, a sensitive, specific and accurate method is needed to provide the necessary data to understand the oxidation status of foods and their acceptability. Peroxide value is generally used as an effective indicator of the degree of oxidation of lipids. The Italian Official Method measures the oxidation value iodometrically and expresses it as meq of O_2 per Kg of fat.

Preliminary experiments indicate that using a glassy carbon electrode on which peroxidase enzyme is absorbed can be a useful means for the determination of peroxide in vegetable oils.

The adsorptive immobilization was accomplished by placing 10 μL of a peroxidase solution (10mg/ml in 0.1M phosphate buffer pH=7) onto the electrode surface which was first polished with a SiC paper. Flow injection experiments were performed at a potential of -0.100 V vs Ag/AgCl and by using chloroform (saturated with phosphate buffer), containing tetrabutyl-ammonium bromide and $2x10^{-3}M$ ferrocene as carrier solution. Figure 2 displays typical recorder outputs for lauryl peroxide obtained using the peroxidase electrode. Work is in progress in our laboratory for the application of this bioelectrode for the determination of peroxide value in vegetable oils.

Other food application of organic -phase biosensors are anticipated.

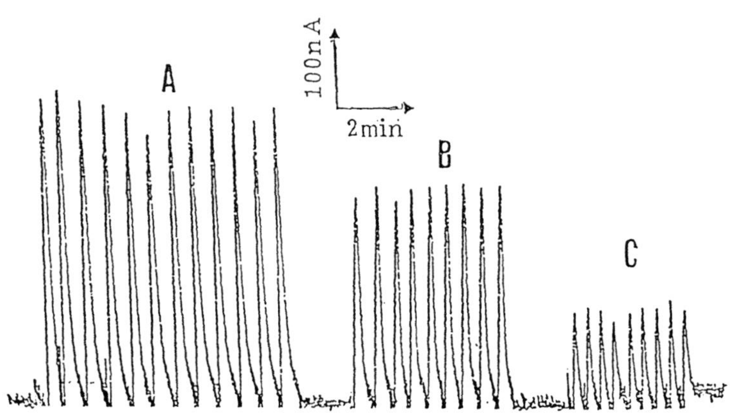

Figure 2. Responses of the peroxidase enzyme electrode to repetitive injections of different amounts of lauryl peroxide: A, $2x10^{-3}$ M; B, $1x10^{-3}$ M; C, 10^{-4} M. Conditions: Flow rate, 0.6 ml/min; Potential -0.1 V vs Ag/AgCl.

IV. References

1) M. Solinas, and A. Cichelli, Riv. Ital. Sost. Grasse, 1981, 10, 159
2) Cevdet Nergir, Kemal Unal, J. of Food Sci., 1991, 56, 79
3) G. Montedoro and M. Servili, J. Agr. Food Chem., 1992, 40, 1571
4) M. Thisdou, G. Papadopulos, D. Boskou, Food Chem.,1992,45, 53
5) J. Wang, J. Reviejo, S. Mannino, Anal. Letters, 1992,25(8),1399
6) S. Mannino, M.S. Cosio, M. Bertuccioli, Italian J. of Food Techn., in press.

IMMOBILIZED BIOMOLECULES FOR DETECTION OF ENVIRONMENTAL POLLUTANTS

K.R. ROGERS
U.S. Environmental Protection Agency
Environmental Monitoring Systems Laboratory
Las vegas, NV 89193-3478

Abstract Biosensors which incorporate immobilized enzymes, receptors, antibodies or microbes have been used to measure compounds of environmental significance as diverse as phenolics, ammonia, formaldehyde, benzo(a)pyrene, PCBs, and pesticides. Although a variety of biosensor assay formats have been reported, there are two basic mechanisms for operation: detection of a substrate or product of biocatalysis, and detection of a stoichiometric binding event. Most reported biosensors use one or the other of these mechanisms. Herein reported, however, are two unique biosensors which use both mechanisms to detect organophosphate (OP) and carbamate pesticides.
Most OP and some carbamate pesticides bind stoichiometrically to, and inhibit acetylcholinesterase (AChE) activity. Consequently, biosensors using this enzyme are able to catalytically amplify the binding event. Two AChE-based biosensors were developed using different enzyme immobilization and signal transduction schemes. The fiber-optic biosensor uses a fluorescein isothiocyanate (FITC) label to report AChE activity. Signal transduction was accomplished by catalysis-induced pH changes in the local environment of the enzyme which changed the optical properties of the FITC probe. The AChE-based light addressable potentiometric (LAP) biosensor measures AChE immobilized to a nitrocellulose membrane. Signal transduction in this case was accomplished using the LAP transducer which is sensitive to small changes in surface potential resulting from AChE-induced pH changes in solution. Both of these biosensors responded similarly to various OP and carbamate pesticides despite the use of dramatically different enzyme immobilization and signal transduction schemes.

INTRODUCTION

The detection of anticholinesterases is of concern to the agriculture industry, regulatory agencies, health care providers and the military. The interest in detection of these compounds is due, in part, to their toxic nature. The acute toxic effects of compounds such as OPs and carbamates are mediated primarily through disruption of cholinergic neurotransmission by inhibition of AChE. The ability to detect these compounds based on enzyme inhibition has resulted in the development of a number

G.G. Guilbault and M. Mascini (eds.), Uses of Immobilized Biological Compounds, 477–487.

of assays based on electrochemical[1], colorimetric[2], radiometric[3] and fluorescence[4] detection of enzyme activity.

Advances in enzyme immobilization technologies as well as miniaturized optical and electrochemical transducers has led to a renewed interest in cholinesterase-based biosensors for the detection of anticholinesterases. Because biosensors can be configured to be small, fast and inexpensive, these devices show the potential to be used in a variety of field applications.

Transducers which have been used in cholinesterase-based biosensors include: potentiometric[5,6,7] and amperometric[8,9,10] electrodes, ion sensitive field effect transistors (ISFETS)[11], piezoelectric crystals[12] as well as optical systems using optode[13] or evanescent wave[14] detectors. Although these biosensors use different enzyme immobilization methods and signal transduction formats, operational characteristics of these devices may in some ways be similar. This paper will discuss results from AChE-based electrochemical (LAP) and optical (evanescent wave) biosensors as well as other reported methods. Operational characteristics common to each will also be discussed.

METHODS

The AChE-based fiber-optic biosensor relies upon the pH dependence in the fluorescence quantum yield of fluorescein[14]. The use of evanescent wave technology allowed the continuous observation of the fluorescence of FITC-labeled AChE immobilized to the surface of the fiber. Because acetylcholine hydrolysis catalyzed by AChE results in the production of protons, the change in observed fluorescence intensity of the fluorescein label can be used to measure enzyme activity. AChE from eel was labeled with fluorescein isothiocyanate (FITC), immobilized to silanized quartz fibers and assayed as previously reported[14] Unless otherwise stated in the results, the enzyme was incubated with inhibitor for 10 min. Immobilized enzyme which had been phosphorylated by exposure to OP pesticides was re-activated using pralidoxime methoiodide (2-PAM).

This work was carried out using a fluorimeter designed and built by ORD Inc. (North Salem, NH). The fiber-optic evanescent fluorosensor was similar in configuration to that previously reported[15](Fig 1).

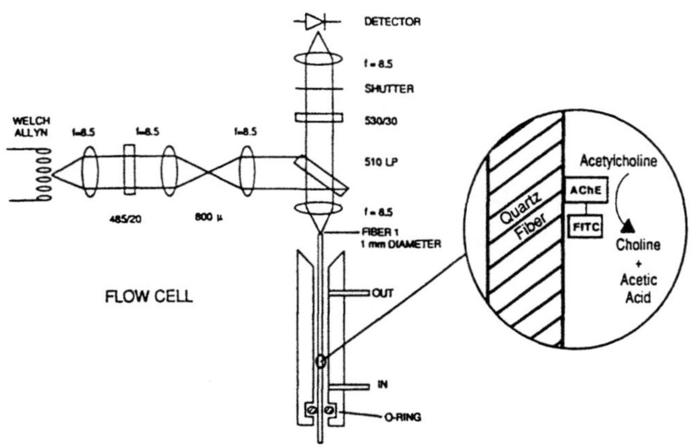

FIGURE 1 Schematic presentation of the optical system. Fiber inset illustrates biochemical model for AChE biosensor.

The light addressable potentiometric (LAP) sensor (Threshold™, Molecular Devices Corp., Menlo Park, CA) uses an insulated semiconductor device that responds to surface potentials (generated by protons) at a silicon nitride surface. The semiconductor produces a transient photocurrent in response to a transient illumination. By placing the semiconductor between the sample and an array of LEDs, pH measurements may be made at separate locations above the sensor. As a result of this configuration, the activity of acetylcholinesterase immobilized to a nitrocellulose membrane and brought in contact with the sensor can be measured in an extremely small volume[16] on the order of 0.6 μL . Acetylcholinesterase was immobilized on a biotinilated nitrocellulose membrane using a strepavidin bridge (Fig 2). The enzyme, inhibited by treatment with anticholinesterases prior to immobilization on the nitrocellulose membrane was assayed as previously reported[5]. Immobilized enzyme which had been phosphorylated by exposure to OP pesticides was re-activated using pralidoxime methoiodide (2-PAM).

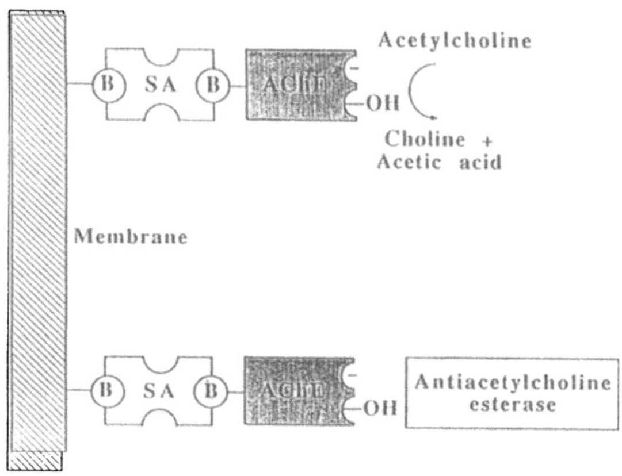

FIGURE 2 Schematic illustration of AChE-based LAP system. B, biotin; SA, strepavidin; AChE, acetylcholinesterase.

RESULTS AND DISCUSSION

For the fiber-optic biosensor, the flow cell allowed the rapid exchange of inhibitors (i.e., OPs and carbamates), analyte (i.e., ACh) and re-activator (i.e., 2-PAM). The biosensor assay was performed by interrupting the flow of the buffered substrate and measuring the pH-induced fluorescence change during a 2 min period.

Interruption of the buffer flow, by turning the pump off, allowed the local pH in the vicinity of the FITC-labeled enzyme to drop resulting in fluorescence quenching (Fig. 3B). Resuming the buffer flow allowed the equilibrium between the surface pH and bulk buffer pH to be re-established. The biosensor response was dependent on the presence of substrate (Fig. 3A) and the assay was very stable and could be repeated numerous times on the same fiber without loss in enzyme activity (Fig. 3C).

Similar to the fiber-optic biosensor, the LAP biosensor response was also dependent on the presence of the substrate ACh. The LAP sensor response (i.e., μV/min) was linear over the assay period (1 min). Also, once the enzyme was immobilized to the nitrocellulose membrane, the membrane could be removed from the reader (i.e., silicon nitride sensor), washed and re-assayed without significant loss in activity.

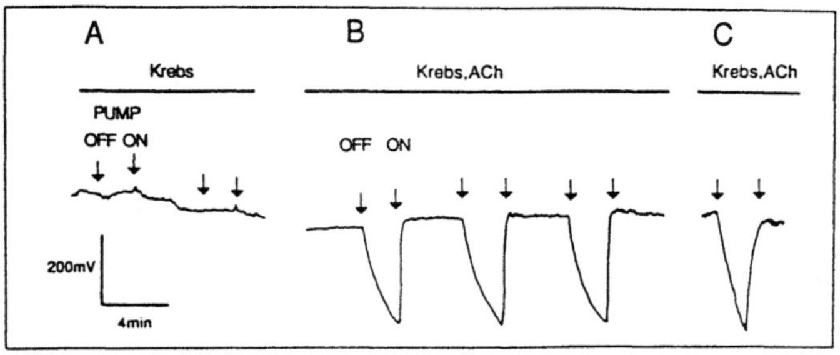

FIGURE 3 Fluorescence response of the AChE-based fiber-optic biosensor. Response of biosensor as a result of (A) buffer alone (response in the absence of acetylcholine, arrows indicate the start or stop of buffer flow); (B) Acetylcholine; and (C) Time (response was measured after 2 hr of continuous operation).

Both quenching of the fluorescence signal (in the case of the fiber-optic biosensor) and the rate of change in surface potential (for the LAP biosensor) varied as a function of the substrate (i.e., ACh) concentration. Furthermore, in both cases, a rectangular hyperbola, indicative of Michaelis-Menten kinetics, was observed. K_{app} values of 0.42 mM and 5.2 mM were observed for AChE immobilized to the fiber-optic and LAP biosensors, respectively (Table 1). It is interesting to note that the herein reported K_{app} for the LAP sensor is similar to the 5.3 mM value reported by Skladal et al.[8] for AChE immobilized to nylon nets and measured using an amperometric biosensor.

In order to determine the kinetic effects of FITC labeling and immobilization of the enzyme to quartz, non-immobilized fluorescent-labeled enzyme (FITC-AChE) and non-labeled enzyme were assayed using a colorimetric method[17]. Kinetic analysis of both enzymes yielded K_{app} values close to 0.05 mM. These values are similar to those reported for non-immobilized AChE from eel by both Karlin[18] and Skladal et al.[8]

TABLE 1 Kinetic Parameters for Free and Immobilized AChE.

Cited Work	K_{app} (mM)
Rogers et al.[14]	0.42 mM (immobilized) 0.05 mM (free)
Rogers et al.[5]	5.2 mM (immobilized)
Skladal et al.[8]	5.3 mM (immobilized) 0.04 mM (free)

Several factors might be expected to influence a change in the K_{app} observed upon immobilization of this enzyme. The affinity of the substrate for the enzyme may have been decreased by conformational changes resulting from chemical modification required for immobilization of the protein. Although the chemically active sites on the enzyme involved in fluorescent-labeling may have been different from those involved in protein immobilization, direct conformational changes appear an unlikely cause for the change in K_{app} since FITC-labeled AChE was kineticaly identical to the unlabeled enzyme. A more likely explanation is an effective decrease in the substrate concentration at the catalytic site due to decreased accessibility of the enzyme to the substrate caused by charge or steric interactions between the substrate, the surface and the enzyme. Both increases and decreases in K_{app}, observed in immobilized enzymes, have been attributed to surface interactions[19].

AChE activity and resulting biosensor responses for both the LAP and fiber-optic biosensors were inhibited by the presence of (slowly reversible) carbamates and (irreversible) OPs. The presence of these two types of inhibitors could be distinguished based on recovery of the biosensor response. For example, the fiber-optic biosensor response which was completely inhibited (i.e., 5% of control activity) by the carbamate, neostigmine (100 μM) recovered to 71% of its control activity after perfusion for 20 min with buffer in the absence of inhibitor. Similar results were also observed for the LAP biosensor response which showed 10% of the control signal after treatment with the carbamate, bendiocarb (3 μM) but recovered to yield 77% of the control activity after perfusion of the membrane for 15 min with buffer in the absence of inhibitor.

In contrast to the carbamates, after treatment with OPs, neither the fiber-optic nor LAP biosensor could be re-activated by perfusion with buffer alone. The fiber-optic biosensor response which was inhibited by the OP, echothiophate (100 μM) required the presence of the nucleophilic oxime, 2-PAM for re-activation. This was also the case for the LAP biosensor which after complete inhibition (i.e., to 4% of control activity) by paraoxon (1 μM), required treatment with 2-PAM to regain 94% of its control response. It was observed for both the fiber-optic LAP biosensors, that the immobilized AChE could be re-activated numerous times without substantial loss of activity and was stable for at least 2 hr after re-activation.

The ability to re-activate these biosensors allowed for a control measurement (i.e., calibration) to precede each determination of anti-cholinesterases. Similar results have been reported for immobilized AChE in a flow-injection analysis method in which the enzyme was inhibited to 22% of the control activity with paraoxon (1 μM) and reactivated to 94% with 2-PAM[1] Although nucleophilic attack on the OP-AChE complex by the oxime 2-PAM resulted in a high degree of reactivation for OP pesticides and therapeutics, this would most likely not be observed for OP "nerve agents". After phosphorylation of the enzyme, compounds such as sarin and soman lose an alkyl group, resulting in enzyme aging, where the enzyme-inhibitor complex is resistant to hydrolysis.

Because AChE is irreversibly inhibited by OP anti-cholinesterases, the detection limits of systems using immobilized AChE will be affected by the amount of time that the enzyme is exposed to the OP. It was observed using the LAP biosensor that the log of the control activity (for any specific concentration of OP) varied linearly with the exposure time (Fig. 4A). Consequently, the inhibition curves and resultant detection limits may be shifted as a function of enzyme exposure time (Fig 4B). This behavior has also been reported by Marty, et al.[9] using an AChE/choline oxidase electrode.

Because of variations in the time of inhibitor exposure afforded by different assay formats, it is not surprising that the detection limits (for a particular OP) would vary somewhat among reported AChE-based biosensors (Table 2). In addition to the dependence on exposure time, the detection limits of a method to OPs may also depend on the amount of enzyme immobilized to the insoluble support. Marty etal. observed that the detection limit for an amperometric biosensor decreased with decreasing AChE immobilized to the electrode[9].

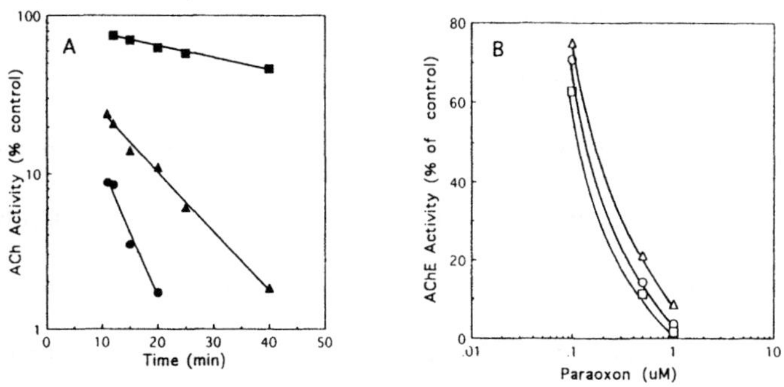

FIGURE 4 Conditions that affect AChE inhibition. (A) Time dependence of paraoxon inhibition of the AChE-based LAP biosensor response (1 μM,●; 0.5 μM,▲; 0.1 μM,■). (B) Dependence of inhibition concentration response curves on incubation time (12 min,△; 15 min,○; 20 min,□; enzyme exposure times).

TABLE 2 Detection Limits Among Various AChE-Based Methods for Paraoxon.

Cited Work	Detection Limit (IC_{10})[a]
Rogers et al.[14]	3.5×10^{-9} M
Rogers et al.[5]	3.0×10^{-9} M
Skladal et al.[8]	4.4×10^{-8} M[b]
Marty et al.[9]	1.0×10^{-8} M
Kumaran et al.[1]	9.0×10^{-8} M

[a]IC_{10} is the molar concentration required to inhibit the control AChE activity by 10 %. This value is being used as a conservative estimate of the method detection limit.
[b]The reported detection limit using butyrylcholinesterase (BChE) was 5.4×10^{-9} M.

It is clear from these limited comparisons that a number of transducers and formats can be used for detection of anti-cholinesterases using AChE-based methods. The more important issue to consider, however, is which of these methods (if any) will emerge into the commercial market? The answer will most likely depend on how well the method can be developed to meet the needs of the potential user. The following issues must first be considered: assay location, target compounds, matrices, assay time, sensitivity, selectivity and data quality objectives. A more detailed discussion of this subject can be found in the literature[20,21]. Another important consideration involves the restrictions of the developer. It is typically the case that instrument development in an academic setting is designed for optimal versatility and performance rather than simplicity and manufacturability. Consequently, in order for these biosensors to become more than laboratory curiosities, it is crucial for biosensor researchers to look forward to the specific requirements of potential applications and the unique restrictions of development, manufacturing and marketing of these devices.

ACKNOWLEDGMENTS

I would like to acknowledge Dr. M.E. Eldefrawi, in whose laboratory this work took place and whose input was vital to this project.

NOTICE

The research described in this report has been supported, in part, by the United States Environmental Protection Agency under Cooperative Agreement No. CR-820460-01-0 with the University of Maryland at Baltimore. It has been subjected to Agency review and has been approved for publication.

REFERENCES

1. S. Kumaran, C. Tran-Minh, Anal. Biochem., 200, 187, (1992).
2. M.F. Leon-Gonzales, A. Townshend, Anal. Chim. Acta, 236, 267, (1990).
3. M.K. Lewis, M.E. Eldefrawi, Anal. Biochem., 57, 588, (1974).
4. G.G. Guilbault, D.N. Kramer, Anal. Chem., 37, 1675, (1965).
5. J.C. Fernando, K.R. Rogers, N.A. Anis, J.J. Valdes, R.G. Thompson, A.T. Eldefrawi, M.E. Eldefrawi, J. Agric. Food Chem., 41, 511, (1993).
6. G. Baum, F.B. Ward, Anal. Chem., 42, 487, (1971).
7. P. Durand, A. David, D. Thomas, Biochim. Biophys. Acta, 527, 277, (1978).
8. P. Skladal, M. Mascini, Biosensors & Bioelectronics, 7, 335, (1992).
9. J.L. Marty, K. Sode, I. Karube, Electroanal., 4, 249, (1992).
10. L. Campanella, R. Cocco, M.P. Sammartino, M. Tomassetti, Sci. Tot. Environ., 123/124, 1, (1992).
11. C. Dumschat, H. Muller, K. Stein, G. Schwedt, Anal. Chim. Acta, 252, 7, (1991).
12. J.N. Ngwainbi, P.H. Foley, S.S. Kuan, G.G. Guilbault, J. Am. Chem. Soc., 108, 5444, (1986).
13. W. Hobel, J. Polster, Fresenius J. Anal. Chem., 343, 101, (1992).
14. K.R. Rogers, C.J. Cao, J.J. Valdes, A.T. Eldefrawi, M.E. Eldefrawi, Fundam. Appl. Toxicol., 16, 810, (1991).
15. T. Glass, S. Lackie, T. Hirschfeld, Appl. Opt., 26, 2181, (1987).
16. L. Bousse, G. Kirk, G. Sigal, Sensors and Actuators, B1, 555, (1990).
17. G.L. Ellman, K.D. Courtney, V. Andres, R.M. Featherstone, Biochem. Pharmacol., 7, 88, (1961).
18. A. Karlin, Biochim. Biophys. Acta, 139, 358, (1967).
19. G.P. Royer, In, Immobilized Enzymes, Antigens, Antibodies, and Peptides, (H. Weetal ed.), p.49, Marcel Dekker, New York, (1975).
20. K.R. Rogers, J.N. Lin, Biosensors & Bioelectronics, 7, 317, (1992).
21. K.R. Rogers, E.J. Poziomek, In, Third International Symposium: Field Screening Methods for Hazardous Wastes and Toxic Chemicals, (in press).

ENZYME INHIBITION SENSOR FOR ORGANOPHOSPHORUS PESTICIDES ANALYSIS

L. Campanella, M. Tomassetti

Dipartimento di Chimica, Università "La Sapienza",
P.le A. Moro, 5, 00185 Roma, Italy

Abstract

The study of the inhibiting action exerted by organophosphorus pesticides on the esterase enzymes is of great interest, due to the potential dangerous activity of pesticides and to their large-scale use in agriculture. Different biosensors for the determination of pesticides were developed by several researchers in the recent years, basing on this behaviour. The biosensor developed by us consists of a Clark oxygen electrode, modified with a dialysis membrane, placed over it, containing immobilised two biological mediators: butyrylcholinesterase and choline oxidase. A comparison with the results obtained by a similar previously published procedure, characterised by butyryl- (or acetyl-) cholinesterase free in solution, is also reported. A new enzyme sensor to determine organophosphorus pesticides is also described, constituted by a hydrogen peroxide amperometric electrode modified by a superimposed nylon membrane containing, chemically immobilised, two biological mediators (butyrylcholinesterase and choline oxidase). Some applications to authentic matrices of environmental or of agricultural interest are reported and discussed.

Introduction

Pesticides are highly toxic compounds as able to cause neurological diseases [1,2].

Nevertheless these compounds are largely used in agriculture, so that their determination is of great interest from both the industrial and the environmental point of view. Very suitable and accurate analysis for the pesticides can be performed by chromatographic (TLC, GC, and HPLC) methods [3,4]; nevertheless analysis time and costs are generally considerable.

Thus, actually, the researches of several authors, aiming to develop a rapid and cheap method for the pesticide analysis, are justified; e.g. organophosphorus pesticides can be determined on the basis of their inhibiting action on some enzymes (i.e. esterases) [5], using spectrometric [6,7] or electrochemical detection [8-11].

Esterase enzymes play an important role in biology firstly because they are responsible for the hydrolysis of the esthers of choline and secondly because, in case of inhibition of their activity, the original state of postsynaptic membranes can not be reestablished. The study of the inhibiting action exerced by pesticides on esterases is important because of the potential danger of such chemicals on the human healty; on the other hand the behaviour of the system esterase-pesticide was largely investigated in the last years in view of the analytical possibilities arising from the inhibiting action exerted by pesticides on esterases. Following this guiding lines, different biosensors for the determination of pesticides finally resulted.

Short review of published papers in this field

Several papers were published by G. Guilbault and other coworkers concerning pesticide analysis. In the year 1965 this author published

G.G. Guilbault and M. Mascini (eds.), Uses of Immobilized Biological Compounds, 489–499.

[12] a continuous fluorometric method employing cholinesterase entrapped in a starch gel matrix on a polyurethane foam pad, for the organophosphorus insecticides detection, as Sarin, Systax, Parathion; in the year 1981 the same author published a research [13] to evaluate a series of copper complexes for the detection of organophosphorus compounds in air, using a coated piezoelectric quartz crystal, while in the year 1986 described a system [14] consisting of a piezoelectric quartz crystals coated with antibodies against Parathion and used for the assay of Parathion in the gas phase.

In the year 1985 C. Tran-Minh largely explained in a review [15] as, the immobilisation of acetylcholinesterase on a glass electrode, gives rise to an enzyme probe sensitive to organophosphorus pesticides, (Malathion, methyl- and ethyl-Parathion, Paraoxon), due to the esterase-enzyme inhibition.

The same author, in collaboration with H. El Yamani et al., described [16] in the year 1988 an automated system for organophosphorus pesticides and carbamates detection, based on the use of a butyrylcholinesterase probe as the element sensitive to the pesticide activity, which is prepared by immobilisation of the enzyme onto a nylon net, using a covalent crosslinking and adapting the nylon net on the sensitive element of a pH electrode. Lastly, in the year 1990, G. Tran-Minh, P.C. Pandey et al. studied [17] and compared acetylcholinesterase sensors, based on a glass Pd/PdO and Ir/IrO_2 electrodes as pH sensors and the acetylcholinesterase enzyme immobilised in acrylamide-methacrylamide hydrazides prepolymer. Employed these sensors for the analysis of very low concentration of Paraoxon, Malathion, or methyl-Parathion.

G. Schwedt et al., in the year 1988 [18], immobilised the enzyme acetylcholinesterase in a matrix of bovine albumine by crosslinking with glutaraldehyde directly at the surface area of a pH electrode and observed that an irreversible inhibition was detectable by phosphoric acid ester insecticides starting from 50 $\mu g\ L^{-1}$ (Dichlorvos) and a reversible inhibition by methyl-Carbamate insecticide was detectable starting from 0.1 $mg\ L^{-1}$ (Aldicarb). In the year 1989 F. Galgani and G. Bedquene published an interesting method [19] for routine detection of organophosphates and carbamates in sea water, using a microplate reader and the inhibition of acetylcholinesterase. These authors claimed that microplate readers used in immuno-chemistry can be used directly for determination of spectrometric assay and the system is particularly well-adapted to routine monitoring of colorimetric and enzymatic assays.

In the year 1989 V. Wollenberger, F. Scheller et al. assembled a good amperometric biosensor [20] by coupling immobilised choline oxidase and acetylcholinesterase membrane with an amperometric electrode for the H_2O_2 analysis, being hydrogen peroxide, which results from the oxidase reaction, anodically monitored; the influence of the reversible inhibitor F^- ion, on this bienzyme-electrode response, was investigated too. In the same year (1989) F. Scheller et al. described an amperometric method [21,22] for the determination of the activity of cholinesterases, based on the anodic oxidation of thiocholine iodide; furthermore both reversible and irreversible inhibitors of the cholinesterase have been quantified by using standardised serum samples. More recently (1991) V. Wollenberger, F. Scheller et al. constructed [23] potentially disposable sensors for inhibitors (as F^-, Butoxycarboxime, Trichlorfon, Dimethoate), with cholinesterase co-immobilised with choline oxidase in a gelatin membrane on a platinum electrode and with cholinesterase immobilised in polyurethane on a thick-film metallized platinum electrode.

Karube et al., in the year 1990, described [24] a biosensor based on two enzymes, acetylcholinesterase and choline oxidase, co-immobilised on a photocrosslinkable polymer and coupled to an hydrogen peroxide amperometric sensor and using a suitable incubation time, for carbamates and organophosphorus pesticides analysis.

In the same year (1990), M.E.L. Gonzales and A. Townshend described in a paper [25] the rapid flow-injection determination of Paraoxon, using acetylcholinesterase chemically immobilised on controlled porosity glass and the spectrophotometric detection, based on the inhibition of the enzymic idrolysis of α-naphtyl acetate and subsequent reaction of the α-naphtol formed with p-nitrobenzendiazonium fluoroborate. Two flow-injection systems are proposed, based on a continuous-flow, or a stopped-flow system.

J. Kulys and E.J. D'Costa described in the year 1991 [26] an interesting amperometric sensor, based on 7,7,8,8-tetracyanoquinondimethane-modified graphite and immobilised butyrylcholinesterase. The amperometric sensor operated at a potential of 100 mV versus Ag/AgCl and was sensitive to inhibition by organophosphates, after an incubation of the sensor by organophosphate solution from 1 to 17 min.

Mascini and coworkers published two papers, respectively in the years 1991 [27] and 1992 [28], concerning organophosphorus insecticides, and carbamates, determination, using an amperometric hydrogen peroxide and two enzymes (esterase and choline oxidase) both covalently immobilised on a preactivated membrane, by adopting generally incubation times of 30, or 120 min. Recently the light addressable potentiometric sensor (LAPS) was shown to be an effective way to measure small changes in pH; indeed, using a LAPS, the enzyme activity (acetylcholinesterase immobilised to a biotinylated cellulose nitrate membrane via a streptavidin bridge) was determined as a function of substrate concentration (i.e. acetylcholine) by K. R. Rogers et al., in 1991 [29]. This biosensor was also evaluated for its potential to detect organophosphate anticholinesterases. A critical assessment, concerning the application of a chemically sensitive field-effect transistor (CHEMFET) and the enzyme acetylcholinesterase-coating layer to detect organophosphorus compounds and pesticides in the environment was made, by J. Janata et al., in 1981 [30]. More recently (1992) Colapicchioni et al. described an ENFET device [31] with immobilised acetylcholinesterase, used for the detection of organophosporus compounds as Paraoxon, that inhibit the enzyme activity proportionally to their concentration.

Main researches in this field, performed by our group

Some years ago (1988), our research group proposed an operating system [32], based on a choline oxidase sensor, as indicating one, consisting of an oxygen Clark electrode, modified by a cellulose triacetate membrane, containing immobilised the enzyme choline oxidase. This sensor was dipped into a buffer solution, containing dissolved the enzyme butyryl- (or acetyl-) cholinesterase; the addition of butyryl- (or acetyl-) choline to this thermostatted buffer solution, results in an oxygen consumption, quantitatively due to the two in series enzymic reactions of hydrolysis and oxidation and monitored by Clark electrode. As butyryl-, (or acetyl-) cholinesterase activity, was inhibited by organophosphorus pesticides, so that, in their presence, less choline was formed and less oxygen consumed, compounds, such as Malathion or Paraoxon, can be determined, by a calibration procedure and by measuring the decrease of the reaction rate, on adding increasing amounts of pesticides.

In this research choline oxidase enzyme was immobilised in a cellulose triacetate membrane, using the technique described in previous papers [33,34]. The thus obtained membrane was superimposed to the head of the Clark oxygen electrode and secured to it by means of a rubber O-ring. The enzyme butyryl-, (or acetyl-) cholinesterase (2, or 1 U cm^{-3} respectively) was used free in solution. The adopted enzymic-amperometric method was based on the following two reactions:

$$\text{butyryl-, (or acetyl-) choline} \xrightarrow[\text{cholinesterase}]{\text{butyryl- (or acetyl-)}} \text{choline + butyric (or acetic)acid}$$

$$\text{choline} + 2\ O_2 + H_2O \xrightarrow[\text{oxidase}]{\text{choline}} \text{betaine} + 2\ H_2O_2$$

The first reaction was catalyzed by butyryl-, (or acetyl-) cholinesterase free in solution, while the second reaction was catalyzed by choline oxidase immobilised in cellulose triacetate membrane. The rate of the first reaction was determined by measuring the rate of choline formation, related, according to the second reaction, to the oxygen consumption, monitored by the Clark electrode. The determination of the organophosphorus compounds was performed by measuring the decrease of the rate of the oxygen consumption, due to the pesticide action inhibiting butyryl-, (or acetyl-) cholinesterase. Really, after having incubated for 45 min the enzyme, in the presence of the pesticide, in glycine buffer solution and, after the stabilisation time of 15 min of the signal of the electrode, dipped in the buffer, the response of the enzyme-choline sensor, under magnetic stirring, was recorded. The variation rate of the choline production was measured from the slope values of the recorded graphs. On performing the experiment in the presence of known pesticide concentrations, a calibration graph was obtained, by which the unknown pesticide concentration, in the analysed sample, was obtained. This procedure, with all its experimental aspects, was already optimised and particularly described in two papers [35, 36] published in 1990.

The disadvantages of the described approach consist in the required incubation time and in the enzyme consumption for each determination.

A development of the method was carried out by us [37] one year later (1991), including the assembly of a sensor with the two enzymes, butyryl- (or acetyl-) cholinesterase and choline oxidase, both immobilised in a dialysis membrane, superimposed to the oxygen sensor. The new method was characterised by a good handiness, shorter operating times (incubation was not requested) and little amounts of enzyme needed. In this method both enzymes were immobilised in a dialysis membrane, overlapping that one of the Clark electrode. The enzyme electrode was stabilised in 25 cm^3 buffer solution of glycine 0.1 mol L^{-1}, pH 8, at 25°C, free of any other reagents and then a suitable amount of choline derivative was added to obtain the final concentration of butyryl-, (or acetyl-) choline equal to 2×10^{-4} mol L^{-1}. A decrease of the dissolved oxygen concentration was immediately observed. After the steady-state has been attained (about three minutes), the organophosphorus pesticide was added and an increase of the oxygen concentration was consequently observed; really oxygen concentration was renewed, corresponding to its limited consumption rate and to its diffusion from the atmosphere into the reaction vessel. This increase, measured after the attainment of the new oxygen gas equilibrium conditions, was proportional to the pesticide final concentration in the buffer solution. This procedure and the sensor response, are schematically given in Fig. 1.

For each measurement it was necessary to repeat this sequence dipping and maintaining the biosensor in the glycine buffer solution, under stirring, for a time of about twenty minutes, between two subsequent measurements. Using a suitable calibration graph, obtained by these measurements, the concentration of the unknown sample was suitably determined.

The used biosensor was obtained in a very easy manner: both enzymes, choline oxidase and butyryl-, (or acetyl-) cholinesterase, were immobilised in a dialysis membrane. To carry out the operation, 250 U of liophilyzed acetyl-, (or butyryl-) cholinesterase and 50 U of liophilyzed choline oxidase, were stratified on a dialysis cellulose membrane damping with glycine buffer solution. The obtained membrane was coupled to the Clark oxygen electrode, as in the case of cellulose triacetate membrane.

The new arrangement of the sensor and the new operating way were well standardised, showing satisfactory results, reported in a paper [37], published in the year 1991. Nevertheless, by the new method, minimum detection limit and linearity range were shifted to higher concentrations (about one decade) in comparison with those ones found in the previous research above described [35]; in addition the lifetime of the sensor was not satisfactory.

For these reasons we recently (1992) worked at the development of a new enzyme sensor, using a hydrogen peroxide electrode as indicating, modified by superimposing to it a nylon membrane, containing the two enzymes (choline oxidase and butyrylcholinesterase) well immobilised [38]. This new sensor shows a longer lifetime and a better response, in comparison with those ones above described; so the determination of pesticide concentration in some environmental matrices as lake water and soil extracts, were carried out by this sensor. Enzymes immobilisation and assembly of this new sensor are as it follows: both the enzymes (choline oxidase and butyrylcholinesterase) were immobilised in nylon Biodyne membrane, with carboxyl groups on the surface, using the classical methods described in literature [39].

With this aim 250 U of lyophilized butyrylcholinesterase and 30 U of lyophilized choline oxidase were stratified on a nylon membrane (0.6 cm diameter) pretreated with 1-ethyl-3(3-dimethylaminopropyl) carbodiimide. The so obtained membrane was then rapidly washed with glycine buffer 0.1 mol L^{-1}, pH 8.0, and superimposed to the head of the hydrogen peroxide electrode and covered with a high mesh nylon net. Finally, membrane and net, were fixed to the electrode by means of a rubber O-ring. The immobilised enzyme specific activities of choline oxidase and butyrylcholinesterase, measured by the same method as reported in a previous paper [40], were respectively, 1.3 and 0.5 μmol(minxcm^2)$^{-1}$ immediately after the preparation and maintained about 50-70% of the original value, after a short stabilisation time, under working conditions.

The analytical method is based on the two reactions above reported. The enzyme electrode is stabilised in 15 mL of glycine buffer solution 0.1 mol L^{-1} pH 8, at 25°C, then a suitable amount of substrate (butyrylcholine) is added, so that the final concentration of butyrylcholine is 2x10^{-4} mol L^{-1}. An increase of the hydrogen peroxide concentration is immediately observed. Steady-state is reached after about two minutes. At this moment the pesticide is added and a decrease of the hydrogen peroxide in solution is consequently

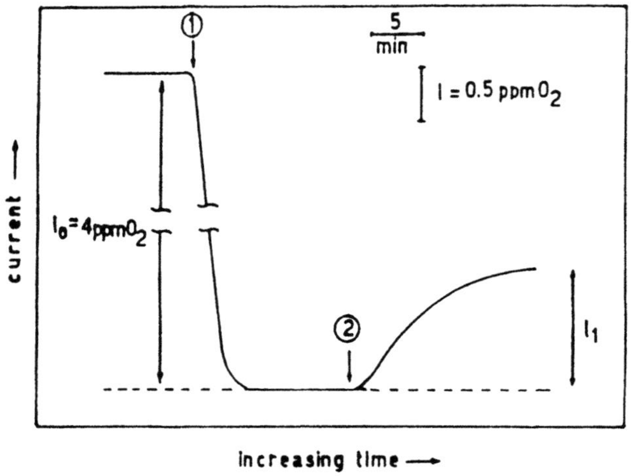

Fig. 1 Malathion determination. Response of the enzyme sensor, with choline oxidase and butyrylcholinesterase both immobilised in dialysis membrane, in glycine buffer 0.1 mol L^{-1}, pH 8:

1) to the addition of butyrylcholine (final concentration 2 x 10^{-4} mol L^{-1});
2) to the addition of Malathion (final concentrations: 1.7 mg L^{-1}.

l_o Variation of O_2 concentration in solution (as $\Delta ppmO_2$) corresponding to the butyrylcholine addition.

l_1 Variation of O_2 concentration in solution (as $\Delta ppmO_2$), corresponding to the Malathion addition.

observed; this decrease, measured at new steady-state conditions (after about 10 min), is proportional to the final pesticide concentration in the buffer solution. By a suitable calibration plot, the unknown concentration can be determined. After each measurement, the sensor is renewed, by an immersion in a stirred concentrated butyrylcholine solution, for about 10 minutes.

The achieved biosensor can be considered satisfactory in order to get long lifetime with sufficient enzyme immobilised activity, but with no increase of cost and complexity. By comparing the results reported in Tables 1 and 2, obtained by this new sensor, with those ones found by two previous developed enzyme sensors [35,37] also summarised in the same tables, we observe that the original method [35], which uses only one immobilised enzyme in membrane and a long incubation time, was however characterised by linearity range shifted by half or one decade to lower concentration values, with respect to the second method, that employs two enzymes immobilised in dialysis membrane and no incubation time. By the last new sensor hereby described all the advantages, reached by the original method (only one enzyme immobilised), were reobtained and improved, but maintaining too all the advantages of the second method (two enzymes immobilised), i.e. no incubation time, simple, cheaper and transportable apparatus.

Table 1

Analytical data for Malathion analysis by the enzyme sensor: (a) choline oxidase immobilised in cellulose triacetate membrane, butyrylcholinesterase in homogeneous solution; (b) two enzymes both immobilised in dialysis membrane; (c) two enzymes both immobilised in nylon functionalised membrane. Butyrylcholine 2 x 10^{-4} mol L^{-1} as substrate).

	(a)	(b)	(c)
- Response time(min)	$\leq$ 10	15	10
- Time of analysis(min)	$\leq$ 70	25	15
- Buffer(mol L^{-1})	glycine 0.1	glycine 0.1	glycine 0.1
- pH	9	8	8
- Working temperature(°C)	25	25	25
- Regression line	y = - 0.24 x + 0.65 (y in $\Delta ppmO_2$ min^{-1};x in mg L^{-1})	y = 0.62 x + 0.09 (y in $\Delta ppmO_2$; x in mg L^{-1})	y = 0.16 x + 0.21 (y in a.u.; x in μg L^{-1})
- correlation coefficient	0.9952	0.9981	0.9983
- Linearity range	0.075 - 0.6 (mg L^{-1})	0.3 - 2.0 (mg L^{-1})	10 - 100 (μg L^{-1})
- Precision on standard solutions(RSD%)	$\leq$ 5.0	$\leq$ 3.2	$\leq$ 2.8
- Inaccuracy on standard solutions by direct method(% values)	-2.0 - + 7.0	- 3.2 - + 4.8	-1.1 - +4.8

Table 2

Calibration graphs and linearity range for different pesticides, obtained by the enzyme sensor (choline oxidase immobilised in cellulose triacetate membrane, butyrylcholinesterase in homogeneous solution and butyrylcholine 2 x 10^{-4} mol L^{-1} as substrate).

	linearity range ($\mu g\ L^{-1}$)	slope ($\Delta ppmO_2\ min^{-1}\ mg^{-1}\ L$)	y axis intercept ($\Delta ppmO_2\ min^{-1}$)	correlation coefficient
Paraoxon	10 - 150	- 0.323	0.061	0.9763
Ethilparathion	150 - 600	- 0.735	1.215	0.9337
Malathion	75 - 600	- 0.243	0.647	0.9952

Calibration graphs and linearity range for different pesticides, obtained by the enzyme sensor (choline oxidase and butyrylcholinesterase both immobilised in dialysis membrane and butyrylcholine 2 x 10^{-4} mol L^{-1} as substrate)

	Linearity range ($mg\ L^{-1}$)	slope ($\Delta ppmO_2\ mg^{-1}\ L$)	y axis intercept ($\Delta ppmO_2$)	correlation coefficient
Paraoxon	0.3 - 1.2	0.72	0.39	0.9981
Ethilparathion	0.2 - 2.0	0.75	-0.002	0.9966
Malathion	0.3 - 2.0	0.62	-0.09	0.9981

Calibration graphs and linearity range for different pesticides, obtained by the enzyme sensor (choline oxidase and butyrylcholinesterase both immobilised on nylon membrane and butyrylcholine 2 x 10^{-4} mol L^{-1} as substrate)

	Linearity range ($\mu g\ L^{-1}$)	slope (a.u. μg^{-1} L)	y axis intercept (a.u.)	correlation coefficient
Paraoxon	5-50	0.18	0.93	0.9999
Ethilparathion	10-85	0.19	-0.02	0.9997
Malathion	10-100	0.16	0.21	0.9983

It is also evident, from the results of Table 3, that the prepared sensor can be satisfactory used for the analysis of authentic matrices of industrial and environmental interest.

Lastly it is worth while to recall the reader's attention to the consideration that our indicating system (in practice the choline enzyme sensor) is characterised by high specific response and works in well buffered solution, so results unaffected from pH change, due for example to the environmental authentic sample to be analysed. This is a great advantage of our sensor in respect to other electrochemical systems which use less specific indicating electrodes [11,16-18,22]. For this reason, in the last years, other authors too [23,24,27,28], studied this problem with the same approach here described.

We are working, at present, on a procedure for entrapping both the two enzymes in cellulose triacetate membrane for the analysis of organophosphorus pesticides and carbamates and on the development of a new "solid state" potentiometric enzyme sensor [41] based on the same esterase enzyme above said, easy to miniaturise and handling.

Table 3

Recovery on authentic matrices of environmental interest, by standard addition method, of Malathion; (a) by the biosensor with choline oxidase immobilised in cellulose triacetate membrane and butyrylcholinesterase in homogeneous solution; (b) by the biosensor with choline oxidase an butyrylcholinesterase both immobilised in the dialysis membrane; (c) by the biosensor with two enzymes immobilised in nylon membrane. (a) and (b) Clark's sensor, (c) H_2O_2 sensor, as indicating electrodes.

Method	Matrix	Recovery values for Malathion addition (% values)	Precision of measurements (as RSD%)
(a)	aqueous extracts of agricultural soil	93.3 - 100.0	≤ 4.0
(b)	"	100.0 - 103.8	≤ 4.0
(c)	"	94.8 - 98.8	≤ 2.7
(a)	lake water	93.3 - 99.2	≤ 4.0
(b)	"	101.7 - 104.6	≤ 4.0
(c)	"	97.8 - 105.0	≤ 3.9

Acknowledgement

Work carried out by the financial support of "Consiglio Nazionale delle Ricerche" (CNR) of Italy, Targets projects in "Biotechnology and Bioinstrumentation" and "Solid State Electronic Materials".

References

1. R.D. O'Brian, Insecticide Action and Metabolism, Academic Press, New York, 1967.
2. G.B. Ansell, S. Spanner, Metabolism, L. Horrocks Raven Press, New York, 1982.
3. J.S. Sherma, Manual of Analytical Quality Control for Pesticides and Related Compounds in Human and Environmental Samples, EPA-600, 2-81-059.
4. J. Sherma, G. Zweig, Pesticides, Anal. Chem., 55 (1983) 57R-70R.
5. G.G. Guilbault, Enzymatic Methods of Analysis, Pergamon Press, Oxford, 1970.
6. T.E. Archer, G. Zweig, Direct colorimetric analysis of cholinesterase inhibiting insecticides with indophenyl acetate, J. Agr. Food Chem., 7 (1959) 179-181.
7. N.V.N. Kumar, M. Ramansundary, Colorimetric determination of methylparathion and oxygen analysis, J. Assoc. Anal. Chem., 63 (1980) 536-538.
8. K.L. Crochet, J.C. Montalvo, Enzyme electrodes system for assay of serum cholinesterase, Anal. Chim. Acta, 66 (1973) 259-269.
9. M. Granmer, A. Peoples, Determination of trace quantities of anticholinesterase pesticides, Anal. Biochem., 55 (1973) 255-265.
10. Z.Y. Yuan, T.M. Zhu, S.Y. Li, X.S. Ji, D.Y. Li, W.B. Wlodarski, Detection of organophosphorous pesticides using an immobilized butyrylcholinesterase membrane and pH-ISFET, 1st World Biosensor's Congress, Singapore, May, 1990.
11. G.G. Guilbault, J. Kristoff, Detection of organophosporus compounds with coated piezoelectric crystal, Anal. Chem, 57 (8), (1985) 1754-1756.
12. G.G. Guilbault, D.N. Kramer, Fluorometric system employing immobilised cholinesterase for assaying anticholinesterase compounds, Anal. Chem., 37 (13), (1965) 1675-1680.
13. G.G. Guilbault, J. Affolter, Y. Tomita, Piezoelectric crystal coating for detection of organophosphorus compounds, Anal. Chem., 53 (1981) 2057-2060.
14. J. Nyeh-Ngwainbi, P.H. Foley, S.S. Kuan, G.G. Guilbault, Parathion antibodies on piezoelectric crystals, J. Am. Chem. Soc., 108 (1986) 5444-5447.
15. C. Tran-Minh, Immobilized enzyme probes for determining inhibitors, Ion Select. Electr. Rev., 7 (1985) 41-75.
16. H. El Yamani, C. Tran-Minh, A. Abdul, D. Chavanne, Automated system for pesticide detection, Sens. Actuat. 15 (1988) 193-198.
17. C. Tran-Minh, P.C. Pandey, S. Kumaron, Studies on acetylcholine sensor and its analytical application based on the inhibition of cholinesterase, Biosens. Biolectr., 5 (1990) 461-471.
18. G. Schwedt, M. Hauck, Reaktivierbare enzymelektrode mit acetylcholinesterase zur differenzierten erfassung von insecticiden in spurenbereich, Fresenius Z. Anal. Chem., 331 (1988) 316-320.
19. F. Galgani, G. Bocquene, A method for routine detection of organophosphates and carbamates in sea water, Environm. Techol. Lett. 10 (1989) 311-322.
20. U. Loffler, U. Wollenberger, F. Scheller, W. Gopel, Amperometric biosensors for the determination of choline and acetylcholine, Fresenius Z. Anal. Chem., 335 (1989) 295-298.
21. R. Gruß, F. Scheller, Amperometric method for the determination of the activity of cholinesterases and detection of the inhibitiors of the enzymes, Fresenius Z. Anal. Chem., 333 (1989) 29-32.
22. R. Gruss, F. Scheller, M.J. Shao, C.C. Liu, Electrochemical determination of cholinesterase activity and indication of its inhibitors using a thick-film metallized platinum electrode, Anal. Lett., 22(5) (1989) 1159-1169.
23. U. Wollenberger, K. Setz, F.W. Sheller, U. Loffler, W. Gopel, R. Gruss, Biosensors for choline, choline esters and inhibitors of choline esterase, Sens.

Actuat. B.4 (1991) 257-260.
24. J.L. Marty, R. Rouillon, K. Sode, I. Karube, Biosensor for carbamates and organophosphorus insecticides, Biosensor 90, Singapore 2-5-1990, Abstracts, Elsevier Sci. Publ. l.d.t., p. 85.
25. M.E. Leon-Gonzales, A. Townshend, Flow-injection determination of paroaxon by inhibition of immobilised acetylcholinesterase, Anal. Chim. Acta, 236 (1990) 267-272.
26. J. Kulys, E.J. D'Costa, Printed amperometric sensor based on TCNQ and cholinesterase, Biosens. Bioelectr., 6(1991) 109-115.
27. M. Bernabei, C. Cremisini, M. Mascini, G. Palleschi, Determination of organophosphorus and carbamic pesticides, with a choline and acetylcholine electrochemical biosensor, Anal. Lett., 24(8), (1991) 1317-1331.
28. G. Palleschi, M. Bernabei, C. Cremisini, M. Mascini; Determination of organophosphorus insecticides with a choline electrochemical biosensor, Sens. Actuat., B, 7 (1992) 513-517.
29. K.R. Rogers, M. Foley, S. Alter, P. Koga, M. Eldefrawi, Light addressable potentiometric biosensor for the detection of anticholinesterases, Anal. Lett., 24(2), (1991) 191-198.
30. J. Janata, R. Robert, J. Huber, R. Cohen, E.S. Kalesar Jr., Chemically sensitive field-effect transistor to detect organophosphorous compounds and pesticides, Aviat. Space Environm. Medicine, 52(11), (1981) 666-671.
31. C. Colapicchioni, A. Barbaro, F. Porcelli, Fabrication and characterization of ENFET-devices for biomedical applications and environmental monitoring, Sens. Actuat., 13,6, (1992) 202-207.
32. L. Campanella, M.P. Sammartino, M. Tomassetti, A simple enzyme probe for detecting pesticides and biological neurotransmitters, European Conference on Biotechnology, Verona (Italy), 7-11-1988, Abstracts, p. 198.
33. L. Campanella, M. Tomassetti, M.P. Sammartino, Enzyme sensor for the determination of choline containing phospholipids in some biological fluids, Analyst, 113 (1988) 77-80 and 113 (1988) 994.
34. L. Campanella, M. Tomassetti, M.P. Sammartino, Enzyme-entrapping membranes for enzyme sensors, Sens. Actuat., 16 (1989) 235-245.
35. M. Achilli, L. Campanella, M.P. Sammartino, M. Tomassetti, Un nuovo metodo per la determinazione enzimatica di pesticidi, Inquinamento, november (11), (1990) 66-70.
36. L. Campanella, M. Achilli, M.P. Sammartino, M. Tomassetti, G. Visco, Determinazione di pesticidi fosforati con un sensore enzimatico a colina ossidasi, Notiz. IRSA (CNR), Mat. Anal. Acque, n. 10 (4), (1990) 29-35.
37. L. Campanella, M. Achilli, M.P. Sammartino, M. Tomassetti, Butyrylcholine enzyme sensor for determining organophosporus inhibitors, Bioelectrochem. Bioenerg., 26 (1991) 237-249.
38. L. Campanella, R. Cocco, M.P. Sammartino, M. Tomassetti, A new enzyme inhibition sensor for organophosphorus pesticides analysis, Sci. Tot. Environ., 123/124 (1992) 1-16.
39. M. Mascini, M. Mazzei, Amperometric sensor for pyruvate with immobilized pyruvate oxidase, Anal. Chim. Acta, 192 (1987) 9-16.
40. M. Mascini, M. Tomassetti, M. Iannello, Determination of free and total cholesterol in human bile samples using an enzyme electrode, Clin. Chim. Acta, 132 (1983) 7-15.
41. L. Campanella, W.B. Wlodarskij, M.P. Sammartino, M. Tomassetti, Solid state sensors for determining organophosphorus pesticides, Biosensor '92, 20-5-1992, "Proceedings" (Suppl.) p. 24.

BIOSENSORS FOR ELECTROCHEMICAL NON INVASIVE ANALYSIS

G.Palleschi*, D.Compagnone*, G.J.Lubrano**, M.H.Faridnia** and G.G.Guilbault***

* DISTAAM, Universita' del Molise Via Tiberio, 21 86100 Campobasso Italy
** Universal Sensors Inc. 5258 Veterans Blv Suite D Metairie La 70006 U.S.A.
*** Dept. of Chemistry, University of New Orleans, New Orleans La 70148 U.S.A.

Abstract Amperometric alcohol, lactate and glucose sensors for non invasive measurements in human saliva, sweat and transbuccal mucosa have been developed and evaluated. These probes resulted in a high degree of selectivity and good stability. The response time was from a few seconds to two minutes making them suitable for analysis in a very short time especially for patients such as diabetics, hemophiliacs, neonates and the elderly.

I. INTRODUCTION

Non invasive analyses are becoming increasingly more important because they can solve problems for patients who have to control daily parameters such as glycemia and urea and in general for people with problems in collecting blood such as hemophiliacs, neonates, elderly.

Many sampling sites are possible in non invasive sensing but most efforts in the past have been directed to saliva, sweat, lacrimal fluid and human skin itself.

Electrochemical biosensors have proven to be specific, selective and easy to use for the determination of metabolites in clinical environmental and food analysis [1]. The advantage of using biosensors is that the sample can be processed immediately after collection in many cases, requires no separation steps and the response time is from seconds to a few minutes.

Moreover, the specificity of the immobilized enzymes often used to assemble biosensors and the selectivity of the electrochemical transducer make this probe suitable for measurements in biological fluids [2].

All of these biosensor features led us to investigate the possibility of measuring alcohol in saliva, glucose in saliva and in transbuccal mucosa, and lactate in saliva and sweat.

Human saliva, as reported by Tenovuo [3], is a complex mixture of different gland secretions mixed with bacteria, leukocytes, sloughed epithelial cells, and crevicular fluid. The concentration of most salivary constituents depend on the flow rate of saliva, consequently the collections of saliva need to be standardized. As a result of bacterial action, the composition of

G.G. Guilbault and M. Mascini (eds.), Uses of Immobilized Biological Compounds, 501–514.

saliva will change on standing. For this reason metabolites present in saliva have to be processed as soon as possible and biosensors fit perfectly this challenge. Earlier studies demonstrated that most therapeutic agents were transferred rapidly from plasma to saliva, and more recently it has been shown that the concentration of drugs in saliva is proportional to the concentration in plasma. The view that, for the most drugs, salivary concentrations reflect unbound drug concentrations in plasma is now accepted [4].

II. APPLICATIONS.

1. MEASUREMENT OF ALCOHOL IN SALIVA

The increase in traffic accidents has enhanced the importance of the determination of alcohol concentration. Impairment of driving ability starts at as low as 50 mg of ethanol per 100 ml of blood [5]. Alcohol analysis is also important in industry for process control, in fermentation and clinical laboratories. Blood alcohol is presently measured with gas chromatographic [6] and enzymatic, automatic or spectrocphotometric methods [7] or breath analyzers [9]. These methods are inaccurate as the last one or require complex and expensive intrumentation and reagents. Jones has studied the saliva/blood alcohol ratio. Results showed a mean ratio of 1.077 (n = 336) with 95% confidence limit of 1.065 and 1.088 so he established a sound basis for non-invasive saliva testing for blood alcohol content.

Several studies published in the literature have shown a relationship between blood alcohol and the saliva alcohol concentration [9, 11].

A painless, non-invasive alcohol measurement probe has been developed herein, based on the measurement of alcohol in saliva with an enzyme based sensor. The sensor is an amperometric hydrogen peroxide electrode that is covered with a gas permeable membrane.

Alcohol Base Sensor

The saliva alcohol device requires a specific, sensitive, reliable alcohol sensor. The two most appropriate base sensors for alcohol electrodes are the oxygen electrode and the amperometric, hydrogen peroxide electrode. When these sensors are coupled with alcohol oxidase, response to alcohol is achieved due to the following reaction:

$$\text{Alcohol} + O_2 \xrightarrow{\text{Alcohol Oxidase}} H_2O_2 + \text{Acetaldehyde}$$

Assay of alcohol is achieved by measuring the decrease in oxygen with an oxygen electrode or by measuring the increase in hydrogen peroxide with a hydrogen peroxide electrode. With either system, the initial rate of reaction within the first 12 seconds, or the steady state current at 1 to 2 minutes after the injection of ethanol can be measured. Because of the potential difference in the oxygen level between the atmosphere and testing matrices (blood, saliva, urine), the hydrogen peroxide system could give the best results, but both systems were investigated in the development of the base sensor. The oxygen measurement system is more selective, and would eliminate electroactive interferences that are a problem in the peroxide based probe.

Assays in Saliva

The response to ethanol in saliva was measured by steady state and rate methods. The applied potential was set to +0.650 V vs Ag/AgCl. The electrode was rinsed with deionized water and placed in 10 mL of PBS until a steady base line was achieved. The electrode was then removed from the buffer and excess liquid was removed with tissue paper. Thirty µL of the sample was injected onto the tip of a 2 cm diameter Teflon rod, forming a droplet. The electrode was lowered over the rod until just touching the droplet, forming a 5 mm diameter contact disk. The maximum rate of response was calculated from the peak height of the first derivative curve. The steady state response was calculated by subtracting the residual current from the steady state current which was obtained within one or two minutes.

Saliva ethanol concentrations were calculated from the calibration curves, and are plotted in Figures 1 and 2 for two subjects as concentration versus time after ingestion.

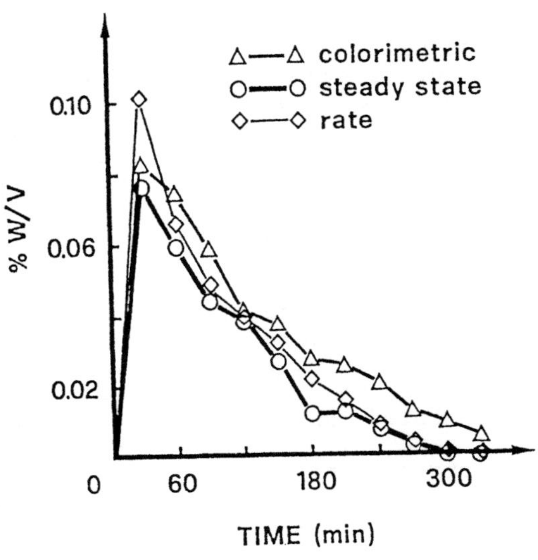

Figure 1: Ethanol metabolism curve is shown for subject A through measurement of saliva alcohol content, (○) = steady state, (◇) = rate, (△) = Colorimetric method.

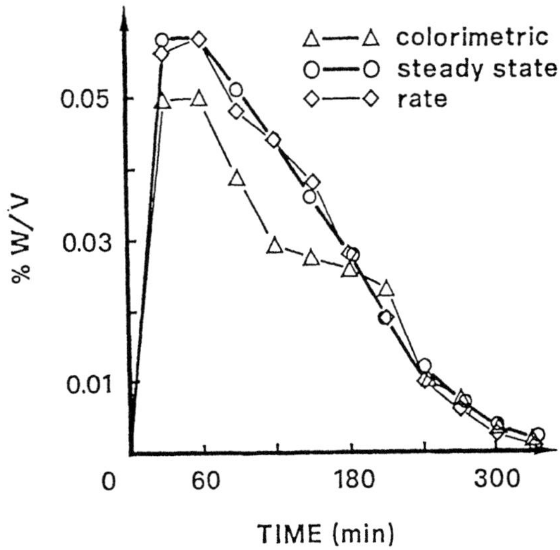

Figure 2: Ethanol metabolism curve is shown for subject B through measurement of saliva alcohol content, (○) =steady state current, (◇) = rate and (△) = Colorimetric method.

The alcohol metabolism curve obtained through analysis of saliva alcohol content resembled closely that reported by many investigators for blood alcohol content [12]: an increase in alcohol concentration in body fluids during the absorption phase and a maximum peak level, followed by a decrease, denoting the elimination of alcohol from the body fluids.

2. MEASUREMENT OF LACTATE IN SALIVA

Our success with the alcohol biosensor led us to investigate the possibility of measuring lactic acid in saliva. A bibliographic search gave very scarce information on this subject. According to a biological handbook [13] lactic acid in human saliva should be around 0.2 mmol/l. Lactate is an important metabolite to be monitored rapidly in people who are sick enough to be in critical care units in hospitals, mainly to prevent heart attacks. Lactate monitors are also used in diabetes control, food analysis and sport medicine to help athletes tailor their training [14-16]. It has already been demonstrated that lactate in blood increases after meals [14, 17], and during physical exercise [18].

An electrochemical lactate probe operates in the following way:

$$\text{Lactate} + O_2 + H_2O \longrightarrow \text{Pyruvate} + H_2O_2$$

This reaction is catalyzed by the enzyme lactate oxidase. The hydrogen peroxide produced is measured at +650 mV with platinum and silver/silver-chloride electrodes. The current change due to the oxidation of hydrogen peroxide is proportional to the lactate concentration in the sample.

Our lactate sensor consists of dual platinum electrodes and a common Ag/AgCl reference electrode assembled in a way that one platinum electrode is an active lactate sensor and the other is inactive to lactate. In this way, changes in background current due to pH, temperature and ionic strength variations, together with current changes due to possible interferences affecting both sensors, can be evaluated and eliminated.

Saliva Lactate Probe

The saliva lactate probe (Fig. 3) was assembled in the following way. A 100 MWCO cellulose acetate membrane was placed on the electrode surface such that the platinum and reference electrodes were completely covered. The enzyme membrane was then cut in two pieces. One piece was placed on one of the platinum electrodes and the other was immersed in boiling distilled water for 1 min, then placed on the other platinum electrode. Finally, the probe was covered with a 0.03 µm polycarbonate membrane and firmly held on with an O-ring.

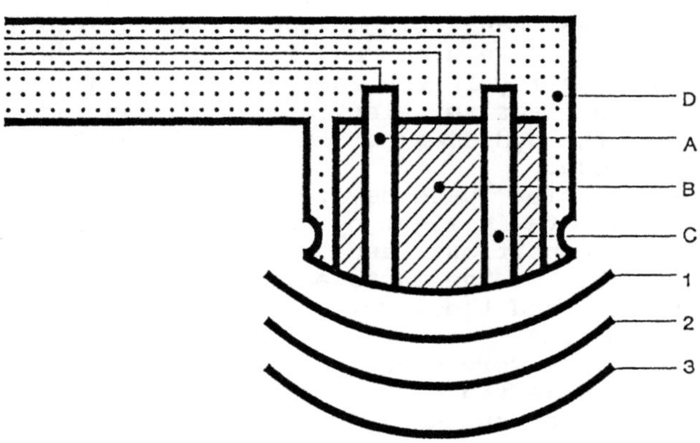

Figure 3: Dual lactate electrode scheme and assembly. D = Plexiglas; A and C = platinum electrodes; B = silver/silver-chloride electrode; 1 = cellulose acetate membrane (100 MWCO); 2 = enzyme membrane; 3 = polycarbonate membrane.

For measurements in saliva, the probe was equilibrated in 4 ml of stirred PBS, then 1 ml of freshly collected saliva was injected and the current change for both electrodes recorded. This dual electrode does not have an internal filling solution, so the electrolytic film on its surface is due to the bulk solution. The PBS used has the same chloride concentration as blood, and chloride in saliva has approximately the same concentration, so the reference potential is not affected by chloride variations. The reference lactate analysis was an enzymatic spectrophotometric method at 340 nm, according to Sigma procedure No. 826-UV.

Saliva Samples

Saliva samples were collected without stimulation among colleagues and students working in the laboratory. The fasting saliva lactate concentration should be ca. 0.2 $mmol \cdot l^{-1}$. Saliva lactate was measured by injecting 1 ml of freshly collected saliva from a subject in the fasting state into 4 ml of PBS. The saliva lactate response (1.8 nA) was calculated by subtracting the inactive electrode response from the active electrode response. This corresponds to 0.21 $mmol \cdot l^{-1}$ lactate in the saliva sample, which is in agreement with the value reported in the literature [13].

The main electrochemical interferences present in saliva were studied to establish their effect on the current background. Ascorbic acid and uric acid gave no current variation when injected at concentrations one and ten times higher than that present in saliva. Hydrogen peroxide (0.01 $mmol \cdot l^{-1}$) gave a current variation of 6 nA in 1.5 min at the active lactate probe and 8 nA in less than 1 min at the inactive probe. The lower response at the active electrode is presumably due to catalase present as an impurity in the enzyme, which consumes the hydrogen peroxide at the active electrode, whereas at the inactive electrode all the hydrogen peroxide is "free" to diffuse to the platinum electrode.

Further studies were devoted to observe possible lactate variations in saliva in eight subjects working in the laboratory, under fasting conditions and after meals. The results are reported in Table 1. The lactate level in saliva varied widely between subjects and the level after a meal increased randomly for all the subjects.

Saliva samples were measured by two different methods. Five samples were collected from subjects in the fasting state and three samples from subjects 30 min after eating. Lactate was measured immediately after sample collection with the electrochemical method and with the spectrophotometric method. The results are reported in Table 2.

TABLE 1

Saliva lactate measurements for eight subjects before and after eating.

Subject No.	Lactate ($mmol \cdot l^{-1}$)	
	Fasting	After meal
1	0.2	0.56
2	0.34	0.50
3	0.70	0.76
4	0.06	0.50
5	0.50	1.20
6	0.40	0.44
7	0.32	0.56
8	0.36	0.40

TABLE 2

Comparison of saliva lactate contents measured by amperometric and spectrophotometric methods

Sample No.[a]	Lactate concentration ($mmol \cdot l^{-1}$)		
	Spectrophotometric	Amperometric	Difference
1	1.20	0.85	-0.35
2	0.12	0.09	-0.03
3	0.76	0.88	0.12
4	0.60	0.64	0.04
5	0.75	0.84	0.09
6	2.50[b]	1.50	-1.00
7	1.00[b]	0.48	-0.62
8	0.60	0.42	-0.18

[a] Samples 1,6 and 7 were collected after a meal. [b] Samples 6 and 7 when collected displayed a brownish and yellowish color, respectively.

Samples taken after a meal displayed the largest difference. Probe results obtained from "clear" saliva in fasting subjects were in good agreement with the results given by the spectrophotometric procedure.

Another study performed with this dual lactate probe was the determination of lactate in saliva from a subject in the fasting state before and after physical exercise. This experiment was done because of the well known lactate increase in blood during exercise when a subject is in the anaerobic mode [27]. Lactate measurements by non-invasive methods will be easier and painless for athletes involved in sports competitions and for people who exercise frequently.

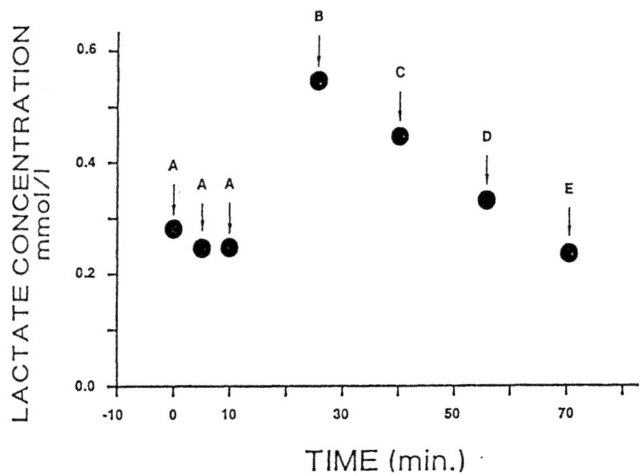

Figure 4: Lactate acidaemic curve. A, Saliva lactate measured while the subject was in the resting state; B, C, D and E, saliva lactate measured immediately, 15, 30, 45 and 60 min, respectively, after physical exercise.

Lactate was measured in a subject in a resting condition until a stable current baseline was reached, then the subject was asked to run at an initial speed of 8 $km \cdot h^{-1}$, increasing every 5 min by 5 $km \cdot h^{-1}$. After 15 min the subject went into the anaerobic mode, stopped running and rested. Saliva was collected three times before the exercise, immediately after the subject stopped running (anaerobic condition) and 15, 30 and 45 min after the exercise. The results are shown in Fig. 7.

Figure 4 shows an increase in saliva lactate when the subject was in the anaerobic mode and a return to the base lactate level after 30 min. Only saliva lactate and not blood lactate was measured because of the non-availability of a micro-scale lactate blood test, but the curve obtained reflects the same trend frequently obtained with continuous monitoring of blood lactate [17]. The lactate probe was calibrated before and after each experiment. The probe showed no decrease in sensitivity and was reproducible.

3. GLUCOSE DETECTION IN SALIVA

Previously, it was mentioned that saliva is an excellent matrix for detection of many substrates and that the number of analytes detectable in saliva are continuously increasing, as more research is carried out in this area. We investigated the applicability of saliva as a matrix for glucose detection.

The glucose probe used was based on a novel H_2O_2 electrode which behaved excellently in terms of stability and reproducibility with no electrochemical interferences due to a new blocking membrane used to protect the platinum electrode surface.

The linear range for glucose was 1-1200 mg/dl and the response time was two minutes for current steady state and 10 seconds for the rate.

Four subjects participated in this study, three males and one female. The men's age ranged between 27 and 39 years and they weighed 72 and 85 Kg; the female subject was 23 years old and weighed 54 Kg. All males were healthy individuals but the female was diabetic and was periodically prescribed insulin by her physician.

All men, according to their weight, drank appropriate amounts of a sugar solution to increase their blood glucose level. Only men participated in the initial part of the study and, to prevent any liability, the female subject was only asked to supply saliva samples before and after insulin injections.

Saliva samples were collected in disposable plastic cups and were immediately analyzed without dilution. The analysis procedure was similar to that described above for detection of glucose in sera samples.

Blood samples were also collected from the subjects finger tips, and were analyzed using an Ames Glucometer. There was no correlation between the saliva and blood glucose content of any of these male subjects. This is perhaps due to the effect of insulin that is released into the blood stream to normalize the blood glucose level when its concentration increases beyond a certain level.

This factor was investigated with the help of the subject who was diabetic. The subject was advised to collect saliva before insulin injections, when the blood glucose content was at its highest, and afterwards, when the glucose level had dropped due to the presence of insulin. The saliva samples were collected in a 5 ml vial and kept at 0 °C until analysis. The subject's blood glucose levels were also registered at the times when saliva samples were collected. The saliva samples were analyzed by the glucose probe and compared to the blood glucose level. A marked increase in saliva glucose content was observed in the absence of insulin when the subject's blood glucose level was the highest. After insulin injection the saliva glucose content had dropped down, comparable to those observed in healthy individuals (Table 3). This illustrates that saliva measurement may be useful in measuring hyperglycemic patients.

TABLE 3

Saliva glucose content of a hyperglycemic patient determined by the glucose biosensor and compared to blood glucose concentration measured colorimetrically.

Samples	Glucose concentration (mg/dl)	
	Blood	Saliva
A1	205	46.0
A2	201	44.9
A3	203	43.3
A4	211	47.6
B1	103	17.3
B2	98	16.3
B3	102	16.5
B4	105	17.9

A(1-4) - Samples were collected before administration of insulin when the patient showed high blood glucose level.

B(1-4) - Samples were collected after the administration of insulin when the patent's blood glucose level had decreased considerably due to the presence of insulin.

The sensor, applied to glucose analysis in undiluted sera and saliva samples, showed very good correlation with that of the spectrophotometric method in sera measurements. Although there was no correlation between saliva and blood glucose content in healthy subjects, there was a good correlation for the hyperglycemic patient. Further studies are required to establish this correlation.

4. MEASUREMENT OF LACTATE IN SWEAT

The use of sweat as an analytical sample dates back almost 20 years. The analysis of sweat for electrolyte concentrations still remains the laboratory "Gold standard" for the diagnosis of cystic fibrosis [28]. Here the chloride concentration in sweat provides greater discrimination between CF and normal populations than does sodium [29].

An excellent correlation between the lactate concentration in sweat to blood was shown many years ago [4], and can form the basis of a useful analytical method.

In the previous section, a dual lactate electrode was introduced that enabled analysis of saliva lactate content. However, due to the large difference in saliva and sweat lactate concentrations, 2 mg/dl and 300 mg/dl, respectively, at resting condition, and due to limited linearity of the probe developed for saliva lactate content measurement (while sufficient for that purpose in this media it was not ideal for measurements carried out in sweat samples), a dilution step would be necessary in use of this biosensor that would have resulted in a decrease in sensitivity of the electrode. Hence a new probe was developed that utilized the enzyme lactate oxidase immobilized on the novel hydrogen peroxide electrode developed by us for glucose detection. The probe performance has been throughly investigated and documented. An electrochemical lactate probe operates according to the following reaction:

$$\text{Lactate} + O_2 + H_2O \xrightarrow{\text{Lactate Oxidase}} \text{Pyruvate} + H_2O_2$$

This reaction is catalyzed by the enzyme lactate oxidase and detection is via a platinum based electrode.

The participants in the study were three males, ages 24-36 years, accustomed to moderate exercise and were not professional athletes.

Sweat samples, 1-2 mL, were collected in 5 mL disposable vials, capped and either immediately assayed or kept at 0 °C until assay. During each experiment, samples were collected from the upper part of legs and arms, areas covering Gluteus maximas and biceps, respectively. To establish a base line at least three samples were collected before exercise, while the subject was at rest. The subjects then ran at a speed of 4-6 miles per hour for 30 minutes. Sweat samples were collected immediately after the run and thereafter at 5 minutes intervals, until the sweat lactate level returned back to normal. The biosensor was calibrated before the analysis and 20 mg/dL L-lactate controls were run throughout the study. One hundred µL of each sample was pipetted into 5.00 mL of stirred PBS and the steady state current and maximum initial rate responses were monitored. The steady state current was measured within two minutes and the initial rate of reaction was measured in less than 10 seconds.

A typical plot of sweat L-lactate vs time is given in figure 5. All subjects had an increase in sweat lactate after the physical exercise and the concentration decreased during the resting period. This finding is in agreement with that reported by many researchers on L-lactate metabolism after a physical exercise [21, 22, 23]. Location of sample collection is crucial and the concentration of L-lactate depends on the type of exercise. The samples collected from subjects' arms and legs showed the same trend, but the former showed much lower baseline levels and peaked at a lower maximum value.

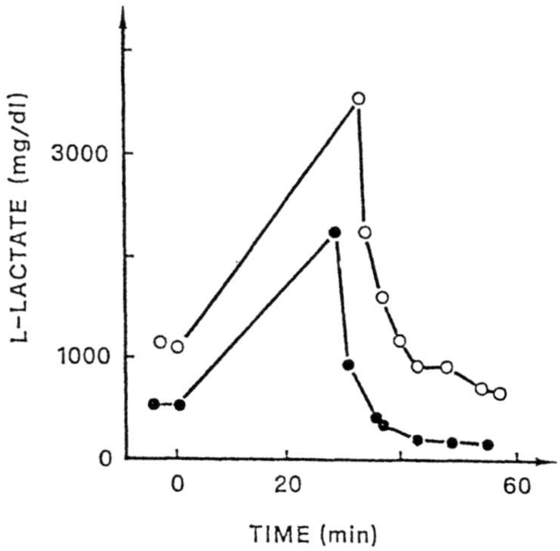

Figure 5: L-lactate metabolism curve presented as sweat L-lactate concentration (mg/dl) vs time (min), samples were collected from the subject's legs (○) and arms (●) before and after a physical exercise.

5. TRANSBUCCAL MUCOSA MEASUREMENTS

Transbuccal fluid is distinct from saliva, and is an ultrafiltrate of blood, with apparently microsolute concentrations that follow these of blood.

Studies performed at Universal Sensors measuring glucose diffusion through the buccal mucosa (TBM) have shown that a relationship exists between the blood glucose level and the glucose from the buccal mucosa. Xienta Inc. [24] has a patent that covers any method for performing in vivo measurements of a substrate, enzyme or other biochemical agents on the skin or mucous membranes of an accessible body surface.

It has been shown that an electrode can be fabricated for direct measurements in the transbuccal mucosa [25]. As shown in Fig. 6, there is a direct correlation between the response of the amperometric glucose oxidase based sensor (rate mode, Curve A) with the results obtained from whole blood analysis (Curve G) using the Ames glucose kit (the Glucometer). A second inactive electrode, designed to cancel out any electroactive impurities present in the TBM, showed no response in the period of the measurement (Curve D). Curve C shows the changes in the steady state. A change occurs, but is not as pronounced as in the rate mode.

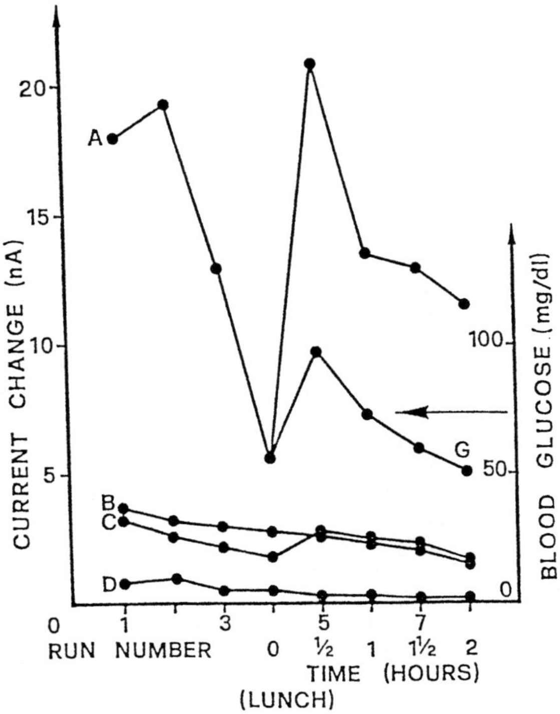

Figure 6: Correlation study in glucose analysis by the TBM biosensor and whole blood analysis. Volunteer 1
A - Rate mode of the TBM biosensor
B - Glucose control
C - Steady state mode of the TBM Biosensor active electode
D - Same, except inactive electrode
G - Glucose in whole blood (Ames Glucometer).

The device underwent clinical trials on over 100 patients at Pharmacontrol Inc., the parent of Xienta Inc., and showed good correlations of TBM to whole blood glucose in about 50% of the patients tested. Unfortunately this non-invasive technique showed too much variation in the remaining patients to be reliable enough to enter the marketplace.

REFERENCES

[1] M.Mascini, and G.Palleschi, Selective Electrode Review 11 (1989) 191.
[2] G.Palleschi, M.A.NabiRanhi, G.J.Lubrano, J.N.Ngwainbi, and G.G.Guilbault, Anal. Biochem., 159 (1986) 114.
[3] Human Saliva, Clinical Chemistry and Microbiology, (Tenovuo, J.O., ed.) CRC Press, Boca Rotan, Florida 1988.
[4] Blood and Other Body Fluids (R.L.Altman, and D.S.Dittmer, eds), Federation of American Societies for Experimental Biology. Washington, D.C. 1961, pp.399-403.
[5] J.D.Bauer, P.G.Ackermann, and G.Toro, Alcohol and the Impaired Driver. Bary's Clinical Laboratory Methods, 7th Ed, Mosby, St. Louis, 1968, pp. 684-687.
[6] A.W.Jones, (1978), J. Stud. Alc. 39, 1931-9.
[7] L.Prencipe, E.Iaccheri, and C.Manzati, Clin. Chem. 483, 486-500 (1987).
[8] M.F.Mason, and K.M.Dubowski, (1976). J. Forensic Sci. 21, 9
[9] A.W.Jones, Clin. Chem. (1979) 25, 1394.
[10] Life Scan Corp., Mountain View, Ca, Manual for the Alcoscan Saliva Alcohol Dipstick.
[11] A.W.Jones, (1979) Scand. J. Clin. Lab. Invest. 39, 199-203.
[12] A.W.Jones, K.A.Jonsson, and L.Vorfeldt, (1989) Clin. Chem. 35, 400-404.
[13] Blood and other body fluids. Dittmer D.S. (Ed.), Federation of American Societies for Experimental Biology, Washington, D.C. p. 400 (1961).
[14] M.Mascini, S.Fortunati, D.Moscone, (1985), Clin. Chem. 31, 451-453.
[15] R.Pilloton, T.N.Nwosu, M.Mascini, (1988), Anal. Lett. 21 727-740.
[16] G.Palleschi, M.Mascini, L.Bernardi, P.Zeppilli, (1990) Medical and biological Engineering and Computing, 28, B25-B28.
[17] M.Mascini, F.Mazzei, D.Moscone, G.Calabrese, and M.Massi Benedetti, (1987), Clin. Chem. 33 591-593.
[18] I.Karube, (1989) Biosensors 14, 3-13.
[19] P.M.Tocci, and R.M.McKey, (1976) Laboratory Conformation of the Diagnosis of Cystic Fibrosis. Clin. Chem. 22, 1841-4.
[20] M.Gleeson, and R.L.Henry, (1991) Sweat Sodium or chloride Clin. Chem. 37, 112.
[21] G.Palleschi, R.Pilloton, M.Mascini, L.Bernardi, A.De Luca, and P.Zeppilli, Biosensors 14, 3-13 (1989).
[22] G.Palleschi, M.Mascini, L.Bernardi, and P.Zeppilli, Biosensors, B25-28 (1990).
[23] N.Shimojo, K.Fujino, S.Kitahashi, M.Nakao, K.Naka, and K.Okuda, Clin. Chem., 37, 1978-1980 (1991).
[24] P.Pugliese, 1978. Apparatus and Methods for Performing In-Vivo Measurements of Enzyme Activities. U.S. Patent, 4, 071, 020.
[25] G.Guilbault, G.Lubrano, G.Palleschi, (1987) Transbuccal Mucosa Biosensor for Direct Assay of Glucose. Universal Sensors Final Report to Pharmacontrol Inc.

FLOW-THROUGH BIOSENSORS IN CLINICAL ANALYSIS

M.D. LUQUE DE CASTRO and J.M. FERNANDEZ-ROMERO
Department of Analytical Chemistry, Faculty of Siences,
University of Córdoba, E-14004 Córdoba, Spain

Abstract An overview of the design and implementation in clinical analysis of flow-through biosensors is presented. A first section is devoted to biosensing systems as predecessors of flow-through biosensors. Then, a systematic study of the continuous manifolds used for implementation of reaction-rate methods based on biosensors is shown. The most recent developments involving transient retention of the reaction product in the flow-cell and the dual immobilization of an enzyme (permanent) and a reaction product (transient) are also commented.

INTRODUCTION

Chemical and biochemical sensors have undergone a spectacular growing in the last few years. A number of monographies devoted to chemical and biochemical sensors based [1,2] or not [3-9] on the use of optical fibers show the exponential development of these devices.

The concept of biosensor has evolved along the last decade: for some authors it is a self-contained analytical device that responds selectively and reversibly to the concentration or activity of chemical species in biological samples. No mention is made in this case of a biologically active material involved in the device; thus any sensor physically or chemically operated in biological samples can be considered a biosensor. Nevertheless, for most of the scientifics working in this area the association of a biological sensing material with a transducer is compulsory, and even a biosensor can be simply defined, in a first approach, as a device that intimately associated a biological sensing element and a transducer.

Our contribution to biosensors is restricted to optical devices that can work in flowing systems, as the best way to automatize steps previous to measurement. Thus, separation processes, dilutions, derivatizing reactions, etc. can be developed into continuous-flow manifolds to fit the features of the sample and/or analyte to the performance of the biosensor.

Biosensors based on continuous systems can be of two main types: Flow-through

G.G. Guilbault and M. Mascini (eds.), Uses of Immobilized Biological Compounds, 515–525.

biosensing systems, in which the biochemical reaction takes place before the sample reaches the flow-cell, and flow-through biosensors, which involve development of either the overall process or only the last step of the biochemical reaction in the flow-cell.

Biosensing systems can be based on the use of a dissolved enzyme continuously circulated continuously through the flow-manifold, into which the sample is inserted, or on circulating the sample along the system and injecting the enzyme into it, thus decreasing consumption of this last. In any case, the best way of using expensive catalysts in continuous systems is to immobilize them on an appropriate support. An enzymatic reactor placed between the injection system and the detector is the most common location for enzymes used in continuous-flow systems.

Flow-through biosensors are among the most promising biosensors as they can solve real analytical problems. For example, a continuous-flow manifold prior to the sensor allows the sample to be modified (manipulated) as required for implementation of a separation step, automatic dilution or preconcentration, a chemical or biochemical reaction, etc. before the final step in the flow-cell, where the catalyst is immobilized on an appropriate support.

The layout for this paper is as follows: An overview of the most interesting approaches to biosensor systems developed by our team, each of which overcomes a specific drawback, will be followed by a description of the different approaches to biosensors we have designed so far, namely:

-Simple biosensors based on the immobilization of dehydrogenases on CPG located in the flow-cell. These were our first steps in this field.

-A systematic study of the different ways of implementing reaction-rate measurements in flow-through biosensors with the aid of an appropriate manifold,

-The ability to accomplish a transient retention of the reaction product on a support packed in the flow-cell. This is an atypical, versatile type of biosensor with a very wide scope of application.

-Finally, the flow-cell can be packed with two different supports, which enable permanent immobilization of a catalyst and subsequent transient retention of the reaction product for monitoring purposes.

BIOSENSING SYSTEMS

The development of biosensing systems was the first step of our research and it was aimed at expanding the field of use of flow-injection (FI) manifolds and to know in deep the behaviour of biochemical systems in flowing approaches for the subsequent implementation of real flow-through biosensors. So, we have designed continuous-flow biosensor systems that involve:

-Placement of an enzymatic reactor in a nested-loop arrangement, thus making possible the development of reaction-rate methods for determination of both substrate concentrations [10] and enzymatic activities [11-13].

-Double halting of the flow to select an appropriate reaction time for each step [14].

-Iterative passage of the sample plug through the reactor and detector, aimed at obtaining multipeak recordings for reaction-rate determination with conventional photometers or fluorimeters by use of immobilized enzymes [11].

-HPLC/continuous systems including an enzyme reactor, with the main purpose to reduce costs by avoiding continuous use of a liquid chromatograph for the analysis of large numbers of samples [15], or to perform continuous post-column monitoring of the reaction rates of analytes isolated by HPLC [16].

BIOSENSORS

However, the final aim of our research in this field was the development of flow-through biosensors. Initially, after several attempts, our group eventually succeeded in establishing the optimal experimental conditions for immobilizing enzymes in the flow-cell of a photometric or fluorimetric detector in order to perform reaction-rate measurements by halting of the flow when the plug reaches the detection zone. The flow-cell used consisted of two dismountable parts. On its inner walls were fixed with silicone controlled-pore glass (CPG) beads previously activated with silane and glutaraldehyde. The two side were glued together. The enzyme, alcohol dehydrogenase, was immobilized by recirculation for four hours. It is interesting to note that the light beam may cross the cell freely, so the absorbance of the solution inside the cell could be measured very readily. This flow-cell was placed in a photometer included in a continuous manifold, where a dual valve allowed simultaneous injection of two volumes of serum or urine sample and co-enzyme into two streams of the same buffer in a merging-zones configuration. When the mixed plug reached the flow-cell, the flow was stopped and a set of absorbance-time data (which was related to the concentration of ethanol in the sample) was acquired by the computer. In this way it was possible to perform photometric reaction-rate measurements by using immobilized enzymes. The main advantage of this approach relative to conventional packed enzyme reactors lies in the elimination of matrix effects by measuring absorbance increments rather than absolute signals [17].

Malate dehydrogenase was also immobilized on CPG for the determination of oxaloacetate in serum by including the biosensor in this continuous flow-system, where the serum sample was injected into a buffered stream and then merged with the coenzyme. The mixture was stopped in the flow-cell to perform fluorimetric reaction-rate measurements.

SYSTEMATIC STUDY ON THE IMPLEMENTATION OF REACTION-RATE MEASUREMENTS

After the above mentioned conventional way of performing reaction-rate measurements, we carried out a systematic study of the different continuous approaches which could be used for this purpose.

The choices were stopped-flow operation and iterative passage of the sample plug through the flow-cell, whether with reversal of the flow direction, or by use of an open-closed circuit.

The biochemical system chosen for this comparative study was the hydrolysis of ß-D-glucuronide compounds (4-nitrophenol-ß-glucuronide specifically). The reaction was catalysed by ß-D-glucuronidase immobilized on CPG, and was monitored at 405 nm via the 4-nitrophenol produced. All these biosensors relay on immobilization of the biocatalyst on the walls of the flow-cell of a photometric detector.

The biosensor based on stopped-flow measurements used a single-channel manifold, where the sample was inserted into a buffer stream and driven to the flow-cell (Fig. 1A), there it was stopped giving rise to a type of recording (Fig. 1B) that shows the extent of development of the enzymatic reaction during the stop interval, after which the plug was sent to waste.

A single-channel manifold was also used for iterative reversal of the flow, which was accomplished by altering the operation of a computer-controlled peristaltic pump. So, after the sample passed through the flow-cell, changing the flow direction results in a further passage and so on (Fig. 1C). The multipeak recording shown in Fig. 1D was obtained as a result.

In the biosensor coupled to an open-closed manifold the sample was inserted into a buffer stream that was passed through the selecting valve, which was then switched in order to close the circuit. The plug was then circulated along the closed circuit, thus repeteadly crossing the flow-cell, where the amount of reaction product, the monitored ingredient increased on each passage and a multipeak recording (Fig. 1F) was thus obtained. Figure 1E depicts the functioning of the open-closed system: as the injected sample enters the circuit the open circuit is converted into a closed one and the plug is recirculated through it. The shape of the multipeak recording obtained is also shown in the slide.

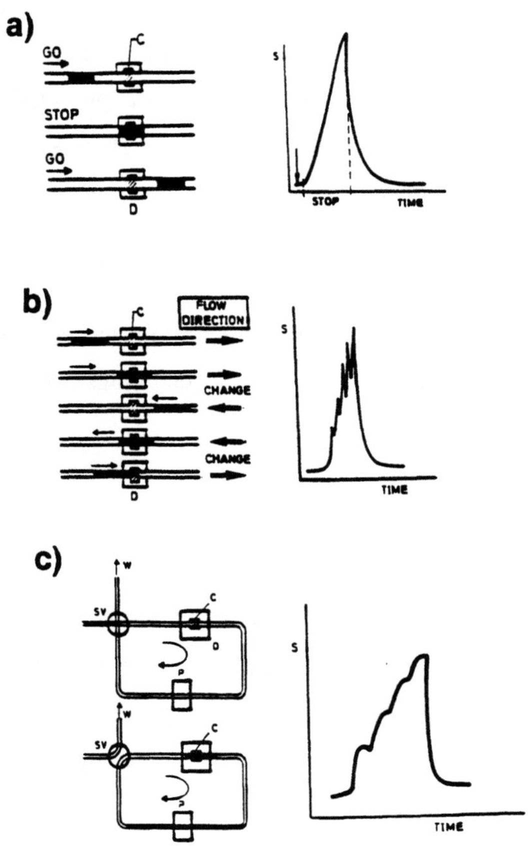

Figure 1. Alternatives to reaction-rate measurements by use of immobilized enzymes (C) in the flow-cell of a photometric detector (D). (a) By stopping the flow; (b) by iterative reversal of the flow direction; (c) by using an open-closed configuration. SV denotes selecting valve, W waste, P pump.

Signal-time data from stopped-flow and multipeak recordings provided by biosensors allow for different types of measurements.

Stopped-flow measurements are based on the expression: $V= (A_2-A_1)/(t_2-t_1)$, where A_2 and A_1 are the absorbances at the ends of the linear- range acquired at times t_2 and t_1, respectively.

Multipeak recordings, either from flow-reversal or open-closed systems, allow for two types of measurements: single-rate measurements, either between consecutive or non-consecutive peaks, and multi-rate measurements, based on the sum of several single measurements.

The methods developed by using these biosensors can only be compared to one another on the basis of the first peak when multipeak recordings and stopped-flow measurements are involved. Of the three types of biosensors, that based on flow reversal is the most sensitive, (almost twice as the sensor based on the open-closed manifold), while the sensitivity of the stopped-flow sensor lies between those of the previous two. On the other hand, the best determinaion limits are provided by the open-closed biosensor. The precision was calculated at two concentration levels and, as can be seen, the biosensor based on stopped-flow is slightly less precise than the rest, yet the difference is not too large. Comparison of the proposed methods in terms of sampling frequency must be done on the basis of all the peaks in the multipeak recordings, since the number of the peak determines the sampling rate achieved. This parameter can be flexibly altered for methods based on multipeak recordings [18].

BIOSENSORS BASED ON THE TRANSIENT IMMOBILIZATION OF THE REACTION PRODUCT

One of our more recent approaches is an uncommon flow-through biosensor in which the enzymatic reaction takes place in the flow-manifold. When the mixture reaches the cell, a support packed within retains the reaction product, providing a transient signal being obtained as the product is concentrated on the support. A constant signal is obtained if the support is saturated with the retained product, but the slope of the rising portion of the concentration curve is always proportional to the product concentration in the stream and, hence to that of the analyte in the sample. After measurement, a valve is switched and an appropriate solution elutes the product from the support, which is thus made ready for retaining the product from the next sample.

The hydrolysis of a ß-D-glucuronide derivative in the presence of ß-D-glucuronidase, with retention of p-nitrophenol was used in order to verify the performance of this sensor, which works as follows:

-The reaction product is retained on a resin packed in the flow-cell (Dowex 1x2) by displacement of the hydroxyl group and the absorbance increment is monitored.

-When the selecting valve is switched, the EDTA contained in the stream displaces the monitored product, which is sent to waste.

-Finally, a new switching of the valve restores passage of the basic stream through the system to regenerate the support, and make it ready for the next determination.

The overall continuous manifold used for the determination of ß-D-glucuronidase activity in serum is shown in Fig. 2. The sample, diluted to an appropriate extent with phosphate buffer, is injected into a phosphate buffer stream and subsequently merged with another stream containing the substrate (p-nitrophenyl-ß-D-glucuronide). The first reaction step, which is catalysed by the enzyme-analyte, takes place along the reactor L1. Then, the plug is merged with a basic stream to enhance the colour of the reaction product, which is retained on the support on passage of the reactant plug through the flow-cell. The transient signal obtained as a result is acquired by the computer. Switching the selecting valve introduces the eluent into the system, so the retained p-nitrophenoxide is sent to waste [19].

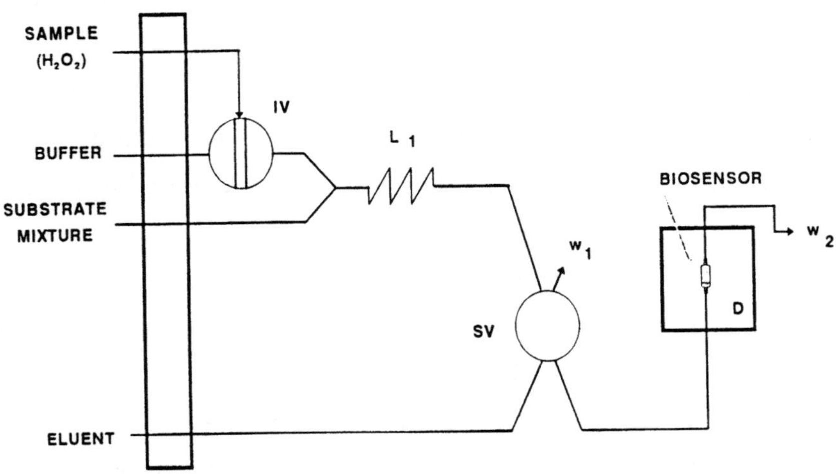

Figure 2. Flow injection manifold for determination of enzyme activity based on immobilization of the reaction product on a support placed in the flow-cell. S denotes sample, B buffer (carrier solution), E reagent (substrate), N basic solution (derivatizing reagent), E eluent, P peristaltic pump, IV injection valve, L reactor, SV switching valve, W waste, D detector.

The main advantage of this biosensor is its wide scope of application. So, it can be used to determine the enzyme activity α- and ß-amilases, α- and ß-galactosidases, glucuronidases, transpeptidases, phosphatases, etc.

BIOSENSORS BASED ON THE DUAL IMMOBILIZATION OF AN ENZYME (PERMANENT) AND A REACTION PRODUCT (TRANSIENT)

The most recent biosensor developed by our team is based on a dual immobilization of a catalyst and a reaction product, both of which retained on materials packed in the flow-cell , but in different ways: the catalyst is immobilized permanently for repeated use, whereas the reaction product is retained temporarily for monitoring purposes, after which it is discarded.

The flow-cell used is a conventional one, in which two layers of packed material can be distinguished: an upper layer of packed CPG containing an immobilized enzyme, and a second layer located in the optical path formed by an exchange resin for retention of the reaction product. Two filters prevent the packed material from being carried away from the flow-cell. The mechanism for reaction/retention and elution in the flow-cell is as follows:

-The substrate molecules enter the flow-cell and undergo the enzymatic reaction on being brought into contact with the catalyst. The product is retained and concentrated on passage through the resin, then being eluted by displacement with an appropriate reagent.

-The enzymatic reaction used was the oxidative condensation of 4-aminophenazone and 3-methyl-N-ethyl-N'-(ß-hydroxiethyl)-aniline in the presence of hydrogen peroxide, catalysed by peroxidase, to obtain the alkyl-aniline cationic derivative according to this reaction.

-The condensation cation formed was retained on a Sephadex carboxymethyl ion-exchange resin. The exchange between the 4-AMHA cation and the proton on the active surface took place as shown in the retention equation.

-The cation was removed from the exchanger by sodium present in the eluent.

The auxiliary manifold for this flow-through biosensor is similar to that shown in Fig. 2.

However, the scope of application of this biosensor can be dramatically expanded if an enzyme reactor containing an oxidase is included in the continuous system. The hydrogen peroxide formed as product of the oxidase-catalysed reaction can be merged with the most appropriate substrate for the next enzymatic step to take place in the flow-cell in contact with the peroxidase immobilized on the CPG. Both photometric and fluorimetric monitoring are possible depending on the nature of the substrate of the second enzymatic step.

This biosensor has been applied to the determination of glucose in serum by using a glucose oxidase reactor between the injection system and the detector in Fig. 2. The procedure is the same as in the previous manifold; that is, injection of the serum sample after appropriate dilution and conversion of glucose into ß-D-gluconic acid and hydrogen peroxide, which reacts with the reaction mixture on passage through the layer of peroxidase-CPG packed in the flow-cell.

The method for the determination of glucose has an excellent linear range at the ppb level (10-200 ng/ml) and good relative standard deviation calculated at three concentration levels: r.s.d. 1.9% (20 ng/ml); 1.2% (50 ng/ml); and 0.9% (150 ng/ml).

In order to check the performance of this last biosensor we applied it to the determination of glucose in different serum samples. An AACC method [20] was used to compare the concentration of glucose provided by this method and the proposed biosensor, and the agreement of both methods was excellent. Also, two additions of standard to aliquots of each saple were performed to check recovery, which was excellent: between 97 and 105% in all instances [21].

The wide scope of application of this approach is a consequence of the generic enzymatic reaction which takes place in the flow-cell. Any reaction yielding hydrogen peroxide can be monitored using the proposed biosensor. The most immediate type of methods to be developed, as shown in the above example, are those involving oxidases. A field of application currently in development is the determination of enzymatic activities, for which a simple change of the continuous manifold makes possible the discrimination of isoenzymes.

FINAL REMARKS

It is necessary to emphasize the unique features of these new approaches to flow-through biosensors. First, compared to conventional probe-type sensors, they enable ready implementation of reaction-rate measurements by using a continuous manifold with halting of the flow iterative passage of the reaction plug through the flow-cell.

The selectivity and/or sensitivity of the associated methods can be enhanced by using an appropriate manifold, by performing kinetic measurements, or by having the concentration step take place in the flow-cell.

Methods based on flow-through biosensor are very ease to automate as the only essential requirements are automatic injection and the use of selecting valves.

Calibration is also very easy process as standards can be inserted into the flowing stream as the required intervals without disturbing the determination sequence.

A single line of the process is sufficient for on-line monitoring of evolving systems.

Some specific advantages of biosensors based on transient immobilization of the reaction product are,

-A high miniaturization capability resulting from the fact that both reaction and retention take place at the same point.

-Enhanced sensitivity as a result of in situ concentration of the product.

-Applicability to a number of analytes, both substrates and biocatalyst, thanks to the universal principle on which they are based.

REFERENCES

1. R.W. Murray, R.E. Dessy, W.R. Heineman, J. Janata and W.R. Seitz, Eds. "Chemical Sensors and Microinstrumentation", Americal Chemical Society, Washington, DC (1989).
2. W. Göpel, J. Hesse and J.N. Zemel, Eds. "Sensors: A Comprehensive Survey" VCH Pub. New York, Vol. I (1991), Vol. II (1992).
3. O.S. Wolfbeis, "Fiber Optic Chemical Sensors and Biosensors", CRC Press, Vols. I and II (1991).
4. D.L. Wise and L.B. Wingard, Jr. Eds. "Biosensors with Fiberoptics", Humana Press, New Jersey (1991).
5. A.P.F. Turner, I. Karube and G.S. Wilson, Eds. "Biosensors: Fundamentals and Applications", Oforx Science Pub. (1987).
6. A.E.G. Cass, Ed. "Biosensors: A Practical Approach", Oxford University Press, Oxford (1990).

7. L.J. Blum and P.R. Coulet, Eds. " Biosensor: Principles and Applications", Marcel Dekker, Inc., New York (1991).
8. F. Scheller and R.D. Schmid, "Biosensors: Fundamentals, Technologies and Applications", VCH Pub. Weinheim (Germany) (1992).
9. F. Scheller and F. Schubert, "Biosensors", Elsevier, Amsterdam (1992).
10. J.M. Fernández-Romero, M.D. Luque de Castro and M. Valcárcel, Anal. Chim. Acta, 274 (1993) 99.
11. Ibid, J. Biotechnolog. 14 (1990) 43.
12. J.M. Fernández-Romero and M.D. Luque de Castro, J. Pharm. Biomed. Anal., 9 (1991) 679.
13. M.D. Luque de Castro and J.M. FErnández-Romero, Anal. Lett., 24 (1991) 749.
14. J.A. Infantes, M.D. Luque de Castro and M. Valcárcel, Anal. Chim. Acta, 242 (1991) 179.
15. A. Membiela, F. Lázaro, M.D. Luque de Castro and M. Valcárcel, Anal. Chim. Acta, 249 (1991) 461.
16. M.D. Luque de Castro and J.M. Fernández-Romero, Anal. Chim. Acta, 263 (1992) 43.
17. P. Linares, M.D. Luque de Castro and M. Valcárcel, Anal. Chim. Acta, 230 (1990) 199.
18. J.M. Fernández-Romero, M.D. Luque de Castro and M. Valcárcel, Sensors & Actuators, 10 (1993) 203.
19. M.J. Fernández-Romero, M.D. Luque de Castro and M. Valcárcel, Anal. Lett., in press.
20. W.R. Faulkner and B. Meites (Eds.) "Selected Methods of Clinical Chemistry", vol. 9, p. 241, American Association for Clinical Chemistry, Washington D.C. 1982.
21. J.M. Fernández-Romero, M.D. Luque de Castro and M. Valcárcel, Anal. Chem., in press.

FLOW INJECTION ANALYSIS: A COMPLEMENTARY OR ALTERNATIVE CONCEPT TO BIOSENSORS

Elo H. Hansen
Chemistry Department A
The Technical University of Denmark, Building 207
DK-2800 Lyngby
Denmark

ABSTRACT. In practical applications biosensors are often forced to operate under less than optimal conditions. Because of their construction, and the physical processes and chemical reactions involved in their operation, compromise conditions are frequently required to synchronize all events taking place. In this communication it is shown Flow Injection Analysis offers itself as complementary facility to augment the performance of biosensors, and in many cases as an attractive alternative. Based on the argument that instead of striving to force different reactions to work under compromise conditions, but that it is infinitely much better to separate them, optimize them individually and let them interact with each other in harmony, selected flow-injection applications are discussed. Initially, comments are made on operating sensors in the FIA mode, while the following sections deal with selected examples of enzymatic assays, aimed either at measuring substrate concentrations, including the use of enzyme amplification schemes, or determining enzyme activies.

INTRODUCTION

As a result of great commercial and academic interests, an impressive amount of manpower at numerous laboratories around the world has been invested in the development of biosensors in recent years. However, considering the concentrated research efforts afforded, the practical results are hereto rather restricted. The reasons and explanations for this lack of apparent success are many, but one of the major ones can probably be ascribed to the requirements that these devices must fulfill in order to meet the desires of their originators and the necessities of real-life applications. Consisiting in essence of a transducing element (electrochemical or optical), covered by an appropriate layer/membrane of an active material (typically an enzyme or an enzyme combination, a biocatalyst, or a custom-made chemical compound), a biosensor must inevitably meet a variety of demands which must be honoured simultaneously. However, considering the diversity of chemical and physical events that can be involved in its function, in addition to satisfying individually dictated reaction parameters, it is not surprising that operational compromises necessarily have to be made for practical synchronization of all the processes.

G.G. Guilbault and M. Mascini (eds.), Uses of Immobilized Biological Compounds, 527–540.

A prime example of such a compromise approach was presented at the previous conference of this series of meetings in 1987. Thus, in one of the lectures [1], a researcher described how he had spent six month to find an appropriate diffusion barrier for a glucose sensor, based on immobilized glucose oxidase in conjunction with amperometric detection of the hydrogen peroxide generated. The sensor was to be used for assaying glucose in human blood samples. However, since the glucose level in such samples are higher than the Michaelis-Menten constant for glucose oxidase, ancillary diffusing impeding measures had to be taken so that first-order-reaction kinetics with respect to glucose could be fulfilled, that is, by restricting the glucose concentration reaching and penetrating into the sensor, it could be "tricked" into monitoring an appararent lower level than that actually present in the sample. After the lecture the author of this paper mentioned that this feat could have been accomplished in six seconds rather than six months, namely by incorporating the glucose sensor, without the diffusion membrane barrier, into a Flow Injection Analysis (FIA) system [2], and simply manipulating the concentration that would be sensed by the detector by adjusting the time that the sample were to be exposed to the detector (for details, see later).

This experience gave this author the impetus to take a closer and somewhat different look at FIA, and to develop it as a complementary facility to or, better, possibly as an alternative to biosensors, that is, to try to demonstrate that instead to strive to force different reactions to work under compromise conditions, it was infinitely much better to separate them, optimize them individually and let them interact with each other in harmony. It is the hope that the examples given in this paper will serve to demonstrate that FIA indeed offers itself as an accomplished tool to effect exactly these task, and thus in several cases constitute an attractive vehicle for augmenting the performance of biosensors, and in many instances advantageously can replace these ones. In the first section comments are made on operating sensors in the FIA mode (re the example above), while the following sections deal with selected examples of enzymatic assays, aimed either at measuring substrate concentrations or determining enzyme activies.

ADVANTAGES OF OPERATING SENSORS IN THE FIA MODE

As mentioned in the introduction, the present author at the previous meering in Florence in 1987 called attention to exploit FIA for taking maximum advantage of (bio)sensors, which, of course, requires further explanation. Indeed, the performance of many sensors used in analytical chemical assays is diffusion controlled. This is particularly true of electrochemical transducers, and therefore FIA, with its strict and reproducible timing of all events within each assay cycle, has proven itself as an efficient means for utilizing these detection devices. Thus in potentiometry it is observed that many ion-selective electrodes operated in the dynamic mode will exhibit kinetic discrimination towards the ion under investigation and interfering species, that is, during the short duration of sample exposure the response of the sensor to the different species might vary considerably, which in turn might be exploited to increase the selectivity and the detection limit of the sensor [3-5].

The very same concept of manipulating the sample exposure time might be extended to apply for more complex sensors such as enzyme sensors, in which a membrane containing one or several immobilized enzymes is placed in front of the active surface of a detection device. The analyte is transported by diffusion into the membrane and here degraded enzymatically forming a product which subsequently can be sensed optically or electrochemically. A condition for obtaining a linear relationship between analyte and signal is, however, that pseudo-first-order reaction conditions are fulfilled, that is, the concentration of converted analyte reaching the detector surface must be much smaller than the Michaelis-Menten constant. Since this constant for most enzyme systems is of the order of 1mM, and many sample matrices (e.g., human sera or blood) contain much higher concentrations, the use of enzyme sensors in static systems often calls for exactly that complicated use of additional membrane layers mentioned previously, aimed at restricting diffusion of analyte to the underlying enzyme layer, or, in case of electrochemical detectors, for chemically modified electrodes where the electron transfer is governed by appropriate mediators. However, in FIA the degree of conversion might simply be adjusted by the *time* the analyte is exposed to the enzyme layer, and therefore the amount of converted analyte can be regulated directly by adjusting the flow rate of the FIA system [6]. This has very elegantly been demonstrated by Petersson [7] using a system for the determinations of glucose or lactate incorporating an amperometric electrode comprizing glucose oxidase and lactate oxidase, respectively. Thus, as illustrated in Fig. 1, by increasing the flow rate from 0.5 to 1.0 ml/min, the linear measuring range could readily be expanded from 20 to 40 mM glucose, ensuring that the system could be applied for physiological measurements. Additionally, the time of exposure might possibly be exploited for kinetic discrimination, advantage being taken of differences in diffusion rates of analyte and interfering species within the membrane layer of the electrode. This was equally well illustrated by Petersson in the afore-mentioned work, where he achieved total spatial resolution of the signals due to glucose and paracetamol, even when the interferent was added at excessive levels. Besides, attention should be drawn to the fact that the operation of a (bio)sensor in the FIA mode ensures a constant monitoring of the sensor itself, that is, while the recorded signal is a measure of the concentration measured, the baseline signal is an appraisal of the stability of the sensor.

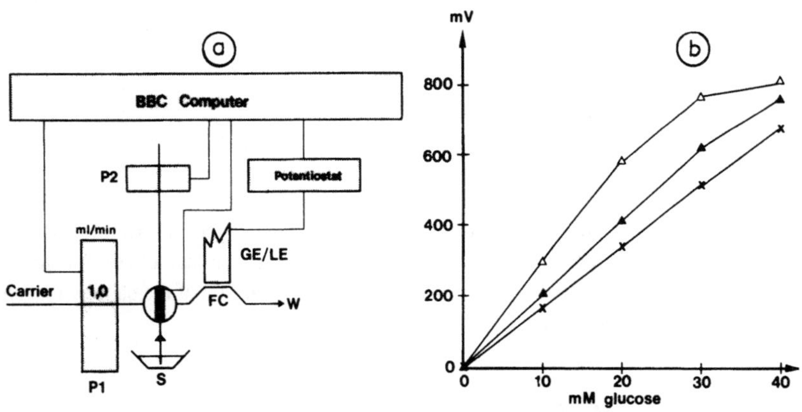

Figure 1 (a) Manifold for the detection of glucose or lactate with an amperometric electrode to detect enzymatically generated hydrogen peroxide. It comprises a syringe pump (P1), peristaltic pump (P2), injection valve, BBC computer, potentiostat and a glucose electrode (GE) or lactate electrode (LE). S - sample injection; FC - flow cell; and W - waste. (b) Calibration graphs for glucose in the concentration range 0 - 40 mM at three different flow rates: (Δ) 0.50, (▲) 0.75, and (x) 1.00 ml/min [7, 11].

Measuring homogeneous sample solutions with an enzyme sensor is technically a relatively uncomplicated task. But when heterogeneous samples, such as undiluted blood, are to be assayed by means of an enzyme sensor incorporated into a flow-injection system, it is a prerequisite that the individual samples at the time they are exposed to the enzyme are undiluted by the carrier stream [8, 9], that is, the dispersion coefficient D of the sample is equal to 1. If this condition is not fulfilled, the analytical readout will be severely influenced by the composition of the matrix of the sample (i.e., in the case of blood, the hematocrit level, the pH value and the buffering capacity). This ostensible paradox, that the sample at the moment of measurement must remain undiluted, yet at the same time necessarily has to be mixed with a reagent contained in the carrier stream in order to promote a given chemical reaction, has found its simple solution in FIA [9], advantage being taken by that a perfectly controlled spatial and kinetic discrimination prevails in the membrane operated under FIA conditions, so that the sensing process, in fact, occurs in two steps (Fig. 2): In the first step, where the injected sample is in contact with the enzyme layer, the sample leaves an "imprint" promoting the degradation of substrate to a non-detectable species. In the next step, when the undiluted sample zone is swept away by the reagent-containing carrier stream, the generated and entrapped species is then converted to a detectable component which is subsequently transferred to the active surface of the sensing element. (In fact, the procedure might be compared to a photographic process: When the picture is taken, its imprint is on the film, but latently, and it becomes visible only after it has been developed via chemical processing). This approach has been successfully implemented for the optical assay of urea in undiluted blood samples [9].

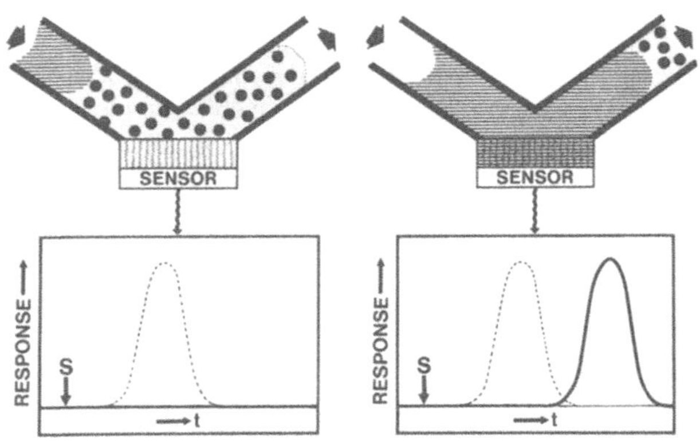

Figure 2. Operation of a latent image sensor during the passage of a heterogeneous sample zone, showing on the left the penetration of the interstitial liquid into the sensor structure so that a latent image is formed (dotted line), and on the right the development of the latent image with the aid of the passing reagent zone, resulting in a recorded response (solid line) [8, 11].

APPLICATION OF IMMOBILIZED ENZYMES. ENZYME AMPLIFICATION SCHEMES

Given the advantages of the selectivity of enzymes towards specific substrates, these catalytic reagents have gained increased significance in several areas of analytical chemistry such as clinical, pharmaceutical, agricultural and industrial applications, and, most recently, process monitoring. Despite the fact that several hundred enzymes of high purity and activity today are commercially available, these reagents are, however, still rather expensive, and once dissolved, many of them exhibit limited stability. Therefore it is imperative that either a high sample frequency, with minimum consumption of enzymes per assay, be maintained, or that the enzymes be immobilized on suitable supports so that they can be used repeatedly, advantage being taken of the often increased stability obtained when they are immobilized [10-12]. This, in addition to the convenience of operation, is, of course, the philosophy for employing immobilized enzymes in biosensors as well.

However, when applied in the FIA mode the use of immobilized enzymes, packed into small column reactors, offers not only the selectivity and the economy and stability gained by immobilization, but they also ensure that strict repetition and hence a fixed degree of turnover from cycle to cycle are maintained. In addition, by obtaining a high concentration of enzyme immobilized within a small volume, the ensuing high activity will facilitate an extensive and rapid conversion of substrate at minimum dilution of the sample, the small dispersion coefficient in turn promoting a lower limit of detection [11]. The enzymes needed for a particular assay might be incorporated either into a single packed reactor, or into sequentially connected reactors in order to process the required sequence of events and also to afford optimal operational conditions for the reactions occurring in the individual reactors. For oxidases, which give rise to the generation of hydrogen peroxide, and also for dehydrogenases, which require the presence of suitable coenzymes such as NADH or NADPH, the oxidized or reduced forms of which are monitored, both optical and electrochemical detection can be used. Thus hydrogen peroxide has been detected optically by coupling to appropriate chromogenic agents, by chemiluminescence via reaction with luminol, or by amperometry, while the coenzymes are usually detected by spectrophotometry or fluorometry or by various electrochemical techniques including (pulsed) amperometry and detection at modified electrodes. For other enzymes, such as the iminohydrolases, which degrade substrates with the ultimate formation of ammonium/ammonia, selective determination might be achieved potentiometrically by an ion-selective electrode, or via gas diffusion followed by optical or electrochemical detection [13], or via chemical derivatization with o-phthalaldehyde and sulfite with ensuing fluorometric quantification [14].

An intriguing application of coimmobilized enzymes is for enzyme amplification reactions (cf. Fig. 3). By this approach advantage is taken by the synergestic operation of appropriately coupled enzyme combinations, so that the product (or a concurrent coproduct/cofactor) generated from the substrate by one enzyme system serves as the substrate for a second system, which in turn leads to the generation of a product (and possibly a coproduct) which constitutes the substrate for the first enzyme system. By repeated recycling, the formation of a product/coproduct beyond the stoichiometric limitation can be effected, and thus a considerable enhancement of sensitivity of either a substrate/product or coproduct might be achieved.

This approach was first reported used in enzyme electrodes for the assay of low concentrations of constituents of biological and biotechnological interest [15-18], and it was claimed that amplification factors of the order of 200 (and in one case even 4100 [17]) could be obtained. Yet so far, all the described biosensors have been characterized by very slow response times - generally of more than 10 min, and often much longer - owing to the diffusion limitations in the enzyme-containing membrane layers, and the necessary compromise between conditions for the enzymatic conversion reactions and the actual detection. Besides, these devices exhibited lack of reproducibility of measurement, and therefore their practical application has been restricted. Following the basic idea of the present communication, an alternative and beneficial avenue would be to separate the enzymatic step and the detection step and optimize them individually. This is readily feasible in a FIA system, where the enzymes can be immobilized on a suitable support and incorporated into a miniaturized column reactor, while quantification of a product/coproduct formed subsequently can be executed by the most appropriate detection device. In this manner, the chemical conversion can be effectively accomplished, advantage being taken of the fact that the radial mass transfer in a packed reactor is much more intense than in a diffusion-controlled membrane, and therefore the conversion within the bulk of the sample zone occurs much more readily.

In order to demonstrate this approach, this research group a couple of years ago embarked upon a comprehensive project aimed at FIA-amplification schemes, a prerogative being that the entire analytical cycle, in order to present itself as analytically interesting in practical applications, could be accomplished within 1 to 1.5 minutes, thus facilitating a sampling frequence of at least 40 samples per hour. The initial systems tested were based on suitable combinations of oxido-reductases, such as lactate oxidase (LOD) and lactate dehydrogenase (LDH) [19, 20], which enzyme entities require consumption of oxygen and NADH and leads to generation of hydrogen peroxide and NAD^+ (Fig. 3). While the operation of enzyme sensors with oxidoreductases almost exclusively is based on measuring electrochemically the consumption of oxygen - which might infer measuring a small decrease in signal on a high background - it is analytically preferable to base the measurement on a species generated. Given the freedom in FIA of selecting the most appropriate detection system, it was opted to assay the hydrogen peroxide formed, which constituent was measured by chemiluminescence via reaction to luminol and hexacyanoferrate(III) [21]. Thus, by optimizing the enzymatic degradation step and the detection step individually it was shown that lactate and pyruvate in a FIA system comprizing a 140 μl column reactor containing an immobilized LOD/LDH-enzyme combination, at a sampling rate of 50-60 s/h, and at amplification factors of ca. 140, could be determined down to 48 and 103 nM, respectively [20].

Inherently, the thus optimized FIA-system should be applicable for assay of a variety of substrates provided that suitable enzyme combinations, such as oxidoreductases, can be found. One such combination is glucose oxidase (GOD) and glucose dehydrogenase (GDH). As seen in Fig. 3, the LOD/LDH and GOD/GDH enzyme systems operate virtually similarly. However, in the glucose system there is a possibility of a side reaction, that is, the preliminarily generated glucono-δ-lactone formed by degradation of glucose might hydrolyse to gluconic acid, and if this happens the cycling will be terminated. This side reaction is very fast, and indeed, Scheller et al. [17], who attempted to use this enzyme combination in a membrane electrode, merely managed to obtain amplification factors of the order of 3-8. However, it was hoped that the combination of convection and diffusion in the column reactor of the FIA system would result in a much better amplification, because it intuitively would be expected that the side-reaction of hydrolysis by means of kinetic discrimination could be minimized. Unfortunately, the results were poor, with amplification factors very similar to those previously obtained by Scheller et al. Viewed in retrospect, there are several plausible explanations for the poor results with this particular system [20], but it was considered that the concept of enzyme amplification should be pursued as a useful technique for the assay of minute concentrations of substrates.

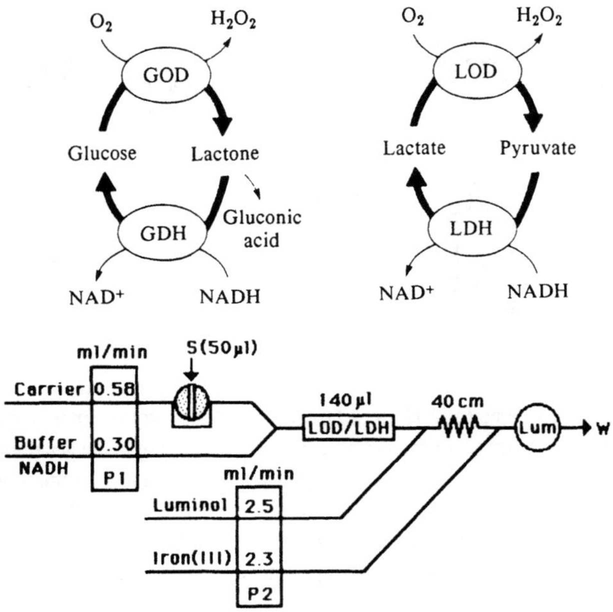

Fig.3. Enzyme amplification in FIA. Amplification reactions for (top, left) glucose by means of glucose oxidase (GOD) and glucose dehydrogenase (GDH); and (top, right) pyruvate and lactate as facilitated by lactate oxidase (LOD) and lactate dehydrogenase (LDH). In the former system, the occurrence of the side reaction leading to the formation of gluconic acid will remove the substrate for the GDH and hence terminate the cycling sequence. Bottom, optimized FIA manifold for the determination of lactate or pyruvate by enzyme amplification with ensuing detection of the hydrogen peroxide generated as facilitated by chemiluminescence (Lum.), yielding amplification factors of ca. 140. S - sample injection (50 µl); LOD/LDH - column reactor (140 µl) containing the two coimmobilized enzymes (LOD/LDH); C - mixing coil (40 cm); and W - waste. Solutions were propelled by two peristaltic pumps, P1 and P2 [10, 20].

In fact, encouraged by the initial results with enzyme amplification schemes, attention was then focused on the assay of ATP (adenosine-5'-triphosphate) via a three-enzyme amplification system as that depicted in Fig. 4 [22]. The assay of low levels of ATP is analytically interesting, because it is inherently an indicator for the activity of somatic cells and bacteria, and therefore it is an intriguing parameter to evaluate not only in medicine, but also in the food industry. As seen from the reaction scheme, ATP is degraded to ADP (adenosine diphosphate) by hexokinase (HK) in the presence of glucose, the ADP being regenerated to ATP by pyruvate kinase (PK) during consumption of phosphoenol pyruvate (PEP) and generation of pyruvate. By the recycling, glucose is transformed into glucose-6-phosphate (G6P), which then by means of the enzyme glucose-6-phosphate dehydrogenase (G6PDH) is converted to gluconate-6-phosphate during consumption of NAD^+ and formation of NADH. Assays based on this reaction sequence have previously been reported by Kirstein et al. [23], who used an enzyme

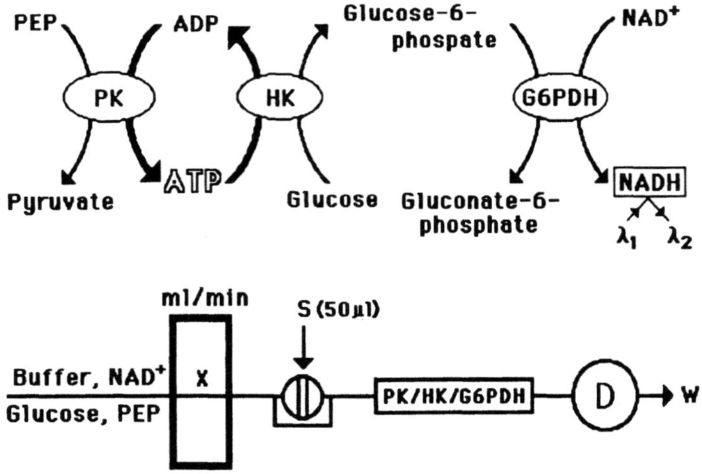

Fig. 4 (top). Amplification scheme for ATP, based on the use of the enzymes pyrovate kinase (PK), hexokinase (HK) and glucose-6-phosphate dehydrogenase (G6PDH), resulting in the generation of NADH which is measured fluorometrically at an excitation wavelength of 340 nm (λ_1) and an emission wavelength of 460 nm (λ_2). Bottom, FIA-manifold used for the assay of ATP by enzyme amplification with ensuing detection by flourescence (D). HK/PK/G6PDH, column reactor (100 µl) containing the three coimmobilized enzymes hexokinase, pyrovate kinase and glucose-6-phosphate dehydrogenase. X - pumping rate (ml/min); S - sample injection; and W - waste [22].

thermistor whereby the heat generated by the enzymatic recycling process could be monitored thermometrically and related to substrate concentration, which obviously is a rather delicate and complicated means of detection, and by Yang et al. [24] who employed amperometric detection by means of a modified electrode. However, this device proved to be very unstable, and therefore it had to be replaced daily. In order to overcome these shortcomings, and again taking advantage of the virtually unlimited options in FIA of chosing the most appropriate means of detection, it was decided to settle for detection of the NADH generated by means of fluorometry. Not only because optical detection generally is simpler to administer, but particularly because fluorometry inherently is a very sensitive detection procedure. The actual FIA system used is shown in Fig. 4, the enzymatic degradation step and the detection step again being completely separated in order to allow their individual optimization. As might be expected, the amplification factor was directly proportional to the residence time of the sample zone within the enzyme reactor, which time might be manipulated by altering the flow rate (X) and in the extreme by performing stopped-flow experiments. As it was shown, amplification factors up to 1000 with this system were readily feasible, permitting to reach a lower limit of detection of the order of 6 nM [22].

For many applications in the food industry, particularly for assaying bacterial activity, it is, however, imperative with even lower limits of detection. Therefore it should be mentioned that recent research activities at this laboratory based on assaying ATP by means of bioluminescence, as facilitated by means of employing a specially designed combination of a flow-cell and a high-sensitivity photomultiplyer, has yielded limits of detection in the pM-range [25]. In this case, soluble enzyme was employed, but by means a FIA manifold permitting recirculation of any unused enzyme solution the system offered a high degree of economy of the precious (luciferase) enzyme, whereby the cost per assay actually could be held lower than that for a system using immobilized enzyme.

MEASUREMENT OF ENZYME ACTIVITIES

Just like assaying substrates there is an increasing need for determining enzyme activities. Especially, because enzymes in industrial scale increasingly are produced by genetic engineering technology, and therefore it is, for instance, of importance to be able to monitor the production process at its various stages by suitable means. And given the complexities of the manipulations and reactions involved to facilitate these types of measurements, the potentials for using biosensors in this area appear rather limited. But FIA presents itself as an attractive vehicle.

For assay of enzyme activities, the operational conditions must be made so that the substrate concentration is made very much higher than the Michaelis-Menten constant. In this manner, the reaction becomes pseudo-zero-order with respect to the substrate, and hence the reaction rate is dirtectly proportional to the concentration of enzyme, or, more precisely, its activity. Most enzymatic reactions also involve the interaction of appropriate cofactors. Thus, oxidases consume oxygen and generate hydrogen peroxide, while dehydrogenases require the presence of $NAD^+/NADP^+$ which in turn are converted to NADH/NADPH. Therefore, the activity of an enzyme can be measured either by determining the substrate/cofactor consumed or the product/cofactor generated. For the dehydrogenases, this task is fairly straightforward, because NADH/NADPH can be monitored both optically and electrochemically. If none of the constituents in an enzyme reaction are directly determinable, one of them might be quantified via an accompanying indicator reaction. For instance, hydrogen peroxide, as generated by the oxidases, can be measured photometrically by coupling to a chromogenic reagent, or by chemiluminescence through reaction with luminol and hexacyanoferrate(III) [11, 20]

The determination of the activity of an enzyme is in practice most often performed by incubating the substrate and the enzyme for an appropriate period of time, whereupon the concentration of one of the constituents is measured - possibly by an accompanying indicator reaction. However, such an approach yields only a single point measurement. And merely assuming that the enzymatic degradation process, as recorded by the change in detector response over a fixed of period of time, necessarily will follow a linear relationship, might, as is depicted in Fig. 5, easily be most hazardous. In fact, it could very well be that the degradation reaction followed a course as the one outlined in the figure, that is, initially comprizing an inherent lag-phase followed by a period during which the signal were to change linearly as a function of time, the process eventually reaching a level where pseudo-zero-order reaction condition with respect to substrate no longer is fulfilled. Hence, the single-point measurement approach would obviously lead to a false answer, and therefore it is inherently much to be preferred to rely on multi-point assays for quantifying the enzyme activity.

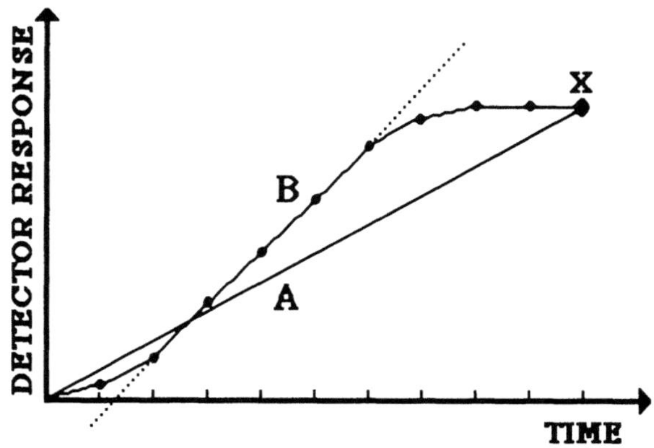

Fig. 5. Single-point (curve A) and multi-point (curve B) determination of the activity an of enzyme as obtained by measurement over identical time spans, that is, from time zero to a time t corresponding to endpoint X. The slope of the linear part of each curve is a measure of the enzyme activity. In curve B is illustrated how the detector output possibly might develop if the degradation process were to be monitored intermittently [27].

Flow injection analysis (FIA) entails exactly such possibilities, namely by applying the stopped-flow approach, where a suitable section of the injected and dispersed sample zone is arrested within the observation area of the detector [2, 26]. Provided that the conditions are manipulated in such a manner that they conform with pseudo-zero-order with respect to substrate and pseudo-first-order with respect to the enzyme, the slope of the reaction rate curve will be directly proportional to the concentration of enzyme and hence to the enzyme activity. Reaction rate measurements in which the the rate of formation (or consumption) of a certain species is measured from a larger number of data points not only improve the reproducibility of the assay, but also ensures its reliability. Thus, the stopped-flow method is an efficient and convenient vehicle for measuring substrates and indeed determining enzyme activities. Furthermore, in FIA an added feature can be exploited, that is, the sample might be stopped at different parts of the the concentration gradient corresponding to different ratios of substrate to enzyme, so that it might be feasible to select exactly that part of the dispersed sample zone where the operational parameters are optimal [11, 26].

However, an absolute condition in order to employ the FIA stopped-flow method for enzyme activity measurements is that the enzymatic reaction either generate or consume a species which can be detected directly (e.g., NADH/NADPH), or that the coupled indicator reaction is much faster than the enzymatic degradation reaction itself (e.g., the use of luminol for the chemiluminescence detection of hydrogen peroxide), because the ensuing chemical reaction must not, of course, become the rate determining step in the overall reaction sequence. Fortunately, most auxillary chemical reaction are generally much faster than the enzymatic degradation procedure, yet this does not always hold true. A prime example of this is the assay of the enzyme activity of cellulase which recently has been studied at this laboratory [27]. Cellulase is an enzyme which selectively degrades cellulose randomly, giving rise to the formation of sugar entities with reducing terminals.

The enzyme is much used for pretreating blue denim materials and as an additive in washing powders, yet it must be administered with utmost care, and for that reason there is a great need for assaying its activity. All available chemical derivatization procedures rely on determining the reducing sugars entities, which can be effected by various means, the most promising avenue being by reaction with p-hydroxybenzoylhydrazine (p-hydroxybenzoic acid hydrazide), which in strongly alkaline solution at elevated temperatures give rise to the formation of a yellow compound which can be measured spectrophotometrically at 410 nm [28, 29] The problem is that not only is the enzymatic degradation reaction slow, but the coupled indicator reaction is very slow indeed.

Various solutions have been attempted to circumvent this problem, the most promising ones being based on employing FIA and separating the enzymatic degradation procedure and the accompanying indicator reaction [30, 31]. Yet, despite their individual ingenuities, all the proposed systems yield essentially only a single-point assay, and therefore they suffer from the drawbacks mentioned previously.

Consequently, and in order to effect the assay of the enzyme activity irrespectively of the coupled chemical reactions and its particular characteristics (absolute reaction rate, and relative reaction rate in respect to that of the enzymatic degradation procedure), this research group decided to take an entirely different approach, that is, not only to try to solve the actual problem of assaying the activity of cellulase, but to attempt to design a generic enzymatic system which inherently would allow to handle any type of samples. Furthermore, it was a prerogative that such a system should be versatile, i. e., it should not

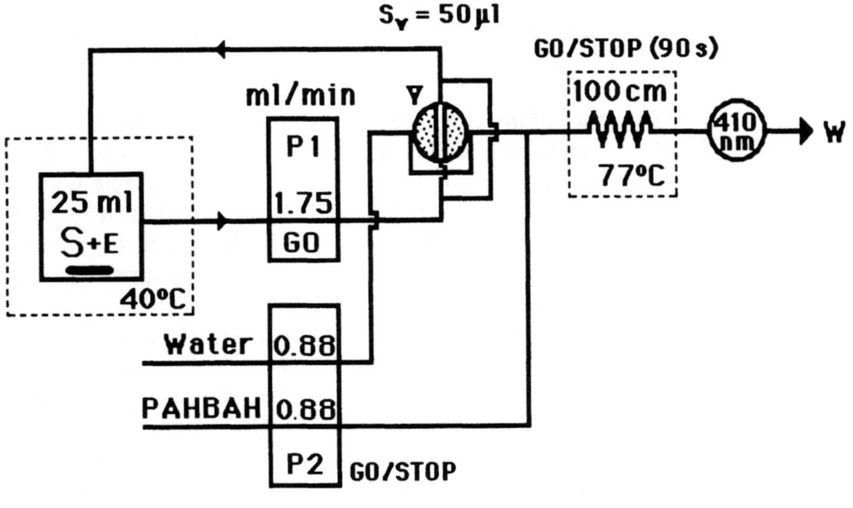

Fig. 6. Integrated system as used for enzyme activity determinations, comprizing essentially two entirely separated subsystem communicating via a common valve (V). The subsystem at left is the enzyme reactor where substrate (S, in large excess) and enzyme (E) react, while the chemical derivatization subsystem is depicted to the right. The operational parameters shown on the figure reflect the ultimately adopted ones for the assay of cellulase with ensuing chromogenic derivatization with p-hydroxybenzoylhydrazine. S_V - sample volume injected [27].

only allow the enzymatic degradation procedure to be completely separated from the chemical detection procedure, but it should also permit each of the processes to be optimized individually, and it should inherently allow multipoint determinations to be performed.

These conditions were met by employing the approach illustrated in Fig. 6 (the parameters depicted show the optimal values for the assay of cellulase). In part based on a concept previously devised at this department for an entirely different purpose [32], it consists in essence of two individually and totally separated subsystems, one for the enzymatic degradation reaction, consisting of a well-stirred reactor vessel, and one for the ensuing chemical indicator reaction, comprizing a FIA manifold, the two subsystems communicating via a common valve. The content of enzyme reactor is via an external loop constantly circulated through pump 1 and the injection valve at such a rate that the solution in the loop at all times is identical to and representative of the solution in the reactor vessel. At preselected times the injection valve is switched to the inject position and a metered sample volume is by pump 2 introduced into the FIA system where it is mixed with suitable reagent(s) and carried towards the detector. During its path through the manifold the sample/reagent mixture might, if required, be arrested in order to gain additional reaction time, heated, or possibly cooled, or manipulated in any other appropriate manner before it is guided into the detection device. Sample injection can be repeated as often as required so that a suitable number of data-points on the concentration/time curve can be recorded, the only constriction being, in fact, the time required for the derivatization chemistry in the FIA manifold. In the present example, in which each sample after withdrawal and mixing with reagent was arrested for 90 s at elevated temperature in order to promote the colour forming reaction (cf. Fig. 6), individual samples were, in order to demonstrate the validity of the system, withdrawn at 3.5 min intervals over a time range of 21.5 min.

The actual performance of the system for the assay of the activity of cellulase is illustrated in Fig 7, which shows the readouts obtained for four different activities of cellulase, that is, 2.0, 1.0. 0.5 and 0.1 U/ml. In each case, the first sample was withdrawn after 30 s, and thereafter samples were taken at 3.5 min intervals over a time range of 21.5 min, that is, totally seven samples were monitored. Clearly, the highest enzyme concentration does not result in a straight line relationship over the entire time range. Possibly primarily because pseudo-zero-order reaction conditions with respect to substrate are not fulfilled at extended times, but maybe also because of the relatively high absorbances recorded. Therefore only the first three points in the response/time relationship were used in the evaluation of the slope, which is the quantitative representation of the enzyme activity. Nor does the 1.0 U/ml sample result in an absolutely straight line relationship over the entire time range, although the first five points (used for evaluation of the slope) appear to fall on a straight line, while the two low enzyme activities conform pertfectly well to the pseudo-zero-order reaction conditions with respect to substrate concentration over the whole time frame. Relative to each other, the slopes were generally in very good agreement with each other. Because the system for practical work is to be used for measuring an unknown activity as based on calibration with known activities, it is important that it is reasonable stable in operation. This was investigated by repeating the set of experiments mentioned above five consecutive times and comparing the results. As it was verified, the relative standard deviation (between runs) was within a range of ca. 3.6-5.7%, depending on the number of measuring points included. For the entire acivity range tested (0.1 - 2.0 U/ml) the precision was generally found to be of the order of ± 3-12%, which for practical routine applications should be adequate.

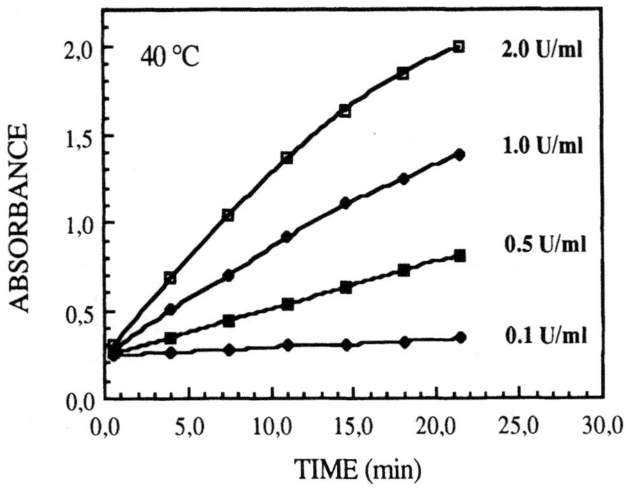

Fig. 7. Signal readouts for four different cellulase enzyme activities in the range 0.1 to 2.0 U/ml at 40 °C. For the two highest cellulase activities the first 3, respective 5, readouts were used for evaluation of the slopes [27].

CONCLUSION

The aim of this communication has not been to cast doubt on the potential validity and applicability of biosensors at the expense of FIA. Rather it has been the intention of the author to demonstrate that biosensors advantageously can be used in FIA, where their performance in many cases can be augmented, and to show that in other instances, where the physical processes and chemical reactions involved in a particular assay are complex, it might be beneficial to replace the compromise working conditions of a biosensor with the optimal ones readily feasible in a FIA system whereby the overall operations can be simplified.

Acknowledgement. The author wishes to express his appreciation to Julie Damms Foundation for partial financial support of the research activities described herein. Thanks are also due to to the North Atlantic Treaty Organization (NATO) for facilitating participation in the Brixen Meeting.

REFERENCES

1 C. C. Feistel, C. F. Strenburg, C. D. Luther and L. L. Gayleard, in "Analytical Uses of Immobilized Biological Compounds for Detection, Medical and Industrial Uses" (G. G. Guilbault and M. Mascini, Eds.), NATO ASI Ser., Ser. C, **226** (1988) 341.

2 J. Ruzicka and E. H. Hansen, "Flow Injection Analysis", 2nd Ed. John Wiley and Sons, Inc., New York, USA, 1988.
3 E. H. Hansen, J. Ruzicka and A. K. Ghose, STI/PUB/**535** (IAEA, Vienna), (1980) 77.
4 M. Trojanowicz and W. Matuszewski, *Anal. Chim. Acta*, **151** (1983) 77.
5 L. Ilcheva and K. Cammann, *Fresenius' Z. Anal. Chem.*, **322** (1985) 323.
6 E. H. Hansen, *Fresenius' Z. Anal. Chem.*, **329** (1988) 656.
7 B. A. Petersson, *Anal. Chim. Acta*, **209** (1988) 231.
8 J. Ruzicka and E. H. Hansen, *Anal. Chim. Acta*, **214** (1988) 1.
9 B. A. Petersson, H. B. Andersen and E. H. Hansen, *Anal. Lett.*, **20** (1987) 1977.
10 E. H. Hansen, *Anal. Chim. Acta*, **261** (1992) 125.
11 E. H. Hansen, *Anal. Chim. Acta*, **216** (1989) 257.
12 E. H. Hansen and H. S. Mikkelsen, *Anal. Lett.*, **24** (1991) 1419.
13 M. T. Jeppesen and E. H. Hansen, *Anal. Chim. Acta*, **214** (1988) 147.
14 M. T. Jeppesen and E. H. Hansen, *Anal. Chim. Acta*, **245** (1988) 89.
15 F. W. Scheller, F. Schubert, R. Renneberg and H.-G. Müller, *Biosensors*, **1** (1985) 135.
16 F. Mizutani, T. Yamanaka, Y. Tanabe and K. Tsuda, *Anal. Chim. Acta*, **177** (1985) 153.
17 F. Scheller, D. Kirstein, L. Kirstein, F. Schubert, U. Wollenberger, B. Olsson, L. Gorton and G. Johansson, *Phil Trans. R. Soc. Lond. B*, **316** (1987) 85.
18 F. Scheller, F. Schubert, D. Pfeiffer, R. Hintsche, I. Dransfeld, R. Renneberg, U. Wollenberger, K. Riedel, M. Pavlova, M. Kühn, H.-G. Müller, P. M. Tan, W. Hoffmann and W. Moritz, *Analyst*, **114** (1989) 653.
19 E. H. Hansen, A. Arndal and L. Nørgaard, *Anal. Lett.*, **23** (1990) 225.
20 E. H. Hansen, L. Nørgaard, and M. Pedersen, *Talanta*, **38** (1991) 275.
21 B. A. Petersson, E. H. Hansen, and J. Ruzicka, J., Anal. Lett., **19** (1986) 649.
22 E. H. Hansen, M. Gundstrup and H. S. Mikkelsen, *J. Biotechnol.* (in press).
23 D. Kirstein, B. Danielsson, F. Scheller and K. Mosbach, *Biosensors*, **4** (1989) 231.
24 X. Yang, D. Pfeiffer, G. Johansson and F. Scheller, *Anal. Lett.*, **2** (1991) 1401.
25 G. Gamborg, "Analytical Applications of Bioluminescence for Quantitative Determination of ATP in FIA-Systems", M.Sc. Thesis (In Danish), Chem. Dept. A, Tech. Univ. Denmark, 1993.
26 E. H. Hansen, in "Analytical Uses of Immobilized Biological Compounds for Detection, Medical and Industrial Uses" (G. G. Guilbault and M. Mascini, Eds.), NATO ASI Ser., Ser. C, **226** (1988) 291.
27 E. H. Hansen and A. Jensen, *Talanta* (in press).
28 M. Lever, *Anal. Biochem.*, **81** (1977) 21.
29 M. Lever, T. A. Walmsley, R. S. Visser and S. J. Ryde, *Anal. Biochem.*, **139** (1984) 205.
30 P. J. Worsfold, I. R. C. Whiteside, H. F. Pfeiffer and H. Waldhoff, *J. Biotechnol.*, **14** (1990) 81.
31 H. F. Pfeiffer, H. Waldhoff, P. J. Worsfold and I. R. C. Whiteside, *Chromatografia*, **33** (1992) 49.
32 C. Ridder, *Anal. Chim. Acta*, **217** (1989) 303.

HEAT SENSITIVE BIOSENSORS

BENGT DANIELSSON, ULRIKA HEDBERG, MORTEN RANK, AND BIN XIE
Pure and Applied Biochemistry, University of Lund,
Box 124, S-221 00 Lund, Sweden.

Abstract The unsurpassed versatility of heat sensitive biosensors is illustrated by various new applications of the flow-calorimetric biosensor system, the enzyme thermistor (ET) developed in our laboratory. The ET measures the reaction heat in a small column containing immobilized biocatalyst. Since most biological reactions are exothermic, the ET has a very broad applicability and numerous assays of clinical as well as biotechnological interest have been studied. The procedure is well suited for monitoring and control of biotechnological processes due to its high operational stability. New results from this area are presented as well as developments on an automated thermometric immunoassay (TELISA). Miniaturized ET-devices are suitable for portable monitoring of, for instance, blood glucose. Other recent work include determinations in organic solvents, direct characterization of immobilized biocatalysts, and monitoring of bioactivities in solution.

I INTRODUCTION

In recent years biosensors have attracted a considerably increased interest due to an extended demand for sensors suitable for specific monitoring and control in biotechnology, especially of continuous processes. There is also a rapidly growing need for sensors in the biomedical field, for decentralized clinical chemistry and for *in vivo* monitoring of, in particular, glucose in attempts to develop systems for diabetes control. In these fields heat sensitive biosensors, such as the enzyme thermistor (ET) have definite advantages. Since biological processes generally are more or less exothermic, the combination of specific biocatalysts with a heat or temperature sensor leads to a biosensor concept with general applicability, which can be used for the determination of a particular compound present in a complex mixture, irrespective of the optical and electrochemical properties of the sample.

There are numerous ways to exploit the calorimetric detection principle in combination with biological materials, not only in metabolite assays, but also for analyses of proteins and other macromolecules as well as for whole cells.[1] In this presentation examples from bioprocess monitoring, measurements of enzymic activities in separation procedures, determinations in organic solvents and work with miniaturized thermal biosensors will be given.

G.G. Guilbault and M. Mascini (eds.), Uses of Immobilized Biological Compounds, 541–548.

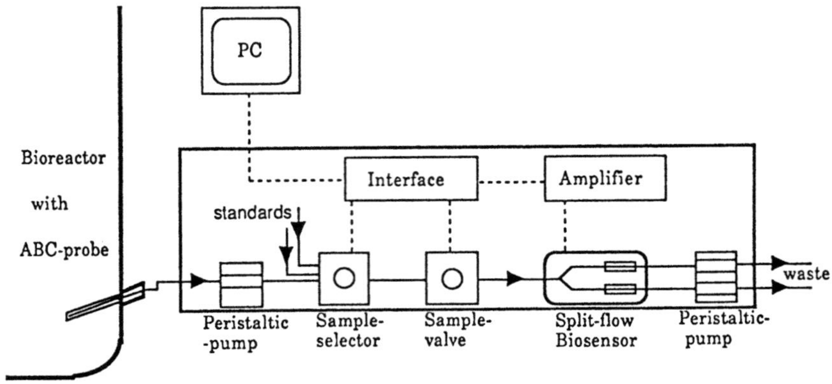

FIGURE 1. Schematic diagram of the set-up using a filtration probe for continuous withdrawal of sample from a fermentor. The filtered sample is led to the FIA-system with the heat sensitive split-flow biosensor. A personal computer was used for controlling the system and for calculation of concentrations.

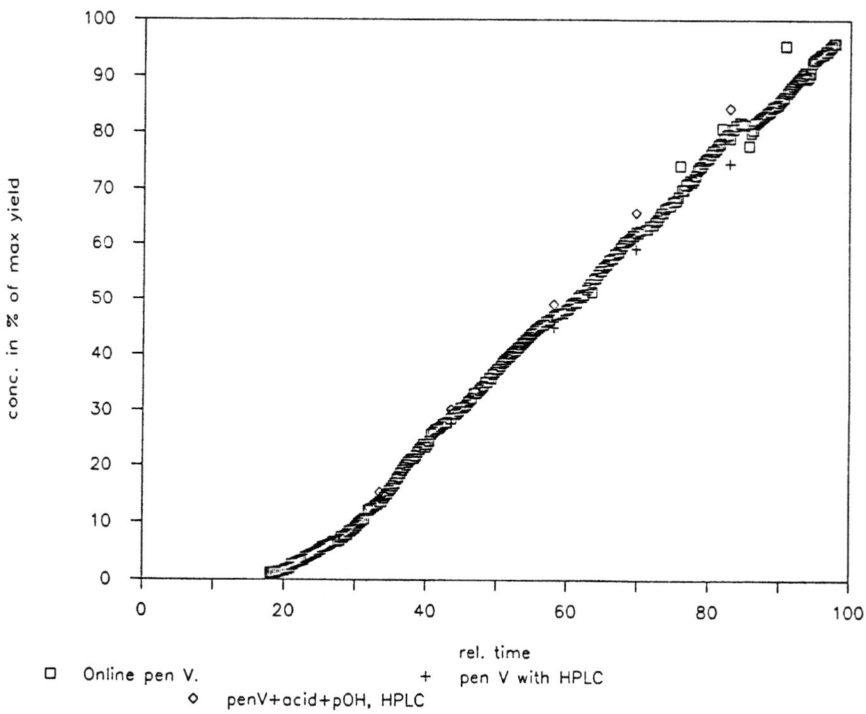

FIGURE 2. On-line monitoring of the penicillin V production in a 160 m^3 bioreactor over 1-2 weeks using an ET-column with penicillin V acylase.

II INSTRUMENTATION

The ET-calorimeter used in most of the studies described herein, with the exception of the studies on miniaturized systems, consists of a carefully thermostated aluminium cylinder, 80 x 250 mm, containing a heat sink with heat exchangers and two column positions, which can be used either with different enzymes for two different assays or with one column acting as a reference column (split-flow). [1] The columns are attached at the end of the thermistor probes and are readily exchangeable. The thermistors are connected to a Wheatstone bridge producing 100 mV/m°C at maximum sensitivity. Commonly used fullscale sensitivities are in the 10-50 m°C range permitting determinations in the 0.01-100 mM range. A large excess of enzyme (10-100 units or more per column, column volume up to 1 ml) bound to a mechanically stable, highly porous support, such as controlled pore glass or Eupergit C ensures high operational stability. A continuous flow of 0.5-2 ml/min through the system is maintained with a peristaltic pump. Samples are introduced with a sample loop valve (0.1-1 ml or smaller). Such small sample volumes will result in a temperature peak proportional to the substrate concentration.

III ON-LINE MONITORING OF BIOPROCESSES

At the implementation of biosensors in process environment the technical difficulties caused by high humidity, varying ambient temperatures up to 40 °C, large temperature variations, water and steam outlets, and continuous vibrations are dealt with. The complex composition of the fermentation media and high and variable concentrations of various components are additional problems.

The equipment was assembled inside a steel cabinet flushed with cool, filtered air to keep the temperature sufficiently constant. The ET was automated and equipped with a pneumatic sampling valve and a sample selector (Fig. 1). The instrumentation was tested in penicillin fermentations at the fermentation pilot-plant and the production plant at Novo-Nordisk (Bagsvaerd and Kalundborg, Denmark). A sample stream of 0.5 ml/min was taken from a tangential flow filtration unit with a 0.1 μ microfilter placed in a sterile loop connected with the fermentor. In later studies sampling has been accomplished using a polypropylene hollow fiber filtration probe (Advanced Biotechnology Corp., Puchheim, Germany). Every 10-30 min a 0.1 ml sample was injected over a time period of 1-2 weeks in the evaluation of the system. The flow through the ET unit (0.9 ml/min) was equally split between the enzyme column (ß-lactamase (EC 3.5.2.6) bound to controlled pore glass) and an inactive reference column containing immobilized bovine serum albumin. The split-flow technique was considerably improved by a newly designed flow divider. The column had to be protected against microbial growth by adding 1 mM sodium azide to the buffer solution.

Penicillin V could be measured during the whole fermentation with the same enzyme column without serious troubles in spite of temperature variations between 20 and 40 °C, high humidity and vibrations. The linear range of penicillin V was 0.05-500 mM. The reference column efficiently compensated for nonspecific heat effects. The results from one experiment is displayed in Figure 2. The concentration determined by the ET is 5-10% larger than the concentration of only penicillin V as determined by HPLC, since ß-lactamase hydrolyzes 6-APA and p-hydroxypenicillin in addition to penicillin V. Isopenicillin N and penicilloic acid are, however, not detected.[2] A more specific assay for penicillin V can be obtained with penicillin amidase (EC 3.5.1.11).[2] Presently, this assay is the preferred one, since it can be used to follow the conversion of penicillin to 6-APA. The reaction is, however, less exothermic.

In addition, measurements have been performed on penicilllin and *Saccharomyces* fermentations using alcohol oxidase for ethanol, glucose oxidase for glucose and lactate oxidase for lactate.[3] In all cases catalase was coimmobilized for better sensitivity and extended linear range. All assays and the total system worked fully satisfactory. Larger variations, could occassionally be seen, especially at the end of the fermentations when the viscosity of the broth was high, caused by improper function of the filtration unit. The general impression is that the ABC filtration probe works better than the tangential flow unit at higher viscosities in smaller fermentors as well as in larger fermentors up to 160 m^3.

IV CHARACTERIZATION OF IMMOBILIZED BIOCATALYSTS

The usefulness of the ET for the characterization of preparations of immobilized enzymes (invertase) was demonstrated in a study where the kinetic constants were directly determined without the need for postcolumn analysis.[4] An extension of thi work allowed for the direct determination of the catalytic activity of immobilized cells as well.[5] *Trigonopsis variabilis* strains selected by mutagenesis for high cephalosporin transforming activity were used in a model system in which the yeast cells were immobilized by crosslinking with homobifunctional reagents or by entrapment in gels. The thermometric signal arising from the activity of one specific dominating enzymic step in the cells can be identified by comparative HPLC-analysis of the reaction mixture. The cephalosporin transforming activity of D-amino acid oxidase isolated from selected yeast strains and immobilized by gel entrapment was identified in the same way. The thermometric signal was found to be proportional to the number of cells as well as to the amount of enzyme (DAAO) immobilized in the ET minicolumn.

Furthermore, the thermometric signal associated with cephalosporin transforming activity of both the immobilized cells and the isolated DAAO was found to be proportional to the height of the glutaryl 7-aminocephalosporanic acid peak in the HPLC chromatogram. Best agreement was obtained when the ET was operated at a flow rate of about 1 ml/min. The $K_{m(app)}$ values correlated in all cases very well with data computed for the (differential) ET-column.

V SUBSTRATE RECYCLING - TELISA

In TELISA (Thermometric Enzyme-Linked ImmunoSorbent Assay) the ET column contains an immunosorbent. In the competitive assay the sample is mixed with enzyme-labelled antigen and the concentration of bound antigen is determined by introduction of a substrate pulse to the enzyme, whereafter the column is regenerated by a pulse of glycine at low pH. The whole cycle takes less than 13 min. The sensitivity is adequate for the determination of hormones, antibodies, etc. produced by fermentation. [6]

If alkaline phosphatase is used as the enzyme-label, it is possible to enhance the sensitivity by using phosphoenolpyruvate as substrate and using a separate detection column in the ET-unit for the determination of the product, pyruvate, by substrate recycling. This is accomplished by the co-immobilized enzymes lactate dehydrogenase (reduces pyruvate to lactate under the consumption of NADH), lactate oxidase (oxidizes lactate to pyruvate), and catalase. By substrate recycling we have obtained several 1000-fold amplification of the sensitivity.[7]

In addition, genetically engineered enzyme conjugates have been used in immunoassays. Thus a human proinsulin/*E. coli* alkaline phosphatase conjugate was used for the determination of insulin or proinsulin. Concentrations lower than 1 μg/ml could be determined in less than 15 min.[8]

VI MEASUREMENTS IN ORGANIC SOLVENTS

Much interest has been paid in recent years to the possibility to run enzymic reactions in organic solvents. In this context it should be valuable to be able to analyze the reaction mixture directly in the organic medium using biosensors. Heat sensitive biosensors such as the ET are of special interest since the temperature response depends on the heat capacity of the system and the specific heat is up to 3 times lower in some organic solvents than in water. In addition the solubility of some enzyme substrates (cholesterol, triglycerides for instance) is higher than in water. It turned out to be possible to design practically useful procedures for analysis in organic solvents, especially since the enzymes involved become stabilized by the immobilization. In cases where the enzymic activity is lost after some time, it is often possible to restore it fully by treatment with aqueous buffer. Since the enthalpy change may be different in organic solvents or in solvent-water mixtures and in pure buffer the temperature response is difficult to predict. A comparison was made of the temperature responses obtained for tributyrin in a buffer-detergent system and in cyclohexane with lipoprotein lipase immobilized on Celite. In the latter case the response was about 2.5 times higher (as would be expected from the specific heats) and linear up to higher concentrations. In other experiments, the increase in sensitivity was found to be much higher.[9]

VII SPECIFIC MONITORING OF SOLUBLE ENZYMES

On-line monitoring of chromatographic separations of enzymes is usually restricted to UV-registration and determination of the pH or conductivity of the mobile phase. The normal procedure is to assay for the component of interest fraction-wise collecting the fractions with the highest concentrations for further purification or concentration. This is time- and labor-consuming and may be fatal to labile components. Specific monitoring of proteins, for instance direct identification of a special enzyme, would greatly facilitate and speed up purification, since it would allow for an eluted enzyme fraction to be taken directly (on-line) to a subsequent purification step. It has been shown that the ET has definite advantages as detector of enzymic activities and may be of interest to mention here even if this instrument arrangement falls outside the strict biosensor concept. For chromatography monitoring the effluent or a suitable aliquot of the effluent from a chromatographic column is mixed with a stream of substrate to the enzyme of interest. The heat registered upon passing the mixture through an empty inert column in the ET-unit is proportional to the enzymic activity. Furthermore, it has been demonstrated that the ET can be utilized in automated, rapid (10-15 min/sample) TELISA monitoring of biomolecules other than enzymes.[10]

Another study demonstrated the control of an affinity-adsorption procedure by the specific enzyme activity signal from an ET.[11] Lactate dehydrogenase (LDH) was recovered from a solution by affinity binding to an N^6-(6-aminohexyl)-AMP-Sepharose gel. The LDH activity signal from the ET was used in a PID controller or a computer to regulate the addition of AMP-Sepharose suspension to the LDH solution. The rapid and precise control of the addition of adsorbent in our model experiments suggests that this technique should be attractive in pilot plant and industrial scale purifications of enzymes.

VIII MINIATURIZED HEAT SENSITIVE BIOSENSORS

A current research project in our laboratory deals with the development of miniaturized FIA-systems with the final goal to construct a complete system by micromachining. The calorimetric measuring principle has been found to work surprisingly well in these small devices. Encouraging results have been obtained with various silicon chip test structures. [12]

As an intermediary step in the development of even more miniaturized constructions, plastic devices in the size of 50 mm long and 15 mm in diameter or smaller have been designed. Due to high sensitivity, small dimensions modest buffer consumption, and good operational stability these devices are very suitable for portable use, for instance for home monitoring of glucose in diabetes. To allow analysis directly on whole blood samples three different approaches have been tested, all of them with good results.

Firstly, the blood cells can be removed by dialysis or filtration. Small coaxial dialysis units were constructed by attaching a 25 mm long 0.2 mm (i. d.) cuprophan hollow fibre inside a 0.5 mm PVC tubing. These units give about 5 % glucose resulting in a linear range of 0-25 mM glucose with a 1.5 x 15 mm glucose oxidase/catalase column, which is adequate for diabetes monitoring.

Secondly, a superporous agarose material developed by Dr P.-O. Larsson at our department was employed as enzyme carrier. This material allows a large number of whole blood samples to be injected on the enzyme column without any sign of clogging. Fig. 3 shows a calibration curve obtained with 20 µl samples injected in a flow of 100 µl/min. For comparison the response for a 10-fold diluted whole blood with glucose added to raise the concentration by 0, 0.45 and 0.9 mM is indicated in the diagram. A suitable measuring range is chosen by proper selection of sample volume/flow rate.

Thirdly, a very small column (0.6 mm x 10 mm) was filled with spherical 125 micron CPG-particles loaded with glucose oxidase/catalase. The spaces between

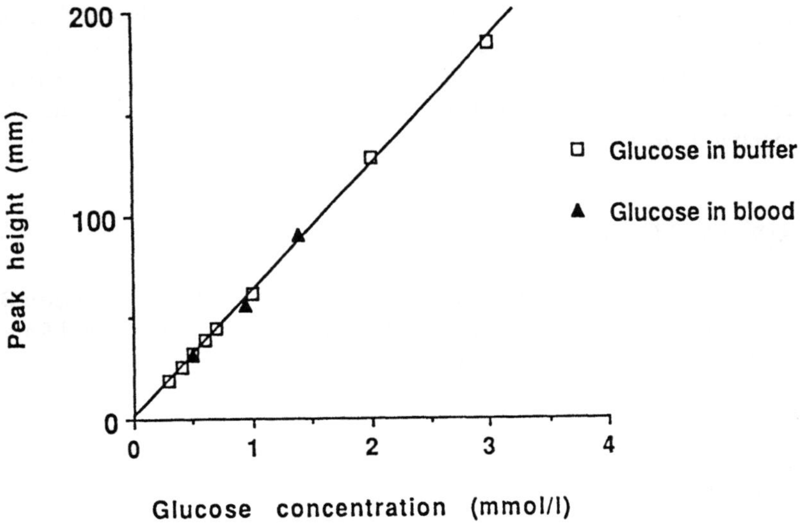

FIGURE 3. Calibration curve for glucose obtained with a miniaturized ET containing glucose oxidase/catalase on superporous agarose using 20 μl samples injected in a flow of 100 μl/min.

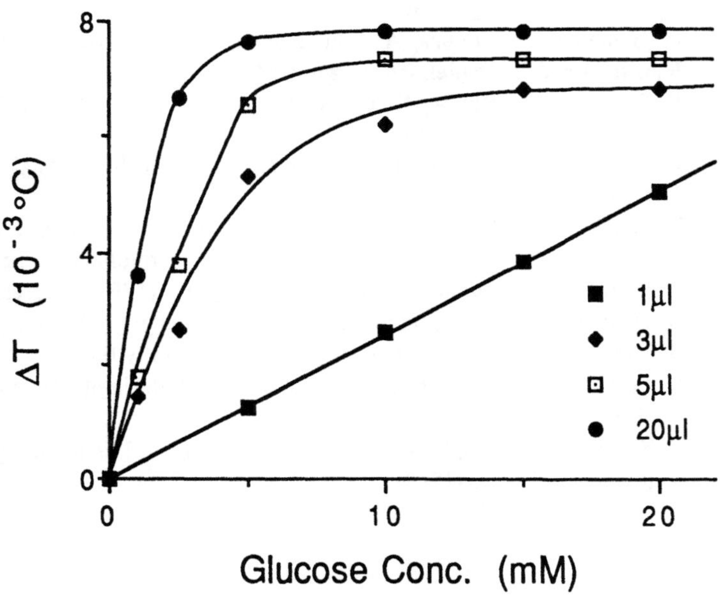

FIGURE 4. Effect of sample volume on the linear range of a glucose sensor with a 0.6 mm x 10 mm CPG-column with glucose oxidase/catalase. The flow rate was 50 μl/min.

these particles are large enough to allow the cells to pass through without being trapped. A sample volume of 1 µl gave a suitable measuring range (1-25 mM) for blood glucose determination (Fig. 4). This type of column can be used with over 100 samples. In conclusion all three approaches have been found to be practically useful for a home monitoring device.

ACKNOWLEDGEMENTS

This work has been supported by grants from the National Swedish Board for Technical Developments and from the Foundation for Biotechnological Research and by Novo-Nordisk A/S, Denmark.

REFERENCES

1. B. Danielsson, J. Biotechnol., 15, 187, (1990).
2. M. Rank, B. Danielsson and J. Gram, Biosensors & Bioelectronics, 7, 631, (1992).
3. M. Rank, J. Gram and B. Danielsson, Industrial on-line monitoring of penicillin V, glucose and ethanol using a split-flow modified thermal biosensor. Anal. Chim. Acta, (1993), in press.
4. V. Stefuca, P. Gemeiner, L. Kurillová, B. Danielsson and V. Báles, Enzyme Microb. Technol., 12 , 830, (1990).
5. P. Gemeiner, V. Stefuca, A. Welwardová, E. Michalková, L. Welward, L. Kurillová and B. Danielsson, Enzyme Microb. Technol., 15, 50, (1993).
6. S. Birnbaum, L. Bülow, K. Hardy, B. Danielsson, and K. Mosbach, Anal. Biochem. 158, 12, (1986).
7. F. Scheller, N. Siegbahn, B. Danielsson, and K. Mosbach, Anal. Chem., 57, 1740, (1985).
8. M. Mecklenburg, C. Lindbladh, Li Hongshan, K. Mosbach and B. Danielsson, Enzymatic amplification of a flow injected thermometric enzyme-linked immunoassay for human insulin, Anal. Biochem., (1993), in press.
9. B. Danielsson and L. Flygare, Sensors and Actuators, B1, 523, (1990).
10. B. Danielsson and P.-O. Larsson, TrAC 9, 223, (1990).
11. L. Flygare, P.-O. Larsson and B. Danielsson, Biotechnol. Bioeng., 36, 723, (1990).
12. B. Xie, B. Danielsson, P. Norberg, F. Winquist, and I. Lundström, Sensors and Actuators B, 6, 127, (1992).

THE MARKET FOR BIOSENSORS
THE STRENGTHS AND THE COMPETITION

Valerie M Owen
SciTec Management Consultants Ltd
1, Otter Close
Bletchley
Milton Keynes MK3 7QP
UK

ABSTRACT

Penetration of the market by biosensors will depend upon their ability to service an application not being reached by alternative techniques or to compete successfully against other technologies.

This paper will attempt to identify the requirements for analytical devices and monitors, with particular emphasis on medical applications. The question of whether a device, incorporating an immobilized biological compound, is likely to be able to compete in this market will be examined. This will be addressed in terms of the technical strengths and weaknesses of biosensors, their development costs, the impact of regulation and legislation, and whether the device can be produced and marketed at a competitive price.

INTRODUCTION

If a new technology is to succeed in the market place it must offer some advantage over existing technologies. It may compete by:–

- being able to perform a function for which there is no alternative existing technology;

- performing better than existing techniques in terms of ease of operation, reliability or cost;

- opening up a new market which has been overlooked or has not previously existed.

The use of the technology must offer some real benefit to the end–user and to the purchaser. It must satisfy a need rather than just a want. A technology is highly unlikely to succeed just because it is an innovative clever piece of science.

G.G. Guilbault and M. Mascini (eds.), Uses of Immobilized Biological Compounds, 549–557.

At present there are few devices which can be used to measure chemicals and biochemicals in the human body in real–time; however, there are some disease states which could benefit from such a level of monitoring. Some of these disease states are sufficiently common or the incidence is increasing so that they represent a market which is already, or shortly will become, large enough to support the development costs of a novel analytical system. These disease states include diabetes and renal failure.

There are still problems with producing reliable analytical results outside the central laboratory. Although there has been great progress in bedside testing for intensive care, at times the results are still questionable. Any technology which is to be used outside the central laboratory and which is going to have a significant effect upon the treatment of a patient needs to be very well researched and developed, and therefore will probably be costly to bring to market. Thus it must be for use in a large market in terms of the number of tests to be performed, and the amount of money which the purchaser will pay for the test. Assisted human conception is a growing market which appears to comply with these criteria.

A market which is only just being developed is for the monitoring of the biochemistry of the athlete. This requires very robust technology which can be used in the field, but the cost per test must be fairly low. Although reliability is desirable it is not as critical as in some clinic or hospital situations. The requirements approach those of a consumer product.

By contrast, although the central laboratory is still the site for the majority of analyses, it is questionable whether novel technology based upon immobilised biological compounds will make a significant impact.

DIABETES

The ability of the patient to monitor blood glucose using dry chemistry strips or biosensors (both technologies involving immobilized biologicals) has led to better control of blood sugar levels. It is estimated that the worldwide market for blood glucose monitoring is between 1,000 million and 2,000 million self–monitoring blood glucose tests per year. Between 1 and 2% of the populations of Europe and North America are diagnosed as diabetic and this percentage is increasing.

Patients find current glucose monitoring techniques laborious and inconvenient especially as it is an invasive procedure. Also as the patient gets closer to normoglycaemia there is a greater risk of hypoglycaemia. This increases the need for continuous monitoring with a hypoglycaemia alarm. The use of such an alarm could be justified on the basis of peace of mind for the patient and economic cost to the community. It has been estimated that hypoglycaemic episodes in Europe cost $150 million per year in terms of cost of accidents, treatment, etc.

The need for continuous short–term blood glucose monitoring could be satisfied, during surgery or intensive care, by a biosensor in an extracorporeal circulation. For longer–term monitoring the biosensor would probably have to be implanted. This would give continuous readings and would relieve the patient of the blood taking and analysis tasks. The process of implanting a biosensor is invasive and, therefore, not without some risk and inconvenience. However, this risk will have to be reduced if we are to reach the ultimate goal of an artificial pancreas based upon a glucose biosensor linked to an insulin pump.

Alternative competing technologies for continuous blood glucose monitoring include transcutaneous measurements using near–infrared light[1,2] or measurements using high frequency electromagnetic radiation to monitor the capacitivity of tissues. It has been reported that there is a correlation between the capacitive effects of blood and changes in blood glucose levels [2].

Advances in transplant technology are leading to transplants of pancreas or, at least, islet of Langerhans cells. Also there is work in progress to predict the onset of Type I diabetes with a view to preventing the destruction of the beta cells and the prevention of diabetes. In the long run these advances could reduce the market for blood glucose monitoring.

RENAL THERAPY

Kidney failure leads to a build up of waste products such as urea and creatinine in the body. Also it has the effect of disturbing the water and electrolyte balance as little or no urine is produced to rid the body of excess water. Treatment by haemodialysis involves linking the patient to a dialysis machine for several hours at two to three day intervals. The trend in haemodialysis is toward high efficiency technology which shortens the time required for the dialysis. This improves the life style of the patient. However, any miscalculation of the urea clearance or slowing of the pump rate will result in less than optimum removal of toxic waste products [3].

There is a need to monitor the removal of the waste products and monitor the fluid and electrolyte balance by measuring the electrolytes (especially pH and K^+). Presently waste product is monitored retrospectively by measuring urea. The demand is for continuous on–line measurement of urea. This is an ideal application for a biosensor. The data obtained could be used to develop computer modelling of the dialysis treatment for each individual patient.

Figure 1 shows the steady rise in the worldwide population of dialysis patients. There are currently about half a million patients being treated by haemodialysis and this number is rising with the increase in the aged population.

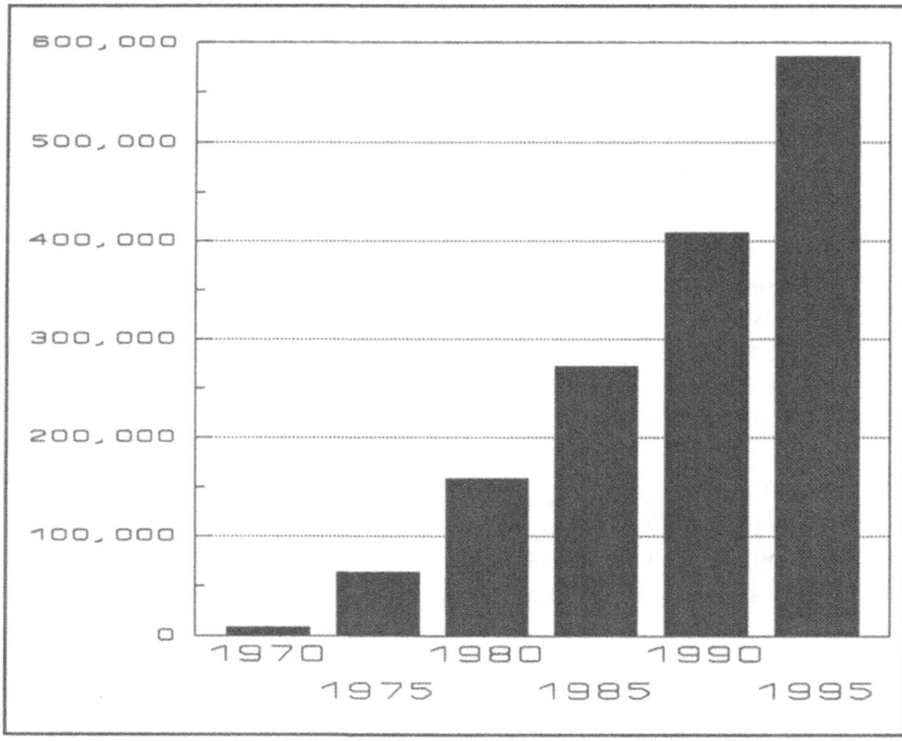

Figure 1 Estimated worldwide population of dialysis patients

The increase in number of patients requiring dialysis therapy means that renal transplant using human kidneys is never likely to keep up with demand. Thus there will always be patients receiving long term artificial dialysis unless animal organ transplants become a large scale reality.

An alterative to the use of haemodialysis is the use of peritoneal dialysis in which the dialysis fluid is introduced into the peritoneal cavity and then changed at the end of a six hour period. This procedure has been monitored using creatinine measurements rather than urea. Once the fluid has been introduced into the peritoneal cavity there is unlikely to be any modifications to the treatment until the fluid is changed some six hours later. Therefore continuous monitoring of the process is unlikely to effect the treatment. At present samples are taken and analysed intermittently. This analysis is carried out using the chemical Jaffé reaction technique; however, the dialysate causes interference with this reaction and a correction factor has to be applied or the more expensive enzymatic method has to be used[4]. The cost of the enzymatic assay could be reduced perhaps with the use of an instrument incorporating immobilized biologicals.

The UK is the only European country in which peritoneal dialysis is used widely, with about half the patients requiring dialysis treatment being treated in this way. The choice of this form of dialysis was due, in the past, to a lack of the expensive haemodialysis machines rather than clinical preference. There used to be problems of infection in patients treated using peritoneal dialysis, but improved catheter connection technology has reduced this risk and may lead to an increase in popularity of peritoneal dialysis on clinical grounds. However, the ultimate determinant of the method of dialysis may be the structure of reimbursement. Accountancy may take precedence over clinical preference!

ASSISTED CONCEPTION

Already dry (bio)chemistry tests have found a large market in the ovulation and pregnancy self–testing market. These tests are all designed to give positive/negative results and not quantitative results. However, there is a growing demand for quantitative testing of the hormones of the reproductive cycle in assisted human conception.

Patients attend clinics to undergo follicular stimulation. They are treated with a series of drugs which stimulate a number of follicles to mature at the same time so that a number of eggs are released simultaneously (superovulation). A side effect of the drug treatment is that the lining of the womb is not developed fully and the eggs may fail to implant. The oocyte must be recovered as it matures so that *in vitro* fertilisation can take place.

The development of the follicles are monitored using a series of measurements including ovarian ultrasonography and measurements of blood oestradiol, progesterone and LH[5]. The short time frames involved mean that the test results must be available quickly so that the patient and operating–room staff can be prepared for oocyte recovery. The requirement for fast results is more pronounced in this clinic situation than in most other clinics as the results of the analysis will involve several people in an activity. Thus it is a market opportunity for biosensor technology which will enable the operator with little analytical training to obtain a reliable result.

The market for assisted conception is growing. It is estimated that around 10–15% of couples in the developed world have an in–fertility problem and the OTC fertility testing market is growing at about 25% per year.

In the UK nearly 10,000 patients underwent *in vitro* fertilization in 1990 and this number is increasing at about 10% per year. In addition further patients are treated using gamete intra–fallopian transfer (GIFT). Each patient requires multiple analyses as typically three urine samples per day are assayed for LH and this number may increase as the follicle matures. Blood samples are also assayed around and following the time of oocyte recovery.

Some form of immunoassay will be involved in testing in these clinics and the choice of a biosensor will depend upon how well this technology can perform in terms of reliability and cost compared to other techniques.

The market will be driven by consumer–demand for babies and an administrative organisational requirement for prompt analytical results.

ATHLETE MONITORING

There is growing interest in the monitoring of athletes during training. Most of this monitoring is concentrated upon discovering how well oxygen is being delivered to the tissues. If the oxygen supply is inadequate the anaerobic threshold is reached and anaerobic glycolysis begins with the production of lactic acid. There is also interest in monitoring the post–training period to examine the rate of recovery. Urine urea and blood creatine kinase are used in these assessments.

The need is for very robust devices which are user–friendly as the requirement is for them to be used by trainers in the field. The cost will be of great importance as it is unlikely that more than a few national– and international– level sporting teams would invest in an instrument which costs more than $200.

CENTRAL LABORATORY

The central laboratory of a large hospital is highly dependent upon automated equipment. Typically the modern clinical chemistry analyser is capable of performing about 50 different tests with about half this menu being available at any one time. The trend is for more tests to become available on the one instrument system as immunoassay techniques become incorporated. The system depends largely upon all the analyses sharing one detection system.

Many of the needs of the central laboratory refer to the handling of the sample rather than the method of analysis. The needs of an analytical system for the central laboratory are listed in Table 1.

Immobilised biologicals offer the prospect of solving the need for a system to be reagentless and inexpensive to run, but probably do not help solve the other needs.

It is estimated that the European market for automated clinical chemistry analysers is about 1,400 systems per year worth about $200 million, while the market for clinical chemistry diagnostic reagents is about $600 million per year.

The time scale for introducing a new immunoassay analytical system to the market is typically over ten years and a cost of $100 million[6]. In the case of the Kodak analysers, which use dry (bio)chemistry technology, the instrument has been in development for some twenty years and is only just reaching the range of tests offered by other analysers. The development costs have certainly been way in excess of $100 million.

Table 1 Needs for an analytical system for the central laboratory

ability to handle:–
whole blood
multianalytes
minimal sample volumes (sample volumes as low as 1μl are already in use)
used for several laboratory disciplines, not just biochemistry
rapid (systems handling 10,000 tests per hour are already available)
inexpensive to buy
inexpensive to run (eg low consumable costs)
low maintenance
infrequent calibration
infrequent quality control
freedom from interference
accurate
precise
reliable
safe when dealing with high risk samples (eg waste containment)
reagentless.

It seems unlikely that another large scale analyser based upon novel immobilised biologicals technology or biosensors could be introduced onto the market. The role of the immobilised biological will probably remain as part of the separation system in immunoassays and nucleic acid analyses.

LEGISLATION AND REGULATION

There is a move toward uniform international standards of regulation and legislation of medical equipment. At one level much of Europe is becoming subject to EC Directives and at the same time there are discussions between the FDA and EC which could eventually lead to harmonisation between Europe and the USA, and even worldwide.

There are three EC Directives (see Table 2) which will have a direct affect upon medical analytical and monitoring technology.

Table 2 EC Directives

Active Implantable Medical Devices Directive (Directive 90/385/EEC)
Medical Devices Directive (Draft Directive 91/C 237/03)
In vitro Diagnostics Medical Devices Directive

The Active Implantable Medical Devices Directive has been adopted already by the European Council and is being implemented in member states. This Directive effects devices which are intended for use in human beings for diagnosis and monitoring. It refers to devices which rely on a source of electrical energy or any other source of energy which is not generated by the human body or gravity. This Directive would presumably affect an *in vivo* glucose sensor and might affect a sensor for monitoring renal dialysis, depending upon where and how the sensor is used.

The Medical Devices Directive will not come into force until July 1995 which means it will be 1997 or later by the time it is fully adopted by member states. This Directive applies to a very wide range of devices from first aid bandages to CT scanners. It will apply to devices, which are not implantable, but which are used to monitor humans. It will not apply to *in vitro* diagnostics. This Directive will presumably cover devices such as transcutaneous glucose monitors.

The *In vitro* Diagnostics Medical Devices Directive is still at an early stage of drafting and EDMA (European Diagnostics Manufacturers Association) is working very actively on this Directive. The Directive is expected to cover laboratory equipment and reagents, self–testing kits and there will be special requirements for diagnostics which have a direct impact upon subsequent treatment, blood grouping, detection of certain infectious diseases. Other candidates for special controls include *in vitro* diagnostics (IVDs) using radioactive isotopes, DNA technology and culture media–based IVDs. Diagnostics for monitoring athletes could come within the scope of this Directive. Diagnostics for use in assisted conception clinics will come under this category and might be subject to some of the special controls.

The effect of EC Directives will be to remove some of the trade barriers across Europe. Devices which meet the requirements will be entitled to carry the 'CE' mark and member states of the EC must not create any obstacle to the placing on the market, or putting into service, of devices bearing the 'CE mark'.

CONCLUSION

If biosensors are to succeed commercially they must do more than compete head–on with existing technology. They must find markets which are not being served by existing technology.

The market they are to reach must be large enough to support the significant development costs. Alternately the development costs must be spread over several devices. For example, a pH sensor could be modified to work in cardiac muscle (during open heart surgery), cardiac chambers (during diagnosis) and routine monitoring in intensive care.

The device must be amenable to manufacture at a price which allows it to be sold profitably. This may be very difficult in a situation in which there is an existing kit or technology which has been manufactured for a number of years and where the capital costs of setting up the manufacturing process have been written off already. This problem has been demonstrated by MediSense, the manufacturers of the ExacTech glucose analyser. Examination of the company accounts at the beginning of this decade show that most of the company revenue was absorbed by the manufacturing costs leaving nothing for marketing, research and development, or profit!

REFERENCES

1. 'Research into noninvasive measurement of blood glucose by near–infrared technology' Rosenthal, R., Clinical Chemistry, 38, 1645, 1992.

2. 'An overview of minimally invasive technologies' Ginsberg, B. H., Clinical Chemistry, 38, 1645, 1992.

3. 'High–efficiency haemodialysis: an overview.' von Albertini, B., Contr. Nephrology, 61, 37–45, 1988.

4. 'Measurement of creatinine in peritoneal dialysis fluid.' Farrell, S.C., Bailey, M.P., Annual of Clinical Biochemistry, 28, 624–625, 1991.

5. 'Endocrine monitoring for assisted human conception.' Howles, C.M., Macnamee, M.C, British Medical Bulletin, 46, 616–627, 1990.

6. 'The future of immunodiagnostics.' Freytag, J.W., Journal of Clinical Immunoassay,14, 239–244, 1991.

REPORT ON BIOSENSOR TECHNOLOGY IN EUROPE

MARCO MASCINI Dip. di Sanita' Pubblica, Epidemiologia e Chimica Analitca Ambientale, Sezione di Chimica Analitica Via Gino Capponi 9, 50121 Florence, Italy

Abstract A Report on Biosensor Technology in Europe has been prepared for EEC. The Report (about 450 pages) includes all recent work done in 15 European nations (+ Japan and USA) and all future projects in this field. Some overviews and statistics have been included to summarize the full text. Conclusions have been presented and submitted at the end of the Report to orient the EEC to include Biosensor programmes in the, IV Framework Programme of Research Activities for the period 1994-1998.

The importance of Biosensor Research is growing enormously and is now appearing at National as well at Community levels. At present Biosensors are included in several Research Programs in different technological areas from Medicine to Environmental, from Food to Biotechnology, because this technology can solve important problems related to Safety, Testing, Monitoring and Process Control.

However the fundamental problems of Biosensor research are common regardless of scientific field or applications and therefore these problems could be better solved by the integration of different actions.

This Report is a survey of the Biosensor Research in Europe and of its strategic importance for National and Community Research Programs.

G.G. Guilbault and M. Mascini (eds.), Uses of Immobilized Biological Compounds, 559–566.

The Report has been prepared by 17 scientists, working experimentally in the field of Biosensors, who can be considered the most expert in their country in Europe, the USA and Japan.

It reports the activity of the European Research in each country in important fields of application, like Clinical Chemistry, Agriculture, Food Chemistry, Environmental Analysis, or Process Control Monitoring; it is an update inventory of this relatively new technology and its application in different fields in Europe.

Important features like "in vivo monitoring", mass production, miniaturization techniques, self monitoring and food testing are highlighted.

The position of Research in Europe is compared to those of the USA and Japan from the point of view of scientific productivity as well as applied technology and achieved results. Europe has a high level of activity in Biosensor research compared to USA and Japan. However Japan has the highest number of patents and the USA is the most important single market. Europe with different rules in each state suffers from lack of interest from industrial partners.

Moreover Biosensor development requires a range of expertise from Analytical Chemistry to Biochemistry, from Electronics to Microengineering to Genetic Engineering, and only in a few organized centers can experts in all disciplines succesfully interact together. In Europe now we assist the formation of small companies where the number of experts and funds involved are limited (there are of course some exceptions). The EC, in concert with national governments, should play an important role in this strategic sector by integrating industrial interest and academic knowledge to obtain better exploitation in the development of Biosensor technology in Europe (See Fig. 1-3).

In the Report the national programmes on Biosensors are reviewed and large differences are apparent in the attitude of the governments in Europe; some nations have large research programmes, others have only limited funds in this area.

In the Report specific recommendations for public policy in Europe at a National and Community level are formulated to develop a Community strategy for Biosensor Research which will help national industries.

The general conclusions of our Reports are:

1) There are many projects involving Biosensors under way in all European countries with a relatively good number of publications in international journals.

2) Industry in Europe does not generally participate to a great extent in Biosensor Programs.

3) The dominance of electrochemical techniques is decreasing and many programs for the future are based on optical techniques.

4) Many groups in Europe have similar projects, but they often do not communicate except through scientific journals

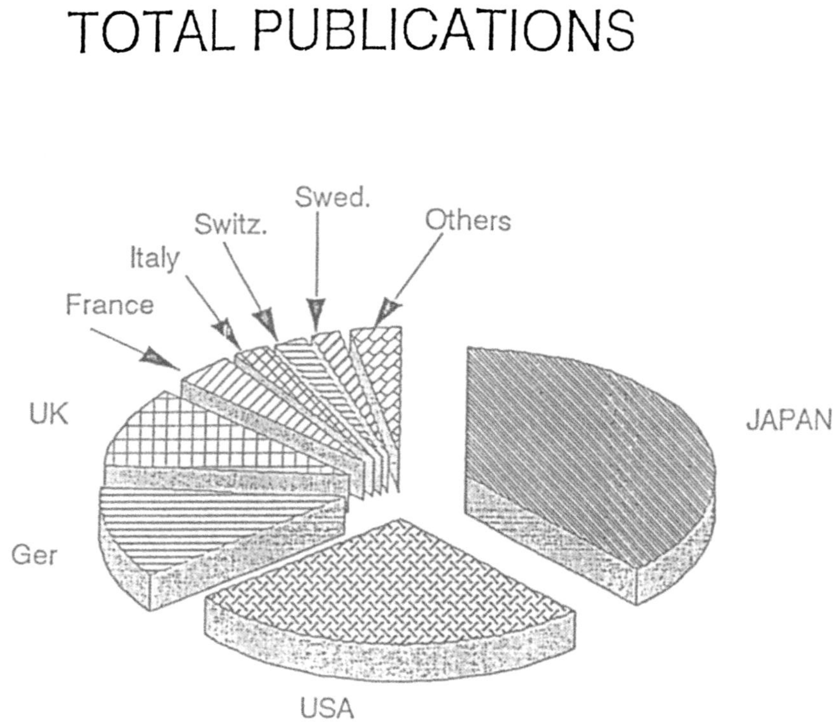

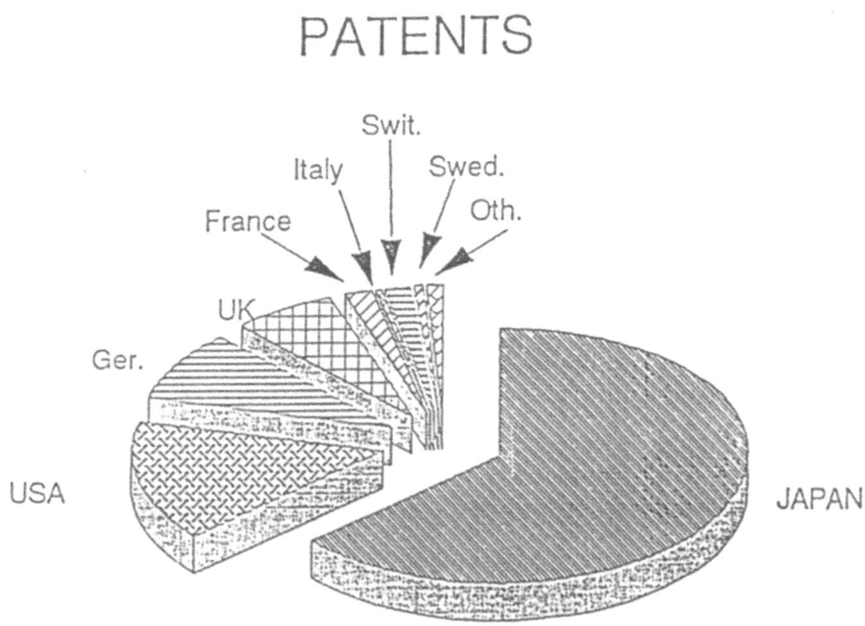

Fig. 1. Distribution of Total Publications and Patents between EEC countries, the USA and Japan

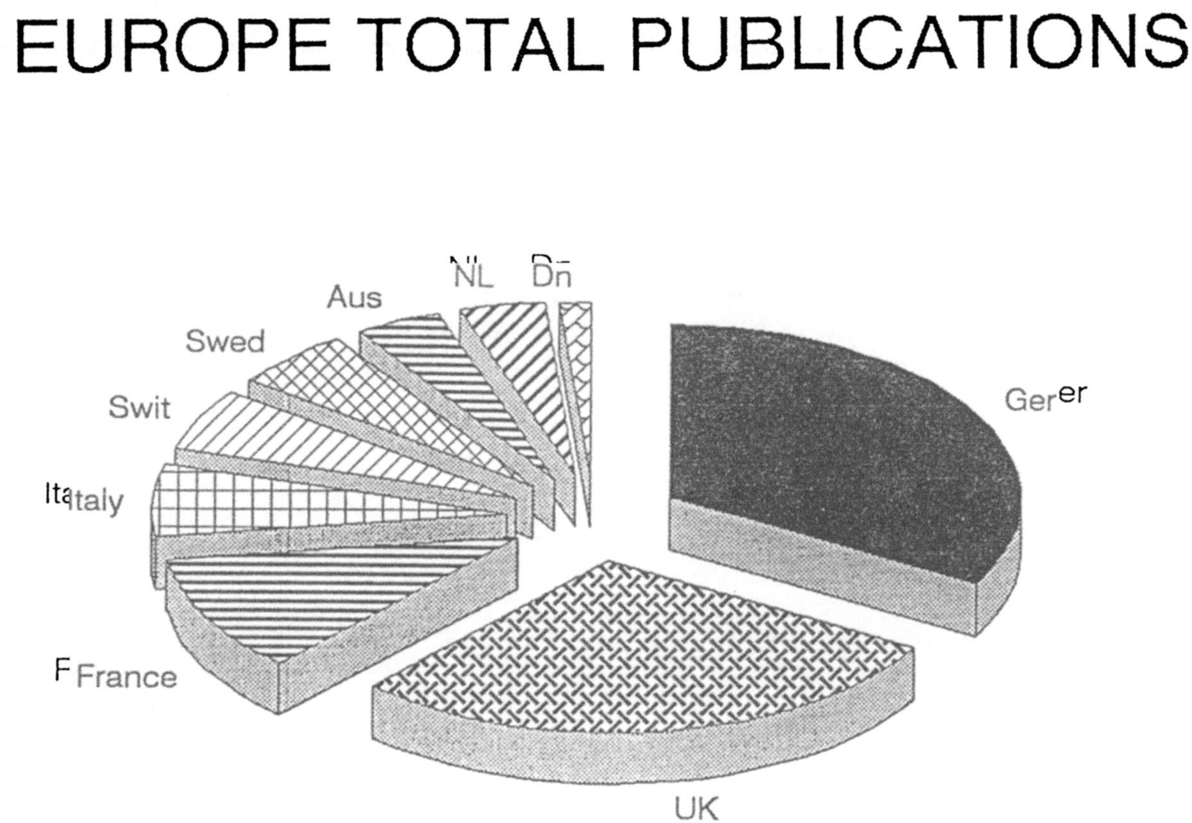

Fig. 2. Distribution of the total publications on Biosensors between the EEC countries

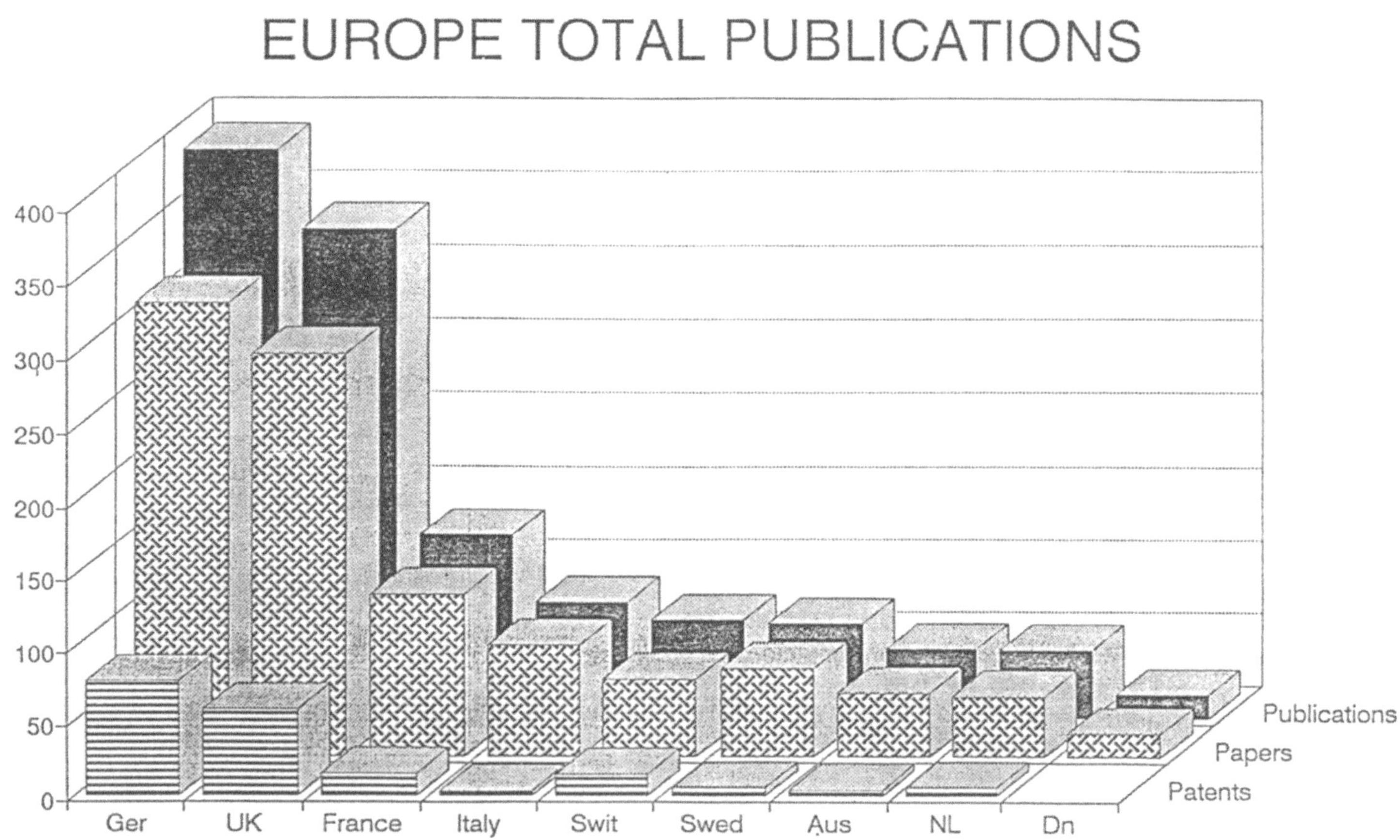

Fig. 3. Distribution of Publications(Papers and Patents) between various countries of the EEC

and formal presentations at International Meetings; the research activity is therefore very dispersed.

5) Biosensor research is not sufficiently focused on "basic" issues such as new immobilization techniques, improved receptors, instability of biological molecules etc.

6) The general request to the EEC is to contribute actively in organizing a specific research program on Biosensors , rather than the existing "service" or "add-ons" to existing programs. This program should fund collaborative research and facilitate exchange of research workers. It should address fundamental issues to assist sensor and biosensor production and technology transfer to industry.

7) The future Biosensor program should be considered independent from Biotechnology and from Microelectronics and it should enable a network between the academic work and industry.

8) The Biosensor program should consider two large areas of research:

- Basic research on Enabling Technologies and, for example, we can cite the following areas for future work
- Miniaturization of Probes
- Fabrication
- Surface Chemistry
- Labelling, Conjugate Chemistry
- Protein Engineering

- New Enzymes
- Stabilization Technology
- Kinetics and Modeling
- Non-invasive Monitoring
- Interfacial Interaction
- Biocompatibility
- Solid State Chemistry
- Surface Technology
- Chemometrics

and

- Target research with applied projects in conjuction with industry like:

- Biosensors for Drinking Water/Water Toxicity
- Biosensors for Nitrate, Phosphate in Sea water
- Biosensors to evaluate Food Freshness
- Biosensors for detection of Infectious Diseases like AIDS, or Cardiovascular Disease
- Biosensors for in vivo Monitoring of Glucose
- Biosensors as Process Controllers
- Maintenance-Free Biosensors
- Innovative Designs for Biosensors.

The Report is free and available from:

Dr. P. de Taxis du Poet, Commission of the European Communities, Directorate General for Science, Research and Development, Joint Research Centre Cube, Rue de la Loi 200, B-1049 Brussels, Belgium

AUTHOR INDEX

Z

SUBJECT INDEX

A

B

C

D

P

R

S

GPSR Compliance

The European Union's (EU) General Product Safety Regulation (GPSR) is a set of rules that requires consumer products to be safe and our obligations to ensure this.

If you have any concerns about our products, you can contact us on ProductSafety@springernature.com

In case Publisher is established outside the EU, the EU authorized representative is:

Springer Nature Customer Service Center GmbH
Europaplatz 3
69115 Heidelberg, Germany

Batch number: 10607320

Printed by Printforce, the Netherlands